FISH BIOMECHANICS

This is Volume 23 in the

FISH PHYSIOLOGY series
Edited by David J. Randall and Anthony P. Farrell
Honorary Editor: William S. Hoar

A complete list of books in this series appears at the end of the volume

FISH BIOMECHANICS

Edited by

ROBERT E. SHADWICK
Canada Research Chair
Department of Zoology
University of British Columbia
Vancouver, British Columbia
Canada

GEORGE V. LAUDER
Alexander Agassiz Professor
Department of Organismic and Evolutionary Biology
Museum of Comparative Zoology
Harvard University
Cambridge, Massachusetts

ELSEVIER

AMSTERDAM • BOSTON • HEIDELBERG • LONDON
NEW YORK • OXFORD • PARIS • SAN DIEGO
SAN FRANCISCO • SINGAPORE • SYDNEY • TOKYO
Academic Press is an imprint of Elsevier

Cover Photo Credit: Paracirrhites forsteri (Schneider, 1801)
Blackstriped hawkfish, Family Cirrhitidae (Hawkfishes),
Photo: Ralph Schill; graphics: Sven Gemballa

Elsevier Academic Press
525 B Street, Suite 1900, San Diego, California 92101-4495, USA
84 Theobald's Road, London WC1X 8RR, UK

This book is printed on acid-free paper. ♾

For all information on all Elsevier Academic Press publications
visit our Web site at www.books.elsevier.com

ISBN-13: 978-0-12-350447-0
ISBN-10: 0-12-350447-3

Printed and bound by CPI Group (UK) Ltd, Croydon, CR0 4YY

Transferred to Digital Print 2011

CONTENTS

4. The Hydrodynamics and Structural Mechanics of the Lateral Line System

Sheryl Coombs and Sietse van Netten

5. Skin and Bones, Sinew and Gristle: The Mechanical Behavior of Fish Skeletal Tissues

Adam P. Summers and John H. Long, Jr.

6. Functional Properties of Skeletal Muscle

Douglas A. Syme

CONTRIBUTORS

The numbers in parentheses indicate the pages on which the authors' contributions begin.

ELIZABETH L. BRAINERD *(1), Biology Department and Program in Organismic and Evolutionary Biology, University of Massachusetts, Amherst, Amherst, Massachusetts**

THEODORE CASTRO-SANTOS *(469), S.O. Conte Anadromous Fish Research Center, USGS-Leetown Science Center, Turners Falls, Massachusetts*

SHERYL COOMBS *(103), Department of Biological Sciences, Bowling Green State University, Bowling Green, Ohio*

ELIOT G. DRUCKER *(369), Museum of Comparative Zoology, Harvard University, Cambridge, Massachusetts*

LARA A. FERRY-GRAHAM *(1), Moss Landing Marine Labs, California State Universities, Moss Landing, California*

SVEN GEMBALLA *(241), Department of Zoology, University of Tübingen, Tübingen, Germany*

ALEX HARO *(469), S.O. Conte Anadromous Fish Research Center, USGS-Leetown Science Center, Turners Falls, Massachusetts*

GEORGE V. LAUDER *(425), Department of Organismic and Evolutionary Biology, Museum of Comparative Zoology, Harvard University, Cambridge, Massachusetts*

JOHN H. LONG, JR. *(141), Department of Biology, Vassar College, Poughkeepsie, New York*

Current address: Department of Ecology and Evolutionary Biology, Brown University, Providence, Rhode Island.

Robert E. Shadwick *(241), Department of Zoology, University of British, Columbia, Vancouver, British Columbia, Canada*

Adam P. Summers *(141), Department of Ecology and Evolutionary Biology, University of California, Irvine, Irvine, California*

Douglas A. Syme *(179), Department of Biological Sciences, University of Calgary, Calgary, Alberta, Canada*

Eric D. Tytell *(425), Department of Organismic and Evolutionary Biology, Museum of Comparative Zoology, Harvard University, Cambridge, Massachusetts*

Sietse van Netten *(103), Department of Neurobiophysics, University of Groningen, Groningen, The Netherlands*

Peter C. Wainwright *(77), Section of Evolution & Ecology, University of California, Davis, Davis, California*

James M. Wakeling *(333), Structure and Motion Laboratory, The Royal Veterinary College, North Mymms, Hatfield, Herts, United Kingdom*

Jeffrey A. Walker *(369), Department of Biology, University of Southern Maine, Portland, Maine*

Paul W. Webb *(281), School of Natural Resources and Environment, University of Michigan, Ann Arbor, Michigan*

Mark W. Westneat *(29, 369), Department of Zoology, Field Museum of Natural History, Chicago, Illinois*

PREFACE

This is the first multi-authored volume on fish biomechanics to appear in over twenty years. In that time the field has grown immensely, with many new experimenters using new experimental techniques to probe questions of how fish work. Consequently, the published literature in fish biomechanics has grown rapidly, and it is time for a comprehensive review and synthesis of the important findings of recent research to update the classic *Fish Biomechanics* volume edited by Paul Webb and Danny Weihs in 1983.

This book begins at the front end of the fish with important biomechanical events that involve the head: breathing and eating. The complexity of head structure is one of the most distinctive and evolutionarily interesting aspects of fishes. The interaction of bones, joints, and muscles of the head is highlighted in Chapter 1 by Brainerd and Ferry-Graham in their review of the mechanics of respiratory pumping. They discuss two-phase (suction and pressure) pumps, as well as ram ventilation and air breathing. The theme of head structure as a set of muscle-powered levers and linkage bars is further elaborated in Chapters 2 and 3, which present detailed accounts of feeding mechanics, a classic illustration of an elegant form and function relationship. Westneat reviews the great diversity of skull morphologies and feeding strategies in fish groups, showing how different kinematic models have been developed, and provides clear illustrations based on high speed videography, as well as discussions of muscle activity patterns associated with feeding activities and their evolutionary relationships. Wainwright then describes how the pharyngeal jaw apparatus, a unique aspect of fish trophic biology, is designed from multiple skeletal elements modified from gill arches. He summarizes recent work on the morphology and the kinematics of pharyngeal jaws based on experimental approaches of cineradiography and sonomicrometry.

Apart from breathing and eating, one of the most important and interesting activities fishes perform is locomotion, and this is broadly the focus of the remainder of the book. Swimming and maintaining hydrostatic equilibrium go hand in hand. In Chapter 4, Coombs and van Netten discuss the structure and biomechanical features of the lateral line system as a collection of flow sensors, and how this system is used to provide information to the fish about the hydrodynamic structure of its environment that aids locomotion and behavior. The body of a fish can be regarded as a complex mechanical structure in which muscles generate forces and movement, while skeletal elements bear the loads and link the internal muscle action to the external resistive fluid medium. In Chapter 5, Summers and Long provide an overview of the engineering principles used to analyse both the static and dynamic mechanical properties of biological materials, and then discuss current data on the mechanical behavior of fish skeletal tissues in the context of the various locomotor modes of fishes. A major focus of research on fish swimming has been the contractile properties of locomotor muscles, most recently advanced by use of the *in vitro* work loop technique to study power production under simulated swimming conditions. In Chapter 6, Syme provides a comprehensive review of the biomechanical properties of skeletal muscle, and shows how studies of isolated muscle have been used to understand the various strategies fish use to power swimming under different conditions. The use of muscle in undulatory swimming is further considered in Chapter 7 where Shadwick and Gemballa describe the structural organization of the lateral myomeres and their connective tissue linkages as the pathway of force transmission along the body. They also discuss body kinematics and muscle dynamics in steady swimming, noting the general trends as well as the exceptions exhibited by the highly specialized tunas and lamnid sharks. The important problem of maintaining both stability and maneuverability is discussed in detail by Webb in Chapter 8, illustrating the elegant biomechanical solutions attained by fishes, and highlighting the importance of this knowledge in biomimetic designs of underwater autonomous vehicles. Wakeling reviews the specific problem of unsteady fast-start maneuvers in Chapter 9, by considering the sequence of events that initiate muscle contraction, bend the body, and generate the hydrodynamic forces that accelerate the fish. The fast-start (c-start) escape response of fishes has been of great importance as a system for understanding the neural control of behavior, and this chapter provides a synthesis of recent advances in the biomechanics of fish escape responses.

Fish pectoral fin function during locomotion has received a great deal of attention in the past twenty years. In Chapter 10, Drucker and his colleagues review a large amount of data on pectoral fin morphology, kinematics, and hydrodynamics, and discuss the ecological implications of different pectoral fin designs. Perhaps the most noticeable feature of fish locomotion is the bending of the body. Lauder and Tytell update classical descriptions of undulatory locomotion with recent experimental data in Chapter 11, where they also discuss new hydrodynamic data from freely-swimming fishes that highlight the importance of three-dimensional effects. Finally, biomechanical approaches are moving out of the laboratory and playing an increasing role in understanding the field behavior of fishes and helping in conservation efforts.

In Chapter 12, Castro-Santos and Haro synthesize a large body of work on the migration and passage of fishes around dams, and describe new tagging technology and bioenergetic models that will guide future efforts in conserving fish stocks.

The editors wish to thank David Randall and Tony Farrell for help and encouragement in the formulation of this volume, and Andrew Richford and Kirsten Funk at Academic Press offices in London and San Diego for shepherding this volume through the publication process. Numerous colleagues provided insightful reviews of chapter drafts, and we thank all the authors for their patience and cooperation throughout this endeavor.

Robert E. Shadwick
George V. Lauder
Vancouver and Boston

1

MECHANICS OF RESPIRATORY PUMPS

ELIZABETH L. BRAINERD
LARA A. FERRY-GRAHAM

I. INTRODUCTION

To facilitate oxygen uptake and carbon dioxide excretion, fishes ventilate their gas exchange surfaces with water or air. Because water and air differ substantially in their density, viscosity, and oxygen content, the biomechanical problems associated with aquatic and aerial ventilation also differ. Nonetheless, aerial and aquatic respiratory pumps do share one biomechanical challenge stemming from the fact that muscles only generate force in the direction of shortening (Brainerd, 1994b). It is a simple matter for muscle contraction to generate positive pressure and force fluid out of a cavity, but respiratory pumps also require an expansive phase to refill the cavity with new fluid. Some biomechanical trickery is necessary for muscle shortening to cause the expansion of a cavity and the generation of subambient pressure. This trickery generally takes the form of a lever system or occasionally elastic recoil, as is described for aquatic and aerial respiratory pumps in Sections II and III below.

Fish Biomechanics: Volume 23
FISH PHYSIOLOGY

DOI: 10.1016/S1546-5098(05)23001-7

The primary biomechanical problems in the design of aquatic respiratory pumps stem from the physical and chemical properties of water: high density (1000 kg m^{-3} for fresh water), high viscosity (1.0×10^{-3} Pa s for fresh water at 20 °C), and low oxygen content (from 0.4% by volume in seawater at 30 °C to 1% by volume in fresh water at 0 °C when in equilibrium with air). To minimize the work of ventilation, the high density of water dictates that the respiratory medium should undergo as little acceleration and deceleration as possible, the high viscosity dictates that fluid velocities should be low, and the low oxygen content dictates that oxygen extraction efficiency should be high. The unidirectional flow, countercurrent gas exchange system of ray-finned and cartilaginous fishes is well designed to meet these requirements (Hughes and Shelton, 1962). Buccal and opercular pumps, as described in Section II, generally work together to produce unidirectional flow of water over the gills, but some interesting cases of momentary flow reversal have recently been discovered (Summers and Ferry-Graham, 2001).

In contrast to water, air has low density (1.2 kg m^{-3} at 20 °C), low viscosity (0.02×10^{-3} Pa s at 20 °C), and high oxygen content (21% by volume). Aerial gas exchange is a primitive characteristic for ray-finned fishes that was lost in basal euteleosts and that has re-evolved at least 38 times and possibly as many as 67 times within acanthomorph fishes (Liem, 1980, 1988; Graham, 1997). Gas exchange organs include lungs, respiratory gas bladders, skin, gills, and various air-breathing organs (ABOs) such as the labyrinth organs of anabantoids (Liem, 1980; Graham, 1997). The biomechanical challenges for aerial respiratory pumps stem from predation risk (because fishes are vulnerable when they go to the surface to breathe and thus must limit their time there), hydrostatic pressure, buoyancy, surface tension, and mechanical conflicts between breathing and feeding. As described in Section III, the solutions to these problems are diverse.

II. AQUATIC RESPIRATORY PUMPS

In fish gills, the exchange of dissolved gases between water and blood occurs on the surface of tiny, leaf-like projections—the secondary lamellae. Water is pumped over the secondary lamellae in a direction opposite to the direction of the blood moving through the vessels of the secondary lamellae (Hughes and Shelton, 1962). This countercurrent flow of water and blood produces much greater oxygen extraction from the water than would be produced by concurrent flow. When the flow is concurrent, water and blood quickly reach diffusion equilibrium and no more oxygen can be extracted. In countercurrent flow, even though diffusion is occurring, the partial pressure of oxygen in the water is always slightly higher than the partial pressure of the oxygen in the blood, allowing extraction of a high percentage of the oxygen from the water. Countercurrent gas exchange results in oxygen partial pressures that are higher in the blood leaving the lamellae and entering the body than in the water exiting the gill slits. Fishes are the only vertebrates that can achieve such high percentages of oxygen extraction from their respiratory medium (Piiper and Scheid, 1992).

The efficiency of countercurrent exchange depends on the ability of the aquatic respiratory pumps to produce unidirectional flow of water over the gills. In both actinopterygian and elasmobranch fishes, unidirectional flow is achieved with a two-phase pump system.

A. Two-Phase Pump in Actinopterygian Fishes

The two-phase pump models of aquatic ventilation come from the pioneering work of G. M. Hughes (Hughes and Shelton, 1958; Hughes, 1960a,b, 1966, 1970, 1978a,b; Hughes and Ballintijn, 1965; Hughes and Umezawa, 1968; Hughes and Morgan, 1973). In Hughes's models, the buccal and opercular cavities are depicted as pistons (Figure 1.1). The movement of a piston to increase or decrease the volume inside a chamber mimics the expansion and compression of the buccal and opercular cavities during normal ventilation. In the two-phase model, the suction pump phase begins with the opercular cavity compressed and just beginning to expand, causing the pressure inside to be much lower than ambient and somewhat lower than the pressure in the buccal cavity (Figure 1.1, stage 1). This expansion of the opercular cavity results in water being drawn into the mouth, over the gills, and into the opercular cavity. At the start of the pressure pump phase, the buccal cavity begins to compress while the opercular cavity continues to expand (Figure 1.1, stage 2). Subsequently, the buccal cavity reaches

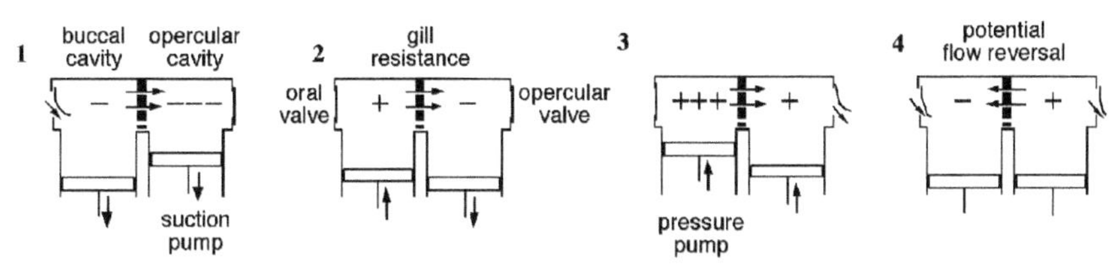

Fig. 1.1. The two-phase pump model of aquatic ventilation as developed by Hughes (1960a,b): stage 1, start of suction pump phase; stage 2, transition from suction to pressure pump; stage 3, pressure pump phase; stage 4, transition from pressure to suction pump phase. During the stage 4 transition, pressure may be momentarily higher in the opercular than in the buccal cavity. Flow reversal may result from the pressure reversal, or adduction of the gill bars may pose enough resistance to block backflow. (Adapted from Ferry-Graham, 1999, Figure 6, p. 1507 and Summers and Ferry-Graham, 2002, Figure 4 p. 96].)

maximal compression before the opercular cavity, thereby maintaining higher pressure in the buccal cavity and maintaining unidirectional flow as water exits the opercular valves (Figure 1.1, stage 3). Just as the pressure pump ends and the suction pump starts again, there is a brief moment of pressure reversal in which opercular pressure is higher than buccal pressure (Figure 1.1, stage 4). This pressure reversal may, in some circumstances, produce brief reversals of flow (see later discussion), but overall the effect of the two-phase pump is to produce flow over the gills that is unidirectional and continuous, albeit highly pulsatile (Hughes, 1960b; Piiper and Schuman, 1967; Scheid and Piiper, 1971, 1976; Malte, 1992; Malte and Lomholt, 1998; Piiper, 1998).

The suction and pressure pumps are powered by abduction and adduction of the opercula, suspensoria, and hyoid apparatus. To generate buccal and opercular expansion and create the subambient pressures of the suction pump, each of these functional units acts as a lever system to convert muscle shortening into abduction of skeletal elements. The motor pattern of the two-phase aquatic respiratory pump is summarized in Figure 1.2 (Liem, 1985). Starting with the pressure phase (P in Figure 1.2) the adductor mandibulae muscle fires (becomes active) to reduce the gape of the mouth, which in many fishes is sealed with a flap-like oral valve that closes in response to superambient pressure in the buccal cavity. Then the geniohyoideus fires to protract and elevate the hyoid apparatus, and the adductor arcus palatini fires to adduct the suspensorium, thereby compressing the buccal cavity. Increased pressure in the buccal cavity drives water across the gills and into the opercular cavity, and at the end of the pressure pump phase, the adductor operculi contracts and water is forced out the opercular valve. At the beginning of the suction pump phase (S in Figure 1.2), the levator operculi fires to open the mouth by a small amount and the levator arcus palatini fires to abduct the suspensorium. After a slight delay, the dilator operculi fires to abduct the operculum, and the pressure in the opercular chamber falls below buccal pressure and water is drawn over the gills. The branchiostegal rays fan out during opercular expansion to maintain the opercular valve seal. Then the adductor mandibulae fires and the pressure phase starts again.

The slight delay between the start of buccal expansion and the firing of the dilator operculi leads to the potential for a momentary pressure reversal (Figure 1.1, stage 4). The available data to date for teleosts suggest that while pressure reversals do occur, concomitant flow reversals likely do not occur (Hughes and Shelton, 1958; Saunders, 1961). Lauder (1984) demonstrated that the gill bars adduct during the pressure reversal, momentarily increasing the resistance between the buccal and opercular cavities. By placing plastic spacers on the gill bars to prevent them from closing fully during normal

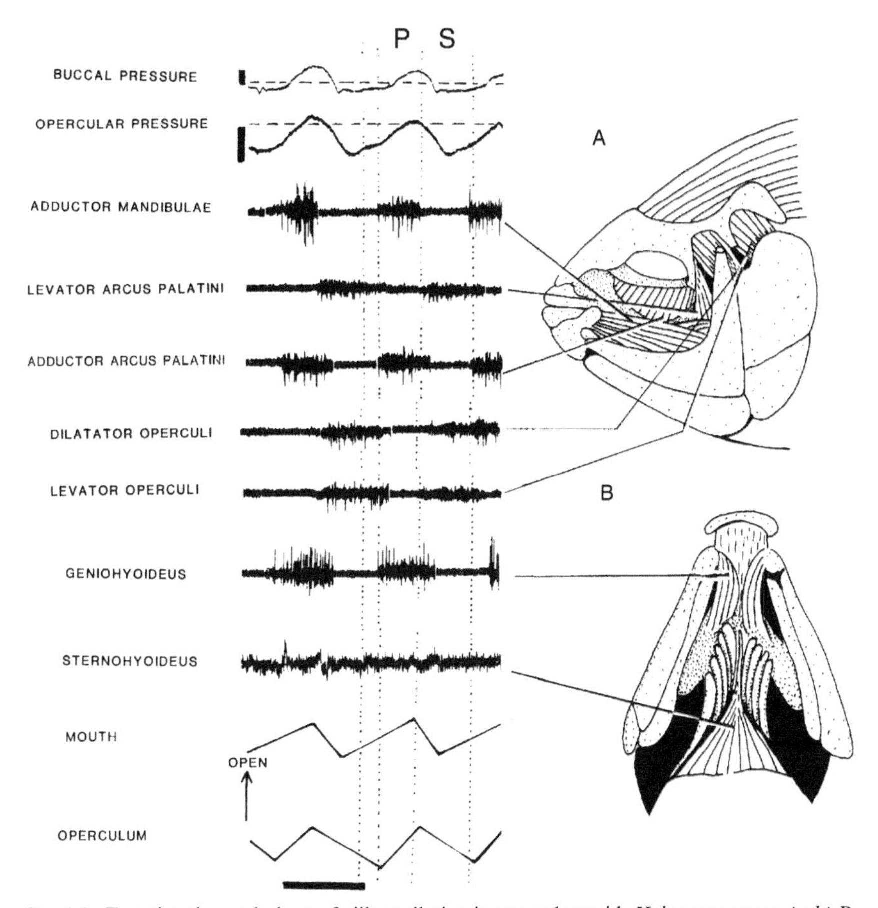

Fig. 1.2. Functional morphology of gill ventilation in an anabantoid, *Heleostoma temmincki.* P, the pressure pump phase (stage 3 of Figure 1.1). Note that buccal pressure always exceeds opercular pressure. S, suction pump phase (stage 1 of Figure 1.1). Note that opercular pressure is lower than buccal pressure. (From Liem, 1985, Figure 11–2, p. 187.)

respiration, Lauder was able to observe flow reversals. When the spacers were absent, flow reversals were not observed (Lauder, 1984).

The two-phase aquatic respiratory pump model has been found to apply to most teleost species studied to date, including the following freshwater fishes: trout *Salmo trutta*, tench *Tinca tinca*, and roach *Leuciscus rutilus* (Hughes and Shelton, 1958); white sucker catfish *Catostomus commersoni* and brown bullhead catfish *Ictalurus nebulosus* (Saunders, 1961); and carp *Cyprinus carpio* (Saunders, 1961; Ballintijn, 1969a,b). Pelagic and semi-pelagic marine species studied include the horse mackerel *Trachurus*

trachurus, herring *Clupea harengus*, whiting *Gadus merlangus*, conger eel *Conger conger*, rockling *Onos mustela*, great pipefish *Syngnathus acus*, and wrasse *Crenilabrus melops* (Hughes, 1960a). Benthic marine species also appear to fit this model: stickleback *Gasterosteus aculeatus* (Anker, 1978; Elshoud, 1978), bullhead sculpin *Cottus bubalis*, butterfly blenny *Blennius ocellaris*, grey gurnard *Trigla gurnardus*, and dragonet *Callionymus lyra* (Hughes, 1960a). Morphological evidence combined with opportunistic observation of live specimens suggests that the two-phase pump is also used by the bowfin *Amia calva* (Liem, 1985) and coelacanth *Latimeria chalumnae* (Hughes, 1995).

Even the morphologically bizarre flatfishes appear to fit this model (Hughes, 1960a; Liem *et al.*, 1985). With one eye having migrated to the opposite side of the head, they rest on the substrate on the "blind side," which can be either the left or the right side of the body. When flatfishes are at rest and buried in mud or sand, the two-phase pump is modified such that water generally exits from only the eyed side (Yazdani and Alexander, 1967; Kerstens *et al.*, 1979). During activity or when exposed to hypoxia, water exits from both sides (Steffensen *et al.*, 1981a; Liem *et al.*, 1985), and during extreme hypoxia, flatfishes will even raise their heads up above the substrate, presumably to reduce the resistance encountered by the exhaled water (Steffensen *et al.*, 1981a).

For reasons that are unclear, some teleosts have gill slits that are restricted to a small hole; the rest of the opercular valve and the branchiostegal rays are covered with skin. Some of the fishes with tiny gill openings are all tetraodontiforms (pufferfishes, triggerfishes, and their allies), some pleuronectiforms (flatfishes), synbranchiform and elopomorph eels, some antennariids (anglerfishes), and some gasterosteiforms (pipefish and seahorses). The puffers, anglerfishes, flatfishes, and seahorses jet water out of their gill openings at the start of locomotion or when handled (Brainerd *et al.*, 1997; E.L.B., personal observation). It is possible that the function of reduced gill slits is to increase the velocity of these water jets, but a more thorough survey of opercular valve morphology and function is needed to draw any firm conclusions.

B. Two-Phase Pump in Elasmobranch Fishes

It was once thought that a countercurrent gas exchange system does not exist in cartilaginous fishes because they often exhibit lower oxygen extraction efficiencies relative to bony fishes (Millen *et al.*, 1966; Piiper and Schuman, 1967). Elasmobranchs differ morphologically from actinopterygian fishes in several ways with regard to respiratory features. Most notably, they have five or more gill slits on each side of the head compared with the single opercular opening in ray-finned fishes. The parabranchial chambers in elasmobranchs, which are homologous with the opercular chamber of actinopterygians, are similarly separated by septa along their length internally. Early work by several authors proposed that the septa, to which the lamellae are attached, might interfere with the flow of water and force concurrent exchange during at least part of the respiratory cycle (Hughes and Shelton, 1962; Piiper and Schuman, 1967). Piiper and Schuman (1967) proposed a "multi-capillary" model, much like gas exchange in birds, to explain the observed partial pressures of oxygen in the blood and the ventilatory water. Further work, however, rejected this view on the grounds that the partial pressure of oxygen in the arterial blood was higher than that of the expired water in the *Scyliorhinus stellaris*, as can only be achieved with a countercurrent gas exchange system (Piiper and Baumgarten-Schumann, 1968). Further investigations support the notion of a countercurrent gas exchanger in elasmobranchs (Grigg, 1970; Scheid and Piiper, 1976; De Vries and De Jager, 1984), and the countercurrent exchange model presently serves to describe gas exchange in all aquatic-breathing fishes, even hagfish (Malte and Lomholt, 1998) and lamprey (Mallatt, 1981, 1996).

The respiratory pump in elasmobranchs is a two-phase pump that is very similar to the actinopterygian two-phase pump (Figure 1.3A) (Hughes, 1960b; Hughes and Ballintijn, 1965). Recent work on several elasmobranch species has demonstrated, however, that flow reversals are only partially prevented by the action of the gill bars, and that flow reversals may be widespread among species and body types (Figure 1.3B) (Ferry-Graham, 1999; Summers and Ferry-Graham, 2001, 2002). It is only with the application of technologies recently made available to biologists that we have been able to observe directly the path and pattern of water flow during ventilation. The pioneers in this field had to rely on pressure recordings taken inside the respiratory chambers to infer patterns of water flow. Further, movements of any pertinent anatomical features, because they are internal, could only be inferred from electromyographic recordings indicating when the muscles were electrically active, but not necessarily performing actual movements. The addition of sonomicrometry to this field has allowed the determination of the physical position of important morphological elements. Sonomicrometry, combined with the use of endoscopy to visualize anatomical elements in action and the movement of the ventilatory water, has confirmed that although the core elements of Hughes's elasmobranch models are correct, small differences exist, at least among the species originally studied and those studied more recently (Ferry-Graham, 1999; Summers and Ferry-Graham, 2001, 2002). The most important of these is the observation that the gill bars do close, but not for the entire duration of the pressure reversal period (Figure 1.4). Thus, water does flow back over the gills and into the oral chamber.

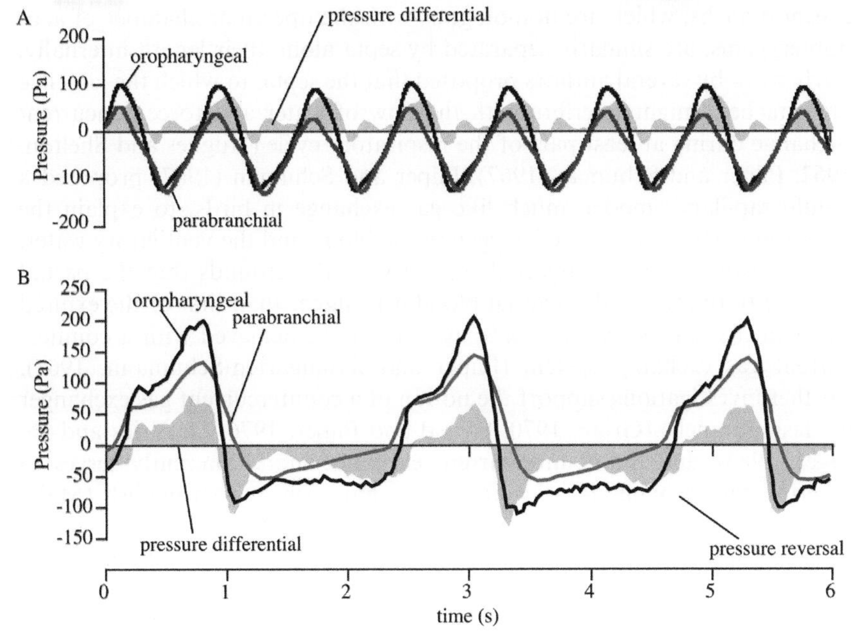

Fig. 1.3. Representative traces showing pressure reversals in (A) *Cephaloscyllium ventriosum* and (B) *Leucoraja erinacea*. The data from *L. erinacea* show much longer pressure reversals (indicated by negative pressure differential). Individuals of *C. ventriosum* also sometimes showed reversals of this magnitude and duration, although they were not as common. *Squalus acanthias* (data not shown) also showed both types of reversal profiles. *L. erinacea* did not exhibit profiles as in (A). Profiles from *C. ventriosum* and *L. erinacea* sometimes lacked a pressure reversal; *S. acanthias* profiles always had a reversal of some nature.

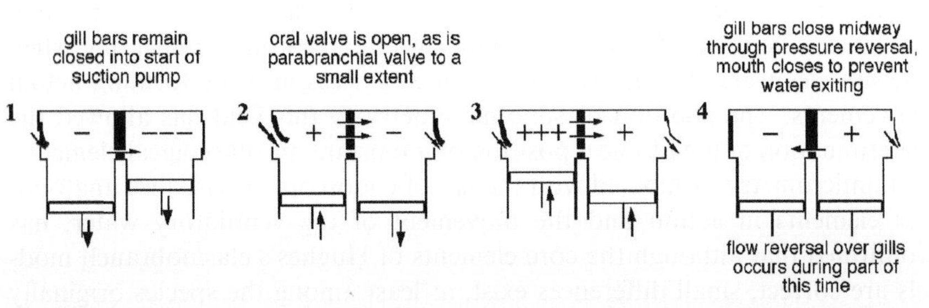

Fig. 1.4. Modifications to the two-phase pump model in elasmobranchs verified by sonometric data and direct observation of anatomical elements and water flow inside the oral and parabranchial chambers using endoscopy (Summers and Ferry-Graham, 2002). Specific modifications are indicated with text on the figures at each time interval. The mouth may be slightly open in stage 4, depending on the species. (Adapted from Ferry-Graham, 1999, Figure 7, p. 1508 and Summers and Ferry-Graham, 2002, Figure 4, p. 96.)

Flow reversals have been difficult to detect since they are typically not apparent externally. Valves normally prevent water from exiting the mouth or entering through the gill slits in most species. Water was never observed exiting the mouth in the swellshark *Cephaloscyllium ventriosum* (Ferry-Graham, 1999; Summers and Ferry-Graham, 2002), and it only rarely exited the mouth in the skates *Leucoraja erinacea* and *Raja clavata* (Hughes, 1960b; Summers and Ferry-Graham, 2001, 2002). Water exited the mouth more frequently in the dogfish *Squalus acanthias*, but not for the entire portion of the pressure reversal period and not during every pressure reversal (Summers and Ferry-Graham, 2002). Water never entered through the gills slits in any species studied. This is likely due to the fact that the reversals are fairly small in nature and short in duration. For example, water did not exit the mouth of most *L. erinacea,* even when the mouth was open and flow reversals were directly observed at the gills (Summers and Ferry-Graham, 2002).

Bidirectional flow has been observed, and tends to be much more obvious, at the spiracles of some elasmobranchs. Spiracles are openings on the dorsal surface of the head that lead directly to the oral chamber and channel water toward the gills. Recent comparative analyses suggest that the spiracle is a derived feature within elasmobranchs (Summers and Ferry-Graham, 2002), but this analysis depends strongly on the placement of the batoids within any given elasmobranch phylogeny, and the position of Batoidea is still in flux (Shirai, 1996; Douady *et al.*, 2003). The presence of the spiracle is not tightly correlated with a benthic habitat, as *C. ventriosum*, a derived carchariniform shark, is largely benthic but lacks spiracles, and *S. acanthias*, a basal squaliform shark, spends much of its time in open water and has fairly large spiracles. However, the use of the spiracle as the exclusive ventilatory aperture has been observed only in benthic species.

Water was seen to enter and exit the spiracle in *L. erinacea* when the skate was resting on the bottom (Summers and Ferry-Graham, 2001), and was also seen on occasion in *R. clavata* in earlier studies (Hughes, 1960b). In contrast, no consistent pattern of exclusive spiracular use was observed in the non-benthic dogfish, *S. acanthias*. Skates tend to rest or even bury themselves in the substrate, and thus the mouth is not or cannot be used to draw in a current of water for respiration during these periods of time. Outflow through the gills may be similarly reduced to prevent stirring up sediment upon discharge. Although distantly related, the sturgeon, *Acipenser transmontanus*, provides some evidence for this notion via the evolution of convergent structures. The sturgeon inhabits and forages in largely silty benthic habitats. Despite its reduced spiracles, enlarged openings on the dorsal regions of the gill slits serve to both draw in and expel water for respiration (Burggren, 1978). Other benthic fishes, such as *C. ventriosum*, in which the spiracles are so reduced that they are presumed to be nonfunctional, have been observed propped up on their pectoral fins or with their neurocrania rotated dorsally during periods of very active buccal pumping, thereby increasing the exposure of the mouth to the surrounding water (L.A.F.G., personal observation).

The physiological consequences of flow reversals, whether the reversals are inadvertent, as during the switch from pressure to suction pump, or apparently deliberate, as in spiracular breathing, may not be as grave as some researchers have suggested. Most species can tolerate large, experimentally induced inefficiencies in gas exchange (Malte, 1992), and it is likely that natural flow reversals decrease as oxygen demand increases and the respiratory pumps work harder.

The kinematics of ventilation in elasmobranchs are highly variable (Hughes, 1960b, 1978a; Hughes and Ballintijn, 1965). Much of this variation may be driven by physiological requirements, such as oxygen demand. For example, increases in ventilatory stroke volume are likely achieved by increases in the compression and subsequent expansion of the oral and parabranchial chambers. When a fish is at rest and the oral and parabranchial chambers are compressed to a lesser degree, the two-pump system can break down. Several scenarios have been documented, ranging from double pressure reversals to a complete failure of the suction pump to operate. Figure 1.5 depicts a scenario in which the pressure reversal is extreme. Sonometric and endoscopic data show that the gill bars are closed during stages 1 and 2 of such sequences, preventing prolonged reversals in water flow. However, water is also not flowing from anterior to posterior, as the suction pump is insufficient to generate flow. Variations of this pattern exist such that pressure reversals are seen at stages 4, 1, and 2 (Summers and Ferry-Graham, 2002), and just 4 and 2 (Ferry-Graham, 1999), whereby the suction pump presumably manages to create some anterior-to-posterior flow between pressure reversals.

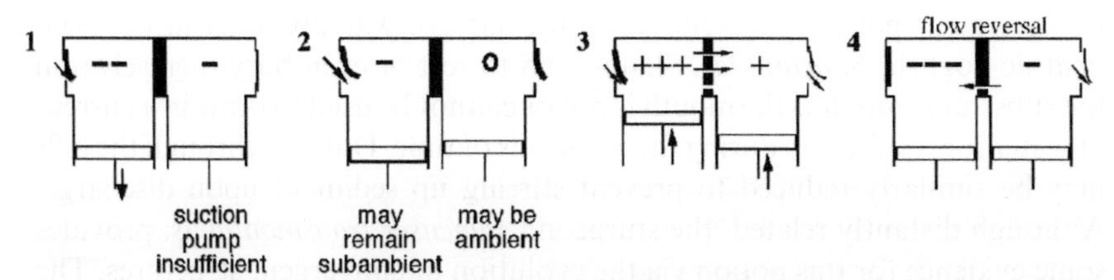

Fig. 1.5. A general scenario depicting a complete failure of the suction pump to generate anterior-to-posterior water flow verified by sonometric data and direct observation of anatomical elements and water flow inside the oral and parabranchial chambers using endoscopy (Summers and Ferry-Graham, 2002). The mouth may be slightly open in stage 4, depending on the species. (Adapted from Ferry-Graham, 1999, Figure 7, p. 1508 and Summers and Ferry-Graham, 2002, Figure 4, p. 96.)

C. Ram Ventilation

During ram ventilation, a respiratory current is generated by the locomotor efforts of the fish. In fast-swimming fishes, water enters the oral cavity and passes over the gills as long as the fish holds its mouth and opercular valves open.

Many fishes are able to buccal pump when needed but switch to ram ventilation at appropriate swimming speeds. Facultative ram ventilation has been documented in paddlefish *Polyodon spathula* (Burggren and Bemis, 1992; Sanderson *et al.*, 1994), sandtiger sharks *Odontaspis* (= *Eugomphodus* or *Carcharias*) *taurus* (von Wahlert, 1964), leopard sharks *Triakis semifasciata* (Hughes 1960b), a variety of salmonids (Roberts, 1978; Steffensen, 1985), several pelagic species such as mackerel *Scomber scombrus*, blue runner *Caranx crysos*, bluefish *Pomatomus saltatrix*, scup *Stenotomus crysops*, and the halfmoon *Medialuna californica* (Roberts, 1975), and sharksuckers *Echeneis naucrates* and remoras *Remora remora* when attached to a fast-swimming shark or aquatic mammal (Muir and Buckley, 1967; Steffensen and Lomholt, 1983; Steffensen, 1985). Interestingly, a number of species, including some that routinely move into open water habitats, never switch to ram ventilation. An apparent inability to perform ram ventilation has been documented in the striped mullet *Mugil cephalus* and in basses and rockfishes of the genera *Paralabrax* and *Sebastes* (Roberts, 1975). In facultative ram ventilation, the switch from buccal pumping to ram ventilation is triggered by a mechanoreceptor that is stimulated by flow velocity (Roberts and Rowell, 1988); benthic fishes may lack this reflex altogether (Roberts, 1978). Switching from active pumping to passive ram ventilating is estimated to save about 10% of the total energy expenditure during high-speed locomotion, although these calculations are only rough estimates (Brown and Muir, 1970; Roberts, 1978; Steffensen, 1985).

In contrast, pelagic fishes such as the scombrids (tuna and mackerel, primarily tuna), istiophorids (sailfish), and xiphiids (swordfish) are obligate ram ventilators. Their branchial anatomy is so severely reduced that they cannot generate a sufficient respiratory current using the buccal pump. There is a great deal of fusion of both the gill filaments and the lamellae in all of these families of fishes as well as in the dolphinfish *Coryphaena hippurus* (Muir and Kendall, 1968). Lamellae on adjacent filaments may be fused to one another along their facing edges, and in some adjacent filaments may even be fused along part of their length. Water passes through small slits or openings where fusion is incomplete. The reason for the fusion is not entirely clear, but it occurs widely among fast-swimming oceanic fish, and there appears to be greater fusion in more-derived species. Possible advantages of fusion include (1) restricting access by parasites to the gill tissues,

(2) increasing the rigidity of the structure so that it does not collapse and can therefore extract the greatest amount of oxygen possible, and (3) reducing the velocity of water flow over the lamellae to increase oxygen extraction (Muir and Kendall, 1968). Interestingly, similar fusion is found in *A. calva*, which lives in stagnant marshes, further suggesting that enhanced oxygen extraction may be a primary function of the fusion (Bevelander, 1934).

D. Gill Ventilation in Lamprey and Hagfish

In the two groups of extant jawless fishes, the anatomy of the respiratory pumps is markedly different from that of gnathostome fishes. Nonetheless, water flow through the oropharynx in lampreys and hagfishes is largely unidirectional and countercurrent gas exchange occurs (Mallatt, 1981, 1996; Malte and Lomholt, 1998).

The respiratory structures of hagfishes consist of pairs of sacs or pouches, anywhere from 6 to 14 depending on the species, that house the gill lamellae. The lamellae are the primary gas exchange surfaces (Malte and Lomholt, 1998). The skin of the hagfish is also quite permeable, but, except when scavenging on carcasses and other large food falls, hagfish are largely buried in the sediment with only their nostrils and tentacles exposed (Steffensen *et al.*, 1984). Water reaches the pouches through afferent ducts originating in the posterior portion of the pharynx and exits through efferent ducts that lead to external gill openings on either side of the animal. In some species, the efferent ducts fuse to form one common opening to the surrounding medium. Water enters the pharynx through the mouth or the nostril and is pumped into the afferent ducts by the action of the velum (Malte and Lomholt, 1998). The velum is a muscular structure situated at the dorsal midline of the rostral portion of the pharynx that serves to contract the chamber and pump water posteriorly. As a result, the flow entering the nostril is pulsatile and the frequency is highly variable, ranging from 0.01 to 1.3 Hz (Steffensen *et al.*, 1984), with the higher frequencies recorded from hagfish under warmer experimental conditions.

Based on anatomical studies, it was long thought that the velum alone was responsible for generating the respiratory current, and hagfish had little ability to alter the path of water once in the head. One of the first studies to examine hagfish anatomy in action was a cineradiographic study (Johansen and Hol, 1960). In this study, the researchers used barium and hypaque contrast agents to follow the path of the respiratory currents in live animals after introducing the contrast agents at either the mouth or the nostril. This foundational, and unequaled, study revealed that hagfish do use pumping of the velum to generate respiratory water flow through the head. However, the gill pouches themselves are muscular and also pump water through the system. Flow is further modified by the active control of sphincters located at both the afferent and efferent ends of the gill ducts. The sphincters open and close rhythmically during normal respiration, but this pattern can be altered as conditions require. The barium solution, for example, rarely entered the gill ducts and instead was routed directly from the esophagus to the gill openings, frequently by extreme expansion of the esophagus. Presumably, overfilling this chamber allowed for the forceful ejection of the offending material through the gill openings, and barium was prevented from entering the gill pouches by the sphincters. If a small amount of barium did enter the pouches, it was ejected back into the esophagus rather than continuing through the efferent gill ducts, where the maintenance of unidirectional flow is assisted by peristaltic-type contractions (Johansen and Hol, 1960). Clearly, hagfish can determine the water quality and/or particle sizes entering the head and alter the path of respiratory water accordingly to avoid contact with gas exchange surfaces.

Similar to hagfish, larval lamprey, or ammocetes, primarily use the action of a velar pump to generate a respiratory current (Rovainen, 1996). Ammocetes are suspension feeders, and thus ventilation and feeding are coupled and rely on a unidirectional current (Mallatt, 1981). The gill pouches are located within the pharynx (Mallatt, 1981), also referred to as the branchial basket (Rovainen, 1996). The velum has flaps that come together to form a seal during contraction, presumably preventing the flow of water back out the mouth. The velum moves posteriorly and the branchial basket contracts to produce an expiratory current, although the contribution of basket compression to expiration seems to be directly and positively related to activity or oxygen demand (Mallatt, 1981; Rovainen, 1996).

The inspiration of water back into the pharynx is powered primarily by elastic recoil of the branchial basket (Mallatt, 1981; Rovainen, 1996). During inspiration, water enters the mouth, passes through the velum and into the pharynx and gill sacs, and then exits via the branchiopores. Valves over the branchiopores reduce the influx of water during expansion of the branchial basket, but Mallatt (1981) noted that they function imperfectly and water is often drawn into the pharynx through the branchiopores during the inspiratory phase.

Mallatt (1981) suggested that the combined action of the velum and the branchial basket in ammocetes is sufficient to generate a two-phase pump as seen in actinopterygians and elasmobranchs. Contraction during expiration forces water laterally over the gill filaments and out the branchiopores and constitutes the first phase of the pumping cycle, the pressure pump phase. Elastic recoil of the basket during inhalation draws water in through the mouth via suction and constitutes the second phase of the pumping cycle. During ventilatory cycles in which only velar pumping is used and contraction of the basket does not contribute to water flow, the suction pump is not sufficient to generate substantial lateral flow across the gills. As noted previously, there is detectable backflow during the suction pump phase where water is drawn in through the branchiopores. This backflow period can be lengthy, persisting for up to half of the complete ventilatory cycle.

During metamorphosis from ammocete larva to adult lamprey, the velum is extensively remodeled. Many adult lamprey are parasitic, feeding by attaching their rasping mouth parts onto the sides of fishes with a sucker-like structure. Therefore, the mouth and anterior portions of the head are largely unavailable for respiration, and water both enters and exits the gill sacs via the external branchiopores. In adults, the velum presumably functions to prevent the rostral flow of water and maintain ventilation separate from feeding, while contraction and elastic recoil of the branchial basket exclusively generate the respiratory current (Mallatt, 1981; Rovainen, 1996).

III. AERIAL RESPIRATORY PUMPS

A. Evolutionary History and Biomechanical Challenges

Lungs are present in basal members of Actinopterygii and Sarcopterygii but not in Chondrichthyes; therefore, it is most parsimonious to conclude that lungs arose in stem osteichthians and have been retained as a primitive character in actinopterygians and sarcopterygians. Within Actinopterygii, paired lungs are present only in Polypteriformes, and an unpaired lung, homologous with paired lungs and termed a gas bladder, is present in other basal actinopterygians (Liem, 1988; Graham, 1997). The pneumatic duct connecting the gas bladder to the pharynx was lost in euteleosts, probably in stem acanthomorphs, and buoyancy control became the primary function of the gas bladder. Thus, the physoclistous swim bladder of euteleosts is homologous with the physostomous gas bladders of basal actinopterygians and with the lungs of tetrapods.

The physostomous gas bladder lost and regained its respiratory function several times in the evolutionary history of basal actinopterygians and teleosts (Liem, 1989b). However, once the pneumatic duct was lost, the swim bladder did not regain its respiratory function in any euteleosts. Instead, various other kinds of air-breathing organs (ABOs) evolved, such as the suprabranchial chambers of *Channa* and *Monopterus*, the branchial diverticulae of *Clarias* and anabantoids, and the stomach and intestinal modifications of some siluriforms (Graham, 1997).

All air-breathing fishes are bimodal or trimodal breathers (Graham, 1997). They retain gills as important sites of CO_2 excretion and ion exchange, and the gills also absorb oxygen when the water is not hypoxic. In addition, the skin is often an important site of gas exchange, both in water (Steffensen *et al.*, 1981b) and when fishes emerge during "terrestrial trespassing" (Liem, 1987). In severely hypoxic water, some air-breathing fishes may actually lose oxygen to the water through their gills and skin if the oxygen derived from air breathing causes the blood to have a higher oxygen tension than the surrounding water. This apparent inefficiency results from the fact that blood from most ABOs flows back to the heart and gills before being redistributed to the rest of the body. This seemingly maladaptive system is one of several lines of evidence that led to the myocardial oxygenation theory (Farmer, 1997), in which selection for increased oxygen delivery to the heart muscle is proposed as a primary selection force in the evolution of air breathing.

Aerial respiratory pumps face biomechanical challenges that result from the interaction of air and water. Within lungs and gas bladders, pressure generated by aerial pumps must overcome the surface tension of the air–liquid interface. However, surface tension is probably quite low, as surfactants are produced by the epithelia of gas bladders and lungs (Liem, 1988). Hydrostatic pressure also affects aerial respiratory pumps. If a fish takes an air breath with its body at an angle with the surface of the water, as is usually the case, then the aerial pump pressure must exceed the hydrostatic pressure at the deepest part of the gas-filled space (Figure 1.6). On the other hand, hydrostatic pressure may also assist breathing by contributing to exhalation.

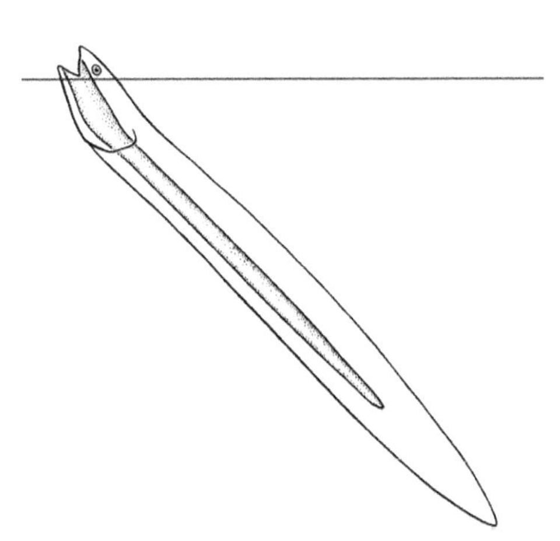

Fig. 1.6. The effect of hydrostatic pressure on air breathing. When a fish approaches the surface at an angle, hydrostatic pressure at the caudal end of the lungs or gas bladder may assist expiration but will also oppose inspiration.

Air breathing strongly affects the buoyancy of fishes, and this coupling between respiration and buoyancy places a constraint on the volume of air that can be held in a gas exchange organ. Fishes are vulnerable to both aerial and aquatic predators when they come to the surface to breathe (Kramer and Graham, 1976). Presumably there is selection to breathe as infrequently as possible, which should favor high-volume gas exchange organs. However, too much air would result in positive buoyancy—a condition that traps fishes on the surface and increases their vulnerability to predators. Therefore, the upper limit on thc size of aerial gas exchange organs is constrained by the need to avoid positive buoyancy.[1] In addition, air-breathing fishes have fine control over their gas volume and manage their buoyancy at slightly negative, neutral, or slightly positive, depending on their behavioral needs at any given moment in time (E.L.B., personal observation). In most cases, total gas volume is probably regulated on the basis of buoyancy, whereas tidal volume and breath frequency vary with metabolic needs.

B. Air Ventilation Mechanics

Unlike ourselves and other amniotes, fishes lack the intercostal and/or diaphragmatic muscles necessary for aspiration breathing. Instead, almost all air-breathing fishes use buccal pump breathing, in which expansions and compressions of the buccopharyngeal cavity ventilate the gas exchange organs (Liem, 1985). As described previously for aquatic ventilation, the hyoid apparatus and suspensorium act as lever systems to convert muscle shortening into buccal cavity expansion, thereby generating subambient pressure and drawing air in through the mouth. As the mouth closes, the hyoid protracts and the suspensorium adducts, generating superambient pressure and forcing air into the gas exchange organ. Aquatic ventilation, suction feeding, and aquatic coughing all involve buccopharyngeal expansion and compression, and the evolution of aerial buccal pumps appears to have occurred by modifying and combining these basic behaviors (McMahon, 1969; Liem, 1980, 1985; Brainerd, 1994a).

In most basal actinopterygian and basal teleost fishes, the respiratory gas bladder is ventilated with a four-stroke buccal pump, named by analogy with the piston movements in four-stroke internal combustion engines (Brainerd *et al.*, 1993; Brainerd, 1994a). A four-stroke air breath begins as the fish approaches the surface and transfers gas from the gas bladder into the buccal cavity. Hydrostatic pressure, elastic recoil of the gas bladder or

[1]One could imagine a scenario in which fishes might experience selection for added bone mass to offset a larger lung, if selection for infrequent air breathing were sufficiently strong. One possible group in which to look for this effect would be the armored catfishes.

body wall, and active expansion of the buccal cavity, thereby sucking gas out of the gas bladder, may all contribute to the transfer phase of expiration (Liem, 1988; Brainerd, 1994a). After gas transfer, the buccal cavity compresses and expired gas is expelled either out the mouth (*Amia*) or out the opercular valves (all others). With the fish still at the surface, the mouth then opens and the buccal cavity expands to inspire fresh air, whereupon the mouth closes and the buccal cavity compresses to pump the fresh air into the gas bladder. Thus, the four strokes of this buccal pump are (1) gas transfer, (2) expulsion, (3) inspiration, and (4) compression (Figure 1.7). Four-stroke breathing has been observed in basal actinopterygians, *Amia* and *Lepisosteus*, and in basal teleosts, *Arapaima*, *Gymnarchus*, *Notopterus*, *Pangasius* (Rahn *et al.*, 1971; Liem, 1988, 1989b; Brainerd, 1994a), and *Megalops* (E.L. B., personal observation).

In contrast to the four-stroke buccal pump of actinopterygians, lepidosirenid lungfishes ventilate their lungs with a two-stroke buccal pump[2] (Bishop and Foxon, 1968; McMahon, 1969; Brainerd *et al.*, 1993; Brainerd,

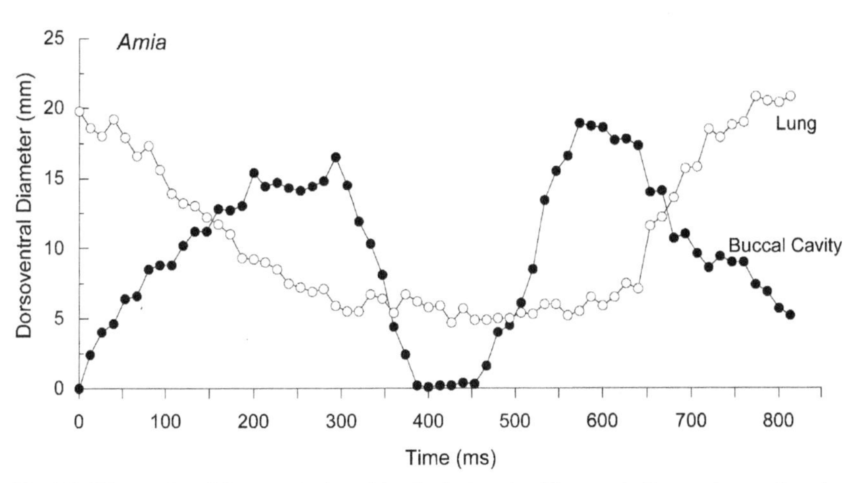

Fig. 1.7. Kinematics of four-stroke breathing in *Amia calva*. Changes in the maximum diameter of the buccal cavity and gas bladder were measured in lateral projection x-ray videos. Note that gas bladder (lung) diameter decreases during the first buccal expansion, and then the buccal cavity compresses to expel all of the expired air. Then the buccal cavity expands to draw in fresh air and gas bladder diameter increases as the buccal cavity compresses for the second time. (From Brainerd, 1994a, Figure 2, p. 291.)

[2]No data are available on air ventilation in the only extant, non-lepidosirenid lungfish, *Neoceratodus*, but observations of an Australian lungfish taking air breaths in a public aquarium suggest that they may use a four-stroke pump (E.L.B., personal observation).

1994a). With the snout of the lungfish protruding slightly from the surface of the water, the mouth opens and the buccal cavity expands to draw in fresh air. While the buccal cavity is expanding, exhalation of air from the lungs begins, driven by hydrostatic pressure, elastic recoil of the lungs and body wall, and possibly the contraction of smooth muscle in the lung walls. Neither buccal suction nor contraction of body musculature contributes to expiration (Figure 1.8). Buccal expansion generally continues beyond the end of expiration, and then buccal compression forces gas into the lungs (Figure 1.9). Because the buccal cavity does not compress after exhalation in two-stroke breathing, expired gas mixes with fresh air in the buccal cavity, and then this mixed gas is pumped into the lungs. In contrast, all of the expired gas is expelled from the buccal cavity in four-stroke breathing before fresh air is inspired and pumped into the gas bladder (Figure 1.7).

The two-stroke buccal pump is present in amphibians as well as in lepidosirenid lungfishes (Brainerd *et al.*, 1993), whereas the four-stroke buccal pump is typical of actinopterygian fishes. This phylogenetic pattern indicates that two-stroke breathing is the ancestral condition for Sarcopterygii, whereas four-stroke breathing is the ancestral condition for Actinopterygii. The ancestral condition for Osteichthyes cannot be determined, because no extant outgroups to Osteichthyes breathe air (Brainerd, 1994a).

The kinematics of the two- and four-stroke buccal pumps resemble kinematics associated with gill ventilation, suction feeding, and aquatic coughing (Brainerd, 1994a). Four-stroke breathing, suction feeding, and

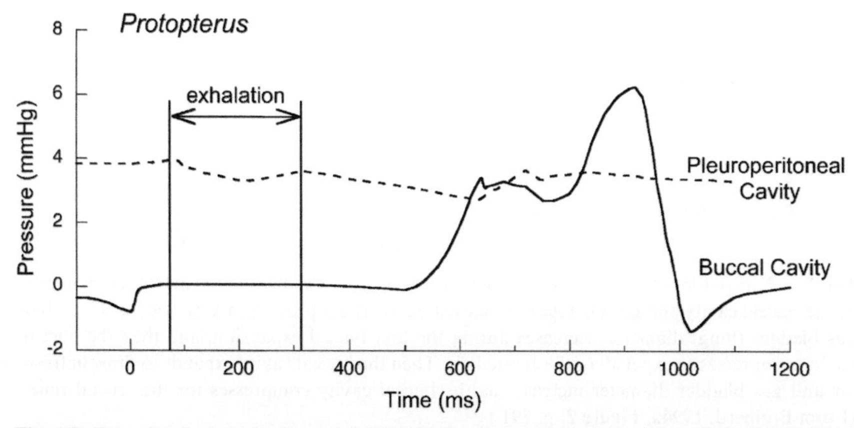

Fig. 1.8. Buccal and pleuroperitoneal (abdominal) pressure during an air breath in *Protopterus aethiopicus*. Note that pleuroperitoneal pressure decreases during exhalation, indicating a slight contribution of body wall elastic recoil to exhalation, but buccal pressure does not decrease, indicating that buccal expansion does not contribute to exhalation. (From Brainerd *et al.*, 1993, Figure 8, p. 176.)

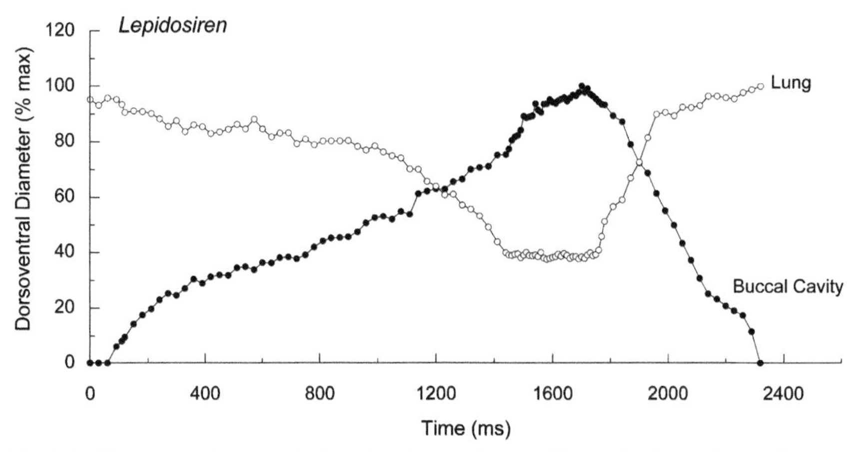

Fig. 1.9. Kinematics of two-stroke breathing in *Lepidosiren*. Changes in the maximum diameter of the buccal cavity and gas bladder (lung) were measured in lateral projection x-ray videos. Note that, in comparison to four-stroke breathing (Figure 7), the buccal cavity expands and compresses only once, and therefore some of the expired air is pumped back into the lung. (From Brainerd, 1994a, Figure 3, p. 293.)

aquatic coughing are all fast movements. The two complete buccal expansion–compression cycles of four-stroke breathing occur in under 1 s, with some fishes completing each cycle in less than 100 ms. The gas transfer and expiration phases may have arisen by modification of the aquatic cough, in which the buccal cavity is expanded with the mouth closed. The inspiration and compression phases may have arisen by modification of the movements associated with suction feeding.

The two-stroke buccal pump of lungfishes more closely resembles the aquatic ventilatory pump in its movements and timing (McMahon, 1969; Brainerd, 1994a). In four-stroke breathing, gill ventilation stops well before each air breath, but in lungfishes, gill ventilation continues as the fish approaches the surface of the water, and the buccal expansion associated with the air breath follows smoothly from the previous gill ventilation cycle (Brainerd, 1994a). The buccal cavity expands more during an air breath than during an aquatic breath, but otherwise the movements are very similar (McMahon, 1969). Aquatic breathing resumes immediately after the buccal compression phase of the air breath, without missing a beat in the aquatic ventilatory rhythm (Brainerd, 1994a).

Although the four-stroke buccal pump is typical for actinopterygians, two alternative ventilatory mechanisms have been described. In polypterid fishes, the patterns of buccal expansion and air transfer are similar to four-stroke, but elastic recoil of the ganoid scale jacket produces subambient pressure in the body cavity whereby air is aspirated into the lungs (Figure 1.10) (Brainerd *et al.*, 1989). Two euteleosts, *Gymnotus* and *Hoplerythrinus*, ventilate their gas bladders in a manner that is completely different from any other actinopterygians (Farrell and Randall, 1978; Liem, 1989b). An air breath starts with a large buccal expansion at the surface of the water (Figure 1.11). Then the fish sinks below the surface and compresses the buccal cavity to pump the air into its esophagus, which expands greatly, and the esophagus gradually empties into the gas bladder through the

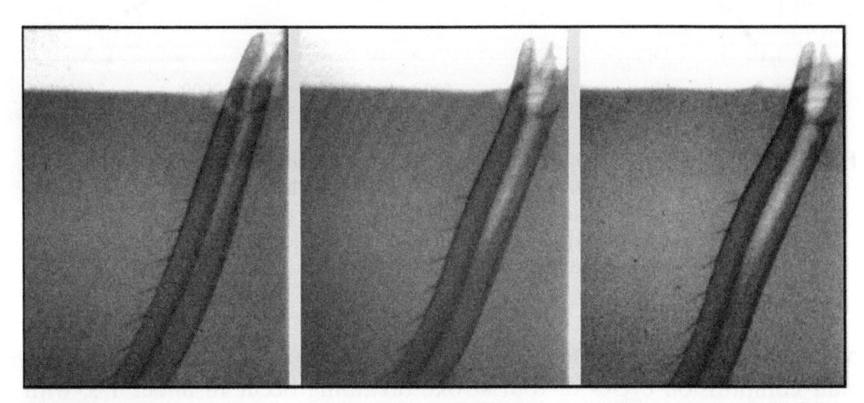

Fig. 1.10. Recoil aspiration in *Polypterus.* Frames from an x-ray video of lung ventilation in *Polypterus senegalis*, lateral projection. The left frame is at the end of expiration, and the middle and right frames show inspiration. Note that the mouth is wide open as the lungs refill with air, indicating that the fish is inhaling by aspiration breathing, rather than buccal pumping (a mouth seal is necessary for buccal pumping).

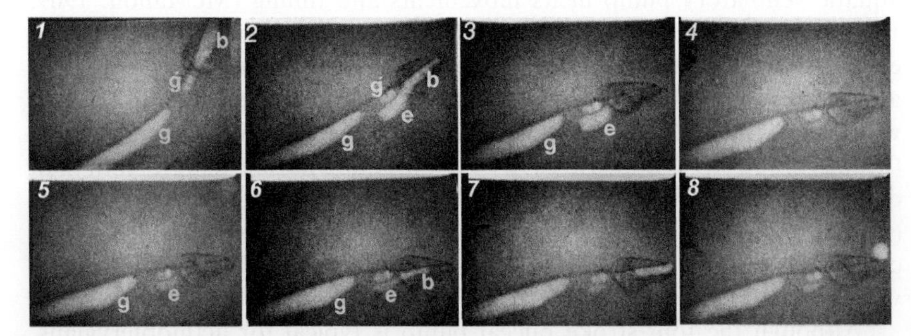

Fig. 1.11. Esophageal pump in *Gymnotus carapo.* Frames from an x-ray video of lung ventilation in lateral projection. Frames 1–4 show inspiration and frames 5–8 show expiration. See text for explanation. Abbreviations: b, buccal cavity; e, esophagus; g, gas bladder; g', anterior chamber of the gas bladder. (Adapted from Liem, 1989b, Figure 8, p. 346.)

pneumatic duct. The fish remains submerged and expiration ensues in reverse of inspiration; gas moves first into the esophagus and then into the buccal cavity and finally is released as bubbles. This mechanism results in relatively small tidal volumes (Figure 1.11), whereas two- and four-stroke breathing and recoil aspiration exchange between 50 and 100% of the gas bladder volume with each breath.

The loss of the pneumatic duct in stem acanthomorphs, presumably through lack of selection for air breathing, apparently produced an evolutionary constraint that prevented the subsequent recruitment of the swim bladder for gas exchange. Nonetheless, air breathing has evolved many times in higher teleosts, most commonly through the use of relatively unmodified buccal, opercular, pharyngeal, and/or branchial surfaces for gas exchange. In these cases, slight modifications of the expansive phase of the aquatic respiratory pump or suction feeding pump are used to draw in a bubble of air at the surface, and then the buccal and/or opercular cavities remain expanded to retain the bubble after submergence (Graham, 1997).

In some teleosts, more elaborate ABOs have evolved. A common theme is the evolution of a suprabranchial chamber (SBC) that may be a relatively simple space dorsal and caudal to the opercular cavity, as in *Monopterus*, or that may contain elaborate structures that increase the surface area for gas exchange, such as the labyrinth organ of anabantoids, the respiratory tree of *Channa*, and the respiratory fans and trees of *Clarias* (Graham, 1997). The dorsal location of the SBC makes biomechanical sense since inspired air will tend to rise up into the chamber and displace the gas or water that is present.

Ventilation of the suprabranchial chamber is accomplished by one of two mechanisms, named monophasic and diphasic by Peters (1978), and renamed triphasic and quadruphasic by Liem to reflect the number of phases recognizable with electromyography and cineradiography (Liem, 1980, 1985, 1989a). Triphasic ventilation is effective when the SBC has both anterior and posterior openings, as in anabantoids. The three phases are as follows: (1) a preparatory phase in which the buccal cavity compresses to expel water, (2) an expansive phase in which the buccal cavity expands to draw in fresh air through the mouth, and (3) a compressive phase in which the buccal cavity compresses to force fresh air into the SBC. The SBC is a rigid structure encased in bone, so the addition of fresh air forces the old gas out of the chamber, thus creating a unidirectional draft of air through the SBC (Liem, 1980).

Muscle activity during the triphasic pump is nearly identical to activity during suction feeding. The levator operculi (LO), levator arcus palatini (LAP), and sternohyoideus (SH) are active during expansion, and the adductor arcus palatini (AAP), adductor mandibulae (AM), and geniohyoideus (GH) are active during compression. One interesting difference is that the dilator operculi (DO) is active during the expansive phase of suction feeding but only becomes active at the end of the compressive phase of triphasic ventilation when bubbles are released through the opercular valve (Liem, 1985).

Quadruphasic ventilation is more complex and is bidirectional (Liem, 1980). The four phases are as follows (Figure 1.12): (1) a preparatory phase in which the buccal cavity compresses to expel water, (2) a reversal phase in which activity in the DO abducts the operculum rapidly, activity in the SH

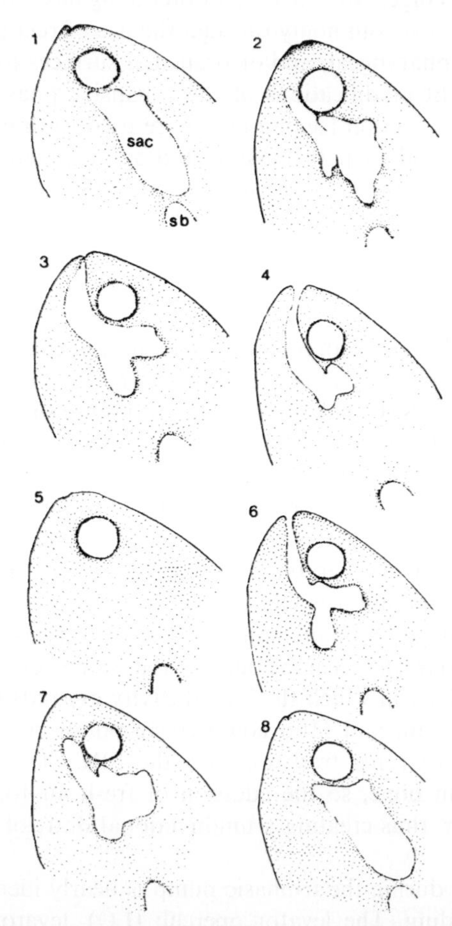

Fig. 1.12. Quadruphasic ventilation of the SBC in an anabantoid, *Heleostoma temmincki.* Drawings traced from an x-ray video of lung ventilation in lateral projection. See text for explanation. Abbreviations: sac, suprabranchial air chamber; sb, swim bladder. (From Liem, 1980, Figure 5, p. 66.)

retracts and depresses the hyoid apparatus, and a current of water is drawn into the posterior opening of the SBC, forcing gas forward through the anterior opening of the SBC and into the buccal cavity whence it is expelled, (3) an expansive phase in which the buccal cavity expands to draw in fresh air through the mouth, and (4) a compressive phase in which the buccal cavity compresses to force fresh air into the SBC. Muscle activity in phases 3 and 4 is identical to suction feeding, including the activity of the DO during expansion. The muscle activity of the reversal phase is identical to muscle activity during the aquatic cough, which is normally used to create a rostrad current of water to clear debris from the gills (Liem, 1980).

Most species of air-breathing fishes with an SBC use either triphasic or quadruphasic ventilation, but anabantoids are able to use both (Liem, 1989a). The quadruphasic pump relies on a current of water for expiration, so this pump works only when fishes are submerged; the triphasic pump works when fishes are in or out of water. Most air breathers that use buccopharyngeal surfaces or an SBC for gas exchange either expel the air bubble before feeding or lose the air bubble in the process of feeding. In anabantoids and clariids, however, air is not lost from the SBC during feeding. Valves separate the SBC from the buccal and opercular cavities, effectively decoupling feeding and air breathing. Liem (1989a) proposed this decoupling as an explanation for the relatively diverse types of food items eaten by anabantoids and clariids, compared to the limited diets of channids and synbranchiforms.

Some air-breathing teleosts, particularly the catfishes and loaches, dedicate parts of the digestive tract to gas exchange. In loricariid and trichomycterid catfishes, part of the stomach is thin walled and highly vascularized, and air breathing has been described for loricariids (Gradwell, 1971). Loricariids release air from their stomachs while resting on the bottom; the air escapes either out the mouth or out from under the operculum. Soon thereafter, the fish darts to the surface and grabs a bubble of air in the buccal cavity and forces it into the stomach. Loaches, family Cobitidae, and armored catfishes in the family Callichthyidae use the intestine for gas exchange. In both groups, the region of the intestine just proximal to the anus is thin walled and vascularized. Armored catfishes have been demonstrated to ventilate the intestine unidirectionally (Gee and Graham, 1978). A fish darts to the surface and grabs a bubble of air, and as it forces the air into the esophagus, a bubble emerges simultaneously from the anus. The armor of the catfish may play a role in this simultaneous expulsion of air. It is highly unlikely that air just pumped into the esophagus travels to the distal end of the digestive tract that quickly, but the armor may limit the total volume of the body to the extent that air forced in the front end increases the pressure in the whole peritoneal cavity, thus forcing air out the anus. Air presumably is then transported by peristalsis to the distal intestine for gas exchange.

IV. FUTURE DIRECTIONS

Work to date has yielded a fairly complete understanding of the functional morphology and basic mechanics of aquatic and aerial respiratory pumps in fishes, but many rich and interesting areas for future research remain. Most of the work reviewed here was done before the experimental techniques of sonomicrometry and endoscopy became available. Application of these techniques to the study of water flow in the pharynx has yielded some unexpected results, such as the discovery of substantial flow reversals during gill ventilation in elasmobranchs (Ferry-Graham and Summers, 1999) and the discovery of crossflow filtration in suspension feeding fishes (Sanderson *et al.*, 2001). Further application of sonomicrometry to quantify shape changes of the pharynx and endoscopy to measure fluid flow could yield the data necessary for the production of more sophisticated and quantitative models of gill ventilation and gas exchange.

Sonomicrometry could also be applied to study the length changes of respiratory muscles during gill ventilation. Most work on whole muscle function has focused on high-performance locomotor activities (reviewed in Biewener, 2002). The study of cranial muscles during gill ventilation could yield information on the behavior of muscles when the strongest selection is likely to act on energetic efficiency rather than on maximizing force or power. This work may also relate to the function of muscles that perform multiple tasks with markedly different performance requirements. The muscles of the gill ventilation pump are also used for suction feeding, a function that presumably requires high power output from the muscles (because the muscles do work to accelerate water into the mouth). Are breathing and suction feeding achieved by different muscle fiber types? How are these fiber types activated? Does the presence of a large volume of inactive fast fibers in a dual-use muscle reduce the energetic efficiency of gill ventilation (due to the inertia and viscosity of the extra muscle mass)? Might this be a source of balancing selection on the size of muscles used for suction feeding?

Finally, as in almost all areas of fish biomechanics, studies of ventilation have focused primarily on adult fishes, with little attention paid to development and ontogeny. At small body sizes, water flow across the gills will be dominated by viscous forces (due to low Reynolds number), which will increase the work of breathing and also decrease the convective transport of oxygenated water to the surfaces of the secondary lamellae. However, this effect is balanced by the efficacy of diffusion over small distances. Small fish larvae absorb oxygen across their body and yolk sac surfaces; only at larger sizes do fish need gills at all. Mathematical modeling, combined with morphological and kinematic data, may provide the most insight into changes in the biomechanics of ventilation over the lifetimes of fishes.

ACKNOWLEDGMENTS

We are grateful to Karel Liem for reading and commenting on an earlier version of this chapter. Thanks to Harvard University Press, Blackwell Publishing, Springer-Verlag GmbH, Thomson Publishing Services, and the Society for Integrative and Comparative Biology for permission to reprint figures. This material is based in part on work supported by the National Science Foundation under Grant Nos. 9875245 and 0316174 to E.L.B. and 0320972 to L.A.F.G.

REFERENCES

Anker, G. C. (1978). Analyses of respiration and feeding movements of the three-spined stickleback, *Gasterosteus aculeatus* L. *Neth. J. Zool.* **28,** 485–523.

Ballintijn, C. M. (1969a). Movement pattern and efficiency of the respiratory pump of the carp (*Cyprinus carpio* L.). *J. Exp. Biol.* **50,** 593–613.

Ballintijn, C. M. (1969b). Muscle co-ordination of the respiratory pump of the carp (*Cyprinus carpio* L.). *J. Exp. Biol.* **50,** 569–591.

Bevelander, G. (1934). The gills of *Amia calva* specialized for respiration in an oxygen deficient habitat. *Copeia* **1934,** 123–127.

Biewener, A. A. (2002). Future directions for the analysis of musculoskeletal design and locomotor performance. *J. Morphol.* **252,** 38–51.

Bishop, I. R., and Foxon, G. E. H. (1968). The mechanism of breathing in the South American lungfish, *Lepidosiren paradoxa*; a radiological study. *J. Zool. (Lond.)* **154,** 263–271.

Brainerd, E. L. (1994a). The evolution of lung-gill bimodal breathing and the homology of vertebrate respiratory pumps. *Amer. Zool.* **34,** 289–299.

Brainerd, E. L. (1994b). Mechanical design of polypterid fish integument for energy storage during recoil aspiration. *J. Zool.(Lond.)* **232,** 7–19.

Brainerd, E. L., Liem, K. F., and Samper, C. T. (1989). Air ventilation by recoil aspiration in polypterid fishes. *Science* **246,** 1593–1595.

Brainerd, E. L., Ditelberg, J. S., and Bramble, D. M. (1993). Lung ventilation in salamanders and the evolution of vertebrate air-breathing mechanisms. *Biol. J. Linn. Soc.* **49,** 163–183.

Brainerd, E. L., Page, B. N., and Fish, F. E. (1997). Opercular jetting during fast starts by flatfishes. *J. Exp. Biol.* **200,** 1179–1188.

Brown, C. E., and Muir, B. S. (1970). Analysis of ram ventilation of fish gills with application to skipjack tuna (*Katsuwonus pelamis*). *J. Fish. Res. Bd. Can.* **27,** 1637–1652.

Burggren, W. W. (1978). Gill ventilation in the sturgeon, *Acipenser transmontanus*: Unusual adaptations for bottom dwelling. *Resp. Physiol.* **34,** 153–170.

Burggren, W. W., and Bemis, W. E. (1992). Metabolism and ram gill ventilation in juvenile paddlefish *Polyodon spathula* (Chondrostei: Polyodontidae). *Phys. Zool.* **65,** 515–539.

De Vries, R., and De Jager, S. (1984). The gill in the spiny dogfish, *Squalus acanthias*: Respiratory and nonrespiratory function. *J. Anat.* **169,** 1–29.

Douady, C. J., Dosay, M., Shivji, M. S., and Stanhope, M. J. (2003). Molecular phylogenetic evidence refuting the hypothesis of Batoidea (rays and skates) as derived sharks. *Mol. Phylogenet. Evol.* **26,** 215–221.

Elshoud, G. C. A. (1978). Respiration in the three-spined stickleback, *Gasterosteus aculeatus* L.: An electromyographic approach. *Neth. J. Zool.* **28,** 524–544.

Farmer, C. (1997). Did lungs and the intracardiac shunt evolve to oxygenate the heart in vertebrates? *Paleobiology* **23,** 358–372.

Farrell, A. P., and Randall, D. J. (1978). Air-breathing mechanics in two Amazonian teleosts, *Arapaima gigas* and *Hoplerythrinus unitaeniatus*. *Can. J. Zool.* **56,** 939–945.

Ferry-Graham, L., and Summers, A. P. (1999). Kinematics of ventilation in the little skate, *Leucoraja erinacea*, as indicated by sonomicrometry. *Bull. Mt. Desert Isl. Biol. Lab.* **38,** 97–100.

Ferry-Graham, L. A. (1999). Mechanics of ventilation in swellsharks, *Cephaloscyllium ventriosum* (Scyliorhinidae). *J. Exp. Biol.* **202,** 1501–1510.

Gee, J., and Graham, J. (1978). Respiratory and hydrostatic functions of the intestine of the catfishes, *Hoplosternum thoracatum* and *Brochis splendens*. *J. Exp. Biol.* **74,** 1–16.

Gradwell, N. (1971). A photographic analysis of the air breathing behavior of the catfish, *Plecostomus punctatus*. *Can. J. Zool.* **49,** 1089–1094.

Graham, J. B. (1997). "Air-Breathing Fishes: Evolution, Diversity, and Adaptation." Academic Press, New York.

Grigg, G. C. (1970). Water flow through the gills of Port Jackson sharks. *J. Exp. Biol.* **52,** 565–568.

Hughes, G. M. (1960a). A comparative study of gill ventilation in marine teleosts. *J. Exp. Biol.* **37,** 28–45.

Hughes, G. M. (1960b). The mechanism of gill ventilation in the dogfish and skate. *J. Exp. Biol.* **37,** 11–27.

Hughes, G. M. (1966). The dimensions of fish gills in relation to their function. *J. Exp. Biol.* **45,** 177–195.

Hughes, G. M. (1970). A comparative approach to fish respiration. *Experientia* **26,** 113–122.

Hughes, G. M. (1978a). On the respiration of *Torpedo marmorata*. *J. Exp. Biol.* **73,** 85–105.

Hughes, G. M. (1978b). Some features of gas transfer in fish. *Bull. Inst. Math. & Appns.* **14,** 39–43.

Hughes, G. M. (1995). The gills of the coelacanth, *Latimeria chalumnae*: A study in relation to body size. *Phil. Trans. Roy. Soc. London B* **347,** 427–438.

Hughes, G. M., and Ballintijn, C. M. (1965). The muscular basis of the respiratory pumps in the dogfish (*Scyliorhinus canicula*). *J. Exp. Biol.* **43,** 363–383.

Hughes, G. M., and Morgan, M. (1973). The structure of fish gills in relation to their respiratory function. *Biol. Rev.* **48,** 419–475.

Hughes, G. M., and Shelton, G. (1958). The mechanism of gill ventilation in three freshwater teleosts. *J. Exp. Biol.* **35,** 807–823.

Hughes, G. M., and Shelton, G. (1962). Respiratory mechanisms and their nervous control in fish. *Adv. Comp. Physiol. Biochem.* **1,** 275–364.

Hughes, G. M., and Umezawa, S. (1968). Oxygen consumption and gill water flow in the dogfish *Scyliorhinus canicula* L. *J. Exp. Biol.* **49,** 557–564.

Johansen, K., and Hol, R. (1960). A cineradiographic study of respiration in *Myxine glutinosa* L. *J. Exp. Biol.* **37,** 474–480.

Kerstens, A., Lomholt, J. P., and Johansen, K. (1979). The ventilation, extraction and oxygen uptake in undisturbed flounders, *Platichthys flesus*: Responses to hypoxia acclimation. *J. Exp. Biol.* **83,** 169–179.

Kramer, D. L., and Graham, J. B. (1976). Synchronous air breathing, a social component of respiration in fishes. *Copeia* **1976,** 689–697.
Lauder, G. V. (1984). Pressure and water flow patterns in the respiratory tract of the bass (*Micropterus salmoides*). *J. Exp. Biol.* **113,** 151–164.
Liem, K. F. (1980). Air ventilation in advanced teleosts: Biomechanical and evolutionary aspects. *In* "Environmental Physiology of Fishes" (Ali, M. A., Ed.), pp. 57–91. Plenum Press, New York.
Liem, K. F. (1985). Ventilation. *In* "Functional Vertebrate Morphology" (Hildebrand, M., Bramble, D. M., Liem, K. F., and Wake, D. B., Eds.), pp. 186–209. Harvard University Press, Cambridge, MA.
Liem, K. F. (1987). Functional design of the air ventilation apparatus and overland excursions by teleosts. *Fieldiana Zool.* **37,** 29.
Liem, K. F. (1988). Form and function of lungs: The evolution of air breathing mechanisms. *Am. Zool.* **28,** 739–759.
Liem, K. F. (1989a). Functional design and diversity in the feeding morphology and ecology of air-breathing teleosts. *Forschr. Zool.* **35,** 487–500.
Liem, K. F. (1989b). Respiratory gas bladders in teleosts: Functional conservatism and morphological diversity. *Am. Zool.* **29,** 333–352.
Liem, K. F., Wallace, J. W., and Whalen, G. (1985). Flatfishes breathe symmetrically: An experimental reappraisal. *Exp. Biol.* **44,** 159–172.
Mallatt, J. (1981). The suspension feeding mechanism of the larval lamprey. *J. Zool. (Lond.)* **194,** 103–142.
Mallatt, J. (1996). Ventilation and the origin of jawed vertebrates: A new mouth. *Zool. J. Linn. Soc.* **117,** 329–404.
Malte, H. (1992). Effect of pulsatile flow on gas exchange in the fish gill: Theory and experimental data. *Resp. Physiol.* **88,** 51–62.
Malte, H., and Lomholt, J. P. (1998). Ventilation and gas exchange. *In* "The Biology of Hagfishes" (Jorgensen, J. M., Lomholt, J. P., Weber, R. E., and Malte, H., Eds.), pp. 223–234. Chapman and Hall Ltd., London.
McMahon, B. R. (1969). A functional analysis of aquatic and aerial respiratory movements of an African lungfish, *Protopterus aethiopicus,* with reference to the evolution of the lung-ventilation mechanism in vertebrates. *J. Exp. Biol.* **51,** 407–430.
Millen, J. E., Murdaugh, H. V., Jr., Hearn, D. C., and Robin, E. D. (1966). Measurement of gill water flow in *Squalus acanthias* using the dye-dilution technique. *Am. J. Physiol.* **211,** 11–14.
Muir, B. S., and Buckley, R. M. (1967). Gill ventilation in *Remora remora*. *Copeia* **1967,** 581–586.
Muir, B. S., and Kendall, J. I. (1968). Structural modifications in the gills of tuna and some other oceanic fishes. *Copeia* **1968,** 389–398.
Peters, H. M. (1978). On the mechanism of air ventilation in anabantoids (Pisces: Teleostei). *Zoomorphologie* **89,** 93–123.
Piiper, J. (1998). Branchial gas transfer models. *Comp. Biochem. Physiol. A* **119,** 125–130.
Piiper, J., and Baumgarten-Schumann, D. (1968). Effectiveness of O_2 and CO_2 exchange in the gills of the dogfish (*Scyliorhinus stellaris*). *Resp. Physiol.* **5,** 338–349.
Piiper, J., and Schuman, D. (1967). Efficiency of O_2 exchange in the gills of the dogfish, *Scyliorhinus stellaris*. *Resp. Physiol.* **2,** 135–148.
Piiper, J., and Scheid, P. (1992). Gas exchange in vertebrates through lungs, gills, and skin. *News Physiol. Sci.* **7,** 199–203.
Rahn, H., Rahn, K. B., Howell, B. J., Gans, C., and Tenney, S. M. (1971). Air breathing of the garfish (*Lepisosteus osseus*). *Resp. Physiol.* **11,** 285–307.

Roberts, J. L. (1975). Active branchial and ram gill ventilation in fishes. *Biol. Bull.* **148,** 85–105.
Roberts, J. L. (1978). Ram gill ventilation in fish. *In* "The Physiological Ecology of Tunas" (Sharp, G. D., and Dizon, A. E., Eds.), pp. 83–88. Academic Press, New York.
Roberts, J. L., and Rowell, D. M. (1988). Periodic respiration of gill-breathing fishes. *Can. J. Zool.* **66,** 182–190.
Rovainen, C. M. (1996). Feeding and breathing in lampreys. *Brain Behav. Evol.* **48,** 297–305.
Sanderson, S. L., Cech, J. J., and Cheer, A. Y. (1994). Paddlefish buccal flow velocity during ram suspension feeding and ram ventilation. *J. Exp. Biol.* **186,** 145–156.
Sanderson, S. L., Cheer, A. Y., Goodrich, J. S., Graziano, J. D., and Callan, W. T. (2001). Crossflow filtration in suspension-feeding fishes. *Nature (Lond.)* **412,** 439–441.
Saunders, R. L. (1961). The irrigation of the gills in fishes: I. Studies of the mechanism of branchial irrigation. *Can. J. Zool.* **39,** 637–653.
Scheid, P., and Piiper, J. (1971). Theoretical analysis of respiratory gas equilibration in water passing through fish gills. *Resp. Physiol.* **13,** 305–318.
Scheid, P., and Piiper, J. (1976). Quantitative functional analysis of branchial gas transfer: Theory and application to *Scyliorhinus stellaris* (Elasmobranchii). *In* "Respiration of Amphibious Vertebrates" (Hughes, G. M., Ed.), pp. 17–38. Academic Press, New York.
Shirai, S. (1996). Phylogenetic interrelationships of neoselachians (Chondrichthyes: Euselachii). *In* "Interrelationships of Fishes" (Stiassney, M. L. J., Parenti, L. R., and Johnson, G. P., Eds.), pp. 9–34. Academic Press, San Diego, CA.
Steffensen, J. F. (1985). The transition between branchial pumping and ram ventilation in fishes: Energetic consequences and dependence on oxygen tension. *J. Exp. Biol.* **114,** 141–150.
Steffensen, J. F., Johansen, K., Sindberg, C. D., Sorenson, J. H., and Moller, J. L. (1984). Ventilation and oxygen consumption in the hagfish, *Myxine glutinosa* L. *J. Exp. Mar. Biol. Ecol.* **84,** 173–178.
Steffensen, J. F., and Lomholt, J. P. (1983). Energetic cost of active branchial ventilation in the sharksucker, *Echeneis naucrates*. *J. Exp. Biol.* **103,** 185–192.
Steffensen, J. F., Lomholt, J. P., and Johansen, K. (1981a). Gill ventilation and O_2 extraction during graded hypoxia in two ecologically distinct species of flatfish, the flounder, *Platichthys flesus*, and the plaice *Pleuronectes platessa*. *Env. Biol. Fish.* **7,** 157–163.
Steffensen, J. F., Lomholt, J. P., and Johansen, K. (1981b). The relative importance of skin oxygen uptake in the naturally buried plaice, *Pleuronectes platessa*, exposed to graded hypoxia. *Resp. Physiol.* **44,** 269–275.
Summers, A. P., and Ferry-Graham, L. A. (2001). Ventilatory modes and mechanics of the hedgehog skate (*Leucoraja erinacea*): Testing the continuous flow model. *J. Exp. Biol.* **204,** 1577–1587.
Summers, A. P., and Ferry-Graham, L. A. (2002). Respiration in elasmobranchs: New models of aquatic ventilation. *In* "Vertebrate Biomechanics and Evolution" (Bels, V. L., Gasc, J. P., and Casinos, A., Eds.), pp. 87–100. Bios Scientific Publishers Ltd., Oxford.
von Wahlert, G. (1964). Passive respiration in sharks. *Naturwissenschaften* **51,** 297.
Yazdani, G. M., and Alexander, R. M. (1967). Respiratory currents of flatfish. *Nature* **213,** 96–97.

2

SKULL BIOMECHANICS AND SUCTION FEEDING IN FISHES

MARK W. WESTNEAT

I. INTRODUCTION

The evolutionary history of feeding biomechanics in fishes is a fascinating story of change in the structure and function of kinetic vertebrate skulls. From batoids to balistoids there is a spectacular diversity of skull form and feeding mechanisms among fishes, from sit-and-wait predators that use high suction forces to engulf their prey, to species that chase their prey during an attack, to fishes that remove pieces of their food using a biting strategy. Major feeding guilds among fishes include piscivores, herbivores, planktivores, detritivores, and molluscivores, and there are many more specialized modes of dietary preference, such as scale eating, parasitism, and consumption of wood. The attempt to explain this diversity has focused research efforts on several complementary themes such as functional and biomechanical analyses of cranial design in fishes (Alexander, 1967; Liem, 1970; Lauder, 1980a), the ecological roles of different feeding modes (Wainwright, 1996; Liem and Summers, 2000), and the evolutionary history of change in structure and function of the skull (Lauder, 1982; Wilga, 2002;

Fish Biomechanics: Volume 23
FISH PHYSIOLOGY

DOI: 10.1016/S1546-5098(05)23002-9

Westneat, 2004). The diversity of fishes and their feeding strategies, the importance of fish feeding to both freshwater and marine ecology, and the wide range of technological tools such as high-speed video, electromyography, sonomicrometry, and fluid mechanics used in feeding studies have coalesced to make fish feeding one of the most fruitful areas of functional and evolutionary morphology.

Several important themes emerge from the deep body of literature on skull mechanisms in fishes and the evolutionary analysis of trends of fish feeding. First, fish skulls are highly kinetic musculoskeletal systems with numerous movable elements (Alexander, 1969; Liem, 1980; Lauder, 1982; Westneat, 1990). The dynamics of skull motion during rapid feeding events have thus been a prime focus of both theoretical and experimental research in biomechanics. Whether they are suction feeding, ram feeding, or biting, fishes use a common set of skeletal structures and muscle systems to feed. What are the fundamental principles of cranial mechanics in fishes, and how has the mechanism of fish skulls been modified in structure and activation in order to achieve appropriate suction, ram, or biting performance during different types of feeding? A central goal of this chapter is to present the basic morphological structure of fish skulls, identify the principles of musculoskeletal biomechanics that transfer force and motion in fish feeding systems, and illustrate modifications of the basic pattern of prey capture that characterize some of the primary feeding modes in fishes.

A second pattern that emerges from the study of feeding in many fish lineages is the widespread use of suction during prey capture as a strategy to transport food into the mouth. Suction feeding is the most common mode of prey capture in teleost fishes (Liem, 1980; Muller and Osse, 1984; Lauder, 1985; Alfaro *et al.*, 2001; Ferry-Graham *et al.*, 2003) and is also used in many sharks and rays (Motta *et al.*, 1997; Wilga *et al.*, 2000). Suction is employed to draw elusive, suspended, or even attached prey into the mouth by rapidly expanding the intraoral cavity and using the viscous aquatic medium as a tool for prey transport. A number of technological and computational approaches such as hydrodynamic modeling and particle imaging have been applied to suction feeding in order to clarify this rapid behavior. A second goal of this chapter is to outline the theoretical and experimental research that has advanced our understanding of hydrodynamic mechanisms of suction generation, the suction profile that occurs during feeding events, and the effect of suction on the prey during feeding in fishes.

A third theme of fish feeding mechanisms, emerging from advances in our understanding of the evolutionary relationships among fish groups, is the phylogenetic pattern of diversity in skull form and function in fishes. Schaeffer and Rosen (1961) and Lauder (1982) outlined the major evolutionary transitions in the function of fish skulls, from the condition found in basal actinopterygians to teleostean designs for cranial kinesis. These studies identified the important musculoskeletal couplings that drive cranial elevation, mouth opening, upper jaw protrusion, and suction generation during feeding. An important conclusion of these investigations is the increasing mobility in fish skulls due to the decoupling of skeletal elements in the skull, allowing multiple linkages to perform parts of the feeding strike independently. More recently, phylogenetic analyses taking a broad look at actinopterygian skull characters have revealed frequent independent origins of such key features as jaw protrusion in fishes (Westneat, 2004). At a finer level of phylogenetic resolution, a number of studies developed a more detailed look at feeding mechanics and ecomorphology in the context of phylogenies of particular lineages such as catfishes (Schaefer and Lauder, 1996), cichlids (Liem, 1978, 1980; Waltzek and Wainwright, 2003), centrarchids (Wainwright and Lauder, 1986, 1992), labrids (Westneat, 1995a; Westneat *et al.*, 2005), and parrotfishes (Streelman *et al.*, 2002). This body of research has shown that the evolutionary history of skull structure, feeding ecology, and prey capture behaviors provides a rich system for exploring historical biomechanics. The final goal of this chapter is to present some of the major events in the evolution of fish feeding mechanics in the context of phylogenetic relationships among fishes.

II. SKULL MORPHOLOGY AND MECHANISMS

The functional morphology of skull mechanisms in fishes has a long and distinguished history (reviewed by Ferry-Graham and Lauder, 2001) and has been an active research field due to interest in the high level of kineticism in fish skulls (Tchernavin, 1953; Alexander, 1967; Osse, 1969; Liem, 1978, 1980; Elshoud-Oldenhave, 1979; Lauder, 1980a, 1982; Westneat and Wainwright, 1989; Westneat, 1991; Waltzek and Wainwright, 2003). Cranial kinesis reaches extraordinary levels in many teleosts because there are 20 or more independently movable skeletal elements in the skull and many more in the pharyngeal apparatus. Detailed illustration and description of skull morphology in fishes can be found in a number of sources (Gregory, 1933; Gosline, 1971; Winterbottom, 1974; Hanken and Hall, 1993; Helfman *et al.*, 1997), but the primary cranial skeletal elements and the key musculoskeletal couplings involved in feeding behavior are presented here.

The skull of actinopterygian fishes is composed of several important movable blocks of bones that function in prey capture (Figure 2.1). The neurocranium (brain case) may be rotated dorsally by epaxial muscles and the pectoral girdle pulled posteriorly by hypaxial musculature. The hyomandibula and the suspensorium (pterygoid series, palatine, and quadrate) are

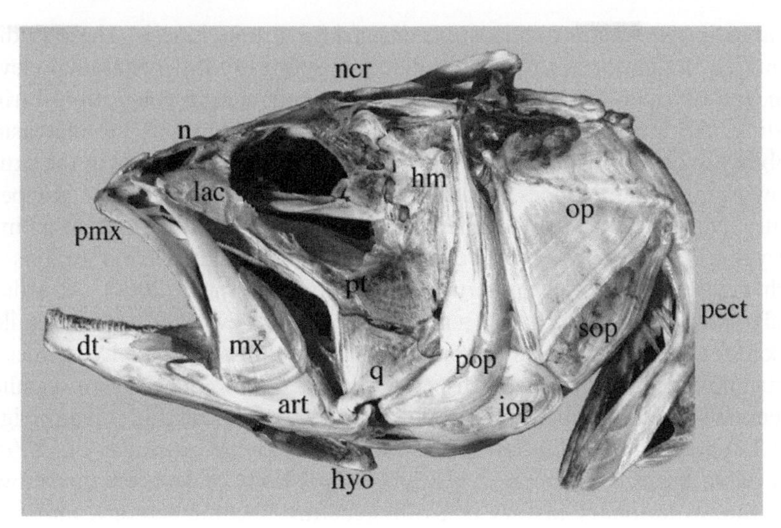

Fig. 2.1. Skull morphology of the large-mouth bass, *Micropterus salmoides*. Abbreviations: art, articular; dt, dentary; hm, hyomandibula; hyo, hyoid; iop, interopercle; lac, lacrimal; mx, maxilla; n, nasal; ncr, neurocranium; op, opercle; pect, pectoral girdle; pmx, premaxilla; pop, preopercle; pt, pterygoid series; q, quadrate; sop, subopercle.

often flared laterally during feeding, driven by adductor and levator arcus palatini muscles. The opercular series (gill cover elements) are capable of posterior displacement via the levator operculi, and lateral motion upon contraction of the dilatator operculi. The upper jaw is formed by the toothed premaxilla and a maxilla that may be toothed or may play a rotational supporting role in forcing the premaxilla forward into a protruded position. The lower jaw is a composite unit formed from the dentary, articular, and angular ossifications that are opened by several possible mechanisms described later. The mandible is powered in the bite by the adductor mandibulae muscles, which are usually subdivided into two or more muscle subdivisions. Last, the hyoid apparatus is the primary element of the floor of the mouth that may move ventrally to enhance the expansion of the buccal cavity during suction feeding.

In their analyses of evolutionary trends in fish skull function, Schaeffer and Rosen (1961) and Lauder (1982) identified several key muscle-tendon-bone connections and skeletal linkages that drive cranial kinesis in ray-finned fishes. Detailed descriptions of the morphology and mechanisms of fish skulls in particular taxa were also presented by Tchernavin (1953), Alexander (1966, 1967), Anker (1974), and Elshoud-Oldenhave (1979). The first experimental analyses involving feeding kinematics and interpretation of musculoskeletal connections were published by Osse (1969) using perch and Liem (1970) using nandid fishes. These detailed and monographic studies on feeding mechanics formed the foundation for many subsequent experimental and theoretical analyses that focused on kinematics, modeling of cranial function, and muscle contraction patterns.

Feeding in fishes is considerably more complex than simply opening the mouth and then closing it around a prey item because of the use of suction, the advent of jaw protrusion, and other kinetic features of the skull. Four main musculoskeletal linkages drive the initial stage of feeding, termed the expansive phase (Lauder, 1983a), by causing cranial elevation, hyoid depression, jaw opening, and jaw protrusion (Figure 2.2). These linkages have been illustrated in a number of ways, including a network approach showing the interconnections between muscles and bones (Figure 2.2A) and a morphological mechanical diagram approach showing functional skeletal groups and the muscles that power their motion (Figure 2.2B).

Cranial elevation is a common feature of fish feeding in which the skull is rotated dorsally around its joint with the vertebral column, raising the roof of the mouth and elevating the snout. This action is driven by the modified epaxial muscles that attach to the rear of the skull on the supraoccipital, exoccipital, and dorsal skull roof in many species. As illustrated by Lauder (1982) (Figure 2.2B), the neurocranium is a relatively simple lever system in which the posteriorly directed force developed by the epaxial muscles is transmitted to the rear of the skull, developing a torque around the craniovertebral joint to lift the skull. This mechanism has been developed as a quantitative lever model by Carroll *et al.*, (2004). Lifting the skull contributes to mouth opening in fishes with the premaxilla firmly attached to the skull, and promotes the production of premaxillary protrusion in fishes with a sliding premaxilla. Cranial elevation is widespread among fishes and tetrapods and is thought to be one of the basal mechanisms for opening the mouth in gnathostomes (Schaeffer and Rosen, 1961).

Although raising the upper jaw is an important part of mouth opening, the majority of increased gape between upper and lower jaws is achieved by rotating the lower jaw ventrally around its joint with the quadrate. The transmission of force to generate this action is surprisingly complex in fishes and may be mediated by several couplings that transmit forces from several ventral muscles to the posteroventral margin of the lower jaw (Schaeffer and Rosen, 1961; Lauder, 1982). The first linkage involves the hypaxial musculature, pectoral girdle, sternohyoideus, hyoid apparatus, and mandible (Figure 2.2). Basal actinopterygian fishes possess a mandibulohyoid ligament, so that retraction of the hyoid posteriorly exerts a posteriorly directed force on the lower jaw for jaw opening. Analysis of feeding in *Polypterus* and *Lepisosteus* by Lauder (1980a) showed that ventral hypaxial muscles stabilize the pectoral girdle to provide an anchor point for the action of the

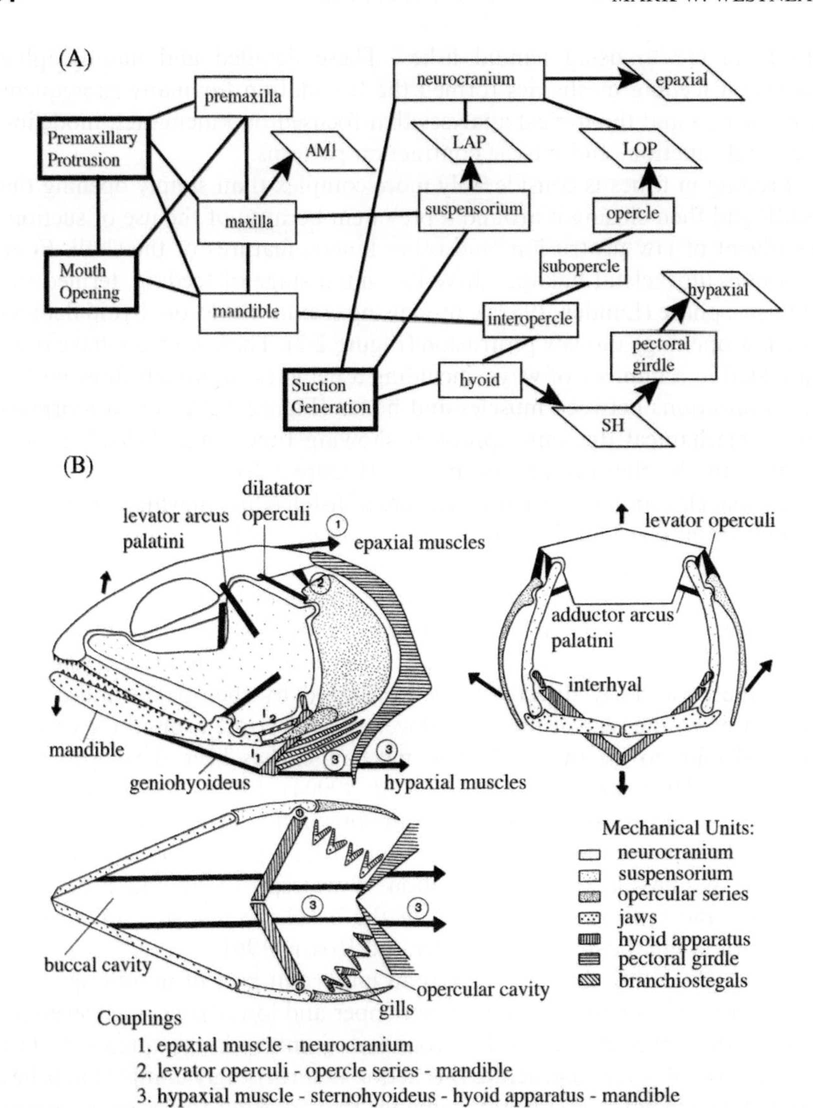

Fig. 2.2. Mechanical couplings and interactions of cranial bones and muscles during feeding in teleost fishes. (A) Network of connections and muscle actions that function in three aspects of feeding (bold boxes): mouth opening, premaxillary protrusion, and suction generation. Bones are in light boxes, muscles in parallelograms. Arrows indicate direction of force transmission of muscles. Muscle abbreviations: AM1, adductor mandibulae subdivision 1; LAP, levator arcus palatini; LOP, levator operculi; SH, sternohyoideus. (Adapted from Liem, 1980; Lauder, 1982.) (B) Three major musculoskeletal couplings in the head of teleost fishes: (1) the epaxial muscle to neurocranium coupling, which causes cranial elevation, (2) the levator operculi muscle to opercle to mandible coupling for lower jaw rotation, and (3) the pectoral girdle-sternohyoideus-hyoid-mandible coupling for hyoid depression and lower jaw rotation. (From Lauder, 1985.)

sternohyoideus muscle. Contraction of the sternohyoideus then pulls the hyoid posteroventrally, and this action is transferred directly to the mandible through the mandibulohyoid ligament. Hyoid retraction thus increases the volume of the mouth for suction and causes lower jaw rotation (Aerts, 1991). This mechanism for jaw depression is present in basal actinopterygians and basal teleosts, but in more derived teleosts (Aulopiformes and above), the insertion of the mandibulohyoid ligament shifts from direct mandibular attachment to insertion on the interopercle, becoming an interoperculohyoid ligament. In these taxa the interopercle is pulled posteriorly and this action is transmitted to the mandible via the interoperculomandibular ligament. Although some basal actinopterygians have only this single mechanism for jaw depression, most teleosts have a second linkage for jaw opening involving the opercular series.

The operculo-mandibular coupling is a second pathway for exerting force on the mandible for jaw depression. The levator operculi muscle originates dorsally on the neurocranium and inserts on the posterodorsal edge of the opercle, enabling it to rotate the opercle posteriorly upon contraction. The subopercle and interopercle in many species are free to move relative to the preopercle, allowing them to be pulled posteriorly by the motion of the opercle. The interopercle has on its anterior tip a robust interoperculomandibular ligament that attaches to the posteroventral margin of the lower jaw and pulls it into an open position. Thus, the force of levator operculi contraction is transmitted through the opercular series linkage to the mandible. This mechanism has been quantitatively modeled as a four-bar linkage mechanism (Muller, 1987; Westneat, 1990) in several fish groups (see Section III). Finally, a third possible jaw opening mechanism involving a hyoid-protractor hyoideus muscle-mandibular coupling (Wilga *et al*., 2000; Adriaens *et al.*, 2001) may aid in lower jaw depression in many species of fishes.

Upper jaw protrusion is a common feature of fish feeding during the expansion phase that allows a fish to move its jaw rapidly forward to contact or engulf the prey. Several linkages have been proposed to mediate upper jaw protrusion in a wide range of taxa from sharks and rays to derived perciforms. The upper jaw (palatoquadrate cartilage) in sharks and rays exhibits protrusion involving anterior sliding along the neurocranium (Motta *et al.*, 1997; Wilga and Motta, 1998; Wilga, 2002), although this protrusion is often not visible until the compressive phase begins. In chondrichthyes, protrusion ability is often associated with a long ethmopalatine ligament (or lack of such a ligament) that allows more anterior excursion of the upper jaw (Wilga, 2002). Extreme ventral protrusion of the jaws in sturgeons is driven by a unique linkage mechanism in which anterior rotation of the hyomandibula is translated into a ventrally directed sliding of the palatoquadrate cartilage (Bemis *et al.*, 1997; Carroll and Wainwright, 2003).

Jaw protrusion has been shown to have arisen independently at least five times in the teleosts (Westneat, 2004), including via the maxillary twisting model of Alexander (1967) for cyprinid fishes, and in loricariid catfishes (Schaefer and Lauder, 1986). Most jaw protrusion in perciform fishes occurs due to premaxillary sliding driven by maxillary rotation, via a four-bar linkage mechanism proposed by Westneat (1990). In addition, there are several unusual and extreme forms of jaw protrusion such as that of *Stylephorus chordatus* (Pietsch, 1978) and *Epibulus insidiator* (Westneat and Wainwright, 1989) that are produced by modified musculoskeletal systems and unique linkages.

Jaw closing occurs in the second half of a feeding event in virtually all fishes, and is powered by the adductor mandibulae muscles. Jaw closing is accomplished primarily by raising the lower jaw, though reversal of cranial elevation and retraction of the premaxilla also serve to bring the upper jaw back toward a closed gape position. The mechanism of lower jaw closing is a relatively simple connection between the adductor muscles and the mandible (Lauder, 1980a; Barel, 1983; Westneat, 2003). The adductors originate on the lateral rim of the preopercle and on the lateral surface of the suspensorium, and insert tendinously on the articular process or on the medial face of the dentary. In many fishes, a third subdivision attaches to the maxilla to retract it and the premaxilla during jaw closing (Alexander, 1967; Liem, 1970). The jaw closing mechanism of fishes has been modeled as a lever system, allowing a wide range of biomechanical calculations of the force, motion, work, and power that are done during jaw closing (Westneat, 2003; Van Wassenbergh *et al.*, 2005). There are a large number of quantitative biomechanical models that have been developed to test the function of the musculoskeletal linkages in fish skulls.

III. BIOMECHANICAL MODELS OF SKULL FUNCTION

The dynamic motion of the teleost skull represents a challenge for the development of biomechanical models, and relatively few actinopterygian fishes have been analyzed in this way. However, the detailed functional morphology of musculoskeletal couplings and linkages in fish skulls described previously has often led to quantitative engineering models of cranial design (Anker, 1974; Lauder, 1980a; Muller, 1987, 1989; Westneat, 1990, 2003; Herrel *et al.*, 2002). Such models for fish skulls have great potential for testing hypotheses of mechanical design in a diversity of fishes, for developing ideas of functional transformation during growth and development, and for examining patterns of evolution in key functionally relevant characters in a phylogenetic context. Most current models for fish skull mechanics are based on the engineering theory of levers and linkages.

The closing mechanism of the lower jaw was first modeled as a simple lever by Barel (1983) in his assessment of suction and biting in cichlid fishes. Although the mechanical advantage of the mandibular lever was not calculated directly in this work, Barel (1983) used regression plots of inlever vs. outlever dimensions to illustrate the alternative strategies of velocity and force transmission for suction feeders and biters, respectively. This model was further developed for jaw opening and closing in labrid fishes (Westneat, 1994), and mechanical advantage was used as a variable for correlation with diet and exploration of evolutionary patterns in the context of a phylogeny (Westneat, 1995a). This research demonstrated that the mechanical advantage of the mandibular lever, and its resultant force profile, was associated with the ecology of feeding in piscivorous and molluscivorous coral reef fishes. The lower jaw as a lever has also been explored in a diversity of fish groups for the purposes of comparing feeding mechanics across taxa and during ontogeny (Wainwright and Richard, 1995; Wainwright and Shaw, 1999; Westneat, 2004; Kammerer *et al.*, 2005). Jaw mechanical advantage is a simple yet broadly informative metric that has also been creatively employed in surveys of fossil forms to estimate patterns of evolution of herbivory in Mesozoic and Cenozoic faunas (Bellwood, 2003) and the mechanics of the jaws of fossil gars (Kammerer *et al.*, 2005).

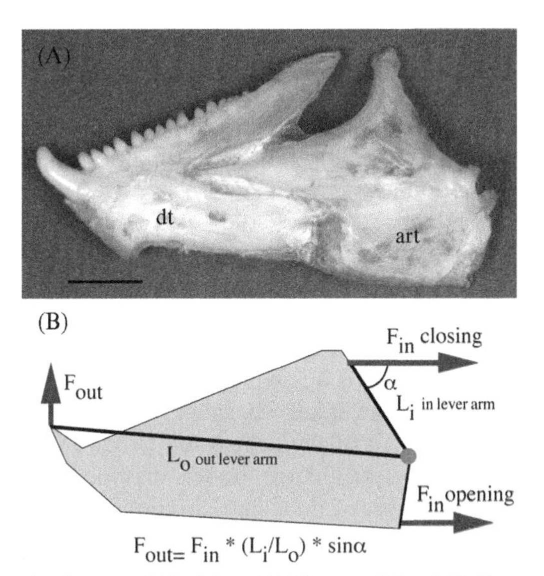

Fig. 2.3. The lower jaw lever model in fishes. (A) The mandible of *Cheilinus trilobatus*, a labrid fish (scale bar = 5 mm). (B) The input forces (F_{in}) for jaw opening and closing can be used in combination with angular insertions of muscles and lever arms to calculate mechanical advantage, bite force, and velocity of jaw closing. (Adapted from Westneat, 1994, 2003.)

The lever mechanics of the lower jaw (Figure 2.3) have also been analyzed as a more complex, dynamic system by including the geometry and properties of the adductor muscles that power jaw closing (Westneat, 2003). Including muscle physiology in the mandibular lever model allows the simulation of muscle mechanics and the force, work, and power of the jaw closing system at each point in the closing cycle. Van Wassenbergh *et al.* (2005) also developed a detailed biomechanical model of jaw closing in catfishes that simulates the semicircular profile of the lower jaw, accounts for the acceleration and deceleration of the mass of the jaw, and estimates the effect of water resistance to jaw closing motions. By including muscle morphology and contraction mechanics, these modeling studies revealed the more complex and dynamic features of the mandibular lever system that go beyond mechanical advantage.

The more complex interactions between movable elements in the skulls of fishes have been analyzed with linkage theory from mechanical engineering. For example, Anker (1974) and Aerts and Verraes (1984) proposed that the operculo-mandibular coupling for lower jaw depression in teleosts may be modeled with a four-bar linkage mechanism in which the neurocranium, opercle, interopercle, and articular of the lower jaw were linked (Figure 2.4). Anker (1978) examined the action of this lower jaw depression mechanism during respiration and demonstrated that respiratory kinematics implicated the opercular linkage rather than the sternohyoideus-hyoid linkage as the driver of low-amplitude jaw motions. In the feeding strike of many fishes, the opercular series begins motion early in the strike cycle, and Aerts *et al.*, (1987) proposed a model in which this linkage plays a role in triggering rapid depression of the lower jaw after it is loaded in tension. Tests of this linkage in labrid fishes (Westneat, 1990) showed a lack of expected correlation between opercular motion and jaw depression, thus rejecting a planar linkage mechanism in some taxa, but this study also noted that accounting for three-dimensional motion of the opercular series may be required for a more complete assessment of the linkage's role in jaw depression. Indeed, recent experiments in which the opercular linkage was surgically disrupted (Durie and Turingan, 2004) have shown that the opercular linkage may be important to jaw-opening mechanisms in some species.

The anterior jaws four-bar linkage mechanism was developed to model maxillary rotation and premaxillary protrusion (Figure 2.4, jaws linkage), and is the most complex set of linkage models developed. Westneat (1990) proposed that the mechanism of maxillary rotation was a four-bar linkage with an additional slider link, composed of suspensorium, palatine, maxilla, coronoid portion of the lower jaw, and sliding premaxilla. The anterior jaws linkage contains a fixed link (Figure 2.4, link a) formed by the neurocranium and suspensorium, whose length equals the distance from the quadrate-articular joint to the palatine-neurocranium joint. Lower jaw depression is

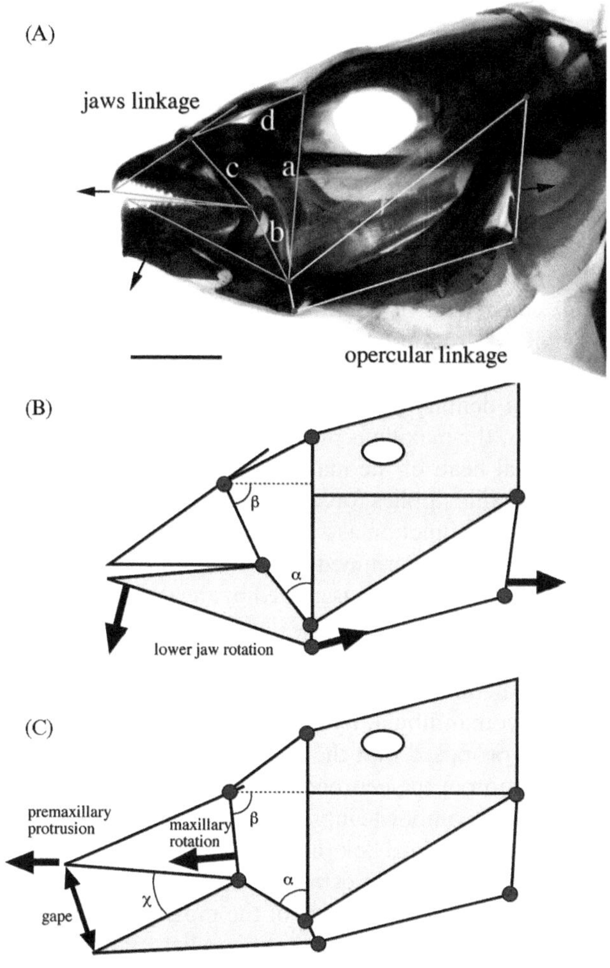

Fig. 2.4. (A) Two four-bar linkages in the feeding mechanism of teleost fishes, illustrated on a cleared and alizarin/alcian stained skull of the labrid fish *Oxycheilinus digrammus*. The opercular linkage (posterior) involves opercular rotation and force transmission through the interopercle to rotate the lower jaw downward. The jaws linkage (anterior links a–d) transmits lower jaw rotation to the maxilla and input for anterior sliding and protrusion of the premaxilla. (B) The input motion that drives the linkage is ventral depression and rotation of the mandible (increase in angle α) that may be transmitted by the opercular linkage. (C) The output motions of the linkage are maxillary rotation (angle β), maxillary displacement ventrally, sliding and protrusion of the premaxilla, and increase in mouth gape and gape angle (angle χ). (Adapted from Westneat, 2004.)

the input motion that drives the anterior jaws linkage. Specifically, the input link of the mechanism is the distance from the quadrate-articular joint to the articular-maxilla articulation (Figure 2.4, link b). As the lower jaw rotates ventrally, the articular and coronoid processes transfer the motion of lower jaw rotation to the ventral shank of the maxilla (Figure 2.4, link c). The coupler link of the chain (Figure 2.4, link d) is the palatine bone and palatomaxillary ligament, or in many fishes the nasal and nasomaxillary ligament, which extend anteriorly from the neurocranium. The anterior process of the palatine provides the dorsal head of the maxilla with a pivot point for rotation.

The output kinematics of the anterior jaws linkage are premaxillary protrusion, maxillary rotation, and increase in gape angle (Westneat, 1990). The output link of the four-bar chain is the maxilla (Figure 2.3, link C), from the maxilla-dentary joint to the joint between maxilla and palatine. The ventral shank of the maxilla is pulled anteriorly by rotation of the lower jaw, while the dorsal head of the maxilla pivots about its articulation with the palatine. The maxilla applies force to the premaxilla, causing protrusion. The premaxilla does not function as a link in a four-bar chain, but is a sliding element whose motion is determined by the action of the four-bar linkage behind it. Premaxillary protrusion is guided by the dorsal head of the maxilla and the dorsal intermaxillary ligament.

The mechanism of hyoid depression was modeled by Muller (1987, 1989), who used four-bar theory to propose a new model based on the connections among the skull, hyomandibula, hyoid, and sternohyoideus muscle (Figure 2.5). Muller (1987) proposed that the pectoral girdle forms a fixed link and that the dorsal rotation of the neurocranium drives anterodorsal rotation of the hyomandibula as a composite input link. The sternohyoideus muscle and urohyal complex form the third link (held rigid in Muller's model). The hyoid is the output link, the anterior tip of which rotates downward rapidly during feeding to create expansion of the floor of the mouth. Muller (1987) was the first to use linkage system simulations to predict cranial kinematics and concluded that the model could explain the rapid, explosive mouth expansion in some species through a mechanism of preloading and triggered release of force. This model was developed as a computer simulation model for labrid fishes by Westneat (1990), who showed that simulation of both cranial elevation and contraction of the sternohyoideus muscle is required to generate kinematic predictions that match real feeding behavior.

Theoretical models of musculoskeletal biomechanics are useful tools in biology for several types of analysis. First, models promote an understanding of the morphological basis of behavior. The ability to quantify feeding mechanisms and predict their actions allows direct interpretation of the structural basis of feeding behavior (Anker, 1974; Barel, 1983; Muller,

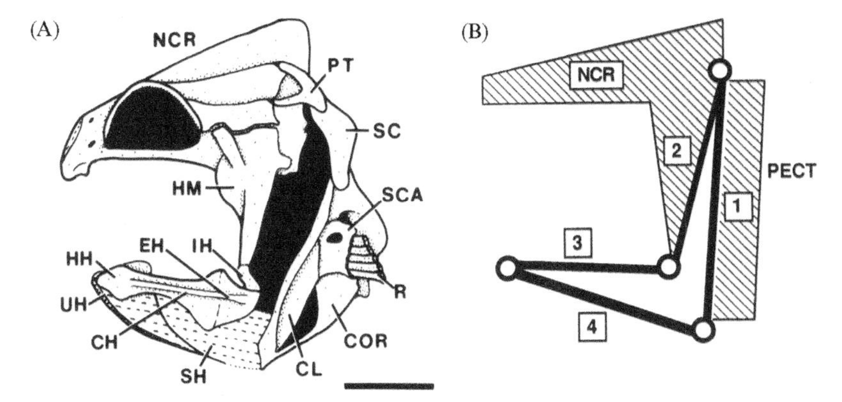

Fig. 2.5. The four-bar linkage model of the hyoid depression mechanism. (A) Morphology of the labrid fish *Cheilinus chlorurus* that forms the hyoid depression mechanism. (B) Mechanical diagram of the hyoid linkage. Links 1–4 are the elements in a four-bar crank chain. Abbreviations: CH, ceratohyal; CL, cleithrum; COR, coracoid; EH, epihyal; HH, hypohyal; HM, hyomandibula; IH, interhyal; NCR, neurocranium; PECT, pectoral girdle; PT, posttemporal; R, radials; SC, supracleithrum; SCA, scapula; UH, urohyal. Scale bar = 1 cm. (Linkage originally proposed by Muller, 1987; figure from Westneat, 1990.)

1987; Westneat, 1990). A second level of analysis is that of examining the mechanics of feeding during ontogeny. Incorporation of mechanical models into developmental studies of cranial function could reveal the direct implications of musculoskeletal development on force and velocity characteristics of growing fish skulls (Wainwright and Shaw, 1999; Kammerer *et al.*, 2005). Finally, the ability to quantify the relationship between mechanism geometry and feeding kinematics in fishes provides a tool for exploring the diversity and evolution of the trophic apparatus of fishes (Wainwright *et al.*, 2004). A key advantage of using quantitative models in biomechanics is comparative power. Models provide a way to quantitatively compare the functional consequences of differences in morphology among species. Simple lever mechanisms in engineering are highly varied and produce a vast number of motions, whereas biological mechanisms such as the jaws of fishes are constrained to certain proportions and actions. Comparing the range of possible feeding mechanisms to the phenotypic range observed in nature could identify features of teleost feeding mechanisms that have constrained the range of natural variation. Finally, evolutionary studies of fish feeding can benefit from the application of functional morphology in a phylogenetic context (Westneat, 1995a, 2004). Functional analyses using biomechanical models may be performed upon fish taxa with known phylogenetic relationships in order to reveal historical patterns of change in feeding mechanics.

IV. SUCTION FEEDING FOR PREY CAPTURE

Suction feeding is the primary mode of prey capture in fishes (Elshoud-Oldenhave, 1979; Liem, 1980; Lauder, 1983a, 1985; Muller and Osse, 1984; Ferry-Graham and Lauder, 2001), a method employed to draw prey into the mouth by using the density of water as a tool for prey transport. Suction feeding behavior has been the focus of numerous studies involving high-speed video analysis of kinematics (Anker, 1978; Grobecker and Pietsch, 1979; Lauder, 1979; Lauder and Shaffer, 1993; Wilga and Motta, 1998; Sanford and Wainwright, 2002), electromyographic study of the motor patterns that drive suction feeding (Osse, 1969; Liem, 1980; Wainwright *et al.*, 1989; Westneat and Wainwright, 1989; Alfaro *et al.*, 2001; Grubich, 2001), and application of a range of modern techniques such as pressure transduction (Lauder, 1983b; Muller and Osse, 1984; Svanbäck *et al.*, 2002), sonomicrometry (Sanford and Wainwright, 2002), and digital particle imaging velocimetry (Ferry-Graham and Lauder, 2001; Ferry-Graham *et al.*, 2003). These studies have discovered the timing of different aspects of cranial kinesis involved in suction feeding, revealed the patterns of muscle contraction underlying suction generation, and measured the hydrodynamics, changing water pressure, and suction velocity during feeding. As a result, we are closer to a full understanding of morphological, behavioral, and biomechanical explanations for one of the fastest and most widespread feeding behaviors among vertebrates.

A. Kinematics

Suction feeding occurs by explosive opening of the mouth and expansion of the oral cavity followed immediately by rapid closing of the jaws. Thus, suction feeding is often described as consisting of two primary phases: expansion and compression (Osse, 1969; Lauder, 1979, 1980a). Lauder (1985) also described a preparatory phase occurring before expansion and an extended process of prey processing and swallowing after compression, though these have not been as well characterized in a range of taxa. The expansive phase consists of many cranial movements that are coordinated in order to increase the volume inside the buccal cavity. These skull motions include neurocranial elevation, opercular rotation and abduction, lateral flaring of the suspensoria, opening of the jaws, jaw protrusion, and hyoid bar depression. These movements create an increase in buccal volume. Since water is virtually incompressible, the rapid buccal expansion produces a drop in water pressure within the buccal cavity (Lauder, 1983b; Lauder and Clark, 1984; Muller and Osse, 1984; Sanford and Wainwright, 2002). With the mouth open and the gill cover and branchial arches restricting water flow into the mouth from the rear, water is drawn at high velocity into the mouth, producing drag on a prey item that may be entrained in the water flow and then captured as the jaws close.

As the kinematic events of the expansive phase reach their maxima, the compressive phase begins (Lauder, 1983a). Jaw closing usually begins the compressive phase, often initiated by dorsal rotation of the mandible to close the gape. In species with upper jaw protrusion, the upper jaw is often capable of remaining protruded during the early compressive phase, even until the jaws snap closed. The neurocranium returns to its rest position, and the laterally flared suspensoria begin to be adducted. The hyoid and floor of the mouth often reach peak depression after the compressive phase begins, but the hyoid begins its role in the reduction of buccal volume shortly thereafter (Lauder, 1979, 1980a). As buccal compression is well underway anteriorly, posteriorly the gill chamber continues to expand. Buccal compression forces water posteriorly into the pharyngeal region, while expansion of the gill apparatus produces lowered pressure posteriorly to maintain a forceful one-way flow of water back across the gills (Lauder, 1983b; Muller and Osse, 1984; Svanbäck *et al.*, 2002). Gill rakers that prevent the escape of food items between the gill arches, while still permitting water to exit, are often present. At the end of the compressive phase, most movable skull components have returned to their closed, rest positions, and if the strike was successful the neurocranium, hyoid, and pectoral girdle may be observed in motions related to prey processing in the pharyngeal region.

The general pattern of suction feeding motions described previously has numerous variations among fish species. First, it should be noted that a large number of fishes deliver a forceful bite with the oral jaws, instead of or in conjunction with some level of suction production. The kinematics and motor patterns of biting are often significantly different than suction feeding (Alfaro *et al.*, 2001). Biting for the purpose of piece removal, as in many sharks, piranhas, and parrotfishes (Figure 2.6), is often performed without employing suction flows. Many fishes such as gar, vampire characins (Figure 2.6C), and many deep-sea forms impale their prey using long, sharp teeth, and in these species suction is normally utilized to assist the biting capture. Most fishes employ suction during feeding.

The kinematics of suction feeding is becoming well documented in a wide range of phylogenetically diverse fishes. Major advances have recently been made in analysis of feeding biomechanics in both sharks (Frazzetta and Prange, 1987; Motta *et al.*, 1997, 2002; Motta and Wilga, 2000) and rays (Wilga and Motta, 1998; Dean and Motta, 2004). A recent example of this research is an analysis of suction feeding in the bamboo shark, *Chiloscyllium plagiosum* (Sanford and Wilga, 2004). As the shark approaches the prey (Figure 2.7), the lower jaw rotates ventrally and the upper jaw is protruded

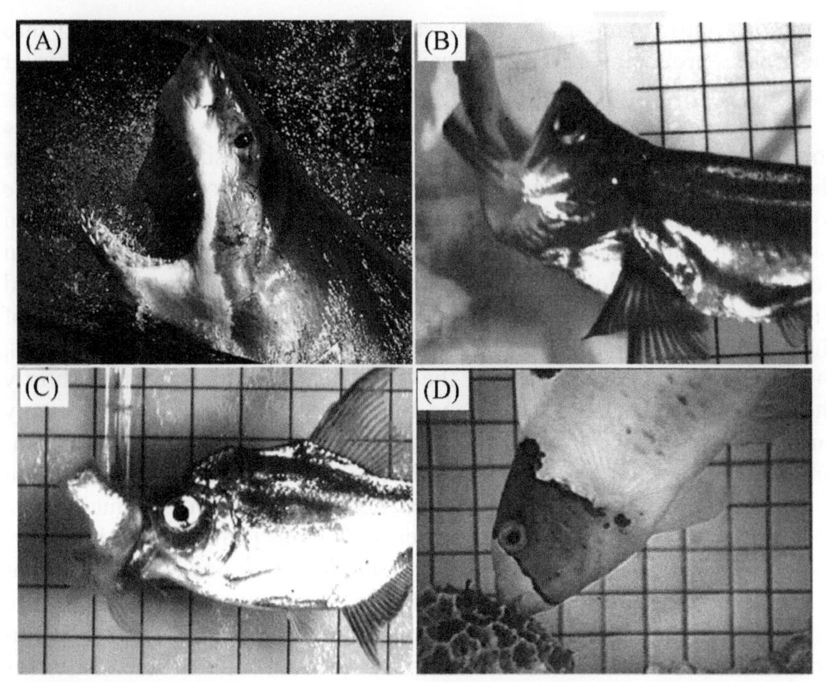

Fig. 2.6. Fishes that deliver a strong bite during feeding. (A) The great white shark, *Carcharodon carcharias*, performing the bite for which it is famous. (Photo from National Marine Fisheries.) (B) The vampire characin, *Hydrolycus scomberoides*, which uses a combination of suction, ventral attack with high cranial rotation, and extremely long mandibular canine teeth to employ an impaling bite. (Photo by M. Alfaro.) (C) The wimple piranha, *Catoprion mento*, uses extreme mandibular rotation, a strong bite, and rasping teeth to feed upon the scales of other fishes. (Photo by J. Janovetz.) (D) The bicolored parrotfish, *Cetoscarus bicolor*, which uses a strong bite to remove algae, detritus, and calcium carbonate from coral surfaces. (Photo by M. Westneat.)

to contact the prey. As the expansive phase reaches its peak with both ventral and lateral expansion of the buccal cavity (Figure 2.7, 26 ms), the prey is seen to travel by suction into the mouth. Compression begins anteriorly with the closing of the jaws, and in late compressive stage (Figure 2.7, 82 ms) the head can be seen to be narrower in lateral profile and the prey has disappeared into the mouth. Many sharks and most batoids employ jaw protrusion and suction feeding; thus, a current area of active research is focused on the mechanics and evolution of these behaviors and how the suction profiles compare to those of other fishes.

Research on the kinematics of basal actinopterygians, including *Amia* (Lauder, 1979, 1980a), *Lepisosteus* (Lauder, 1980a), paddlefishes (Sanderson

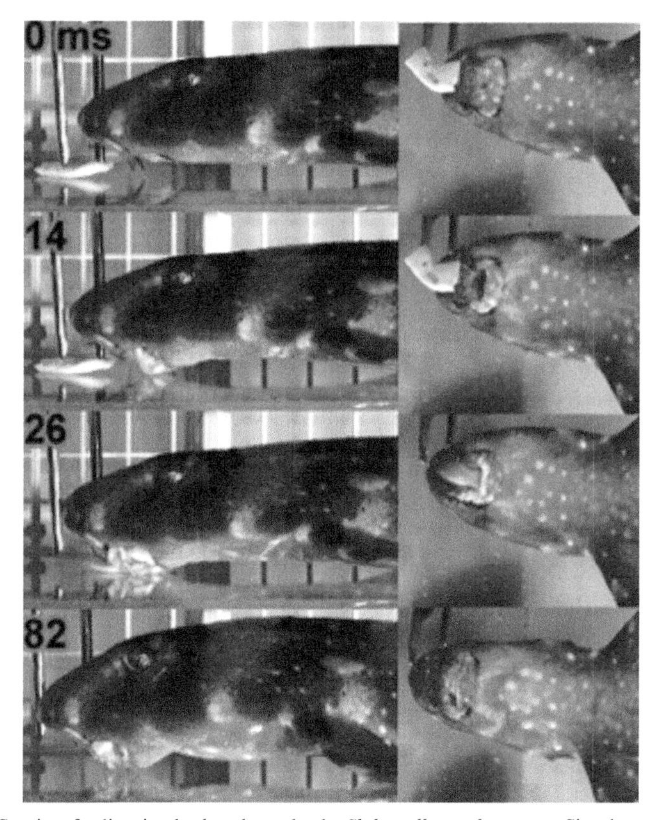

Fig. 2.7. Suction feeding in the bamboo shark *Chiloscyllium plagiosum*. Simultaneous lateral (left) and ventral (right) views show the role of lateral expansion in prey capture by suction. At time 0 the shark approaches the prey, expanding the head through time 26 ms, at which point the prey is being sucked into the buccal cavity. Feeding is complete by time 82 ms, when the jaws are fully closed and still protruded, and the compressive phase has begun. (From Sanford and Wilga, 2004.)

et al., 1991, 1994), and sturgeon (Carroll and Wainwright, 2003) has clarified the feeding motions of these important lineages whose behavior represents the outgroup states with which teleostean mechanisms are compared. Feeding modes and the mechanisms of suction production in these taxa are diverse, with the behavior of the bowfin, *Amia calva*, being particularly well characterized. Lauder (1980a) and Muller and Osse (1984) showed that the feeding kinematics of bowfin (Figure 2.8) consist of cranial elevation and jaw rotation accompanied by extreme anterior rotation of the maxilla. Lauder (1980a) also showed that the opercular coupling plays a role in jaw depression

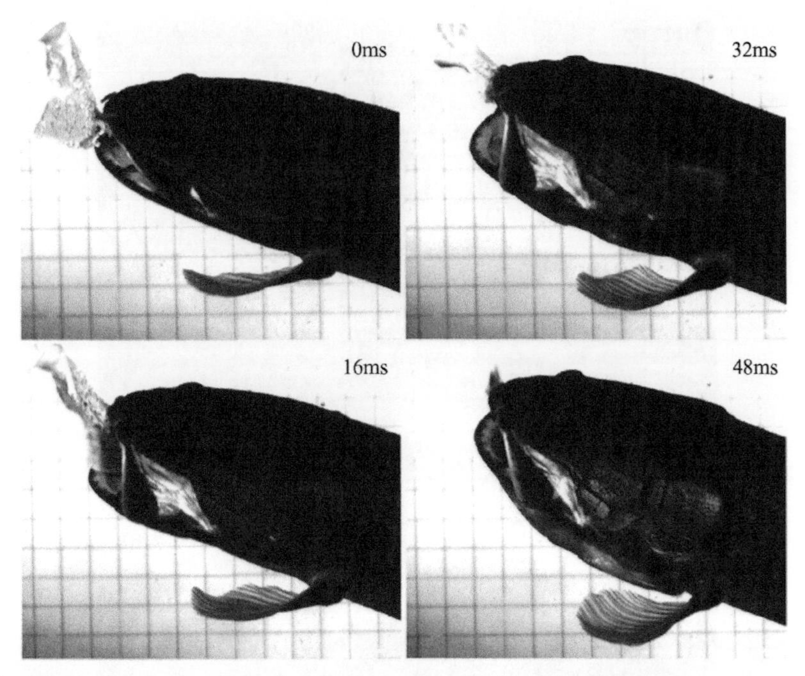

Fig. 2.8. Suction feeding in the bowfin *Amia calva*. At time 0 cranial elevation and mouth opening have just begun and contact with the prey item has occurred. At 16 ms the maxilla is observed in anteriorly rotated position and expansion of the head has begun. At time 32 ms the prey is being sucked into the mouth, after which (time 48 ms) the jaws close on the food item. Hyoid depression reaches a maximum (seen at time 48 ms) after peak gape and then the compressive phase begins. (Photos courtesy of J. Grubich.)

in this species, and noted the late action of hyoid depression (Figure 2.8, 48 ms) after the jaws are nearly closed.

Several species in groups that arose near the base of the teleost radiation have been the subjects of feeding studies involving detailed kinematics, including tarpon (Grubich, 2001), *Salmo* (Lauder, 1979), cyprinids (Sibbing, 1982; Callan and Sanderson, 2003), and characins (Lauder, 1981; Janovetz, 2002). Lauder (1979) showed that salmonids share the high degree of maxillary rotation seen in *Amia*, and maxillary rotation has since been shown to be an important aspect of suction feeding in regard to forming a suction tube with the mouth. Janovetz (2002) demonstrated that suction is sometimes used to a high degree in serrasalmine fishes such as pacus that are known for their strong bite, which is capable of cracking nuts and seeds. The tarpon, *Megalops atlanticus*, is a spectacular suction feeder, often feeding at the surface with a large cranial expansion (Grubich, 2001). The expansive phase in tarpon (Figure 2.9) is rapid and differs from many other suction feeders in that opercular abduction is nearly synchronous with lateral expansion of the suspensorium rather than occurring later in the strike cycle (note Figure 2.9, 16 ms, ventral view). Grubich (2001) concluded that variation in lateral expansion kinematics and muscle recruitment patterns that drive this motion may be an important unexplored source of suction variability in teleosts.

The majority of kinematic data on suction feeding has been collected on perciform fishes such as leaf-fishes (Liem, 1970), basses and sunfishes (Nyberg, 1971; Lauder, 1983b; Wainwright and Lauder, 1992; Svanbäck *et al.*, 2002), cichlids (Liem, 1978, 1980; Wainwright *et al.*, 2001; Waltzek and Wainwright, 2003), and labrids (Westneat, 1990, 1994; Ferry-Graham *et al.*, 2002). The large-mouth bass, *Micropterus salmoides*, is an important "model" species in the study of suction feeding (Grubich and Wainwright, 1997; Sanford and Wainwright, 2002; Svanbäck *et al.*, 2002), and video frames of feeding (Figure 2.9) show that this species is a classic suction feeder. Similarly, the kinematic profile of feeding in piscivorous wrasses (Figure 2.10) is typical of suction feeding across many perciform families (Westneat, 1990, 1994). The bass and the wrasse, like most perciform fishes, usually complete the suction feeding sequence in 50–100 ms, although some species are faster and the duration of a strike within the same individual may vary with motivation or satiation (Svanbäck *et al.*, 2002). Opercular rotation and cranial elevation begin the strike (Figure 2.9, 8 ms) and both the skull and opercle typically rotate through angles of up to 10 degrees (Figure 2.10A and B). Profiles of cranial kinesis often show a synchronous peak in many variables at the point of prey capture (Figure 2.9, 16–24 ms), in which jaw depression, maxillary rotation, gape, and gape angle all show their maxima (Figure 2.10). Premaxillary protrusion may also peak synchronously with other variables, or maximal protrusion may be delayed until 5–10 ms later, and the upper jaw may remain protruded during early jaw closing (Figure 2.9, 24–32 ms). The floor of the mouth is pushed ventrally by hyoid depression (Figure 2.9, 48 ms), which peaks as much as 10 ms after peak values in most skull motions (Figure 2.10H). Ventral views (Figure 2.9) show that lateral flaring of the suspensorium returns to rest position later than the jaws and skull.

The impressive diversity of suction feeding fishes includes many interesting variations on this general theme of cranial kinematics, including changes in timing and magnitude of cranial motions. Among the goals of future research are to analyze the mechanics of these unusual mechanisms and the evolutionary changes that have produced them. Some fishes have extraordinarily fast suction feeding, such as the antennariid anglerfishes (Grobecker and Pietsch, 1979) that perform a complete suction feeding event in 10–15 ms. The flatfishes (flounders and relatives) have a unique cranial developmental process resulting in both eyes on one side of the head, and an

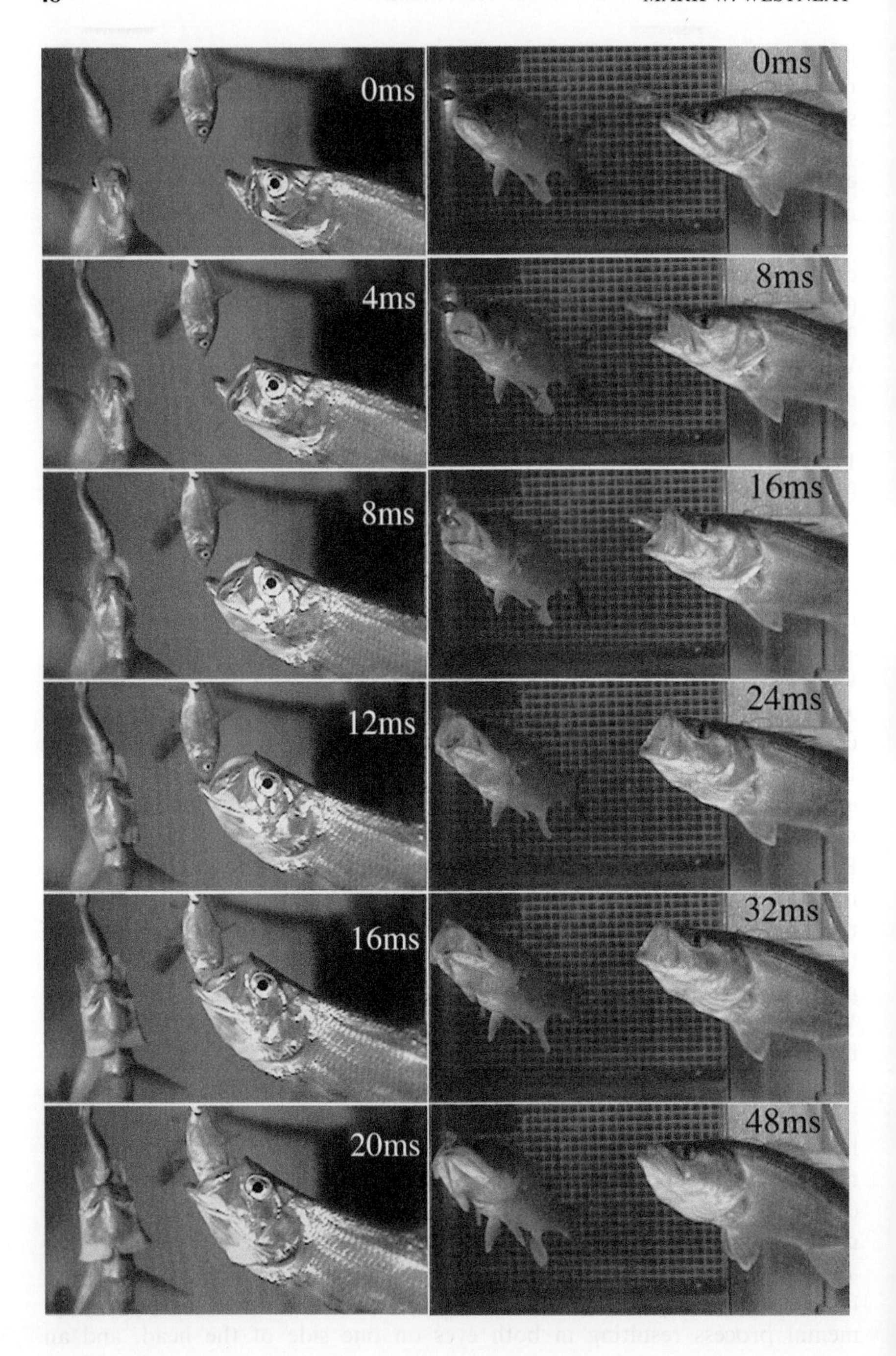
0ms
4ms
8ms
12ms
16ms
20ms
0ms
8ms
16ms
24ms
32ms
48ms

asymmetrical musculoskeletal system with asymmetrical suction feeding kinematics (Gibb, 1996, 1997). Extreme and unusual jaw protrusion kinematics distinguishes many lineages, such as the remarkable ventral protrusion of the gerreids, the dorsally directed protrusion of the leiognathids, and the bizarre anterior extension of the head in the cichlid *Petenia splendida* (Wainwright *et al.*, 2001) and the sling-jaw wrasse *Epibulus insidiator* (Westneat and Wainwright, 1989). The sling-jaw protrudes both upper and lower jaws over 65% of head length (Figure 2.11) in the most extreme jaw protrusion among fishes. This behavior is characterized by unique features such as high maxillary, interopercular, and quadrate rotation that swing the lower jaw anteriorly via a unique linkage system (Westneat, 1991). Extraordinary suction feeding behaviors and the biomechanics that underlie them are promising systems for future research on development, functional morphology, and evolution of function.

B. Motor Activity Patterns of Suction Feeding

The cranial motions employed for prey capture in fishes are powered by the contraction of 20 or more cranial muscles, most of them bilaterally symmetrical pairs. Most muscles in the head that participate in suction feeding are activated by motor neurons descending to both the left and right sides from the 4th (trochlear), 5th (trigeminal), or 7th (facial) cranial nerves. The sequence of contraction of the cranial muscles, the duration of their activity, and the intensity of their contraction provide information about the neural activity that occurs during feeding. For most fish feeding studies, however, we know little about neural activity directly, or about brain function and sensory/motor feedback during feeding. Rather, most biologists depend upon recording of voltage fluctuations due to action potentials within the muscle using electromyography (EMG). By recording and measuring voltage patterns (electromyograms) of multiple muscles, researchers are able to compile a motor activity pattern (MAP) for suction feeding in a particular individual or species. MAPs are used with kinematics for assessing the functional role of muscles in driving feeding kinematics (Osse, 1969;

Fig. 2.9. Suction feeding in the tarpon *Megalops atlanticus* (left) and the largemouth bass *Micropterus salmoides* (right). Simultaneous lateral (left) and ventral (right) views show the role of lateral expansion in prey capture by suction. From time 0 to 8 ms in both species cranial elevation and mouth opening occur before contact with the prey item. At 16 ms the maxilla is observed in anteriorly rotated position and expansion of the head is near peak. For *M. salmoides*, at time 24 ms the prey is being sucked into the mouth, after which (time 48 ms) the jaws close on the food item. Hyoid depression reaches a maximum (seen at time 48 ms) after peak gape and then the compressive phase begins. (High-speed video photos courtesy of J. Grubich.)

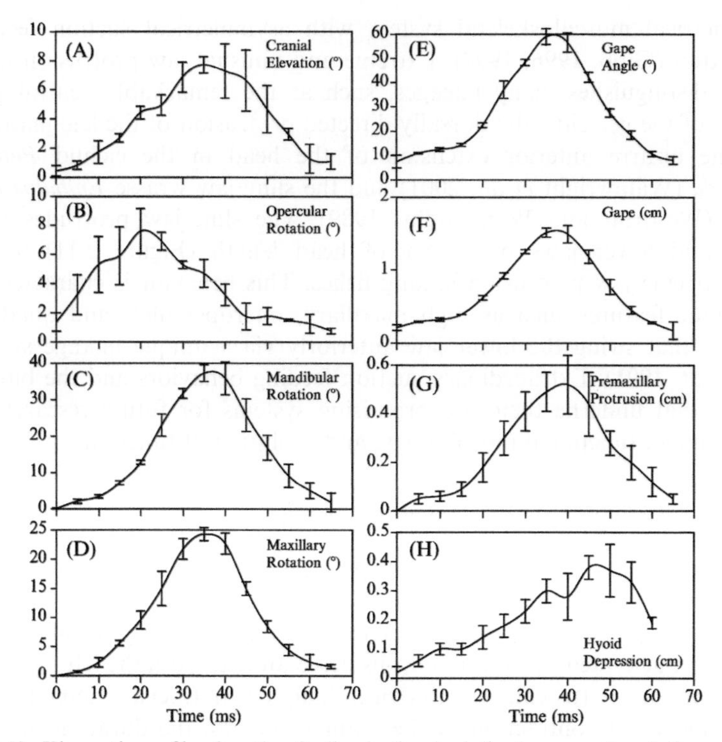

Fig. 2.10. Kinematic profile of suction feeding in the cheek-lined wrasse, *Oxycheilinus digrammus.* Rotational motions of (A) cranial elevation, (B) opercular rotation, (C) jaw depression, (D) maxillary rotation, and (E) gape angle. Distance variables include (F) gape distance, (G) premaxillary protrusion, and (H) hyoid depression. (Adapted from Westneat, 1990.)

Liem, 1978; Lauder, 1979), comparing motor profiles between species, or testing for intraspecific differences between behaviors (Sanderson, 1988; Wainwright and Lauder, 1992; Grubich, 2001). Motor patterns are also used in conjunction with other techniques such as pressure measurements and hydrodynamic techniques to produce a more complete picture of feeding biomechanics.

In combination with basic anatomical origin and insertion data, EMG data provide the clearest evidence for the functional role of muscle in feeding behavior. For example, a MAP of a feeding event (Figure 2.12) shows a fairly typical pattern of suction feeding. The epaxial muscles that attach to the neurocranium are specialized portions of myomeres that cause cranial elevation. EMG recordings (Figure 2.12) show that the epaxial muscle is active early in the strike cycle to raise the head and begin the expansive phase. Also

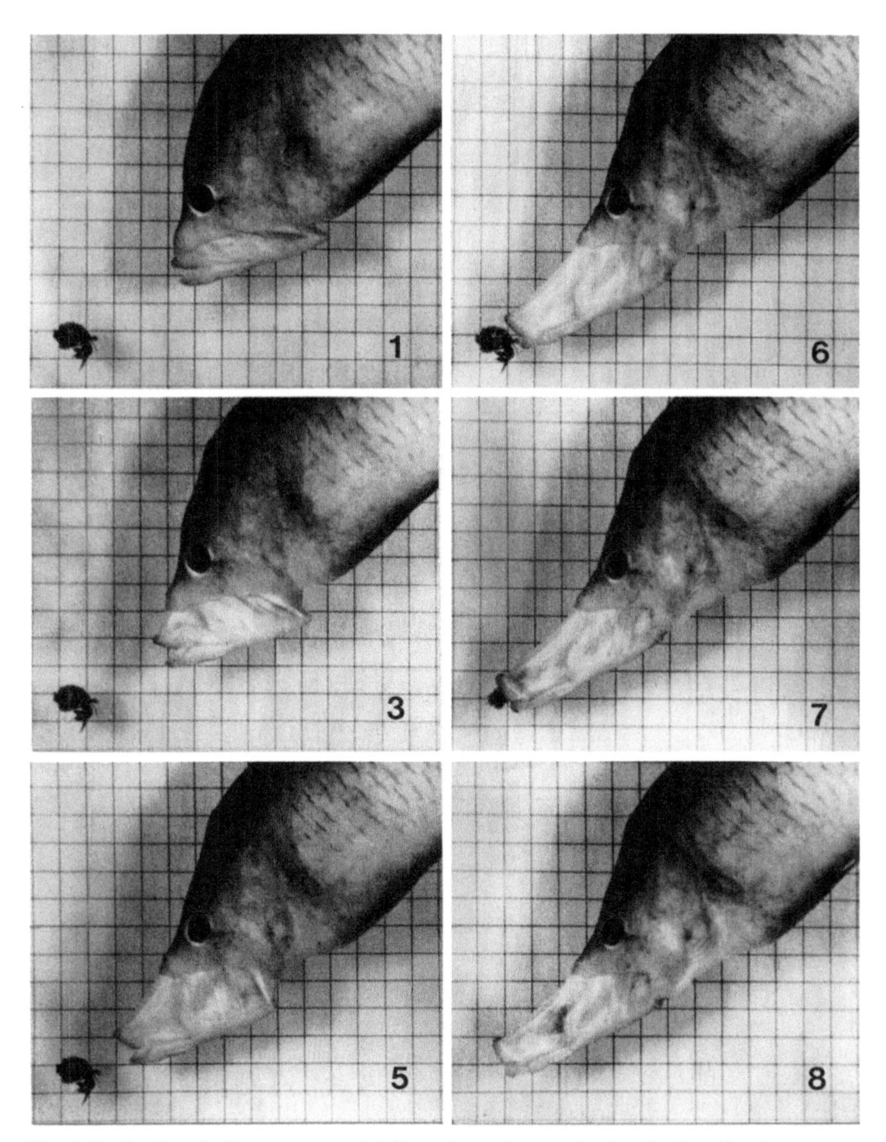

Fig. 2.11. Suction feeding accompanied by extreme jaw protrusion in the sling-jaw wrasse, *Epibulus insidiator*. Frames 1, 3, 5, 6, 7, and 8 from a high-speed film (200 frames/s) of the strike of *E. insidiator*. Successive frames are 0.005 s apart. Note rotation of quadrate, maxilla, and interoperculo-mandibular ligament. Suction is apparent in frames 6, 7, and 8. Grid size = 1 cm^2. (From Westneat and Wainwright, 1989.)

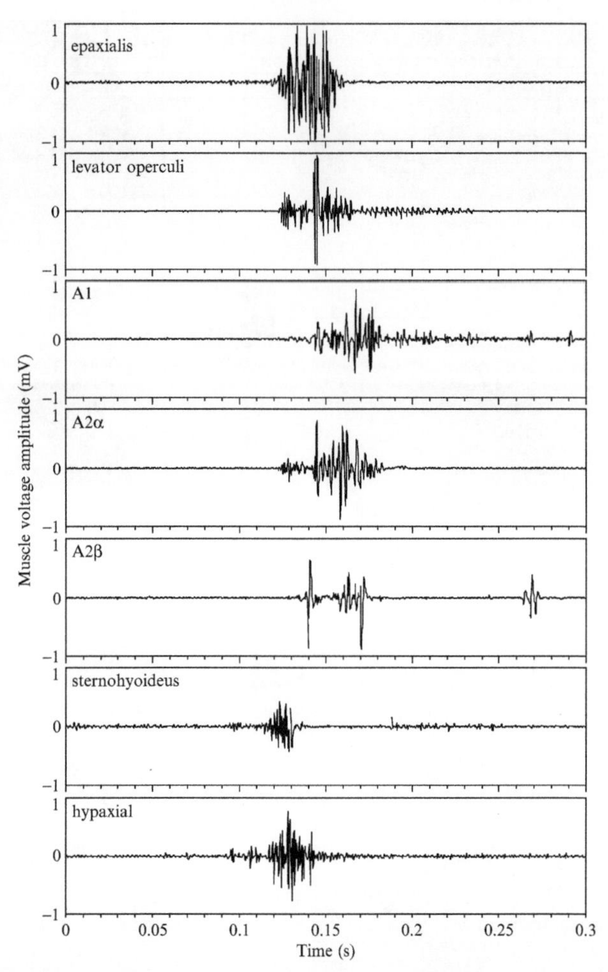

Fig. 2.12. Example of a motor activity pattern (MAP) consisting of EMG traces from seven muscles during suction feeding in the vampire characin *Hydrolycus scomberoides*. Jaw opening muscles showing early activity during expansion phase include epaxialis, hypaxialis, levator operculi, and sternohyoideus. Jaw closing muscles of the adductor mandibulae complex (A1, A2) are subsequently active during compression phase. Later activity of A1, A2β, and sternohyoideus are associated with prey processing (Westneat and Alfaro, unpublished data).

active early in the feeding cycle are the hypaxial, sternohyoideus, and levator operculi muscles (Figure 2.12), which function in jaw depression and buccal cavity expansion. The levator operculi often begins activity slightly earlier than the other expansive phase muscles, whereas the sternohyoideus often begins somewhat later. Also initiating their activity in the expansive phase are the levator arcus palatini, which expands the suspensorium laterally, and the dilatator operculi, which flares the gill cover. Firing last and repeatedly during chewing, the adductor mandibulae begins the compression phase of suction feeding as it closes the jaw. The adductors are almost always subdivided into components with different origins and insertions, which may show different patterns of activation and duration (Figure 2.12).

Comparison of MAPs for cranial muscles during suction feeding across a diversity of fishes has shown that motor patterns are largely conserved among teleost fishes and basal actinopterygians (Wainwright *et al.*, 1989; Friel and Wainwright, 1998; Alfaro *et al.*, 2001). This conserved motor pattern is likely due to the common features of feeding in an aquatic medium, including the basic need to rapidly open and subsequently close the mouth, as well as the fact that suction feeding requires an anterior–posterior water flow. In some cases, the conservation of motor control is maintained even though major changes in function have occurred. For example, Wainwright *et al.*, (1989) showed that major features of feeding MAPs were conserved across basal actinopterygians and even basal tetrapods. Westneat and Wainwright (1989) examined motor patterns in the sling-jaw wrasse and found that the unique mode of prey capture nevertheless utilizes a motor pattern of the major cranial muscles that is similar to other suction feeders. Grubich (2001) found that motor patterns are also conserved in tarpon, which rely on suction combined with body velocity during ram-suction feeding. These examples show that, although changes in behavior are accompanied by some changes in muscle timing or duration, fairly large-scale changes in function can be driven by relatively similar MAPs of the musculature.

On the other hand, despite the overall pattern of motor conservation, a number of studies have discovered changes in onset times and duration of feeding muscle activity that are functionally relevant and may be involved in the diversification of feeding strategies. A classic example is the discovery of "modulatory multiplicity" in cichlid fishes by Liem (1978, 1979). Liem found several different feeding modes within species and used extensive EMG recordings of multiple muscles to show that unique MAPs were responsible for powering a diverse feeding repertoire within species. Alfaro and Westneat (1999) showed that parrotfish feeding EMGs show qualitative and quantitative changes as compared to their close labrid relatives. Friel and Wainwright (1998) analyzed motor patterns in the Tetraodontiformes (triggerfishes, pufferfishes, and relatives), a group in which feeding strategies are diverse, including blowing, suction, and biting, and the adductor mandibulae muscles have undergone extensive subdivision. Their data showed that the MAPs of subdivided muscles are significantly different than the patterns seen in more ancestral unsubdivided forms; this is likely to be an influence on functional versatility and evolutionary diversification in the tetraodontiform radiation (Friel and Wainwright, 1999).

In a broad analysis of motor activity patterns, Alfaro *et al.*, (2001) compared MAPs of suction feeding predators to biting MAPs seen in parrotfishes and piranhas that do not rely on suction in their attacks focused on piece removal of larger fish or sessile (algal) prey. This study showed that the MAPs of biters involved changes in the presence or absence of activity in some muscles and modification of duration times but not the temporal orders of muscle contraction patterns. Biting species showed low levels, or absence, of activity in the epaxial and sternohyoideus muscles, and the adductor mandibulae initiated activity late in the strike cycle compared to suckers (Figure 2.13). This pattern results in less overlap of expansive and compressive phase muscles in biters as compared to suction feeders. Suction feeding performance clearly places a premium on rapid motion and fast muscle contractions to enhance water flow and maximize the probability of capturing evasive prey. These priorities are the underlying source of relatively similar or conserved motor patterns of suction feeding. Within these general constraints, suction feeders often vary in the fine details of muscle activity patterns. A major frontier of future work will be to assess the interactions between muscular contraction patterns and the role of body size, mouth gape, and the kinematic variability of suction feeding across major lineages and feeding guilds in fishes.

C. Suction Feeding Pressure Changes and Hydrodynamics

Experimental approaches to the measurement of pressure changes in the mouth cavity and the dynamics of fluid flow near the head have led to key insights into the mechanisms of suction feeding. Pressure transducers of increasing sophistication have been used to record intraoral pressures during feeding for the purpose of assessing the timing of low pressure and the impact that pressure change has on water flow through the mouth and gill cavity. Studies that have used pressure transducers to measure buccal pressure change (Alexander, 1969, 1970; Lauder, 1980b, 1983b; Van Leeuwen and Muller, 1983; Muller and Osse, 1984; Norton and Brainerd, 1993; Nemeth, 1997) show that water pressure decreases sharply as the mouth opens, creating a flow of water and the prey item into the mouth. Pressure change is highly variable but may drop as low as 50–100 kPa (nearly 1 atm) below ambient pressure, although most feeding events result in drops of just 10–20 kPa. Lauder (1983b, 1985) examined pressure change in both the opercular and the buccal cavity and found that pressures in the opercular cavity may be just one-fifth the magnitude of the subambient buccal

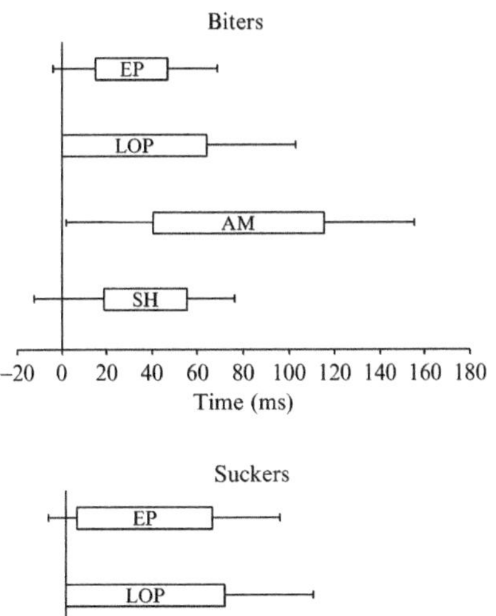

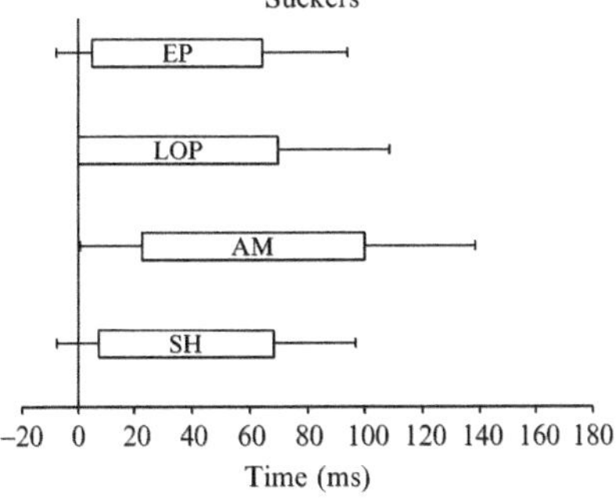

Fig. 2.13. Box plot summary of aquatic feeding MAPs. Shown are the average patterns of activity for five cranial muscles for six biting species and five suction-feeding species. Rectangles represent mean muscle duration with error bars to the right indicating one standard error of the mean. Mean onset times relative to the LOP are indicated by the distance of each bar to the line to the left with error bars to the left indicating one standard error of the mean. Abbreviations: A1-A3, adductor mandibulae subunit; EP, epaxialis; LOP, levator operculi; SH, sternohyoideus. (From Alfaro *et al.*, 2001.)

pressure, and concluded that gill arches play a role in segregating the opercular cavity from the buccal cavity as the mouth opens (Lauder, 1983a,b; Lauder *et al.*, 1986).

Sanford and Wainwright (2002) used sonomicrometry (employing ultrasonic frequencies to measure the distance between implanted crystals) in combination with pressure recordings in order to correlate pressure changes with measurements of buccal cavity expansion. Using synchronized data relating buccal chamber expansion speed to pressure changes through time, Sanford and Wainwright (2002) showed that the point of greatest subambient pressure production occurs early in the strike cycle when the rate of buccal expansion is maximal (Figure 2.14). They were able to show a tight correlation between kinematics and intraoral pressure that was particularly

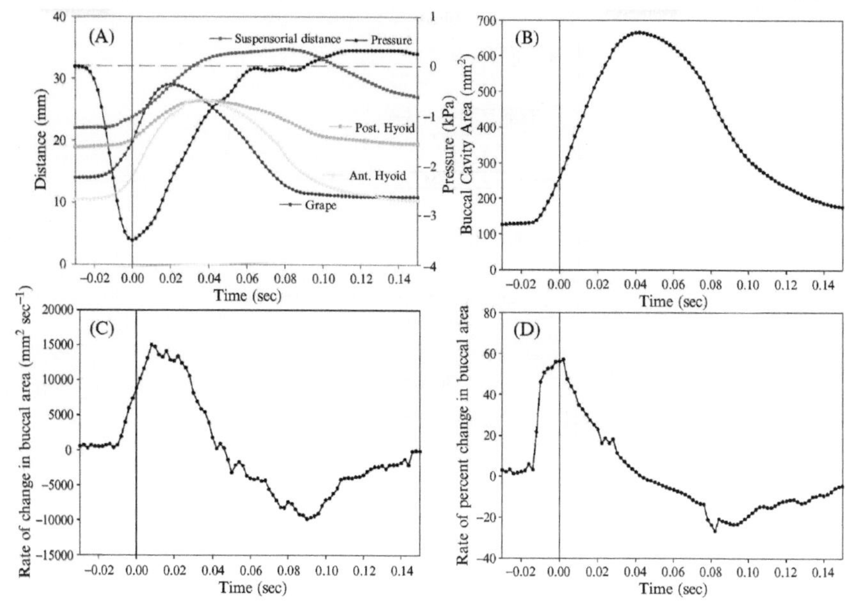

Fig. 2.14. Timing of buccal cavity expansion and pressure change in the large-mouth bass, *Micropterus salmoides*. (A) Representative kinematic profile of buccal cavity variables and pressure during the expansion phase of suction feeding. Abbreviations: Ant., anterior; Post., posterior. (B) Buccal cavity area plotted against time for the same sequence as in A. (C) Rate of change in buccal area for the same sequence as in A. (D) Rate of change in buccal area divided by buccal area for the same sequence as in A. Time zero ($t0$, solid vertical line) represents the time of peak subambient pressure. Note the very early time of peak subambient pressure. (From Sanford and Wainwright, 2002.)

meaningful because the sonomicrometry enabled precise internal kinematic assessment of the expanding volume of the pharynx.

A series of early studies compared the pressure changes observed during suction feeding with the predictions of hydrodynamic models based on buccal cavity expansion. Muller and Osse (1978, 1984), Muller *et al.*, (1982), and Van Leeuwen and Muller (1983, 1984) modeled the increase in volume of the skull as an expanding cone, and used cranial kinematics from a range of species to calculate expanding cone dimensions. These projects effectively combined multiple data sets from modeling, kinematics, and pressure measurements to compare biomechanical models with actual feeding parameters and to demonstrate the suction effectiveness of species with large mouth cavities and small gapes. Recent authors (De Visser and Barel, 1996, 1998; Bouton *et al.*, 1998) have developed complementary models that combine buccal morphology and kinematics to examine correlations with suction forces. De Visser and Barel (1996) and Bouton *et al.*, (1998) demonstrated correlations between hyoid morphology and motion with feeding strategy, suggesting that suction feeders will possess a hyoid with more closely aligned left and right sides that will maximize speed. Similarly, both Muller (1987) and Westneat (1990) used linkage modeling approaches and showed that the kinematic transmission of the hyoid four-bar linkage generally is higher in suction feeders, indicating increased speed of hyoid motion. These modeling studies are important contributions to the basic functional morphology of fish feeding and are central to the effort to develop a complete biomechanical understanding of suction feeding. Continued development of such models, particularly in the direction of three-dimensional linkage modeling, will help to integrate data on kinematics, pressure change, and flow visualization results from particle imaging.

The dynamics of water flow around the head and into the mouth during suction feeding is of critical importance to an understanding of feeding biomechanics. This is so because the rate of flow, the entrainment of prey by drag, and the forward extent of flow (the predator's "reach") are critical to parameters such as the predator–prey distance during feeding and the timing of peak suction force relative to cranial kinematics. Fluid dynamics of rapid events are difficult to quantify, but the technical challenges of assessing these rapid flows are now being overcome using flow visualization techniques. Early particle imaging studies of feeding were done by Muller and Osse (1984), who used polystyrene beads in the water to visualize flow, and Lauder and Clark (1984), who cleverly used the neutrally buoyant aquatic organism *Artemia* as particles to visualize buccal flows during feeding, in combination with high-speed films. The data from these studies enabled the authors to calculate the first empirically derived suction flow rates and estimate the distances at which suction feeding would be effective.

More recently, Ferry-Graham and Lauder (2001) and Ferry-Graham *et al.*, (2003) utilized digital particle image velocimetry (DPIV) to more fully quantify flow fields around the head of a suction-feeding fish. DPIV involves shining lasers through the water to illuminate tiny reflective particles, whose motion is recorded on video, usually in one or more planes of interest relative to the body axis of the fish. Using DPIV with bluegill sunfish feeding, Ferry-Graham and Lauder (2001) generated detailed velocity vector patterns for flow during suction feeding, showing that water flows toward the mouth from a large spherical volume around the head (Figure 2.15). Ferry-Graham *et al.*, (2003) used similar data to show that flow velocity in front of the mouth increased rapidly to peak within 20 ms of the onset of the strike, and occurred at the time that the prey entered the mouth during capture. They also discovered the presence of a bow wave in front of the fish that causes the highest flows toward the mouth to be some distance anterior

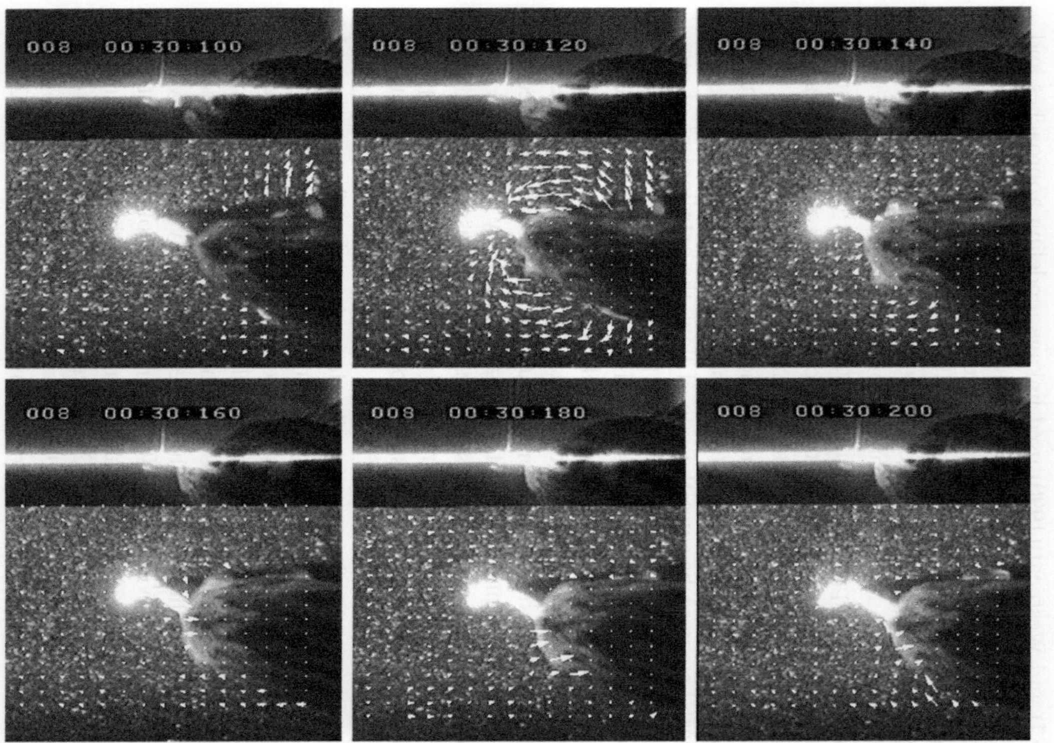

Fig. 2.15. Velocity vector profiles generated by a bluegill sunfish (*Lepomis macrochirus*) feeding on a worm piece held suspended by forceps within a horizontal laser light sheet. The velocity vectors (white arrows) are superimposed on high-speed video images from the same feeding event at the same points in time. Times are indicated on each field (min:s:ms). Peak gape occurred at 140 ms, 40 ms after the onset of the strike (at 100 ms). Note that the velocity vectors pointed to the right at 160 and 180 ms result from water attached via drag forces to the retreating bluegill, and not suction-generated for prey capture. (From Ferry-Graham and Lauder, 2001.)

to the mouth opening, rather than immediately in front of the open mouth. This research confirmed that the timing of the strike relative to predator–prey distance is critical for successful entrainment of the prey in the water flow, and thus successful capture. Flow visualization represents a key source of information with which biologists can test the impact of kinematics of a wide range of species, mouth shapes, or developmental stages on the flow field and suction performance of fishes.

V. ECOMORPHOLOGY OF FISH FEEDING

Skull form and function have provided biologists with an excellent system with which to explore the link between anatomical or biomechanical traits and the ecology of fishes. The field of ecological morphology has emerged from the attempt to explain the links between anatomy, biomechanics, and the performance of ecologically relevant behaviors such as feeding (Wainwright and Reilly, 1994; Motta *et al.*, 1995b). Wainwright (1996) clearly laid out the process whereby an ecomorphologist first searches for morphological variables that affect variability in biomechanics or performance, followed by the determination that the performance character influences resource use. Morphology can play a profound role in fish feeding ecology due to physical constraints that set limits on performance variables such as bite force, jaw velocity, and suction flow. Such constraints can affect prey capture efficiency or success, and a number of studies on fishes (e.g., Gatz, 1979; Mittelbach, 1984; Wainwright, 1987, 1996; Norton, 1991; Turingan, 1994; Turingan *et al.*, 1995) have demonstrated that these limitations can in turn determine patterns of prey use in the environment, the role of predators in shaping community structure, and the biogeographic distribution of species.

The ecomorphology of fish feeding has been studied in a wide range of freshwater fishes (Keast and Webb, 1966; Gatz, 1979; Barel, 1983; Winemiller, 1991; Wainwright and Lauder, 1992; Winemiller *et al.*, 1995) and marine taxa (Motta, 1988; Sanderson, 1990; Turingan, 1994; Clements and Choat, 1995; Luczkovich *et al.*, 1995; Motta *et al.*, 1995a; Norton, 1995; Westneat, 1995b). Many studies have employed the search for correlations between broad sets of morphological characters and ecological variables to determine the strength of ecomorphological associations, but the intermediate step of showing the functional role of morphology in feeding performance and prey use (Wainwright and Reilly, 1994) has become increasingly common. For example, Wainwright and Lauder (1986, 1992) and Wainwright (1996) examined the role of pharyngeal jaw function and oral jaw biomechanics in the ecomorphology of centrarchid fishes (basses and sunfishes). This family has been an excellent group for the study of feeding mechanics and ecomorphology due to their similar body sizes, wide range of feeding habits including piscivory, planktivory, and molluscivory (Mittelbach, 1984), and well-known ecological roles in North American fresh waters. This work showed that the differences in feeding habits of the closely related bluegill (*Lepomis macrochirus*) that eats zooplankton and pumpkinseed sunfish (*L. gibbosus*) that eats snails can be explained by the morphology and performance of the pharyngeal apparatus in producing the bite force necessary to crush snails. Several key morphological and functional attributes such as larger pharyngeal muscles and a unique pharyngeal muscle MAP enable the pumpkinseed to exploit a food resource that bluegill are biomechanically constrained from utilizing. This divergence may locally reduce competition in areas where both species are present (Mittelbach, 1984). In addition, the largemouth bass preys upon them both at certain sizes using its impressive suction-feeding ability, thus restricting smaller ontogenetic stages of both species to shallow, littoral habitats (Werner and Hall, 1976). Wainwright (1996) summarized how this body of work clearly demonstrates the interaction of feeding biomechanics of each species with the availability of prey and the threat of predation to shape important features of community ecology.

The study of ecomorphology is also an inherently comparative exercise, requiring the integration of phylogenetics for the broader analysis of evolutionarily correlated changes in mechanics, performance, and ecology (Westneat, 1995b). The preceding sunfish example is particularly compelling because a phylogeny was used to show the historical pattern of divergence among the fishes that were interacting ecologically based on feeding mechanics. A wide range of phylogenetic comparative methods is now available for integrating functional and ecological data into a phylogenetic framework and searching for patterns of character association (Felsenstein, 1985; Maddison, 1990; Garland *et al.,* 1999; Pagel, 1999; Martins *et al.*, 2002). To study the phylogenetic history of feeding mechanics and dietary ecology, Westneat (1995a,b) analyzed the correlated evolution of oral jaw linkage mechanisms and feeding habits in cheiline labrids (Figure 2.16). The dietary data were categorized as hard prey or evasive prey (Figure 2.16A), whereas the biomechanical characters used were calculations of lever or linkage function, either mechanical advantage or kinematic transmission (KT) of a linkage (Figures 2.16B and 2.16C). Using methods for phylogenetic analysis of these features as both discrete and continuous characters, this research was able to show a statistically significant association between dietary preference and jaw biomechanics while accounting for the phylogenetic pattern of relatedness among species.

Research on the ecomorphology of fish feeding is now burgeoning as investigators are increasingly combining experimental functional morphology

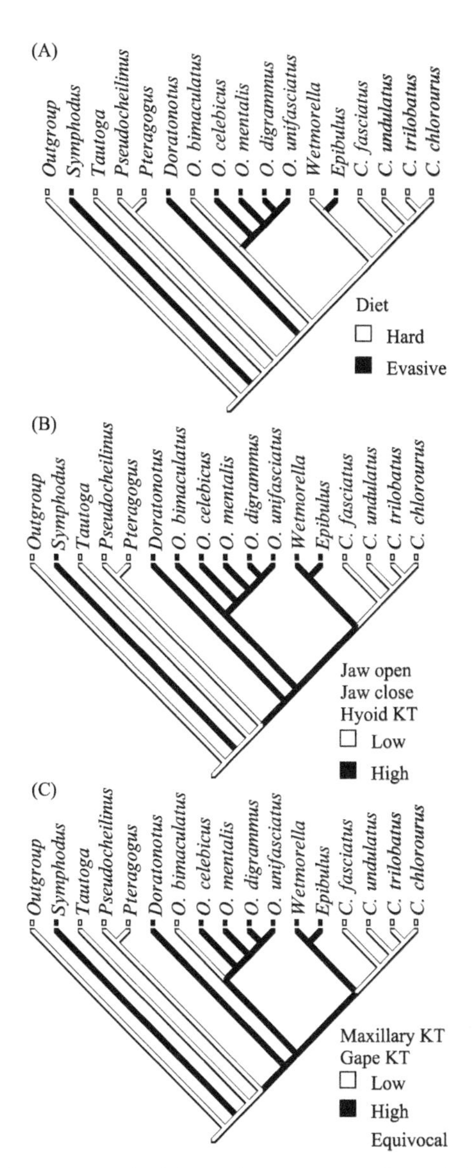

Fig. 2.16. Phylogenetic mapping of dietary habits and biomechanical characters of the oral jaws of cheiline labrid fishes, using gap-coded characters. (A) Dietary habits categorized as hard prey (gastropods, bivalves, echinoderms, etc.) or evasive prey (fishes, evasive decapods, etc.). (B) phylogenetic pattern of estimates of jaw velocity, expressed as displacement advantage of jaw opening and closing levers and the kinematic transmission coefficient of hyoid depression. (C) Oral jaw linkage mechanics expressed as the kinematic transmission of maxillary rotation and gape angle. (From Westneat, 1995a.)

with ecology in a well-resolved phylogenetic context. Of necessity, most studies emphasize their strength with either more detailed ecology or more thorough feeding biomechanics, and the creative students that can combine in-depth data sets from both areas will likely be successful in demonstrating clear ecomorphological links with evolutionary relevance. For the ecomorphology of fish feeding, the focus will likely remain on the link between the skull mechanics of the predator and the escape or defense mechanisms of the prey, and the fluid mechanics events and bite characteristics that lead up to ingestion of the prey. It is at the critical moment of prey capture that natural selection is most likely to have an effect in molding morphology and predator–prey ecology (Ferry-Graham and Lauder, 2001). Feeding ecomorphology in fishes has so far surveyed a limited number of fish groups, and future work could productively explore a wider range of fish clades such as the incredibly diverse gobies, blennies, and other reef fish clades of marine systems and the cyprinids, catfishes, and electric fishes of fresh waters. Similarly, ecomorphologists are likely to discover new links in functional ecology by expanding into other feeding guilds and feeding strategies, including filter feeding and planktivory (van den Berg *et al.*, 1994; Sanderson *et al.*, 2001) and detritivory (Choat *et al.*, 2002), as well as additional work on oral jaw durophagy (Turingan, 1994; Hernandez and Motta, 1997) and pharyngeal processing (Wainwright, 1987; Vanderwalle *et al.*, 1995). It should also be noted that the locomotor system plays a critical role in fish feeding, and the coordination of feeding by the locomotor system (Rand and Lauder, 1981; Rice and Westneat, 2005) is a relatively unexplored axis of diversity important to ecomorphology.

Future ecomorphology also has the potential to integrate other areas of biological endeavor to generate a more complete picture of the role of feeding interactions in evolution (Liem and Summers, 2000). For example, incorporating quantitative genetic information regarding the heritability of features such as jaw dimensions that are biomechanically relevant (Albertson *et al.*, 2003; Streelman *et al.*, 2003) provides the potential to clarify the genetic basis of functionally important traits, how fast they can evolve, and the consequences of phenomena such as population bottlenecks and hybridization events to functional diversity. Ontogenetic studies of trophic morphology and feeding mechanics in fishes (Galis *et al.*, 1994; Cook, 1996; Hunt von Herbing *et al.*, 1996; Hernandez, 2000; Hernandez *et al.*, 2002) have been shown to have clear links to ecology and distribution patterns, and this information provides a potential link to the large body of work on gene expression and the regulation of development in the skull and pharyngeal arches of fishes (Kimmel *et al.*, 2001, 2003; Hunter and Prince, 2002). Linking the role of developmental regulatory genes to skull formation and the biomechanical consequences of changes in development is a key frontier in our understanding of evolutionary diversification of fish feeding systems. The integration of biomechanical, developmental, and genetic data with research on population biology, community ecology, and phylogeny is now an attainable goal that can advance our understanding of the links between morphology and ecology through biomechanics.

VI. PHYLOGENETIC PATTERNS OF FEEDING IN FISHES

The inclusion of phylogenetic information is the hallmark of evolutionary biomechanics, which has the objective of integrating functional morphology with the plethora of recent and emerging phylogenetic hypotheses of relationships among organisms. By analyzing functional traits as characters in the framework of evolutionary trees, biologists can identify the points of origin of functional novelties, compare the variability of functional traits among clades, look for patterns of correlation between mechanics and other features such as ecology, and demonstrate the patterns of evolution of important mechanisms at the level of species, family, and higher phylogenetic levels. Fish feeding mechanics has played a central role in the integration of biomechanics with phylogenetics.

Phylogenetic patterns of change in fish feeding mechanisms have been analyzed primarily at the family and subfamily levels. The first study to join fish jaw mechanisms with phylogenetic information was Liem's (1970) work on the leaf-fishes, the Nandidae. Although a full comparative phylogenetic analysis of feeding was not presented in this monograph, Liem set the stage for explicit phylogenetic analysis of feeding biomechanics by integrating anatomical description, kinematic studies, and a phylogeny of the Nandidae to reveal the diversity of jaw protrusion mechanisms in this group of fishes with extraordinary kinesis in the skull.

A number of evolutionary feeding studies have used phylogenetic mapping of functional and morphological characters onto a phylogeny to reveal the patterns of evolution of the characters of interest in diverse groups of fishes. For example, Lauder (1982) inferred the evolution of the functional design of the feeding mechanism in actinopterygians by mapping functional features onto phylogenies of those groups. Schaefer and Lauder (1986, 1996) mapped the mechanisms of the oral jaws of loricariid catfishes onto a cladogram for the group to identify three major steps in the evolution of the jaw mechanism. Wainwright and Lauder (1992) synthesized a series of previous studies on the phylogeny, ecology, and functional morphology of the centrarchid fishes to identify evolutionary correlates of the novel dietary habit of snail crushing among the sunfishes. Studying coral reef fishes, Streelman *et al.* (2002) analyzed feeding evolution in the context of a

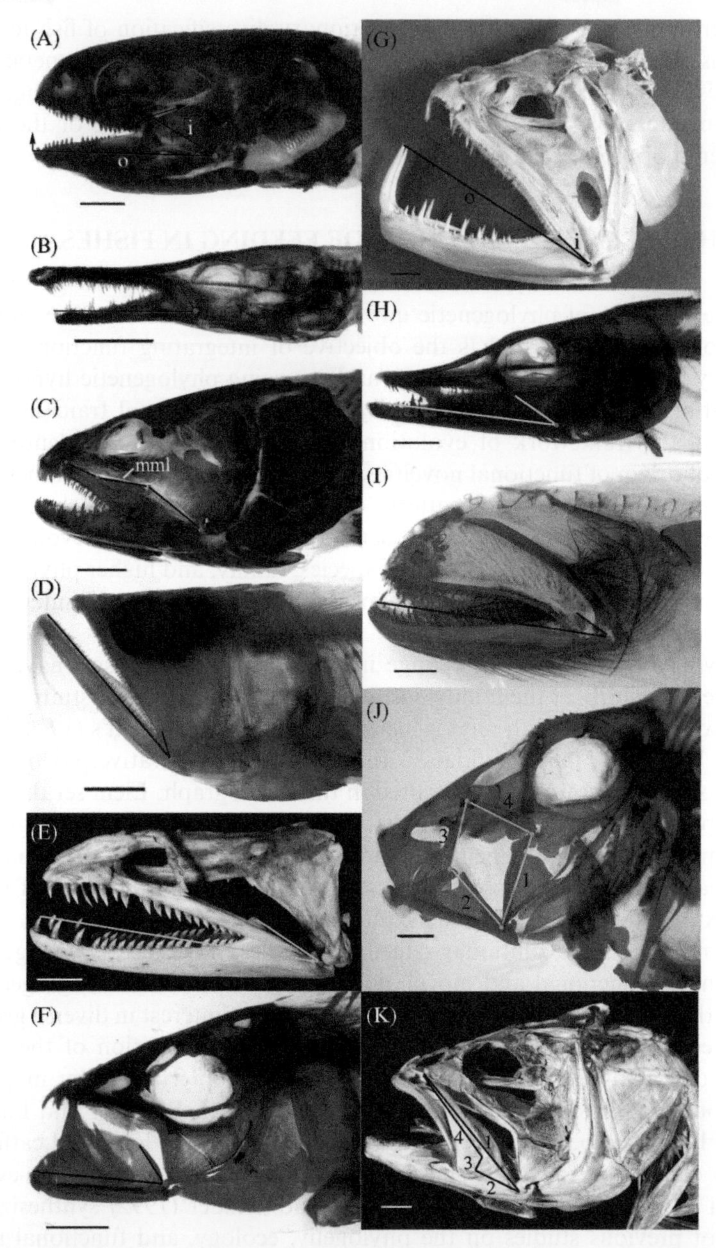

Fig. 2.17. Skull diversity, mandibular lever variation, and linkage structure in actinopterygian fishes. (A) The bichir, *Polypterus senegalus*, illustrating a simple mandibular lever with input (i) and output (o) lever arms. (B) Lever dimensions of the alligator gar *Atractosteus spatula*. (C) The bowfin, *Amia calva*, illustrating the three movable elements in the four-bar linkage

parrotfish phylogeny showing patterns of jaw mechanics as a function of seagrass or coral reef habitat type, and Ferry-Graham *et al.*, (2001) showed the evolutionary patterns of feeding in butterflyfishes.

This species-level research attains the important goal of revealing evolutionary trends in functional morphology of feeding in specific, functionally intriguing fish families. However, only a limited number of particularly interesting fish groups among the huge radiation of actinopterygian fishes have been examined in this way, and broader studies that integrate higher-level phylogenetics with the biomechanical modeling of force and motion in the skull are rare. To attain this broader evolutionary perspective, several recent studies have taken the first steps toward an explicitly phylogenetic framework for jaw biomechanics in fishes. Wilga *et al.*, (2001) combined information on protrusion mechanics in sharks with motor pattern data from jaw muscles and a higher-level phylogeny of elasmobranchs (Shirai, 1996) to reveal the sequence of evolutionary change or character conservation at successive nodes in the tree. Wilga (2002) took a similar approach to tracing the evolution of jaw suspension and feeding mechanisms, focusing primarily on elasmobranchs but expanding the phylogenetic perspective to include most vertebrates. Wilga's research showed an evolutionary increase in mobility of the feeding system that is associated with a diversification of feeding modes among sharks.

The spectacular diversity of skull morphology and function among ray-finned fishes (Figure 2.17) has been explored with the goal of illustrating phylogenetic patterns of jaw function, including changes in the lever and linkage designs found in the skulls of actinopterygian fishes (Westneat, 2004). This work presented the morphological diversity of actinopterygian skulls from the perspective of biomechanical modeling and traced phylogenetic changes in skull mobility to identify key events in the origin and evolution of the linkages that mediate jaw protrusion (Figure 2.18).

A large range of jaw mechanisms is revealed as one moves from *Polypterus* to Perciformes (Westneat, 2004). Many independent evolutionary

---

for maxillary rotation; mml, maxillomandibular ligament. (D) Lever dimensions of the arawana, *Osteoglossum bicirrhosum*. (E) Lever dimensions of the moray eel, *Gymnothorax javanicus*. (F) Lever dimensions of the clupeid *Sardinella aurita*. (G) Skull of the vampire characin, *Hydrolycus scomberoides*, with inlever (i) and outlever (o) labeled. (H) Lever dimensions of the northern pike, *Esox lucius*. (I) Lever dimensions of the bombay-duck, *Harpadon nehereus*. (J) The rosy dory, *Cyttopsis rosea*, the earliest clade to show an anterior jaws four-bar linkage with a rotational palatine that powers protrusion. (K) Skull of the large mouth bass, *Micropterus salmoides*, with diagram of four-bar linkage for maxillary rotation; 1, fixed link; 2, articular input link; 3; maxillomandibular ligament coupler link; 4, maxillary output link. Scale bar = 5 mm. (Modified from Westneat, 2004.)

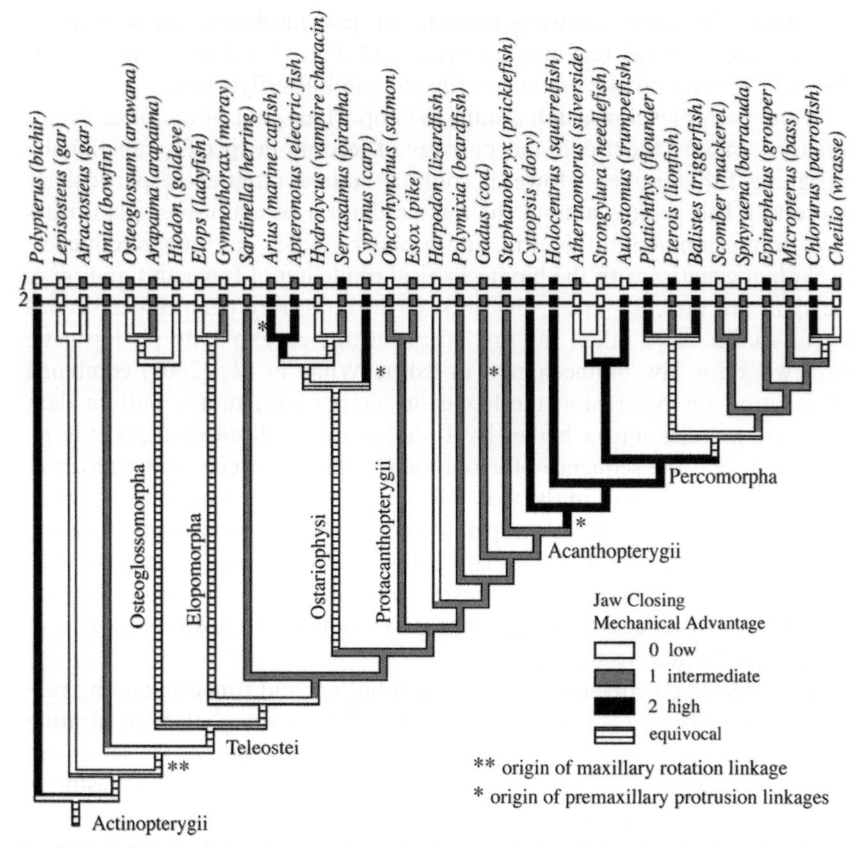

Fig. 2.18. Evolution of jaw lever mechanical advantage in 35 species of actinopterygian fishes. Phylogeny is a composite tree based on Coates (1999), Nelson (1994), Stiassny *et al.* (1996), and Johnson and Patterson (1993). Two characters are optimized on the phylogeny: character 1 is jaw opening mechanical advantage (with tip states only shown) and character 2 is jaw closing mechanical advantage (with tip states and branch shading illustrated). Low MA jaws emphasize velocity and high MA jaws transmit relatively more force. Phylogenetic origin of the maxillary rotation linkage (**) and four of at least five origins of premaxillary protrusion (*) are indicated on the tree topology. (From Westneat, 2004.)

transitions occur from feeding systems with high force transmission to those specialized for speed of jaw motion. Each major group of actinopterygians appears to have members with fast and members with forceful jaw mechanics, suggesting that convergent evolution of jaw function is likely to be the rule at both higher levels and species levels of generality (Figure 2.18). This diversity of mechanical design and emergent pattern of convergence is likely driven by the alternative requirements for force and speed associated with a strategy for biting versus strategies for suction feeding (Alfaro *et al.*, 2001) and the high frequency of switching between these strategies among species (Westneat, 1995a).

The evolution of kinesis in the jaws of fishes is a story of increasing mobility and the origin of linkage mechanisms enabling maxillary and premaxillary motion at several different points in actinopterygian phylogeny (Figure 2.18). A linkage for maxillary rotation originated in Amiiformes and persists throughout many teleost species. This mechanism operates by transferring ventral mandibular rotation into anterior maxillary rotation via a ligamentous coupling that was described by Lauder (1979, 1980a). Kineticism of the maxilla is the precursor to premaxillary protrusion, which is a hallmark of feeding and evolutionary diversity in fishes.

Premaxillary protrusion mechanisms have evolved at least five times among major groups of ray-finned fishes (Westneat, 2004) (Figure 2.18). Using the phylogenetic hypotheses of Johnson and Patterson (1993), it can be shown that these independent evolutionary events are functionally convergent, in that upper jaw protrusion occurs, but the musculoskeletal mechanics underlying each type of protrusion mechanism are different. Many of the cyprinid fishes (minnows and carps) have upper jaw protrusion, via mechanisms described by Alexander (1966) that involve a rotational and twisting maxilla. The loricariid catfishes have independently evolved a highly mobile premaxilla associated with algae scraping (Schaefer and Lauder, 1986). Subsequent clades on the actinopterygian tree such as salmonids, esocids, and aulopiform and stomiiform fishes lack upper jaw protrusion, but most retain a mobile maxilla. A novel, highly kinetic mechanism of maxillary and premaxillary protrusion evolved in the lampridiform fishes at the base of the acanthomorphs (Figure 2.18), in which both maxilla and premaxilla are largely free of the neurocranium and are pulled forward and ventrally during jaw opening. This is found in an extreme form in *S. chordatus*, which has one of the most extremely protrusible mouths among fishes (Pietsch, 1978).

Farther up the actinopterygian tree, *Polymixia* lacks premaxillary protrusion, and this genus occupies a key position as the sister-group to a major sister-pair of large clades, the Paracanthopterygii and Acanthopterygii. At least one origin of upper jaw protrusion occurs in each of these lineages. Most paracanthopts lack premaxillary protrusion, but some cods have protrusion, and the anglerfishes of the family Antennariidae have fairly extensive jaw protrusion used in conjunction with their dorsal fishing lures. The most widespread anterior jaws linkage used for premaxillary protrusion in fishes is that of the acanthopterygians. This mechanism appears to be absent or reduced in *Stephanoberyx* (at the base of Acanthopterygii according to Johnson and Patterson, 1993) but arose initially in

Zeiformes (the dories), which have a rotational palatine link that frees the maxilla to translate and rotate, and an ascending premaxillary process that enables the premaxilla to slide anteriorly and downward. This mechanism is present in some form in most acanthopterygians and most major groups of percomorphs, and has been modified in many lineages to enhance upper jaw motion.

Future work on the phylogenetic patterns and evolution of fish feeding mechanisms should include more species-level analyses of genera or families as well as higher-level phylogenetic surveys of jaw function. Current broad-scale surveys have included only exemplar members of families or even orders of fishes, leaving a compelling need for more intensive taxonomic character sampling of key features of musculoskeletal design, kinematics, motor patterns, and hydrodynamics.

VII. SUMMARY AND CONCLUSIONS

The study of feeding biomechanics in fishes is at an exciting point in its history because progress in animal function is ultimately accomplished through synthesis of ideas and techniques from multiple areas. Fish feeding biomechanics enjoys the position of being a well-developed area of functional morphology equipped with solid morphological background, extensive experimental data, and clear evolutionary interest. As a result, this field is poised to take advantage of rapid advances in technology, developmental biology, genetics, and understanding of phylogenetic relationships. Investigators can readily pursue the experimental biomechanics approaches described here on a group of fishes for which there is fresh phylogenetic information, or in a model system such as zebrafish or medaka for which there is a strong developmental, genetic, or neurobiological literature. Some of the frontiers in this area that offer exciting prospects for research include the exploration of biomechanics in some of the poorly characterized groups of fishes using our growing toolbox of techniques, increasing the sophistication and utility of biomechanical models of skull function, and bridging feeding mechanics to genetics, development, and evolution.

REFERENCES

Adriaens, D., Aerts, P., and Verraes, W. (2001). Ontogenetic shift in mouth opening mechanisms in a catfish (Clariidae, Siluriformes): A response to increasing functional demands. *J. Morphol.* **247,** 197–216.

Aerts, P. (1991). Hyoid morphology and movements relative to abduction forces during feeding in *Astatotilapia elegans* (Teleostei: Cichlidae). *J. Morphol.* **208,** 323–345.

Aerts, P., and Verraes, W. (1984). Theoretical analysis of a planar bar system in the teleostean skull: The use of mathematics in biomechanics. *Ann. R. Soc. Zool. Belg.* **114,** 273–290.

Aerts, P., Osse, J. W. M., and Verraes, W. (1987). Model of jaw depression during feeding in *Astatotilapia elegans* (Teleostei: Cichlidae): Mechanisms for energy storage and triggering. *J. Morphol.* **194,** 85–109.

Albertson, R. C., Streelman, J. T., and Kocher, T. D. (2003). Directional selection has shaped the oral jaws of Lake Malawi cichlid fishes. *Proc. Natl. Acad. Sci. USA* **100,** 5252–5257.

Alexander, R. M. (1966). The functions and mechanisms of the protrusible upper jaws of two species of cyprinid fish. *J. Zool. (Lond.)* **149,** 288–296.

Alexander, R. M. (1967). The functions and mechanisms of the protrusible upper jaws of some acanthopterygian fish. *J. Zool. (Lond.)* **151,** 43–64.

Alexander, R. M. (1969). Mechanics of the feeding action of a cyprinid fish. *J. Zool. (Lond.)* **159,** 1–15.

Alexander, R. M. (1970). Mechanics of the feeding action of various teleost fishes. *J. Zool. (Lond.)* **162,** 145–156.

Alfaro, M. E., and Westneat, M. W. (1999). Biomechanics of parrotfish feeding: Motor patterns of the herbivorous bite. *Brain Behav. Evol.* **54,** 205–222.

Alfaro, M. E., Janovetz, J., and Westneat, M. W. (2001). Motor control across trophic strategies: Muscle activity of biting and suction feeding fishes. *Am. Zool.* **41,** 1266–1279.

Anker, G. (1974). Morphology and kinematics of the head of the stickleback, *Gasterosteus aculeatus*. *Trans. Zool. Soc. Lond.* **32,** 311–416.

Anker, G. (1978). Analyses of respiration and feeding movements of the three-spined stickleback, *Gasterosteus aculeatus* L. *Neth. J. Zool.* **28,** 485–523.

Barel, C. D. N. (1983). Towards a constructional morphology of cichlid fishes (Teleostei, Perciformes). *Neth. J. Zool.* **205,** 269–295.

Bellwood, D. R. (2003). Origins and escalation of herbivory in fishes: A functional perspective. *Paleobiology* **29,** 71–83.

Bemis, W. E., Findeis, E. K., and Grande, L. (1997). An overview of Acipenseriformes. *Env. Biol. Fish* **48,** 25–71.

Bouton, N., van Os, N., and Witte, F. (1998). Feeding performance of Lake Victoria rock cichlids: Testing predictions from morphology. *J. Fish. Biol.* **53** (Suppl.), 118–127.

Callan, W. T., and Sanderson, S. L. (2003). Feeding mechanisms in carp: Crossflow filtration, palatal protrusions, and flow reversals. *J. Exp. Biol.* **206,** 883–892.

Carroll, A. M., and Wainwright, P. C. (2003). Functional morphology of prey capture in the sturgeon, *Scaphirhynchus sapidus*. *J. Morphol.* **256,** 270–284.

Carroll, A. M., Wainwright, P. C., Huskey, S. H., Collar, D. C., and Turingan, R. G. (2004). Morphology predicts suction feeding performance in centrarchid fishes. *J. Exp. Biol.* **207,** 3873–3881.

Choat, J. H., Clements, K. D., and Robbins, W. D. (2002). The trophic status of herbivorous fishes on coral reefs 1: Dietary analyses. *Marine Biol.* **140,** 613–623.

Clements, K. D., and Choat, J. H. (1995). Fermentation in tropical marine herbivorous fishes. *Physiol. Zool.* **68,** 355–378.

Coates, M. I. (1999). Endochondral preservation of a Carboniferous actinopterygian from Lancashire, UK, and the interrelationships of primitive actinopterygians. *Phil. Trans. R. Soc. Lond. B* **354,** 435–462.

Cook, A. (1996). Ontogeny of feeding morphology and kinematics in juvenile fishes: A case study of the cottid fish *Clinocottus analis*. *J. Exp. Biol.* **199,** 1961–1971.

Dean, M. N., and Motta, P. J. (2004). Anatomy and functional morphology of the feeding apparatus of the lesser electric ray, *Narcine brasiliensis* (Elasmobranchii: Batoidea). *J. Morphol.* **262,** 462–483.

De Visser, J., and Barel, C. D. N. (1996). Architectonic constraints on the hyoid's optimal starting position for suction feeding of fish. *J. Morphol.* **228,** 1–18.
De Visser, J, and Barel, C. D. N. (1998). The expansion apparatus in fish heads, a 3-D kinetic deduction. *Neth. J. Zool.* **48,** 361–395.
Durie, C. J., and Turingan, R. G. (2004). The effects of opercular linkage disruption on prey capture kinematics in the teleost fish *Sarotherodon melanotheron. J. Exp. Zool.* **30,** 642–653.
Elshoud-Oldenhave, M. J. W. (1979). Prey capture in the pike-perch, *Stizostedion lucioperca* (Teleostei, Percidae): A structural and functional analysis. *Zoomorphology* **93,** 1–32.
Felsenstein, J. (1985). Phylogenies and the comparative method. *Am. Nat.* **125,** 1–15.
Ferry-Graham, L. A., and Lauder, G. V. (2001). Aquatic prey capture in fishes: A century of progress and new directions. *J. Morphol.* **245,** 99–119.
Ferry-Graham, L. A., Wainwright, P. C., Hulsey, C. D., and Bellwood, D. R. (2001). Evolution and mechanics of long jaws in butterflyfishes (Family Chaetodontidae). *J. Morphol.* **248,** 120–143.
Ferry-Graham, L. A., Wainwright, P. C., Westneat, M. W., and Bellwood, D. R. (2002). Mechanisms of benthic prey capture in labrid fishes. *Mar. Biol.* **141,** 819–830.
Ferry-Graham, L. A., Wainwright, P. C., and Lauder, G. V. (2003). Quantification of flow during suction feeding in bluegill fishes. *Zoology* **106,** 159–168.
Friel, J. P., and Wainwright, P. C. (1998). Evolution of motor patterns in tetraodontiform fishes: Does muscle duplication lead to functional diversification? *Brain. Behav. Evol.* **52,** 159–170.
Friel, J. P., and Wainwright, P. C. (1999). Evolution of complexity in motor patterns and jaw musculature of tetraodontiform fishes. *J. Exp. Biol.* **202,** 867–880.
Frazzetta, T. H., and Prange, C. D. (1987). Movements of cephalic components during feeding in some requiem sharks (Carcharhiniformes: Carcharhinidae). *Copeia* **1987,** 979–993.
Galis, F., Terlouw, A., and Osse, J. W. M. (1994). The relation between morphology and behavior during ontogenetic and evolutionary changes. *J. Fish. Biol.* **45** (Suppl. A), 13–26.
Garland, T., Jr., Midford, P. E., and Ives, A. R. (1999). An introduction to phylogenetically based statistical methods, with a new method for confidence intervals on ancestral states. *Am. Zool.* **39,** 374–388.
Gatz, A. J., Jr. (1979). Community organization in fishes as indicated by morphological features. *Ecology* **60,** 711–718.
Gibb, A. C. (1996). Kinematics of prey capture in *Xystreurys liolepis*: Do all flatfish feed asymmetrically? *J. Exp. Biol.* **199,** 2269–2283.
Gibb, A. C. (1997). Do flatfish feed like other fishes? A comparative study of percomorph prey capture kinematics. *J. Exp. Biol.* **200,** 2841–2859.
Gosline, W. A. (1971). "Functional Morphology and Classification of Teleostean Fishes." University of Hawaii Press, Honolulu.
Gregory, W. K. (1933). Fish skulls: A study of the evolution of natural mechanisms. *Trans. Am. Phil. Soc.* **XXIII,** 75–481.
Grobecker, D. B., and Pietsch, T. W. (1979). High-speed cinematographic evidence for ultrafast feeding in antennariid anglerfishes. *Science* **205,** 1161–1162.
Grubich, J. R. (2001). Prey capture in actinopterygian fishes: A review of suction feeding motor patterns with new evidence from an elopomorph fish, *Megalops atlanticus. Am. Zool.* **41,** 1258–1265.
Grubich, J. R., and Wainwright, P. C. (1997). Motor basis of suction feeding performance in largemouth bass, *Micropterus salmoides. J. Exp. Zool.* **277,** 1–13.
Hanken, J., and Hall, B. K. (1993). "The Vertebrate Skull." University of Chicago Press, Chicago.

Helfman, G. S., Collette, B. B., and Facey, D. E. (1997). "The Diversity of Fishes." Blackwell Science, Malden, MA.
Hernandez, L. P. (2000). Intraspecific scaling of feeding mechanics in an ontogenetic series of zebrafish, *Danio rerio*. *J. Exp. Biol.* **203,** 3033–3043.
Hernandez, L. P., and Motta, P. J. (1997). Trophic consequences of differential performance: Ontogeny of oral jaw-crushing performance in the sheepshead, *Archosargus probatocephalus* (Teleostei, Sparidae). *J. Zool.* **343,** 737–756.
Hernandez, L. P., Barresi, M. J. F., and Devoto, S. H. (2002). Functional morphology and developmental biology of zebrafish: Reciprocal illumination from an unlikely couple. *Integr. Comp. Biol.* **42,** 222–231.
Herrel, A., Adriaens, D., Verraes, W., and Aerts, P. (2002). Bite performance in clariid fishes with hypertrophied jaw adductors as deduced by bite modeling. *J. Morphol.* **253,** 196–205.
Hunt von Herbing, I., Miyake, T., Hall, B. K., and Boutilier, R. G. (1996). Ontogeny of feeding and respiration in larval Atlantic cod *Gadus morhua* (Teleostei, Gadiformes): Function. *J. Morphol.* **227,** 37–50.
Hunter, M. P., and Prince, V. E. (2002). Zebrafish Hox paralogue group 2 genes function redundantly as selector genes to pattern the second pharyngeal arch. *Dev. Biol.* **247,** 367–389.
Janovetz, J. (2002). Functional morphology of feeding in pacus, silver dollars, and piranhas (Teleostei: Serrasalminae). Ph.D. Thesis. University of Chicago, IL.
Johnson, G. D., and Patterson, C. (1993). Percomorph phylogeny: A survey of acanthomorphs and a new proposal. *Bull. Mar. Sci.* **52,** 554–626.
Kammerer, C. F., Grande, L., and Westneat, M. W. (2005). Comparative and developmental functional morphology of the jaws of living and fossil gars (Actinopterygii: Lepisosteidae). *J. Morphol.* in press. (DOI: 10. 1002/jmor. 10293).
Keast, A., and Webb, D. (1966). Mouth and body form relative to feeding ecology in the fish fauna of a small lake, Lake Opinicon, Ontario. *J. Fish. Res. Bd. Can.* **23,** 1845–1874.
Kimmel, C. B., Miller, C. T., and Moens, C. B. (2001). Specification and morphogenesis of the zebrafish larval head skeleton. *Dev. Biol.* **233,** 239–257.
Kimmel, C. B., Ullmann, B., Walker, M., Miller, C. T., and Crump, J. G. (2003). Endothelin 1-mediated regulation of pharyngeal bone development in zebrafish. *Development* **130,** 1339–1351.
Lauder, G. V. (1979). Feeding mechanisms in primitive teleosts and in the halecomorph fish *Amia calva*. *J. Zool. (Lond.)* **187,** 543–578.
Lauder, G. V. (1980a). Evolution of the feeding mechanism in primitive actinopterygian fishes: A functional anatomical analysis of *Polypterus*, *Lepisosteus*, and *Amia*. *J. Morphol.* **163,** 283–317.
Lauder, G. V. (1980b). The suction feeding mechanism in sunfishes (*Lepomis*): An experimental analysis. *J. Exp. Biol.* **88,** 49–72.
Lauder, G. V. (1981). Intraspecific functional repertoires in the feeding mechanism of the characoid fishes *Lebiasina*, *Hoplias*, and *Chalceus*. *Copeia* **1981,** 154–168.
Lauder, G. V. (1982). Patterns of evolution in the feeding mechanism of actinopterygian fishes. *Am. Zool.* **22,** 275–285.
Lauder, G. V. (1983a). Food capture. *In* "Fish Biomechanics" (Webb, P. W., and Weihs, D., Eds.), pp. 280–311. Praeger, New York.
Lauder, G. V. (1983b). Prey capture hydrodynamics in fishes: Experimental tests of two models. *J. Exp. Biol.* **104,** 1–13.
Lauder, G. V. (1985). Aquatic feeding in lower vertebrates. *In* "Functional Vertebrate Morphology" (Hildebrand, M., Bramble, D. M., Liem, K. F., and Wake, D., Eds.), pp. 210–229. Harvard University Press, Cambridge, MA.

Lauder, G. V., and Clark, B. D. (1984). Water flow patterns during prey capture by teleost fishes. *J. Exp. Biol.* **113,** 143–150.

Lauder, G. V., and Shaffer, H. B. (1993). Design of feeding systems in aquatic vertebrates: Major patterns and their evolutionary interpretations. *In* "The Skull, Vol. 3. Functional and Evolutionary Mechanisms" (Hanken, J., and Hall, B. K., Eds.), pp. 113–149. University of Chicago Press, Chicago.

Lauder, G. V., Wainwright, P. C., and Findeis, E. (1986). Physiological mechanisms of aquatic prey capture in sunfishes: Functional determinants of buccal pressure changes. *Comp. Biochem. Physiol. A* **84,** 729–734.

Liem, K. F. (1970). Comparative functional anatomy of the Nandidae (Pisces: Teleostei). *Fieldiana Zool.* **56,** 1–166.

Liem, K. F. (1978). Modulatory multiplicity in the functional repertoire of the feeding mechanism in cichlid fishes. I. Piscivores. *J. Morphol.* **158,** 323–360.

Liem, K. F. (1979). Modulatory multiplicity in the feeding mechanism in cichlid fishes, as exemplified by the invertebrate pickers of Lake Tanganyika. *J. Zool. (Lond.)* **189,** 93–125.

Liem, K. F. (1980). Adaptive significance of intra- and interspecific differences in the feeding repertoires of cichlid fishes. *Am. Zool.* **20,** 295–314.

Liem, K. F., and Summers, A. P. (2000). Integration of versatile functional design, population ecology, ontogeny and phylogeny. *Neth. J. Zool.* **50,** 245–259.

Luczkovich, J. J., Norton, S. F., and Gilmore, R. G., Jr. (1995). The influence of oral anatomy on prey selection during the ontogeny of two percoid fishes, *Lagodon rhomboides* and *Centropomus undecimalis*. *Env. Biol. Fish* **44,** 79–95.

Maddison, W. P. (1990). A method for testing the correlated evolution of two binary characters: Are gains and losses concentrated on certain branches of a phylogenetic tree? *Evolution* **44,** 539–557.

Martins, E. P., Diniz-Filho, J. A., and Housworth, E. A. (2002). Adaptation and the comparative method: A computer simulation study. *Evolution* **56,** 1–13.

Mittelbach, G. G. (1984). Predation and resource partitioning in two sunfishes (Centrarchidae). *Ecology* **65,** 499–513.

Motta, P. J. (1988). Functional morphology of the feeding apparatus of ten species of Pacific butterflyfishes (Perciformes, Chaetodontindae): An ecomorphological approach. *Env. Biol. Fish.* **22,** 39–67.

Motta, P. J., and Wilga, C. D. (2000). Advances in the study of feeding behaviors, mechanisms and mechanics of sharks. *Env. Biol. Fish.* **20,** 1–26.

Motta, P. J., Clifton, K. B., Hernandez, P., and Eggold, B. T. (1995a). Ecomorphological correlates in ten species of subtropical seagrass fishes: Diet and microhabitat utilization. *Env. Biol. Fish* **44,** 37–60.

Motta, P. J., Norton, S., and Luczkovich, J. (1995b). Perspectives on the ecomorphology of bony fishes. *Env. Biol. Fish* **44,** 11–20.

Motta, P. J., Tricas, T. C., Hueter, R. E., and Summers, A. P. (1997). Feeding mechanism and functional morphology of the jaws of the lemon shark *Negaprion brevirostris* (Chrondricthyes, Carcharhinidae). *J. Exp. Biol.* **200,** 2765–2780.

Motta, P. J., Hueter, R. E., Tricas, T. C., and Summers, A. P. (2002). Kinematic analysis of suction feeding in the nurse shark, Ginglymostoma cirratum (Orectolobiformes, Ginglymostomatidae). *Copeia* **2002,** 24–38.

Muller, M. (1987). Optimization principles applied to the mechanism of neurocranium levation and mouth bottom depression in bony fishes (Halecostomi). *J. Theor. Biol.* **126,** 343–368.

Muller, M. (1989). A quantitative theory of expected volume changes of the mouth during feeding in teleost fishes. *J. Zool. (Lond.)* **217,** 639–661.

Muller, M., and Osse, J. W. M. (1978). Structural adaptations to suction feeding in fish. *In* "Proceedings of the ZODIAC Symposium on Adaptation," pp. 57–60. Pudoc, Wageningen, The Netherlands.

Muller, M., and Osse, J. W. M. (1984). Hydrodynamics of suction feeding in fish. *Trans. Zool. Soc. Lond.* **37,** 51–135.

Muller, M., Osse, J. W. M., and Verhagen, J. (1982). A quantitative hydrodynamical model of suction feeding in fish. *J. Theor. Biol.* **95,** 49–79.

Nelson, J. S. (1994). "Fishes of the World" (3rd edn.). John Wiley & Sons, New York.

Nemeth, D. H. (1997). Modulation of buccal pressure during prey capture in *Hexagrammos decagrammus* (Teleostei: Hexagrammidae). *J. Exp. Biol.* **200,** 2145–2154.

Norton, S. F. (1991). Capture success and diet of cottid fishes: The role of predator morphology and attack kinematics. *Ecology* **72,** 1807–1819.

Norton, S. F. (1995). A functional approach to ecomorphological patterns of feeding in cottid fishes. *Env. Biol. Fish* **44,** 61–78.

Norton, S. F., and Brainerd, E. L. (1993). Convergence in the feeding mechanics of ecomorphologically similar species in the Centrarchidae and Cichlidae. *J. Exp. Biol.* **176,** 11–29.

Nyberg, D. W. (1971). Prey capture in the largemouth bass. *Am. Midl. Nat.* **86,** 128–144.

Osse, J. W. M. (1969). Functional morphology of the head of the perch (*Perca fluviatilis* L.): An electromyographic study. *Neth. J. Zool.* **19,** 290–392.

Pagel, M. (1999). The maximum likelihood approach to reconstructing ancestral character states of discrete characters on phylogenies. *Syst. Biol.* **48,** 612–622.

Pietsch, T. W. (1978). The feeding mechanism of *Stylephorus chordatus* (Teleostei: Lampridiformes): Functional and ecological implications. *Copeia* **1978,** 255–262.

Rand, D. M., and Lauder, G. V. (1981). Prey capture in the chain pickerel, *Esox niger*: Correlations between feeding and locomotor behavior. *Can. J. Zool.* **59,** 1072–1078.

Rice, A. N., and Westneat, M. W. (2005). Coordination of feeding and locomotor systems in the Labridae. *J. Exp. Biol.* In Press. **43,** 980.

Sanderson, S. L. (1988). Variation in neuromuscular activity during prey capture by trophic specialists and generalists (Pisces: Labridae). *Brain Behav. Evol.* **32,** 257–268.

Sanderson, S. L. (1990). Versatility and specialization in labrid fishes: Ecomorphological implications. *Oecologia* **84,** 272–279.

Sanderson, S. L., Cech, J. J., and Patterson, M. R. (1991). Fluid dynamics in suspension-feeding blackfish. *Science* **251,** 1346–1348.

Sanderson, S. L., Cech, J. J., and Cheer, A. Y. (1994). Paddlefish buccal flow velocity during ram suspension feeding and ram ventilation. *J. Exp. Biol.* **186,** 145–156.

Sanderson, S. L., Cheer, A. Y., Goodrich, J. S., Graziano, J. D., and Callan, W. T. (2001). Crossflow filtration in suspension-feeding fishes. *Nature* **412,** 439–441.

Sanford, C. P. J., and Wainwright, P. C. (2002). Use of sonomicrometry demonstrates the link between prey capture kinematics and suction pressure in large mouth bass. *J. Exp. Biol.* **205,** 3445–3457.

Sanford, C. P. J., and Wilga, C. D. (2004). The effects of suction generation on prey capture in bamboo sharks. *Integ. Comp. Biol.* **44,** 634.

Schaefer, S. A., and Lauder, G. V. (1986). Historical transformation of functional design: Evolutionary morphology of feeding mechanisms in loricarioid catfishes. *Syst. Zool.* **35,** 489–508.

Schaefer, S. A., and Lauder, G. V. (1996). Testing historical hypotheses of morphological change: Biomechanical decoupling in loricarioid catfishes. *Evolution* **50,** 1661–1675.

Schaeffer, B., and Rosen, D. E. (1961). Major adaptive levels in the evolution of the actinopterygian feeding mechanism. *Am. Zool.* **1,** 187–204.

Shirai, S. (1996). Phylogenetic relationships of Neoselachians (Chondrichthyes: Euselachii). pp. 9–32. *In* "Interrelationships of Fishes" (Stiassny, M. L. J., Parenti, L. R., and Johnson, G. D., Eds.). Academic Press, New York.

Sibbing, F. A. (1982). Pharyngeal mastication and food transport in the carp (*Cyprinus carpio*): A cineradiographic and electromyographic study. *J. Morphol.* **172,** 223–258.

Stiassny, M. L. J., Parenti, L. R., and Johnson, G. D. (Eds.) (1996). "Interrelationships of Fishes." Academic Press, New York.

Streelman, J. T., Alfaro, M., Westneat, M. W., Bellwood, D. R., and Karl, S. A. (2002). Evolutionary history of the parrotfishes: Biogeography, ecology and comparative diversity. *Evolution* **56,** 961–971.

Streelman, J. T., Webb, J. F., Albertson, R. C., and Kocher, T. D. (2003). The cusp of evolution and development: A model of cichlid tooth shape diversity. *Evol. Dev.* **5,** 600–608.

Svanbäck, R., Wainwright, P. C., and Ferry-Graham, L. A. (2002). Linking cranial kinematics and buccal pressure to suction feeding performance in largemouth bass. *Physiol. Biochem. Zool.* **75,** 532–543.

Tchernavin, V. V. (1953). "The Feeding Mechanisms of a Deep Sea Fish *Chauliodus sloani* Schneider." British Museum, N.H., London.

Turingan, R. G. (1994). Ecomorphological relationships among Caribbean tetraodontiform fishes. *J. Zool. (Lond.)* **233,** 493–521.

Turingan, R. G., Wainwright, P. C., and Hensley, D. A. (1995). Interpopulation variation in prey use and feeding biomechanics in Caribbean triggerfishes. *Oecologia* **102,** 296–304.

van den Berg, C., van den Boogaart, J. G. M., Sibbing, F. A., and Osse, J. (1994). Implications of gill arch movements for filter-feeding: An x-ray cinematographical study of filter-feeding white bream (*Blicca bjoerkna*) and common bream (*Abramis brama*). *J. Exp. Biol.* **191,** 257–282.

van Leeuwen, J. L., and Muller, M. (1983). The recording and interpretation of pressures in prey-sucking fish. *Neth. J. Zool.* **33,** 425–475.

van Leeuwen, J. L., and Muller, M. (1984). Optimum sucking techniques for predatory fish. *Trans. Zool. Soc. Lond.* **37,** 137–169.

Van Wassenbergh, S., Aerts, P., Adriaens, D., and Herrel, A. (2005). A dynamic model of mouth closing movements in clariid catfishes: The role of enlarged jaw adductors. *J. Theor. Biol.* **234,** 49–65.

Vanderwalle, P., Saintin, P., and Chardon, M. (1995). Structures and movements of the buccal and pharyngeal jaws in relation to feeding in *Diplodus sargus*. *J. Fish. Biol.* **46,** 623–656.

Wainwright, P. C. (1987). Biomechanical limits to ecological performance: Mollusc-crushing by the Caribbean hogfish, *Lachnolaimus maximus*. *J. Zool. (Lond.)* **213,** 283–297.

Wainwright, P. C. (1996). Ecological explanation through functional morphology: The feeding biology of sunfishes. *Ecology* **77,** 1336–1343.

Wainwright, P. C., and Lauder, G. V. (1986). Feeding biology of sunfishes: Patterns of variation in prey capture. *Zool. J. Linn. Soc. Lond.* **88,** 217–228.

Wainwright, P. C., and Lauder, G. V. (1992). The evolution of feeding biology in sunfishes. *In* "Systematics, Historical Ecology, and North American Freshwater Fishes" (Mayden, R. L., Ed.), pp. 472–491. Stanford Press, Stanford, CA.

Wainwright, P. C., and Richard, B. A. (1995). Predicting patterns of prey use from morphology in fishes. *Env. Biol. Fish.* **44,** 97–113.

Wainwright, P. C., and Reilly, S. M. (1994). "Ecological Morphology: Integrative Organismal Biology." University of Chicago Press, Chicago.

Wainwright, P. C., and Shaw, S. S. (1999). Morphological basis of kinematic diversity in feeding sunfishes. *J. Exp. Biol.* **202,** 3101–3110.

Wainwright, P. C., Sanford, C. P. J., Reilly, S. M., and Lauder, G. V. (1989). Evolution of motor patterns: Aquatic feeding in salamanders and ray-finned fishes. *Brain Behav. Evol.* **34,** 329–341.
Wainwright, P. C., Ferry-Graham, L. A., Waltzek, T. B., Carroll, A. M., Hulsey, C. D., and Grubich, J. R. (2001). Evaluating the use of ram and suction during prey capture by cichlid fishes. *J. Exp. Biol.* **204,** 3039–3051.
Wainwright, P. C., Bellwood, D. R., Westneat, M. W., Grubich, J. R., and Hoey, A. S. (2004). A functional morphospace for labrid fishes: Patterns of diversity in a complex biomechanical system. *Biol. J. Linn. Soc.* **82,** 1–25.
Waltzek, T. B., and Wainwright, P. C. (2003). The functional morphology of jaw protrusion among neotropical cichlids. *J. Morphol.* **257,** 96–106.
Werner, E. E., and Hall, D. J. (1976). Niche shifts in sunfishes: Experimental evidence and significance. *Science* **191,** 404–406.
Westneat, M. W. (1990). Feeding mechanics of teleost fishes (Labridae: Perciformes): A test of four-bar linkage models. *J. Morphol.* **205,** 269–295.
Westneat, M. W. (1991). Linkage biomechanics and evolution of the jaw protrusion mechanism of the sling-jaw wrasse, *Epibulus insidiator*. *J. Exp. Biol.* **159,** 165–184.
Westneat, M. W. (1994). Transmission of force and velocity in the feeding mechanisms of labrid fishes (Teleostei, Perciformes). *Zoomorphology* **114,** 103–118.
Westneat, M. W. (1995a). Feeding, function, and phylogeny: Analysis of historical biomechanics in labrid fishes using comparative methods. *Syst. Biol.* **44,** 361–383.
Westneat, M. W. (1995b). Systematics and biomechanics in ecomorphology. *Env. Biol. Fish.* **44,** 263–283.
Westneat, M. W. (2003). A biomechanical model for analysis of muscle force, power output and lower jaw motion in fishes. *J. Theor. Biol.* **223,** 269–281.
Westneat, M. W. (2004). Evolution of levers and linkages in the feeding mechanisms of fishes. *Integr. Comp. Biol.* **44,** 378–389.
Westneat, M. W., and Wainwright, P. C. (1989). Feeding mechanism of *Epibulus insidiator* (Labridae: Teleostei): Evolution of a novel functional system. *J. Morphol.* **202,** 129–150.
Westneat, M. W., Alfaro, M. E., Bellwood, D. R., Wainwright, P. C., Fessler, J. L., Grubich, J., Clements, K., and Smith, L. L. (2005). Local phylogenetic divergence and global evolutionary convergence of skull biomechanics in reef fishes of the family Labridae. *Proc. Roy. Soc.* **272,** 993–1000.
Wilga, C. D. (2002). A functional analysis of jaw suspension in elasmobranchs. *Biol. J. Linn. Soc.* **75,** 483–502.
Wilga, C. D., and Motta, P. J. (1998). Feeding mechanism of the Atlantic guitarfish *Rhinobatos lentiginosus*: Modulation of kinematic and motor activity. *J. Exp. Biol.* **201,** 3167–3184.
Wilga, C. D., Wainwright, P. C., and Motta, P. J. (2000). Evolution of jaw depression mechanics in aquatic vertebrates: Insights from Chondrichthyes. *Biol. J. Linn. Soc.* **71,** 165–185.
Wilga, C. D., Heuter, R. E., Wainwright, P. C., and Motta, P. J. (2001). Evolution of upper jaw protrusion mechanisms in elasmobranchs. *Am. Zool.* **41,** 1248–1257.
Winemiller, K. O. (1991). Ecomorphological diversification in lowland freshwater fish assemblages from five biotic regions. *Ecol. Monogr.* **61,** 343–365.
Winemiller, K. O., Kelso-Winemiller, L. C., and Brenkert, A. L. (1995). Ecological and morphological diversification in fluvial cichlid fishes. *Env. Biol. Fish.* **44,** 235–261.
Winterbottom, R. (1974). A descriptive synonymy of the striated muscles of the Teleostei. *Proc. Acad. Nat. Sci. Phil.* **125,** 225–317.

3

FUNCTIONAL MORPHOLOGY OF THE PHARYNGEAL JAW APPARATUS

PETER C. WAINWRIGHT

I. INTRODUCTION

No living group of vertebrates rivals teleost fishes in diversity. They make up about half of all living vertebrate species and they show stunning morphological, functional, and ecological variety. Fishes live in nearly every aquatic habitat that has been invaded by metazoans, from the deep sea to high altitude torrential streams. As with any diverse group, it is useful to ask which functional systems underlie such staggering evolutionary success. One such axis of diversity in fishes is their feeding biology. There are fishes that feed on virtually every available food, and this is associated with an equal range of functional specializations for capturing and processing these foods. Much of the functional diversity seen in fish feeding systems lies in the mechanics of prey capture that involves the oral jaws and buccal cavity (Wainwright and Richard, 1995; Wainwright and Bellwood, 2002). But an often overlooked element of fish trophic diversity lies in the functioning of a second set of jaws, the pharyngeal jaw apparatus (PJA), which is used

Fish Biomechanics: Volume 23
FISH PHYSIOLOGY

DOI: 10.1016/S1546-5098(05)23003-0

primarily in separating food from unwanted material and a variety of forms of prey manipulation and processing behaviors.

Fish trophic diversity is impacted by the PJA at two distinct levels. First, the presence of a second set of jaws in the feeding system promotes overall trophic diversity by increasing the range of musculo-skeletal specializations for feeding. The PJA can be thought of as an additional independent axis of morphological diversity that fish lineages have explored during evolution (Yamaoka, 1978). The structural independence of the oral and pharyngeal jaws permits potential autonomy in their evolution, and because the roles of prey capture and processing are potentially decoupled, the degree of specialization of each system is less constrained by the need to maintain secondary functions (Liem, 1973). As a result of this separation of functional role, the oral jaws of some fishes are mechanically specialized for the generation of suction or of gripping benthic prey to remove them from their holdfast, while some of the more extreme modifications to the PJA involve its use in crushing shells, grinding food, and winnowing edible material from unwanted debris, functions not often seen in the oral jaws of these fishes. Independent evolution of the oral and PJA has increased the range of fish feeding abilities and hence their feeding ecology.

The second way in which the PJA influences overall fish trophic diversity comes about because this system is itself structurally complex. The system involves a core group of 12 prominent skeletal elements and is influenced by at least another 15. A similarly large number of muscles cross each joint in the system and provide the potential for intricate movements and in some cases awesome biting forces. The shape and organization of the bones are diverse and the attachment sites and sizes of muscles are highly variable, making for functional diversity that is only partly documented at present (Winterbottom, 1974; Sibbing, 1982; Lauder, 1983b; Wainwright, 1988; Grubich, 2000). Indeed, the functional diversity of the PJA may be far greater than what is seen in the oral jaw system. In a recent survey of 130 species of labrid fishes it was discovered that the mass of the levator posterior muscle, a prominent muscle of the PJA, ranged 500-fold across species as compared to a 10-fold range in the adductor mandibulae and sternohyoideus muscles, two prominent oral jaws muscles (Wainwright *et al.*, 2004). This result was found *after* accounting for body size!

In this chapter I review our understanding of the functional morphology of the PJA in perciform fishes. My aim is to emphasize what is known about the mechanisms of PJA action and to describe some examples of particularly notable functional innovations. Although much of what is covered applies very widely across teleosts, I focus on perciform fishes because this is where the majority of research has been concentrated. By focusing on this group of fishes I will omit a discussion of an excellent series of studies on the cyprinid, *Carpio carpio,* by Sibbing and his colleagues (Sibbing, 1982, 1988; Sibbing *et al.*, 1986; Sanderson *et al.*, 1994) and highly innovative work on mechanisms of suspension feeding (Sanderson *et al.*, 1994, 2001; Cheer *et al.*, 2001).

II. THE PHARYNGEAL JAW APPARATUS OF PERCIFORM FISHES

A. Overview and Anatomy

The PJA is located immediately rostral of the esophagus, suspended from the neurocranium dorsally and bounded posteriorly and ventrally by the pectoral girdle (Figure 3.1A). The muscles and skeletal elements are modified components of the branchial arches (Figure 3.1B). Except where indicated in the descriptions that follow, the bones and muscles of the PJA are bilaterally paired. The lower jaw is formed by tooth plates that are often fused to the fifth ceratobranchial (Nelson, 1967). These bones are oriented anteroposteriorly and converge medially at their anterior end to attach by ligaments to the basibranchials and by muscles to the fourth ceratobranchials and the pectoral girdle. The upper jaw is formed by tooth plates that are variably fused to one or more pharyngobranchial bones. In most perciform taxa the third pharyngobranchial is the largest and most dominant of these, with contributions from a reduced fourth pharyngobranchial (Nelson, 1967; Wainwright, 1989a). A functionally important second element of the upper jaw is the fourth, and often the third, epibranchial (Figures 3.1 and 3.2). These bones form an arch dorsal and lateral to the pharyngobranchial and articulate with the latter through a rounded cartilaginous end (Wainwright, 1989a; Galis and Drucker, 1996; Grubich, 2000).

These jaw elements are stabilized by muscular connections among them and to the larger skeletal elements that surround them (Figures 3.1A and 3.2). The fifth ceratobranchials are connected ventrally and posteriorly to the pectoral girdle by the pharyngocleithralis internus and externus muscles, and anteriorly to the hyoid bar by the protractor hyoideus muscle. The transversus ventralis muscle connects the left and right fifth and fourth ceratobranchials ventrally, helping to stabilize the lower jaw elements into a single functional structure. A small adductor branchialis muscle connects the posterior tip of the ceratobranchials to the epibranchial of the same arch. An obliquus posterior muscle also connects the fifth ceratobranchial dorsally to the fourth epibranchial. This muscle plays an important role in the PJA by providing a ventrally directed force on the epibranchial. The pharyngobranchials are connected dorsally to the neurocranium by levator interni muscles and posteriorly to several anterior vertebrae by the retractor dorsalis muscle.

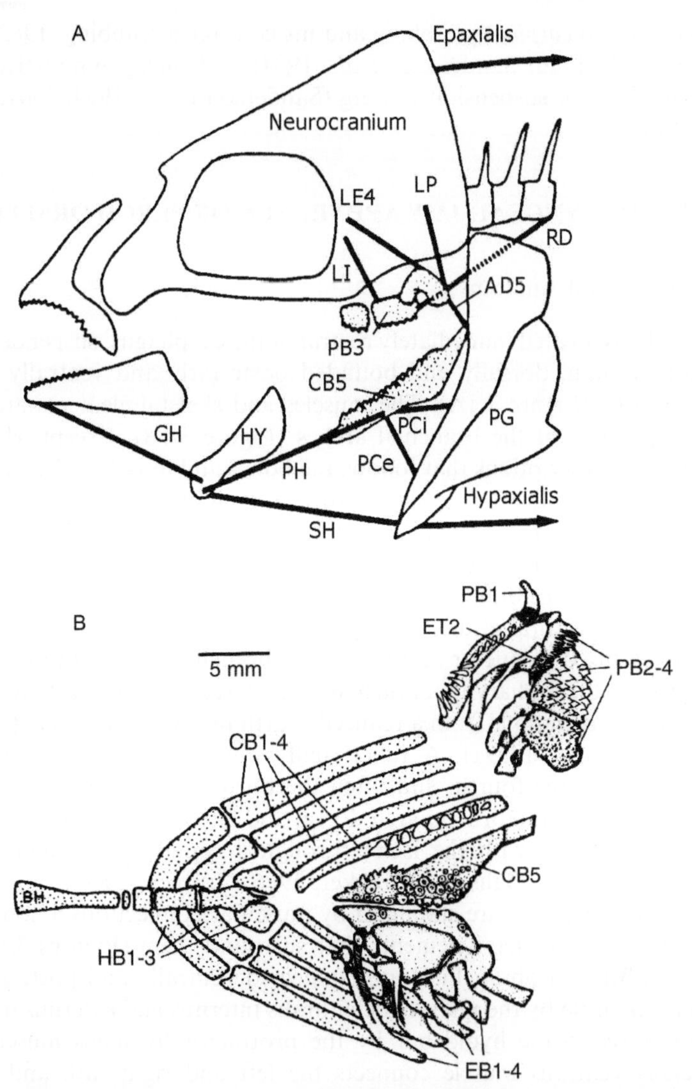

Fig. 3.1. (A) Schematic diagram of the pharyngeal jaw apparatus in teleost fishes with the connections of major muscles indicated by thick black lines. The PJA is positioned at the posterior end of the pharynx immediately anterior to the esophagus and is connected by muscles to structures in this region. (B) Dorsal view of the skeletal elements of the branchial arches in *Haemulon sciurus*. The lower pharyngeal jaw is formed by the paired fifth ceratobranchial and the upper jaw by pharyngobranchials 3 and 4. Abbreviations: AD5, m. fifth adductor branchialis; BH, branchiohyoideus; CB, ceratobranchial; EB, epibranchial; ET2, epibranchial tooth plate; GH, m. geniohyoideus; HB, hyobranchial; HY, hyoid bar; LE4, m. fourth levator exernus; LI, m. levator internus; LP, M. levator posterior; PB, pharyngobranchial; PCe, m. pharyngocleithralis externus; PCi, m. pharyngocleithralis internus; PG, pectoral girdle; PH, m. protractor hyoideus; RD, m. retractor dorsalis; SH, m. sternohyoideus. (Reproduced with permission from Wainwright, 1989a.)

There is also an obliquus dorsalis muscle that connects the pharyngobranchial and epibranchial dorsally. Levator externi muscles connect each epibranchial to the neurocranium dorsally. The levator posterior muscle also connects the fourth epibranchial to the neurocranium.

B. Function in the PJA

Motion of the elements of the oral jaws can be directly observed in most taxa, but the location of the PJA in the pharynx makes observing movement more challenging. However, two approaches, cineradiography and sonomicrometry, have permitted visualization of pharyngeal jaw movement. These approaches have yielded important insights into how the pharyngeal jaws move in several perciform taxa (Aerts *et al.*, 1986; Liem and Sanderson, 1986; Vandewalle *et al.*, 1992, 1995). In conjunction with interpretations of the mechanisms of action in the PJA from anatomy and electromyography, these methods have made it possible to develop a picture of the basic patterns of movement in the PJA and the musculo-skeletal basis of those movements.

A mechanism of action of the PJA was initially identified in the perciform group Haemulidae (Wainwright, 1989a) and subsequently extended to the Centrarchidae and Sciaenidae (Galis and Drucker, 1996; Grubich, 2000). I have observed the anatomical elements of this mechanism in most perciform taxa that I have examined and numerous other teleosts (e.g., Carangidae, Girrelidae, Hexagrammidae, Lethrinidae, Lutjanidae, Percidae, Pomacanthidae, Serranidae, Scorpaenidae, Tatraponidae). Although the mechanism has never been formally mapped onto a phylogeny of actinopterygian fishes, its apparent presence in Osteoglossomorphs and *Amia* suggests that it may be at least as old as the teleosts.

The mechanism implicates the epibranchial bone as a key element in the mechanism for depression of the upper jaw bones (Figure 3.2). Several muscles are oriented such that they can flex the joint between the pharyngobranchial (the upper jaw bone) and the epibranchial. If this joint is flexed while the midpoint of the shaft of the epibranchial is constrained or even pulled ventrally by the fifth adductor branchialis and the obliquus posterior muscles, then the subsequent rotation of the epibranchial bone presses ventrally on the dorsal surface of the upper jaw bone, depressing it (Figure 3.2). The medial margin of the pharyngobranchial is typically connected loosely to the neurocranium by connective tissues, so that this mechanism actually causes a biting action in the PJA in which the lateral margins of the upper jaw are pressed ventrally toward the lower jaw (Figure 3.2A). The joint between the epibranchial and pharyngobranchial can be flexed directly by the obliquus dorsalis muscle, and if the midpoint of the epibranchial shaft

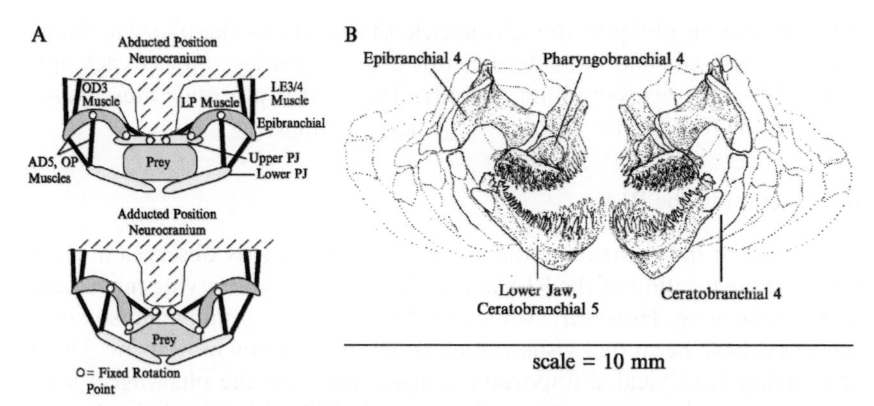

Fig. 3.2. (A) Schematic representation of the mechanism of upper pharyngeal jaw depression in posterior view. Skeletal elements of the jaws are represented by shading and muscles indicated by thick black lines. Joints and rotation points are indicated with small circles. Contraction of the LE, LP, and OD muscles, in concert with stabilization from the AD5 and OP muscles, results in flexion of the joint between the pharyngobranchial and the epibranchial, resulting in depression of the lateral margin of the pharyngobranchial. (Modified after Wainwright, 1989a.) (B) Diagram of the pharyngeal jaw bones of *Lepomis punctatus* in posterior view for comparison with the schematic model in A. Abbreviations as in Figure 3.1 and OD, m. obliquus dorsalis; OP, m. obliquus posterior.

is constrained, the epibranchial can be rotated about this point by action of the levator posterior and fourth levator externus muscles that connect the lateral margin of the epibranchial to the neurocranium.

The significance of this mechanism is that it provides forceful adduction of the PJA. The importance of forceful adduction is clear in the case of behaviors such as mollusc crushing (Lauder, 1983a; Wainwright, 1987), but adduction also can be employed in concert with other actions, most notably sheering of the upper and lower jaws (Vandewalle *et al.*, 1992, 1995). Posterior and anterior translation of the upper jaws can be facilitated by the retractor dorsalis and levator interni muscles, respectively. As we shall see, studies have revealed that a major feature of pharyngeal jaw function in generalized perciform taxa is the combined motion of the upper jaw in both the anterior-posterior axis and the dorsal-ventral axis.

C. Movement Patterns of the PJA

Among generalized perciform fishes, previous studies have documented aspects of pharyngeal jaw movement patterns only in the Serranidae (Vandewalle *et al.*, 1992) and the Sparidae (Vandewalle *et al.*, 1995), while movements have been inferred from muscle activity patterns and anatomy in the Nandidae (Liem, 1970), Haemulidae (Wainwright, 1989a), Centrarchidae (Lauder, 1983b; Galis and Drucker, 1996), and the Sciaenidae (Grubich, 2000). In the serranid, *Serranus scriba*, during routine pharyngeal transport behavior the upper jaw moves in a cyclic pattern that includes anterior-posterior and dorsal-ventral excursions of similar magnitude (Figure 3.3; (Vandewalle *et al.*, 1992). At the start of each cycle the upper jaw (the pharyngobranchial) moves posteriorly and ventrally until it meets the lower jaw. During the recovery stroke the upper jaw moves dorsally before also recovering anteriorly, so that the overall cycle does not involve the jaw exactly retracing its path (Figure 3.3). Lower jaw motion is more restricted than upper jaw movement and occurs mostly in the anterior-posterior axis. The lower jaw cycle involves posterior retraction that peaks shortly before the upper jaw reaches its most posterior and ventral position.

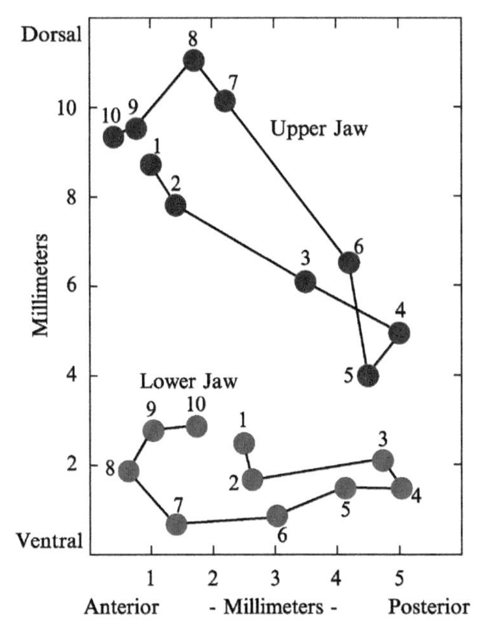

Fig. 3.3. Two-dimensional movement, in lateral view, of the upper and lower jaws of *Serranus scriba* illustrating a typical pattern of jaw movement during pharyngeal transport in generalized perciform fishes. Upper jaw motion involves simultaneous depression and retraction. Note that the upper jaw has greater motion in the dorsal-ventral axis than the lower jaw. Data were collected from radio-opaque markers implanted in pharyngeal jaw bones; the relative positions of the upper and lower symbols in this graph do not reflect their positions with respect to each other. Numbers adjacent to points indicate homologous points in time. Points are separated by 40 ms. (Redrawn from Vandewalle *et al.*, 1992.)

I present unpublished data in Figures 3.4, 3.5, 3.7, and 3.8 on pharyngeal jaw motion from three other perciform taxa, the cabezon *Scorpaenichthys marmoratus* (Cottidae), largemouth bass *Micropterus salmoides* (Centrarchidae), and the lingcod *Ophiodon elongates*, a member of the Hexagrammidae. From these data, two major points can be emphasized in relation to the observations made previously. First, all taxa were capable of a variety of pharyngeal jaw kinematic patterns, including sheering between upper and lower jaw and adduction with retraction as described for *Serranus*. Second, previously unrecognized movement in the medial-lateral axis was sometimes substantial (Figure 3.7).

As with *Serranus*, the upper jaw of *S. marmoratus* begins the cycle with posterior and ventral movement that culminates in a period when the upper and lower jaws adduct against the prey item (Figure 3.4). There is considerable variation in the pattern from cycle to cycle, with one of the primary differences being whether the upper and lower jaws are moving in the same

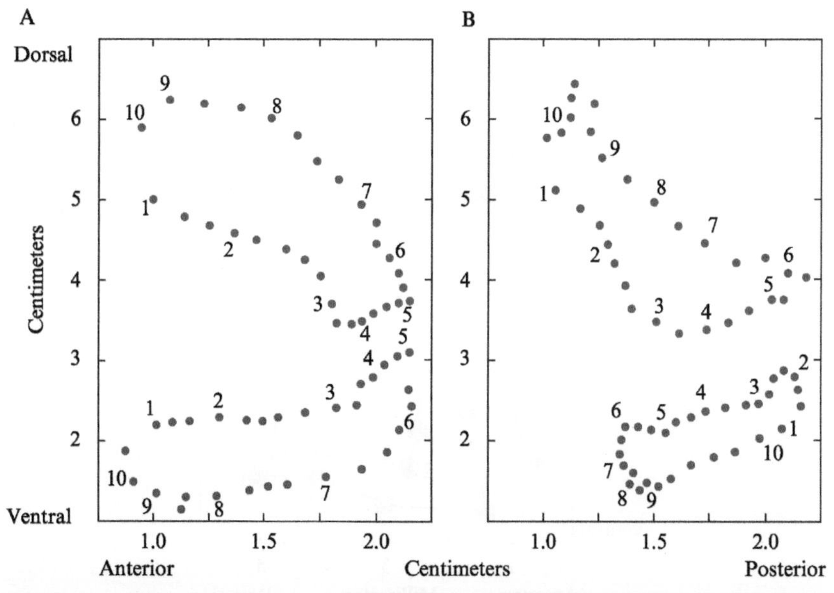

Fig. 3.4. Two-dimensional later view kinematics of the upper and lower pharyngeal jaws in the cabezon, *Scorpaenichthys marmoratus*. Data were generated using sonomicrometry with crystals placed on the jaws and to fixed structures surrounding the jaws that allowed the reconstruction of three-dimensional movement. Here the data are reconfigured to show the motion in two dimensions. (A) Single sequence that shows simultaneous adduction and retraction of both the upper jaw and lower jaw. (B) Sequence from the same feeding bout that shows sheering action of the upper and lower jaws. Note that the positions of the upper and lower jaws in this graph are not meant to represent their position relative to one another. Numbers adjacent to points indicate homologous points in time. Points are separated by 10 ms.

direction together or are moving against each other in a sheering action (compare Figure 3.4A and B). A second point of variation is that the upper jaw often depresses rapidly before moving posteriorly (Figure 3.4B). In these cycles the lower jaw reaches its most posterior position earlier than the upper jaw, and the upper jaw moves ventrally and then posteriorly, raking the prey against the less mobile lower jaw. In the recovery stroke, both the upper and lower jaw move away from their point of adduction before being protracted into anterior positions that form the widest gape between the jaws during the cycle. During cycles when the jaws move in a sheering motion the movement orbit of the lower jaw is smaller than during cycles of simultaneous retraction. The capacity to show sheering motions and simultaneous retraction was also found in *Micropterus* and *Ophiodon*.

A slightly different picture is seen in the largemouth bass, *Micropterus salmoides* (Figure 3.5). During rhythmic pharyngeal transport behavior in this species, the upper jaw undergoes relatively minor ventral excursion but travels about three times further in the posterior direction. As with *Serranus* and *Scorpaenichthys*, *Micropterus* shows both sheering and simultaneous depression and retraction of the jaws. Published data on the sparid, *Diplodus sargus*, illustrate sheering in this species (Figure 3.6) as well as simultaneous retraction (Vandewalle *et al.*, 1995).

In my recordings from *Scorpaenichthys*, I tracked motion of the medial margin of the pharyngobranchial and found that it showed very little ventral or medial movement during the adduction phase of the cycle, in marked

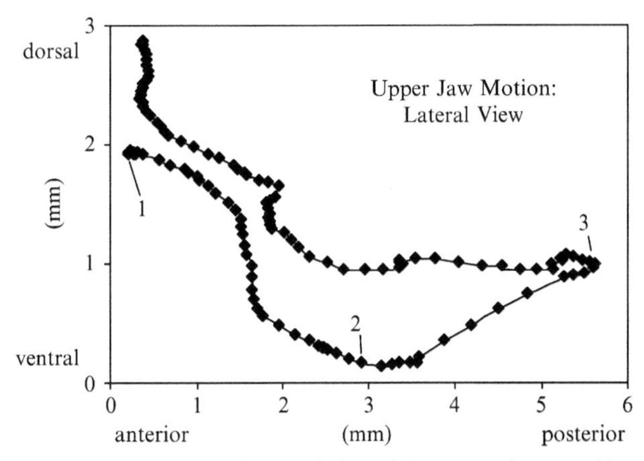

Fig. 3.5. Two-dimensional movement in lateral view of the upper pharyngeal jaw in a 207 mm *Micropterus salmoides*. Data were generated with sonomicrometry. Crystals were sutured to the jaw bones of the fish and to several non-moving structures in the pharynx and buccal cavity to determine movements in two dimensions. Note that in this species, there is considerable anterior posterior motion of the upper jaw, in addition to movement in the dorsal-ventral axis.

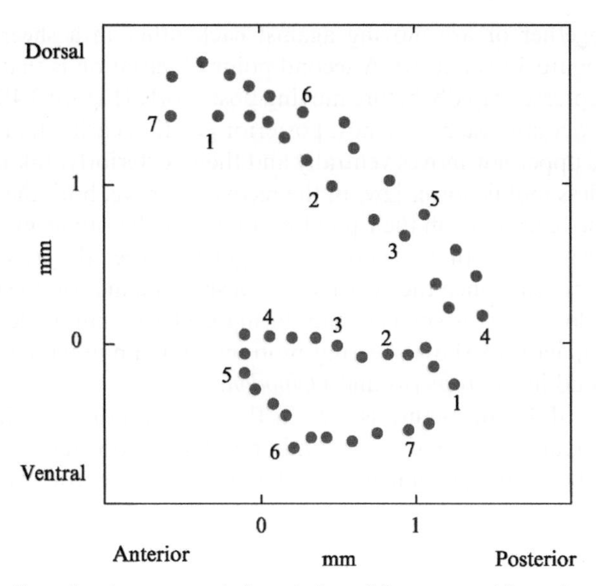

Fig. 3.6. Two-dimensional movement in lateral view of the upper and lower jaws in a 110 mm *Diplodus sargus* (Sparidae). This sequence illustrates sheering between the jaws during the depression and retraction of the upper jaw. (Redrawn from cineradiographic observations presented in Vandewalle *et al.*, 1995.)

contrast to the lateral margin of the pharyngobranchial (Figure 3.7). This may be interpreted in the light of the working model of pharyngeal jaw function (Figure 3.2). The epibranchial depresses the lateral margin of the upper jaw elements, but the medial section of the pharyngobranchial is expected to be relatively stationary during this motion. The medial movement of the lateral margin of the upper jaw appears to reflect the rotation of the pharyngobranchial about its medial region so that the lateral margin swings in an arc.

In generalized perciform fishes the left and right pharyngeal jaws are not constrained to move only in the dorsal-ventral and anterior-posterior axes. Data from *Serranus* (Vandewalle *et al.*, 1995) and *Scorpaenichthys* show that the ventral movement of the upper pharyngeal jaw is associated with movement toward the midline of the pharynx (Figure 3.7). It is important to recognize that this pattern is based on tracking movements of the lateral margin of the pharyngobranchial bone, and thus much of this motion is probably due to the way in which the pharyngobranchial rotates when it is depressed. However, there may also be additional lateral motion in the entire pharyngobranchial bone involved. In *Ophiodon elongates*, a highly piscivorous species of hexagrammid common on temperate rocky reefs along

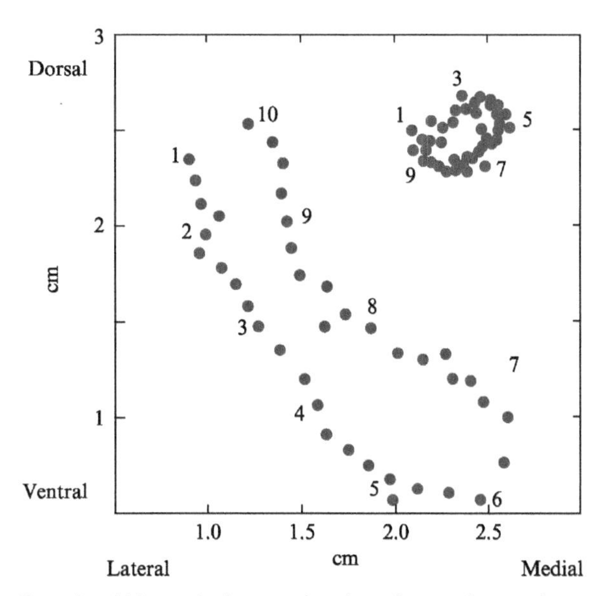

Fig. 3.7. Two-dimensional kinematics in posterior view of two points on the upper pharyngeal jaw (the pharyngobranchial) in a 450 mm *Scorpaenichthys marmoratus*. Data were generated with sonomicrometry by attaching crystals to the lateral (blue) and medial (red) edge of the pharyngobranchial and to fixed structures in the pharynx that allowed resolution of motion in two–dimensions. Note that the pharyngobranchial appears to rotate about its medial edge while the lateral margin undergoes considerable excursions, ventrally and medially. Compare this pattern to the model shown in Figure 3.2. Points are separated by 20 ms.

the coast of western North America, the medial-lateral motion of the lower pharyngeal jaw can be extensive (Figure 3.8). Lateral motion occurs while the PJA is being protracted, such that the jaws are protracted while being strongly abducted in both the dorsal-ventral axis and laterally. This behavior was most apparent in this species when the fish was fed very large prey items. It appears that strong abduction during jaw protraction may aid in moving the jaws to a more anterior position on the prey before beginning the next cycle of retraction and adduction.

Finally, the left and right sides of the PJA may move in phase, as is most common, or they may move out of phase (Liem, 1970; Lauder, 1983b; Vandewalle *et al.*, 1992). The structurally decoupled status of the right and left sides of the system in generalized perciform fishes permits some independent movements in the system and may allow greater dexterity and fine control of prey.

In summary, pharyngeal jaws movements are diverse and take place in three dimensions. It appears that in generalized perciform fishes the orbit of

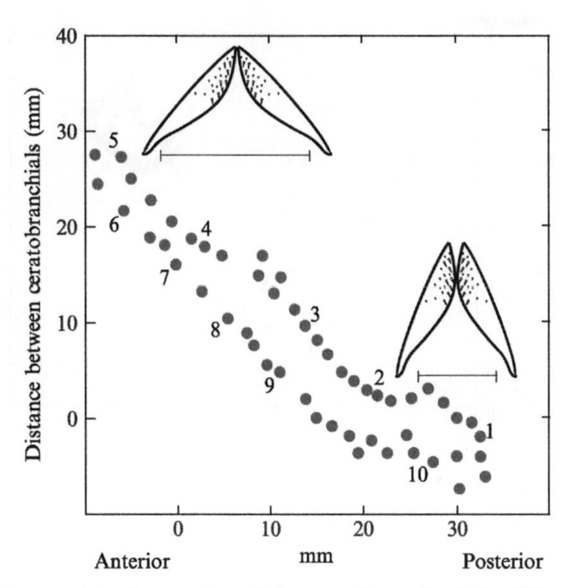

Fig. 3.8. Simultaneous lateral spreading of the posterior ends of the fifth ceratobranchials during protraction of the lower jaw in a 580 mm lingcod, *Ophiodon elongates*. Data were generated with sonomicrometry by attaching crystals to the posterior tips of the fifth ceratobranchials and to fixed structures in the pharynx that allowed resolution of motion in the anterior posterior axis. Points are separated by 20 ms.

motion of the upper jaw is normally greater than that of the lower jaw. During the rhythmic pharyngeal transport behavior that dominates pharyngeal sequences, the upper jaw sweeps from an anterior-dorsal position to a posterior-ventral position. The upper jaw meets the lower jaw in this posterior-ventral region of its orbit, and the relative motion of the lower jaw at this time indicates that either the jaws are being adducted or the upper jaw is moving posteriorly as the lower jaw is moving anteriorly, creating a sheering action. As the jaws are protracted during the recovery stroke they are abducted. This action may involve considerable lateral spreading of the lower jaw bones in preparation for the subsequent cycle.

D. Motor Control of PJA Action

A considerable literature exists on the muscle activity patterns of the PJA in generalized perciform fishes (Lauder, 1983a; Wainwright, 1989a,b; Grubich, 2000). My aim in this section is to describe the major patterns of muscle activity that have been described by various workers. This review is slanted to accomplish two primary goals: (1) to interpret available motor pattern data in light of the data on movement patterns, and (2) to emphasize the extent to which motor patterns appear to be similar across diverse taxa. Among generalized perciform taxa, electromyographic data from the PJA muscles have been reported for members of the Centrarchidae (Lauder, 1983a), the Haemulidae (Wainwright, 1989a,b), and the Sciaenidae (Grubich, 2000). Although we presently lack data from synchronized EMG and kinematics in the PJA, it is possible to identify the probable basis of actions such as sheering and retraction with adduction.

A similar pattern of motor activity is seen during pharyngeal transport behavior in several generalized perciform taxa (Figure 3.9). The activity pattern is characterized by initial onset of the fourth levator externus, almost simultaneously with the onset of activity in the levator posterior. The retractor dorsalis muscle is activated during the middle 50% of the LE4 burst. The relative onset of the retractor dorsalis with respect to the LE4 and levator posterior is quite variable among cycles of activity. The levator interni muscles and the second levator externus, both protractors of the upper jaw, are out of phase with the retractor dorsalis (Wainwright, 1989a). The fifth branchial adductor and obliquus posterior are active together, at the time of the retractor dorsalis. The obliquus dorsalis muscle that flexes the joint between the epibranchial and the pharyngobranchial is active simultaneously with the levator posterior. Given the anatomical interpretations of the

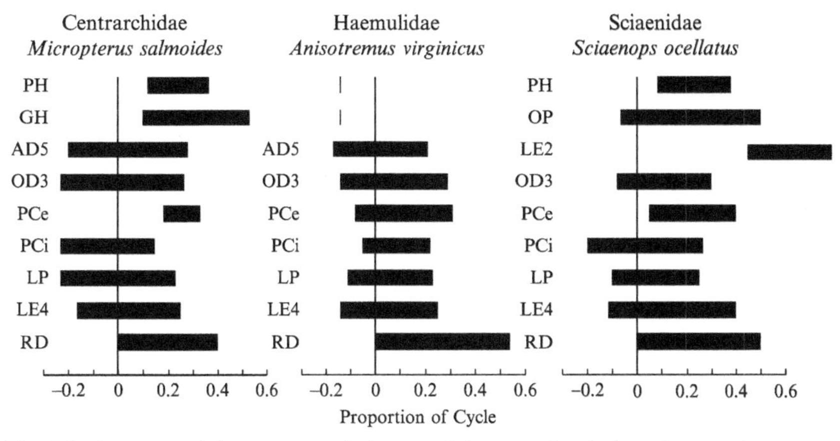

Fig. 3.9. Average activity patterns of pharyngeal jaw muscles during pharyngeal transport behavior in representatives of three generalized perciform groups. Activity is expressed as a proportion of the duration of a single cycle of pharyngeal activity, measured as onset of the retractor dorsalis until onset of the subsequent burst. Muscle abbreviations are as in Figure 3.1. Note that activity patterns in the three taxa are broadly similar. (Data from *Micropterus* are previously unpublished personal observations. Data from *Anisotremus* are redrawn from Wainwright, 1989a, and data from *Scaienops* are redrawn from Grubich, 2000.)

functions of these muscles, these motor activity patterns are consistent with the expected motor basis of the kinematic patterns described previously. Upper jaw depression is caused by the combined activity of the fifth adductor branchialis/obliquus posterior, the obliquus dorsalis, the fourth levator externus, and the levator posterior. Upper jaw retraction is caused uniquely by contraction of the retractor dorsalis. Protraction of the upper jaw is caused by the levator interni and possibly by the second levator externus.

Interestingly, lower pharyngeal jaw motor patterns are more variable than the upper jaw muscles, and can be more difficult to summarize simply. The activities of the pharyngocleithralis externus (PCe) and internus (PCi) muscles are usually out of phase with each other (Figure 3.9). When active, the PCi is activated simultaneously with the fourth levator externus and therefore functions during the posterior-ventral power stroke of the upper jaw. In contrast, the PCe muscle is typically active out of phase with these muscles and appears to function during abduction and recovery of the lower pharyngeal jaws. However, the PCe often shows a second burst of activity that is in phase with all of the PJA adductors (Lauder, 1983a,b; Wainwright, 1989a,b; Grubich, 2000). This activity burst may function to stabilize the lower jaw against the pectoral girdle during more forceful cycles of activity. Overall, the PCe functions to strongly abduct the lower jaw during the recovery stroke of the jaws, analogous to the inferred function of the levator interni muscles of the upper jaw. The lower jaws are protracted by the pharyngohyoideus muscle and also by the geniohyoideus muscle. The latter functions in this context to protract the hyoid apparatus toward the mandibular symphysis, which pulls the entire group of lower branchial structures anteriorly. These muscles can be active singly or together and may or may not be active while the upper jaw depressors are active (Figure 3.9).

While the motor pattern seen during pharyngeal transport behavior is similar in the generalized perciform taxa that have been studied, all taxa show additional behaviors and motor patterns associated with prey capture, buccal manipulation of prey, swallowing behavior, and in some taxa, winnowing and prey crushing. The overall picture that emerges is that the PJA is capable of a wide range of actions that is matched by diversity in motor control. Nevertheless, the general motor pattern during pharyngeal transport behavior tends to be largely conserved among groups of perciforms.

III. INNOVATION IN THE PHARYNGEAL JAW APPARATUS

Much as the oral jaw apparatus has undergone reorganization and functional specialization within various groups of perciform fishes, so too has the PJA. In this section I discuss two major modifications of the PJA that have received considerable attention. First, I review studies of the functional basis of pharyngeal jaw durophagy, or the modifications associated with feeding on very hard-shelled prey. This specialization is noteworthy because it has evolved many times within perciform fishes and the mechanical demands associated with the specialization are quite clear. Mollusc crushing has provided an excellent system for studies of convergent evolution. Second, I review our understanding of the labroid pharyngeal jaw apparatus, the most famous of all teleost pharyngeal jaw innovations. This modification is particularly noteworthy because it was proposed to have a major effect on the trophic diversification of the fishes that possess the innovation, particularly cichlid fishes (Liem, 1973; Friel and Wainwright, 1999).

A. Durophagy

Specialized feeding on molluscs and other very hard-shelled prey types has evolved repeatedly within generalized perciform fishes. In some taxa the prey are crushed by oral jaw biting (Palmer, 1979; Norton, 1988; Hernandez and Motta, 1997; Friel and Wainwright, 1999), and in a few others holes are punched in the shell, allowing digestive juices access to soft parts of the prey after they are swallowed (Norton, 1988). However, in the majority of instances of molluscivory the prey items are crushed in the PJA, and the functional specialization involves being able to exert high forces during jaw adduction (Lauder, 1983a). Crushing strength constrains mollusc predation. This is indicated by ontogenetic studies that have shown in different groups that the youngest, and hence weakest, individuals in the species are not able to crush hard prey and do not eat them (Wainwright, 1988; Osenberg and Mittelbach, 1989; Huckins, 1997). Both within and between species, there is a strong correlation between the strength of the PJA and the percent of the diet made up by hard-shelled prey (Wainwright, 1987, 1988).

Durophagus taxa have larger pharyngeal jaw adductor muscles and enlarged jaw bones when compared to closely related taxa that do not crush hard prey (Lauder, 1983a; Grubich, 2003). Enlarged muscles have higher cross-sectional area and can generate higher stresses, while the enlarged skeletal components are able to resist the higher loads. Within the centrarchid genus *Lepomis* two species are specialized mollusc predators, *Lepomis microlophus* and *L. gibbosus*. The PJAs of these two species are greatly hypertrophied relative to their congeners, including muscular (Lauder, 1983a) and skeletal modifications (compare Figure 3.2B with Figure 3.10). All of the elements of the PJA that are expected to bear loads during jaw adduction are enlarged and the teeth have a wider, "molariform" shape. There is also buttressing of the ventral side of the neurocranium, suggesting

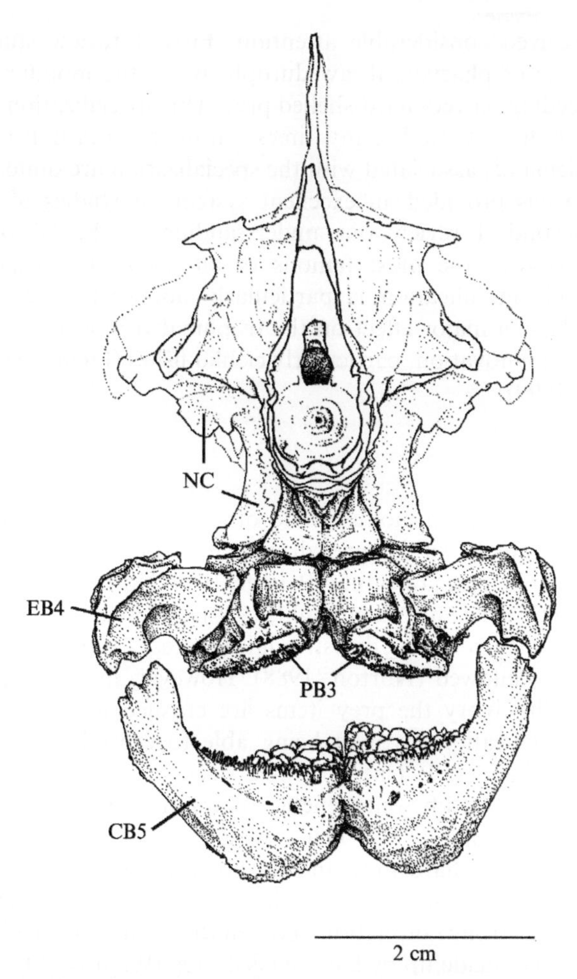

Fig. 3.10. Posterior view of the pharyngeal jaw apparatus in a 210 mm redear sunfish, *Lepomis microlophus*, a mollusc-crushing predator. Note that the skeletal elements of the PJA are greatly hypertrophied relative to those seen in trophically generalized taxa such as *Lepomis punctatus*, shown in Figure 3.1B. Abbreviations as in Figure 3.1 and NC, neurocranium.

that increased loads are transmitted through the upper jaw bones to the neurocranium (Figure 3.10). The muscles that show the greatest hypertrophication are the levator posterior, LE4, and the obliquus dorsalis (Lauder, 1983a; Wainwright *et al.*, 1991), all major muscles involved in adduction.

Grubich (2003) has documented skeletal and muscular hypertrophication in molluscivorous sciaenids and carangids. In both groups, muscles and bones are hypertrophied, although there tend to be unique elements of the specialization in each group. For example, in the carangid *Trachinotus*, the protractor pectoralis is one of the most hypertrophied muscles. This muscle connects the neurocranium to the pectoral girdle and acts to protract the latter. Girdle protraction probably acts to stabilize and protract the lower jaw during prey crushing.

Because mollusc shell crushing probably involves applying increasing forces against a stiff shell, it can be expected that the muscular contractions during crushing are at times purely isometric. Movement patterns of the PJA during mollusc crushing have not been directly observed, but it is well known that molluscivorous *Lepomis* use a derived pattern of muscle activity during crushing that is characterized by long simultaneous bursts of activity in all PJA muscles (Lauder, 1983c). Grubich (2000) found a similar pattern in the sciaenids, in which the black drum exhibited crushing motor patterns very similar to those that have been reported in *Lepomis*. Interestingly, even the trophically generalized sciaenid, *Sciaenops ocellatus*, used this crushing motor pattern when feeding on relatively hard-shelled prey.

B. The Labroid PJA

Monophyly of the Labroidei (Cichlidae, Labridae, Pomacentridae, and Embiotocidae) was proposed initially based on pharyngeal anatomy (Kaufman and Liem, 1982), and this hypothesis was further developed with additional characters (Stiassny and Jensen, 1987). Members of these four groups of perciform fishes share a derived condition of the PJA that has three major features: the lower jaw elements are fused into a single structure, the lower jaw is suspended in a muscular sling that runs from the neurocranium to the posterior muscular arms of the two fused fifth ceratobranchials, and the upper jaw elements have a diarthrotic articulation with the underside of the neurocranium (Figure 3.11). The functional implications of this suite of modifications are primarily that the system appears to be better suited to strong adduction. This is facilitated by direct muscular connection between the neurocranium and the lower jaw, but also by the fused elements of the lower jaw (Liem, 1973; Galis and Drucker, 1996).

In a widely cited series of papers the labroid PJA was proposed to be an important innovation that facilitated the radical evolutionary success and diversity found in the members of this group of perciform fishes (Liem, 1973, 1978, 1979). Liem's hypothesis was that this condition of the PJA allows labroids to process a wider range of prey types and may permit a greater range of jaw behaviors, and because the PJA and oral jaw systems are largely decoupled from each other the evolutionary potential of the labroid

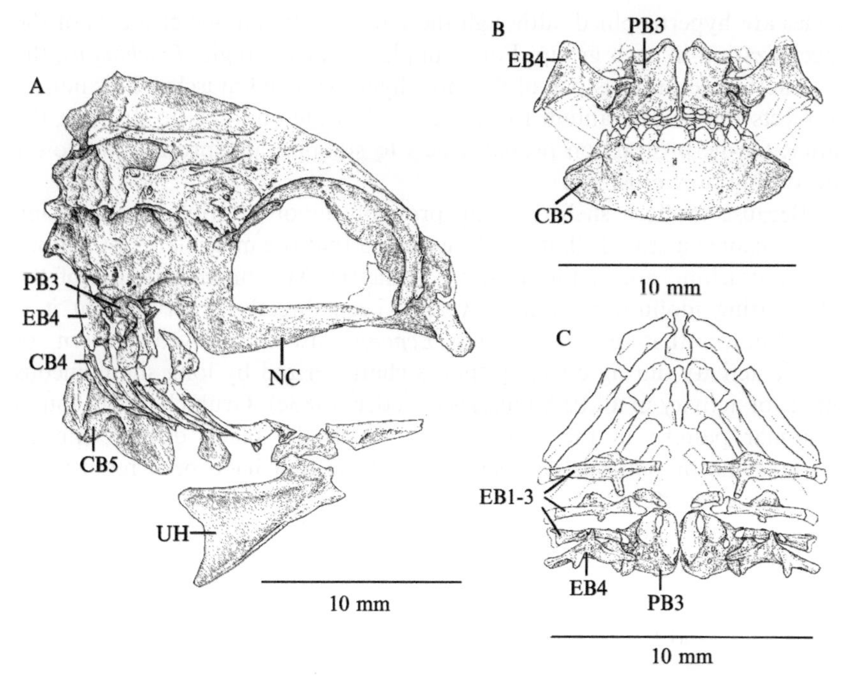

Fig. 3.11. Diagrams of the pharyngeal jaw apparatus in three labrid species to illustrate the labroid condition. (A) Lateral view of the neurocranium and branchial structures in *Bodianus axillaris* to show the position of the PJA at the posterior end of the pharynx. (B) Posterior view of the PJA in *Halichoeres garnoti*. Note the lower jaw bones are fused into a single robust element. (C) Dorsal view of *Cheilinus chlorourus* showing the development of the joint where the upper jaw contacts the underside of the neurocranium. Abbreviations as in Figure 3.1 and UH, urohyal.

feeding system is particularly high. Unfortunately, there are not enough comparative data on jaw function or trophic diversity to rigorously test each of these predictions, although there are some compelling circumstantial data. In the sections below I review what is known about the functioning of the labroid PJA and how this innovation distinguishes these fishes from generalized perciforms. The results are surprising. The labroid PJA appears to exhibit a range of behaviors very similar to that seen in generalized perciform fishes, and because the lower jaw is fused there is less possibility of motion in the medial-lateral axis. While the labroid PJA appears to confer a more efficient and powerful bite, there is no evidence that it is behaviorally or functionally more versatile than that found in generalized taxa that lack the specializations.

1. Morphology

Functional patterns have been inferred from morphology in several labroid taxa, primarily cichlids (Aerts, 1982; Galis, 1992, 1993; Galis and Drucker, 1996), but also the pomacentrids (Galis and Snelderwaard, 1997), embiotocids (DeMartini, 1969; Laur and Ebeling, 1983), and labrids (Liem and Greenwood, 1981; Clements and Bellwood, 1988; Claes and De Vree, 1989, 1990; Gobalet, 1989; Monod *et al.*, 1994; Bullock and Monod, 1997). The chief functional distinctions between the labroid PJA and that of generalized perciforms are that (1) muscles that connect the neurocranium to the fused lower pharyngeal jaw are positioned to directly adduct the lower jaw, and (2) at least in most labrids (P.C.W., personal observations) and possibly in some cichlids (Galis and Drucker, 1996), the upper and lower pharyngeal jaws can move independently. The joint between the medial end of the fourth epibranchial and the dorsal surface of the pharyngobranchial in these taxa has been modified into a sliding joint that allows the pharyngobranchials to move anteriorly and posteriorly, supported dorsally by their articulation to the neurocranium, independent of motion of the lower jaw. The retractor dorsalis and the internal levators appear to effect these actions. Morphological positions indicate that the lower jaw can be strongly adducted by the muscular sling, can be protracted by actions of the pharyngohyoideus and geniohyoideus, and can be abducted and retracted by the pharyngocleithralis externus and internus. It is interesting to note that none of the primary PJA muscles appear to have altered their basic function in labroids, as compared to the generalized condition discussed previously. Even the levator externus and levator posterior, which attach variably on the lower pharyngeal jaw instead of the epibranchial, act to adduct the jaws. The distinction appears to be that in labroids jaw adduction is accomplished mainly by actions of the lower jaw, whereas in generalized taxa adduction is accomplished mostly by depression of the upper jaw.

2. Kinematic Patterns

Direct observations of pharyngeal jaw motion using cineradiography have confirmed most of the anatomically based interpretations. Data from cichlids (Claes and De Vree, 1989, 1990, 1991), embiotocids (Liem, 1986), and labrids (Liem and Sanderson, 1986) reveal that the lower jaw undergoes the largest excursions during prey processing behaviors (Figure 3.12). Sheering actions between the upper and lower jaws are frequent in these taxa, although all authors report that the upper and lower jaws can also move in concert as they swing in the anterior-posterior axis. Thus, observations confirm a large degree of independence in movement of the upper and lower pharyngeal jaws. However, it is not clear that these taxa show an advanced

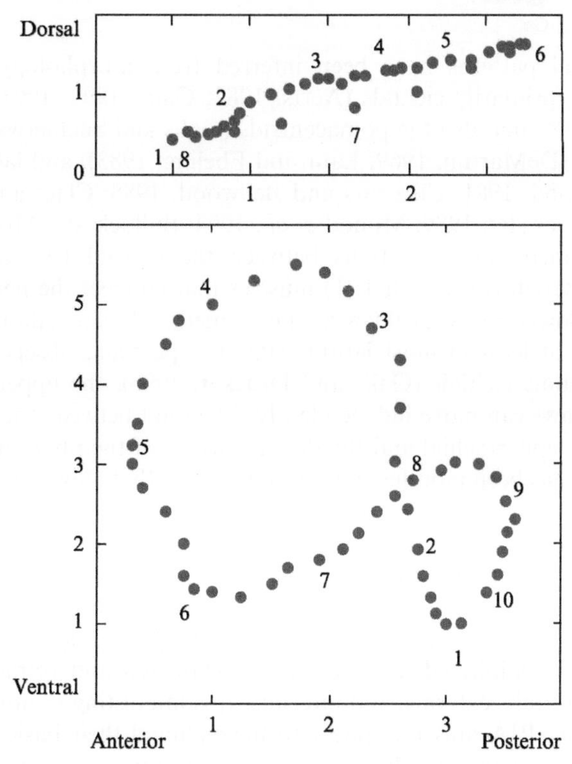

Fig. 3.12. Two-dimensional kinematic pattern of the upper (blue dots) and lower (red dots) pharyngeal jaws in a 200 mm *Oreochromis niloticus*. Note that the movement of the lower jaw is considerably more extensive in the dorsal-ventral axis than is the upper jaw. However, the upper jaw matches the lower jaw in the anterior-posterior excursions. Note sheering action of lower and upper jaw. Units are arbitrary, following Claes and DeVree (1991). (Redrawn from Claes and De Vree, 1991, based on cineradiographic observations.)

level of independence in jaw motion as compared to the generalized perciform condition, in which sheering actions and considerable independence of motion have also been found (Figures 3.5 and 3.6).

3. Motor Patterns

Muscle activity patterns of the labroid PJA are surprisingly unmodified relative to those found in generalized perciforms (compare Figures 3.9 and 3.13). This probably reflects the general conservation of overall muscle function noted previously for the two groups rather than any constraint on the nervous system for generating variation in motor activity (Wainwright,

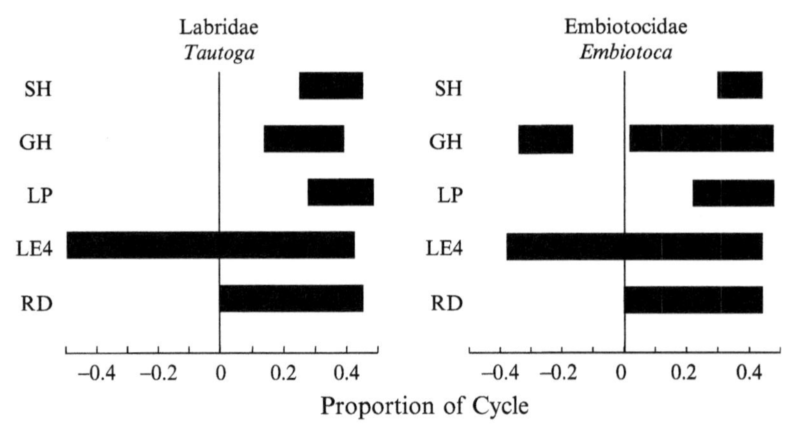

Fig. 3.13. Average pharyngeal muscle activity patterns in a representative embiotocid and a labrid. Note similarity in the activity pattern of the two species. (Redrawn from Liem, 1986; Liem and Sanderson, 1986.)

2002). One distinct modification that is seen in labroids is a very early onset of activity in the fourth levator externus (Figure 3.13). This has been proposed to represent a recovery or preparatory period when this muscle protracts the jaws (Liem, 1986; Liem and Sanderson, 1986; Claes and De Vree, 1991). The interpretation of this pattern is made difficult because the LE4 muscle attaches both to the fourth epibranchial in most labroids and to the lower pharyngeal jaw in labrids, many cichlids, and many embiotocids (Liem, 1986). As in generalized perciform taxa, muscles appear to be active during an adduction phase that includes retraction by the retractor dorsalis, or active during a recovery phase (e.g., the pharyngocleithralis externus). As is seen in other perciform groups, the activity patterns of muscles that control the lower jaws are especially variable and can show a pattern consistent with sheering actions of the upper and lower jaws or of synchronized retraction and adduction (Liem, 1986; Liem and Sanderson, 1986).

4. Labroid Diversity

Comparative studies of morphological and functional diversity across groups of perciform have not been published, so it is not possible to rigorously assess the hypothesis that labroids exhibit greater functional diversity than other groups. As an initial exploration of this area, Figure 3.14 presents data on the diversity of the size of the levator posterior muscle in 154 labrid species and 20 centrarchid species. Variance is a useful measure of diversity, because it describes the spread of a variable around its average value and it does not scale with sample size (Foote, 1997). This comparison reveals that

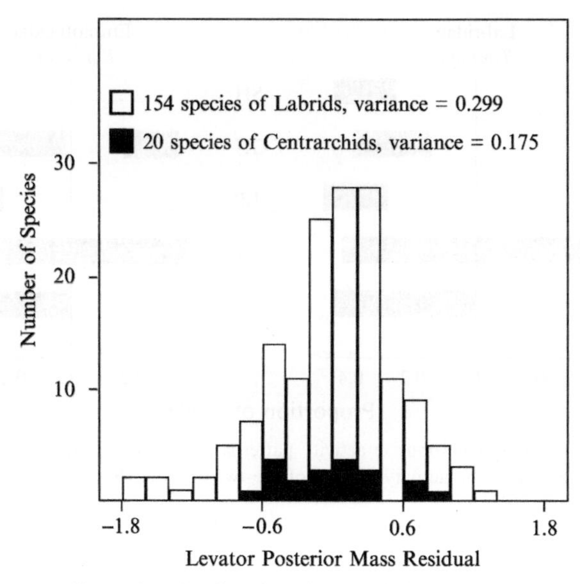

Fig. 3.14. Histogram illustrating the diversity of mass of the levator posterior muscle in 154 labrid species and 20 centrarchid species. Data from all 174 species were fit to a log-log regression on body mass and residuals were calculated to remove body size effects. The histograms are of the residuals from that regression and show that variance of levator posterior mass in centrarchids, a group of generalized perciforms, is about 60% of that in labrids. Since variance is expected to be independent of sample size, this observation suggest that labrids have greater diversity in the size of the levator posterior muscle than do the centrarchids.

levator posterior muscle diversity in centrarchids is about 60% of that seen in labrids, providing modest support for the expectation that labrids are more diverse than centrarchids.

IV. SUMMARY

The pharyngeal jaws have been a more difficult nut to crack than the oral jaws, primarily because they are buried deep inside the head and cannot be observed directly. But anatomical observations and data collected with cineradiography and sonomicrometry have helped in the development of functional models and documentation of how the jaws are used. One of the biggest surprises from this body of work is the lack of evidence to support the expectation that the labroid pharyngeal jaws show greater versatility and are used in a wider range of behaviors than the jaws of generalized perciform taxa. One remaining interpretation of the functional enhancement gained with the labroid condition is that proposed by Galis and Drucker (1996), who suggested that biting forces are more efficiently transferred to the prey in labroids than in generalized taxa. Thus, the advantage may be in the strength of the bite. This point raises the specter of a remaining serious challenge for students of the pharyngeal jaws: how to measure performance. The only PJA performance trait that has been both modeled (e.g., Galis, 1992) and measured is biting strength (Wainwright, 1987; Osenberg and Mittelbach, 1989). Without clear performance metrics upon which to compare taxa, it will not be possible to fully understand the implications of the diversity seen in teleost pharyngeal jaw systems.

ACKNOWLEDGMENTS

I thank Justin Grubich, Darrin Hulsey, and Lara Ferry-Graham for many conversations over the years on pharyngeal jaw function and diversity. Ian Hart prepared the anatomical diagrams shown in Figures 3.2B, 3.10, and 3.11. Funding was provided by National Science Foundation grant IBN-0076436.

REFERENCES

Aerts, P. (1982). Development of the musculs levator externus IV and the musculus obliquus posterior in *Haplochromis elegans* Trewavas, 1933 (Telostei: Cichlidae): A discussion on the shift hypothesis. *J. Morphol.* **173,** 225–235.

Aerts, P., Devree, F., and Vandewalle, P. (1986). Pharyngeal jaw movements in Oreochromis niloticus (Teleostei, Cichlidae)—Preliminary results of a cineradiographic analysis. *Ann. Soc. Roy. Zool. Bel.* **116,** 75–81.

Bullock, A. E., and Monod, T. (1997). Cephalic myology of two parrotfishes (Teleostei: Scaridae). *Cybium* **21,** 173–199.

Cheer, A. Y., Ogami, Y., and Sanderson, S. L. (2001). Computational fluid dynamics in the oral cavity of ram suspension-feeding fishes. *J. Theor. Biol.* **210,** 463–474.

Claes, G., and De Vree, F. (1989). Effects of food characteristics on pharyngeal jaw movements in two cichlid species pisces perciformes. *Ann. Soc. Roy. Zool. Bel.* **119,** 7.

Claes, G., and De Vree, F. (1990). Pharyngeal jaw movements during feeding in *Haplochromis burtoni* cichlidae. *Bel. J. Zool.* **120,** 15.

Claes, G., and De Vree, F. (1991). Cineradiographic analysis of the pharyngeal jaw movements during feeding in *Haplochromis burtoni* Gunther 1893 pisces cichlidae. *Bel. J. Zool.* **121,** 227–234.

Clements, K. D., and Bellwood, D. R. (1988). A comparison of the feeding mechanisms of two herbivorous labroid fishes, the temperate *Odax pullus* and the tropical *Scarus rubroviolaceus*. *Aust. J. Mar. Freshwater Res.* **39,** 87–107.

DeMartini, E. E. (1969). A correlative study of the ecology and comparative feeding mechanism morphology of the Embiotocidae as evidence of the family's adaptive radiation. *Wasmann J. Biol.* **27,** 117–247.

Foote, M. (1997). The evolution of morphological diversity. *Ann. Rev. Ecol. Syst.* **28,** 129–152.

Friel, J. P., and Wainwright, P. C. (1999). Evolution of complexity in motor patterns and jaw musculature of tetraodontiform fishes. *J. Exp. Biol.* **202,** 867–880.

Galis, F. (1992). A model for biting in the pharyngeal jaws of a cichlid fish: *Haplochromis piceatus*. *J. Theor. Biol.* **155,** 343–368.

Galis, F. (1993). Interactions between the phayrngeal jaw apparatus, feeding behavior, and ontogeny in the cichlid fish, *Haplochromis piceatus*: A study of morphological constraints in evolutionary ecology. *J. Exp. Zool.* **267,** 137–154.

Galis, F., and Drucker, E. G. (1996). Pharyngeal biting mechanics in centrarchid and cichlid fishes: Insights into a key evolutionary innovation. *J. Evol. Biol.* **9,** 641–670.

Galis, F., and Snelderwaard, P. (1997). A novel biting mechanism in damselfishes (Pomacentridae): The pushing up of the lower pharyngeal jaw by the pectoral girdle. *Neth. J. Zool.* **47,** 405–410.

Gobalet, K. W. (1989). Morphology of the parrotfish pharyngeal jaw apparatus. *Am. Zool.* **29,** 319–331.

Grubich, J. R. (2000). Crushing motor patterns in drum (Teleostei: Sciaenidae): Functional novelties associated with molluscivory. *J. Exp. Biol.* **203,** 3161–3176.

Grubich, J. R. (2003). Morphological convergence of pharyngeal jaw structure in durophagus perciform fish. *Biol. J. Linn Soc.* **80,** 147–165.

Hernandez, L. P., and Motta, P. J. (1997). Trophic consequences of differential performance: Ontogeny of oral jaw-crushing performance in the sheepshead, *Archosargus probatocephalus* (Teleostei, Sparidae). *J. Zool. (Lond.)* **243,** 737–756.

Huckins, C. J. F. (1997). Functional linkages among morphology, feeding performance, diet, and competitive ability in molluscivorous sunfish. *Ecology* **78,** 2401–2414.

Kaufman, L., and Liem, K. F. (1982). Fishes of the suborder Labroidei (Pisces: Perciformes): Phylogeny, ecology, and evolutionary significance. *Breviora (Museum of Comparative Zoology, Harvard University)* **472,** 1–19.

Lauder, G. V. (1983a). Functional and morphological bases of trophic specialization in sunfishes (Teleostei: Centrarchidae). *J. Morphol.* **178,** 1–21.

Lauder, G. V. (1983b). Functional design and evolution of the pharyngeal jaw apparatus in euteleostean fishes. *Zool. J. Linn. Soc.* **77,** 1–38.

Lauder, G. V. (1983c). Neuromuscular patterns and the origin of trophic specialization in fishes. *Science* **219,** 1235–1237.

Laur, D. R., and Ebeling, A. E. (1983). Predator-prey relationships in surfperches. *Environ. Biol. Fish* **8,** 217–229.

Liem, K. F. (1970). Comparative functional anatomy of the Nandidae. *Fieldiana Zool.* **56,** 1–166.

Liem, K. F. (1973). Evolutionary strategies and morphological innovations: Cichlid pharyngeal jaws. *Sys. Zool.* **22,** 425–441.

Liem, K. F. (1978). Modulatory multiplicity in the functional repertoire of the feeding mechanism in cichlid fishes. I. Piscivores. *J. Morphol.* **158,** 323–360.

Liem, K. F. (1979). Modulatory multiplicity in the feeding mechanism in cichlid fishes, as exemplified by the invertebrate pickers of Lake Tanganyika. *J. Zool.* **189,** 93–125.

Liem, K. F. (1986). The pharyngeal jaw apparatus of the Embiotocidae (Teleostei) – a functional and evolutionary perspective. *Copeia* **1986,** 311–323.

Liem, K. F., and Greenwood, P. H. (1981). A functional approach to the phylogeny of the pharyngognath teleosts. *Am. Zool.* **21,** 83–101.

Liem, K. F., and Sanderson, S. L. (1986). The pharyngeal jaw apparatus of labrid fishes—A functional morphological perspective. *J. Morphol.* **187,** 143–158.

Monod, T., Hureau, J. C., and Bullock, A. E. (1994). Cephalic osteology of two parrotfish (Scaridae: Teleostei). *Cybium* **18,** 135–168.

Nelson, G. J. (1967). Gill arches of some teleostean fishes of the families Girellidae, Pomacentridae, Embiotocidae, Labridae, and Scaridae. *J. Nat. Hist.* **1,** 289–293.

Norton, S. F. (1988). Role of the gastropod shell and operculum in inhibiting predation by fishes. *Science* **241,** 92–94.
Osenberg, C. W., and Mittelbach, G. G. (1989). Effects of body size on the predator-prey interaction between pumpkinseed sunfish and gastropods. *Ecol. Monogr.* **59,** 405–432.
Palmer, A. R. (1979). Fish predation and the evolution of gastropod shell sculpture: Experimental and geographic evidence. *Evolution* **33,** 697–713.
Sanderson, S. L., Cech, J. J., and Cheer, A. Y. (1994). Paddlefish buccal flow velocity during ram suspension feeding and ram ventilation. *J. Exp. Biol.* **186,** 145–156.
Sanderson, S. L., Cheer, A. Y., Goodrich, J. S., Graziano, J. D., and Callan, W. T. (2001). Crossflow filtration in suspension-feeding fishes. *Nature (Lond.)* **412,** 439–441.
Sibbing, F. A. (1982). Pharyngeal mastication and food transport in the carp (*Cyprinus carpio*): A cineradiographic and electromyographic study. *J. Morphol.* **172,** 223–258.
Sibbing, F. A. (1988). Specializations and limitations in the utilization of food resources by the carp, *Cyprinus carpio*: A study of oral food processing. *Environ. Biol. Fish* **22,** 161–178.
Sibbing, F. A., Osse, J. W. M., and Terlow, A. (1986). Food handling in the carp (*Cyprinus carpio*): Its movement patterns, mechanisms and limitations. *J. Zool. (Lond.)* **210,** 161–203.
Stiassny, M. L. J., and Jensen, J. (1987). Labroid interrelationships revisited: Morphological complexity, key innovations, and the study of comparative diversity. *Bull. Mus. Comp. Zool.* **151,** 269–319.
Vandewalle, P., Havard, M., Claes, G., and De Vree, F. (1992). Movements of the pharyngeal jaw during feeding in *Serranus scriba* (Linneus, 1758) (Pisces, Serranidae). *Can. J. Zool.* **70,** 145–160.
Vandewalle, P., Saintin, P., and Chardon, M. (1995). Structures and movements of the buccal and pharyngeal jaws in relation to feeding in *Diplodus sargus*. *J. Fish Biol.* **46,** 623–656.
Wainwright, P. C. (1987). Biomechanical limits to ecological performance: Mollusc-crushing by the Caribbean hogfish, *Lachnolaimus maximus* (Labridae). *J. Zool. (Lond.)* **213,** 283–297.
Wainwright, P. C. (1988). Morphology and ecology: The functional basis of feeding constraints in Caribbean labrid fishes. *Ecology* **69,** 635–645.
Wainwright, P. C. (1989a). Functional morphology of the pharyngeal jaws in perciform fishes: An experimental analysis of the Haemulidae. *J. Morphol.* **200,** 231–245.
Wainwright, P. C. (1989b). Prey processing in haemulid fishes: Patterns of variation in pharyngeal jaw muscle activity. *J. Exp. Biol.* **141,** 359–376.
Wainwright, P. C. (2002). The evolution of feeding motor patterns in vertebrates. *Curr. Opin. Neurobiol.* **12,** 691–695.
Wainwright, P. C., and Bellwood, D. R. (2002). Ecomorphology of feeding in coral reef fishes. *In* "Coral Reef Fishes. Dynamics and Diversity in a Complex Ecosystem" (Sale, P. F., Ed.), pp. 33–55. Academic Press, Orlando.
Wainwright, P. C., and Richard, B. A. (1995). Predicting patterns of prey use from morphology with fishes. *Environ. Biol. Fish* **44,** 97–113.
Wainwright, P. C., Osenberg, C. W., and Mittelbach, G. G. (1991). Trophic polymorphism in the pumpkinseed sunfish (*Lepomis gibbosus* Linnaeus): Effects of environment on ontogeny. *Funct. Ecol.* **4,** 40–55.
Wainwright, P. C., Bellwood, D. R., Westneat, M. W., Grubich, J. R., and Hoey, A. S. (2004). A functional morphospace for the skull of labrid fishes: Patterns of diversity in a complex biomechanical system. *Biol. J. Linn. Soc.* **82,** 1–25.
Winterbottom, R. (1974). A descriptive synonymy of the striated muscles of the Teleostei. *Proc. Acad. Nat. Sci. Phila.* **125,** 225–317.
Yamaoka, K. (1978). Pharyngeal jaw structure in labrid fishes. *Publ. Seto Mar. Biol. Lab.* **24,** 409–426.

4

THE HYDRODYNAMICS AND STRUCTURAL MECHANICS OF THE LATERAL LINE SYSTEM

SHERYL COOMBS
SIETSE VAN NETTEN

I. INTRODUCTION

The lateral line system is a primitive vertebrate sensory system, found exclusively in aquatic, anamniotic vertebrates (cartilaginous and bony fishes, as well as some aquatic amphibians). It is closely associated with a suite of octavolateralis sensory systems, which include the vestibular and auditory organs of the inner ear and the electro- and mechano-sensory lateral line systems (Figure 4.1). The term "octavolateralis" is derived from the cranial nerves that innervate these systems and enter the brain in close proximity to one another. The paired organs of the inner ear are innervated by different

Fish Biomechanics: Volume 23
FISH PHYSIOLOGY

DOI: 10.1016/S1546-5098(05)23004-2

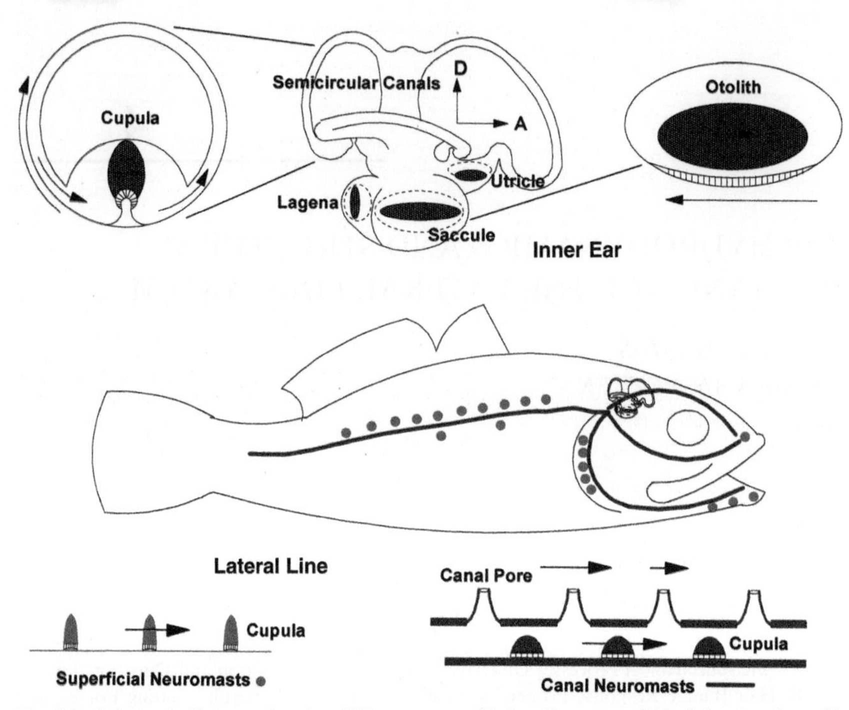

Fig. 4.1. Schematic diagram showing different types of hair cell sensors within fish octavolateralis systems, including the semicircular canals and otolithic endorgans (saccule, lagena, and utricle) of the inner ear and the superficial and canal neuromasts of the lateral line. Adapted from Fig. 32.1 in Platt *et al.* (1989) with kind permission from Springer-Science and Business Media.

branches of the eighth (octavo) cranial nerve, whereas the spatially distributed, multiple-organ systems of the electro- and mechano-sensory lateral line are innervated by as many as five different lateral line cranial nerves (Northcutt, 1989). At the heart of all mechanically driven octavolateralis systems is a tiny hair cell (micrometers in diameter) that functions as a mechano-electrical transducer (Figure 4.2). A displacement of the hair bundle at the apical surface of each cell causes a change in the electrical potential across the cell membrane (Figure 4.3). This, in turn, results in a pattern of rapidly changing (action) potentials across the membrane of the innervating nerve fiber, which is relayed to the brain as the neural "code" of the sensory input.

The peripheral structures surrounding, overlying, or otherwise coupled to the hair cell bundles vary widely among the octavolateralis suite of senses (Figure 4.1), resulting in fundamental differences in how the mechanical energy is transmitted to the hair cells. For example, hair cells enclosed in

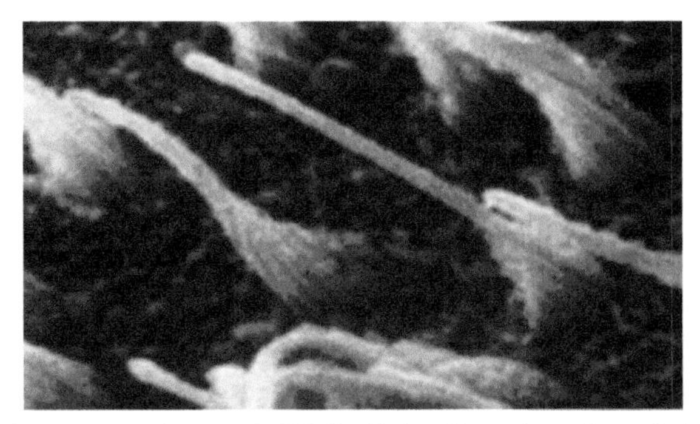

Fig. 4.2. Scanning electron micrograph (SEM) of hair cells on the surface of a lateral line canal neuromast from the mottled sculpin, *Cottus bairdi*. The axis of hair cell polarization and thus the directional response properties of each cell (see Figure 4.3) are determined by the direction of stepwise increase in stereociliary length toward the elongated kinocilium. The kinocilium and tallest stereocilia are always on one of two opposite sides of the cell, resulting in a single axis of hair cell polarization that is parallel to the long axis of the canal.

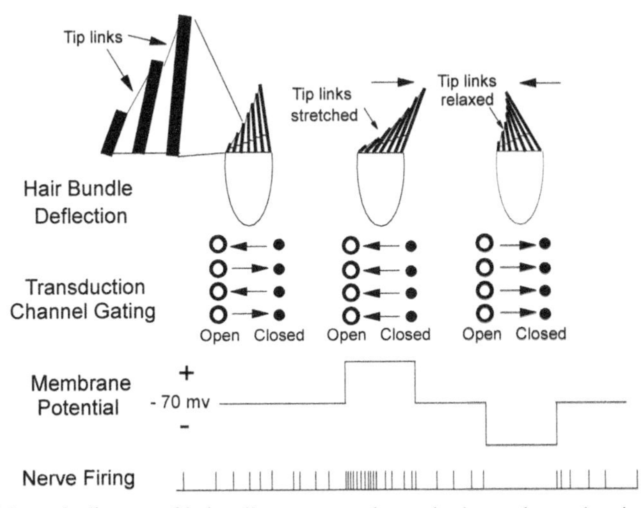

Fig. 4.3. Schematic diagram of hair cell structure and steps in the mechano-electrical transduction process. In the absence of imposed deflection, ambient noises, Brownian motion, and stochastic channel flicker cause the random opening and closing of transduction channels. Minute fluctuations in the resting membrane potential (–70 mV) cause a low level of spontaneous firing activity in the nerve fiber. A deflection of the hair cell's ciliary bundle in the direction of the longest stereocilium causes increased tension in the tip links, leading to an opening of the transduction channels, a depolarization of the membrane potential, and an increase in the firing rate of nerve fibers carrying information from the hair cells to the brain. Deflection in the opposite direction causes a relaxation of tip link tension, leading to the closing of transduction channels and a decrease in the neural firing rate.

the semicircular canals of the inner ear respond to the angular accelerations of the canal fluids, conveying information about angular accelerations, yaw (x), roll (y), and pitch (z), of the fish's body. Those in the otolithic organs of the inner ear (the saccule, lagena, and utricle) are mass-loaded with a single, calcareous stone in a fluid-filled chamber. Density and thus inertial differences between the otolith and the underlying ciliary bundles render these otolithic organs sensitive to linear accelerations of the fish's body. Last, but not least, the spatially distributed hair cell organs in the lateral line system normally respond to net movements between the fish and the surrounding water or the spatial nonuniformities in the surrounding flow field. Furthermore, the frequency response of lateral line organs and whether or not they respond to flow velocity or flow acceleration depend on whether the organs are exposed superficially on the skin surface or are enclosed in fluid-filled canals just below the skin surface. From a neurobiological perspective, the biomechanical and specifically hydrodynamic principles by which these differences arise are thus critical determinants of how the different systems function and the kinds of information that can be extracted and encoded by the nervous system (Kalmijn 1988, 1989; Denton and Gray, 1989; Braun *et al.*, 2002).

Octavolateralis sensory functions are also intricately linked to the biomechanical effects of whole-body or body part movements of fish, which can stimulate one or more octavolateralis systems of a nearby receiving fish to indicate the presence, identity, location, or intention of a predator, prey, or mate. Likewise, the biomechanical consequences of self-motion provide mechano-sensory information about the animal's three-dimensional position in space and its relationship to other entities in the environment. Conversely, self-generated motions can also interfere with an animal's ability to detect signals from exogenous sources like other fishes. As a consequence, biomechanical filters to reduce noise interference and enhance signal detection are common features of octavolateralis systems. Finally, a fish's ability to maneuver and stabilize its position in the face of environmental disturbances will depend, in part, on octavolateralis information about the temporal and spatial characteristics of the disturbance.

The overall goal of this chapter is to summarize some of the key biomechanical and hydrodynamic features of the lateral line system—especially as they pertain to the extraction and encoding of information relevant to the lives of fishes. Because the lateral line is a essentially an array of flow sensors, fluid dynamics plays a particularly significant role in shaping the response properties of this particular octavolateralis system. In addition, the lateral line system shares some of the same structural mechanics of other octavolateralis systems at the level of individual hair cells. Thus, the terms biomechanical and hydrodynamic are used here to distinguish between processes that depend primarily upon the properties of solid structures (e.g., stiffness, elasticity) and those that depend upon the properties of fluids (e.g., viscosity). We begin with a general overview of the structure, function, and behavioral use of the lateral line system. Next, we examine the role of hair cell micromechanics in the mechano-electrical transduction process and the directional specificity of the hair cell response. This is followed by a discussion of how the structural interface between lateral line sense organs and the surrounding water helps to shape the response properties of the system and the kind of information that is encoded. We conclude by identifying unanswered questions and key areas in need of further research.

II. GENERAL FUNCTION, STRUCTURE, AND ORGANIZATION

A. Behavioral Significance

Lateral line function, being somewhat intermediate between the senses of touch and hearing, has been aptly described as "touch at a distance" (Hofer, 1908; Dijkgraaf, 1963). With this sense, fish can "feel" water movements ranging from large-scale river currents to the minute disturbances created by planktonic prey. Likewise, hydrodynamic disturbances can be either biotic (e.g., a nearby swimming fish) or abiotic (e.g., ambient currents) in origin. The lateral line has been experimentally implicated in a number of different behaviors, including (1) schooling (Partridge and Pitcher, 1980), (2) prey detection (e.g., Hoekstra and Janssen, 1985), (3) courtship and spawning (Satou *et al.*, 1994), (4) rheotaxis (Montgomery *et al.*, 1997), and (5) station holding (Sutterlin and Waddy, 1975). In a more general sense, the lateral line system is also thought to form hydrodynamic images of the surroundings, much as the visual system forms visual images. This can be accomplished in both active and passive ways to detect both stationary and moving bodies. *Active* hydrodynamic imaging is analogous to the ability of dolphins or bats to echolocate objects in the environment. Instead of producing ultrasonic sounds, however, fishes produce a flow field around their bodies as they swim through the environment. Thus, they can use their lateral line system to detect distortions in this self-generated flow field due to the presence of stationary objects (Dijkgraaf, 1963; Hassan, 1985). Blind cavefishes, which rely heavily on their mechano-senses for exploration of the environment, are able to gather information about the fine spatial details of objects as they glide past them. For example, they can discriminate between two grates when the spatial interval of the grates differs by as little as 1 mm (Hassan, 1986). Fishes can also form *passive* hydrodynamic images of both moving and stationary bodies by detecting the currents generated by other moving bodies (e.g., another fish) or the distortions caused by stationary bodies in ambient currents of abiotic origins (e.g., a rock in a stream).

B. Neuromast Structure and Organization

Anywhere from less than 50 to more than 1000 hair cells are grouped together into a single sense organ or neuromast in the lateral line system. Neuromasts are located on the head and body of adult fishes either superficially on the skin surface or just under the skin in fluid-filled canals (Figure 4.1). Lateral line canals are typically distributed along the trunk, above and below the eye, across the top of the head, and along the edge of the preopercle and lower jaw. In bony fishes, each canal typically contains several, more or less equally spaced canal neuromasts, each of which is normally located between two openings (pores) in the overlying canal wall and skin surface. The location and number of superficial neuromasts in adult fishes vary widely among different species of fish—ranging from just a few neuromasts distributed into distinct groups or lines at several stereotypical locations (e.g., around the nares, behind the eye, and adjacent to different canal lines) to literally thousands of neuromasts distributed all over the head and body surface, as found in many characid (e.g., blind cavefish) and cypriniform species (see reviews by Coombs *et al.*, 1988; Webb, 1989).

Both superficial and canal neuromasts contain two groups of oppositely oriented hair cells that are spatially intermingled (Figure 4.2). The orientation of the hair cell is determined by the height asymmetry of the hair bundle. That is, individual stereovilli within the bundle increase their lengths in a stepwise fashion—in the direction of a single eccentrically placed and elongated kinocilium (Figure 4.2) (Flock, 1965a,b). This anatomical polarization determines the directional response properties of the cell such that bending of the stereovilli toward the kinocilium results in an excitatory response (depolarization of the hair cell membrane and an increase in the electrical discharge of fibers carrying information from the hair cells to the brain), and bending in the opposite direction results in an inhibitory response (hyperpolarization of the cell membrane and a decrease in the firing rate of afferent fibers) (Figure 4.3). Bending in a direction orthogonal to this axis results in no response, and intermediate directions of bending results in responses that are a cosine function of the bending direction.

In canal neuromasts, the axis of hair cell orientation is predominantly parallel to the long axis of the canal (e.g., Flock, 1965b; Kelly and van Netten, 1991) (Figure 4.2), such that fluid motion in one direction along the canal will excite roughly half of the hair cells while simultaneously inhibiting the other half. In contrast, water motion past superficial neuromasts is, for the most part, not similarly constrained, although the various pits, grooves, and papillae surrounding superficial neuromasts in some species may function to channel flows in different directions. The axis of hair cell orientation on different superficial neuromasts has not been carefully documented for most fish species.

Information from oppositely oriented hair cells is transmitted to the brain in separate channels. That is, each superficial or canal neuromast is innervated by a minimum of two afferent fibers, one that innervates hair cells of one polarity and another that innervates hair cells of the opposite polarity (e.g., Murray, 1955). The functional significance of this particular organization is yet to be fully understood, largely because we have so little data on the central connectivity of inputs from oppositely oriented hair cells. Nevertheless, information about the overall direction of an oncoming current appears to be encoded by superficial neuromasts, as judged by their critical importance to rheotactic behavior in the absence of vision (Montgomery *et al.*, 1997). Likewise, the pattern of local flow directions and amplitudes inside lateral line canals appears to convey useful information about moving sources (Coombs *et al.*, 1996, 2000; see also Section IV.E for further detail). Hair cells and their afferent fibers are also innervated by efferent fibers from the brain. The efferent system tends to reduce lateral line sensitivity just before and during self-generated movements of the fish (reviewed in Roberts and Meredith, 1989).

C. Neuromast Response Properties and Functions

Differential movement between the animal and the surrounding water, or, in other words, water flowing over the skin surface, generally results in stimulation of the lateral line system (Denton and Gray, 1983b; Kalmijn, 1988). The specific stimulus dimension to which a particular lateral line neuromast responds, however, can vary in a number of important ways, depending on the structural interface between the neuromast and the surrounding water (Figure 4.4). Relative to flow along the skin surface, superficial neuromasts tend to show responses that are largely proportional to flow velocity, whereas canal neuromasts respond in proportion to flow acceleration . The biophysical principles by which these differences arise are described in further detail in Section IV, but it is worth pointing out here that this rather simple dichotomy is complicated by the fact that canal neuromasts are initially "born" as superficial neuromasts in larval fish. Only later in development do they invaginate into the dermis and become enclosed in canals (reviewed in Webb, 1989a, 2000). Although the biomechanical properties of a larval superficial neuromast destined to become a canal neuromast may not differ drastically from a superficial neuromast in an adult

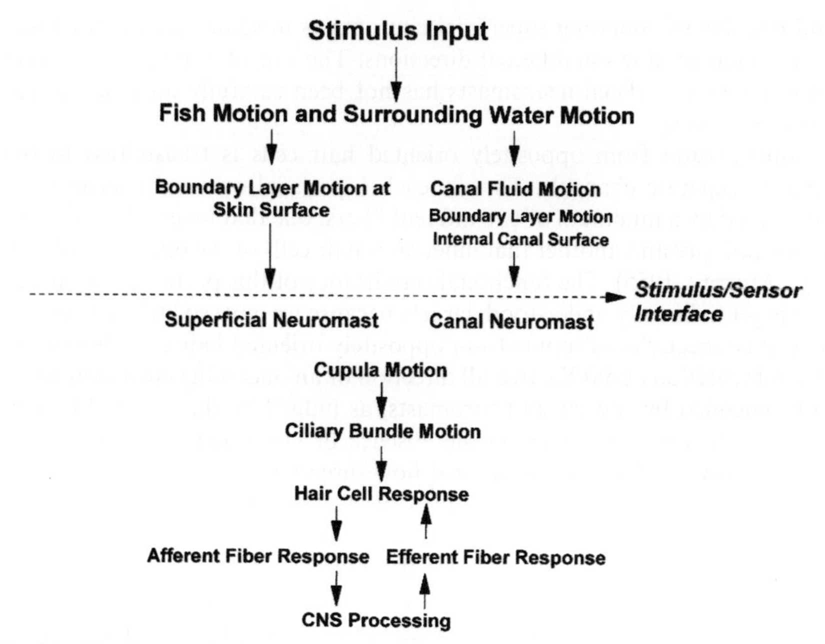

Fig. 4.4. Stimulus transduction pathway in the lateral line system showing the different possible levels of biomechanical/hydrodynamic filter action and signal processing.

(all structural features being equal), the two categories of superficial neuromasts may nevertheless differ in other, fundamental ways (e.g., in their neural connections and pathways in the brain; see discussion in Coombs *et al.*, 2001a).

In the majority of species, canal walls are formed by rigid structures such as bone or scale and open to the environment through a series of pores. In some species, however, canal pores are absent and canal walls are compliant at one or more body location. Such is the case for the ventral portion of the mandibular canal in stingrays (and other batoid species). In this case, lateral line organs appear to respond to tactile stimulation or depression of the canal wall—presumably caused by benthic prey buried in the substrate (Maruska and Tricas, 2004). Cephalic lateral line canals in clupeids have additional structures (compressible air cavities) impinging on their compliant walls (Blaxter *et al.*, 1981). In sprats, in which the mechanical relationship between an air-filled bulla in the cranial cavity and cephalic lateral line canals has been studied closely, fluid displacements in the canals are proportional to pressure on the air cavity; hence, flow velocity in the canal is related to the rate of change of pressure, which in turn is proportional to the rate of change of the acceleration of the source surface (Denton and Gray, 1988). Under these circumstances, the lateral line system can respond to both water acceleration and rate of change in acceleration, and appears to be designed to detect the earliest signs of change—an ability that correlates well with the exquisite schooling maneuverability of these fishes.

The responses of lateral line neuromasts to surrounding flow can also be differentiated in the frequency domain. Frequency tuning in the lateral line system has typically been measured with a dipole stimulus source (sinusoidally vibrating sphere) (e.g., Harris and van Bergeijk, 1962; Kroese *et al.*, 1978). When described in the frequency domain, all octavolateralis systems, including the lateral line, can be subdivided into at least two submodalities: low-pass channels that respond best to the lower end of the frequency range of detection (superficial neuromasts) and high-pass channels that respond best to the higher end of the range (canal neuromasts) (Coombs and Montgomery, 2005). One important outcome of this frequency partitioning is the utility of high-pass channels for improving signal-to-noise ratios when fishes need to detect low-amplitude, high-frequency signals of interest (e.g., prey) in the presence of pervasive high-amplitude, low-frequency noises (e.g., ambient currents, self-generated respiratory flows) (Montgomery *et al.*, 1994). The high-pass tuning characteristics of canals and their dependence on internal canal diameter and geometry are discussed in further detail in Sections IV.B and IV.C.

III. HAIR CELL MICROMECHANICS

A significant challenge to mechano-sensory systems in general is how responsiveness to rapid events can be accomplished—especially in the presence of DC stimuli (i.e., sustained, unidirectional flows). Although some hair cells show tonic sensitivity to sustained stimuli (e.g., superficial neuromast hair cells to DC flows [Voigt *et al.* 2000]), the majority of hair cells are, in effect, AC-coupled, meaning that they respond best to time-varying changes in amplitude. Indeed, hair cells have a remarkable ability to respond to high-frequency signals (in the ultrasound range for the auditory systems of echolocating bats and dolphins) and to phase-lock (respond at the same phase during each sinusoidal cycle) to signals up to ~1000 Hz, despite viscous damping by surrounding fluids. In addition to the directional characteristics of hair cells, displacement sensitivity in the nanometer range—good enough to detect Brownian motion—is another remarkable property of hair cells. The micro-mechanical processes underlying these abilities are described in the following two sections.

A. Micromechanical Processes Underlying Hair Cell Transduction and Directionality

Hair cell transduction channels are transmembrane proteins with pores located on the apical surface of the hair cell. These mechanically sensitive channels allow charged ions to cross the hair cell membrane. The opening and closing of the pore modulate the flow of transmembrane current, which is driven by a significant electrochemical gradient across the membrane (Hudspeth *et al.*, 2000). Although the exact proteins associated with hair cell transduction channels have yet to be thoroughly identified (but see Sidi *et al.*, 2003; Corey *et al.*, 2004), the channels appear to be rather nonselective, permitting primarily K^+ (the most prominent cation in extracellular fluids surrounding the hair cell) and also Ca^{2+} ions to enter the hair cell. This induces small, graded potential differences across the cell membrane, which are then transformed into action (non-graded) potentials (spikes) by afferent nerves that contact the hair cells. The summed receptor potentials across many hair cells can be recorded extracellularly as small microphonic potentials, which in the lateral line system are twice the stimulus frequency because of the bidirectional polarization of the hair cells (Figure 4.2) (Flock, 1965b). Microphonic responses have subsequently been used to study mechanoelectrical transduction in hair cells of wild-type (e.g., Corey and Hudspeth, 1983a; Wiersinga-Post and van Netten, 1998, 2000; Corey *et al.*, 2004) and mutant animals (Nicolson *et al.*, 1998; Sidi *et al.*, 2003).

At the core of each stereovillum is a large number of actin filaments that effectively make the stereovilli behave as rigid rods pivoting around their insertion points in the apical surface (de Rosier *et al.*, 1980; Flock and Orman, 1983). Deflection of a hair bundle causes differential (shearing) motion of the stereovilli that is most likely controlled by tiny filaments serving as mechanical links between individual stereovilli (see inset, Figure 4.3). Several types of links (e.g., lateral and tip) have been identified, primarily from studies on inner ear hair cells in mammals and birds (Osborne *et al.*, 1984; Pickles *et al.*, 1984; Goodyear and Richardson, 2003). Lateral or side links are oriented in parallel with the apical membrane and are thought to structurally organize the hair bundle so that the stereovilli may slide with respect to each other if the hair bundle is deflected (Pickles *et al.*, 1989; Geisler, 1993; Pickles, 1993). Filamentous links at the very tips of the stereovilli are implicated in the transduction process and have been found in nearly all types of hair cells investigated, including those of the lateral line organ (Pickles *et al.*, 1991; Rouse and Pickles, 1991). Information on the ultrastructure (Kachar *et al.*, 2000) and molecular constituents (Siemens *et al.*, 2004; Söllner *et al.*, 2004) of these important links has also become recently available.

Tip links run nearly parallel to the stereociliar axis (see Figure 4.3), connecting the tips of shorter stereocilia to their taller neighbors. Tip links are therefore selectively tensioned when a hair bundle is deflected along the hair cell's axis of best sensitivity (Pickles *et al.*, 1984) (Figure 4.2), meaning that they are in a favorable position to transfer bundle-deflecting forces into those that open the hair cell's transduction channels. When the transverse bundle displacement, X, is small relative to bundle height, L, (i.e., $X \ll L$), the stretching distance of the tip links between adjacent stereociliary, x, is approximately equivalent to a fraction, γ, of the bundle displacement:

$$x = \gamma X = (d/L)X, \tag{1}$$

where d is the shortest distance between adjacent stereocilia (i.e., along a line drawn parallel to the apical surface of the hair cell (Howard and Hudspeth, 1988).

It is now generally accepted that transduction channels, when activated via increased tension in the tip links, increase their probability of being open. Evidence in support of this view comes from studies showing that (1) tip link destruction abolishes the transduction process, whereas tip link regeneration restores it (Zhao *et al.*, 1996) and (2) transduction currents are localized to the hair bundle's tip (Hudspeth, 1982; Denk *et al.*, 1995).

B. Hair Bundle Response Properties: Restorative Forces, Molecular Gating Forces, and Nonlinearities

The transducer apparatus that is engaged by the movement of the hair bundle appears to have a reciprocal effect on the mechanical action of the hair bundle. This molecular-level effect governs crucial parameters such as the operational range and accuracy of the transduction process. These parameters are discussed after first giving a description of the mechanical properties of hair bundles.

Mechanical properties of hair bundles have been measured *in vitro* in the ear and lateral line of a variety of organisms in response to stimulation by micro-fluid jets (Flock and Orman; 1983; Saunders and Szymko, 1989; Kros *et al.*, 1992; Géléoc *et al.*, 1997) or small microfibers (Strelioff and Flock, 1984; Crawford and Fettiplace, 1985; Howard and Ashmore, 1986; Howard and Hudspeth, 1988; Russell *et al.*, 1992). Combination of simultaneous measurement of exerted force and evoked submicrometer bundle displacement has allowed the determination of a hair bundle's mechanical impedance. Apart from indications of a resistive component (Howard and Ashmore, 1986), which most likely reflects the process of adaptation to sustained stimuli (Eatock *et al.*, 1987; Hudspeth and Gillespie, 1994), the mechanical impedance consists mainly of a reactive or stiff component. This dominant elastic property of a hair bundle is clearly of functional importance, as it causes the bundle to restore its equilibrium position after a deflection in the excitatory direction. Different values, depending on hair cell type, have been found for the stiffness component, but most are on the order of 1 mN/m (e.g., van Netten, 1997). Estimates of hair bundle stiffness have also been obtained *in vivo* from submicroscopic displacement measurements of lateral line neuromasts with similar results (van Netten and Kroese, 1987).

Detailed measurements of hair bundle stiffness on individual hair cells (Howard and Hudspeth, 1988; Russell *et al.*, 1992; Geleoc *et al.*, 1997; Ricci *et al.*, 2002) have revealed the presence of two components: (1) a linear (deflection-independent) stiffness component, which is most likely related to the passive restoring force originating from the stereociliar ankle region (pivots) and the lateral links, and (2) a nonlinear (deflection-dependent) component. The latter is present only in the direction of hair cell sensitivity and is therefore associated with the hair cell's transducer apparatus, as described below.

A combination of experimental results and theoretical considerations on the nonlinear component of hair bundle mechanics has led to a concise biophysical description of mechano-electrical transduction as a stochastic molecular gating mechanism, termed the gating spring model (Corey and Hudspeth, 1983b; Howard and Hudspeth, 1988). In this model, a deflection of stereocilia in the direction of the tallest cilia is thought to increase the tension in the stereociliary tip links, which act on the elastic "gating springs" to open the ionic transduction channels (Figure 4.3). When the hair bundle is deflected in the opposite direction, the tip link tension is reduced, leading to the closure of transduction channels. Because the elastic coupling forces on the transduction channel are instantaneous and the rate of transition between open and closed states depends on a relatively low energy barrier (on the order of the thermal energy, as defined later), relatively rapid changes in tip link tension can be registered (Corey and Hudspeth, 1983b).

The mechanical responsiveness or sensitivity of the transduction gating process can be characterized by the elementary gating force, Z, which is related to the force required to open a single transduction channel. At most, there are only one or two of these channels per stereocilium (e.g., Denk *et al.*, 1995). The values of Z in saccular, lateral line, vestibular, and cochlear hair cells, as sensed at their hair bundles tips, are all found in the range of 100–500 fN (Howard and Hudspeth, 1988; van Netten and Khanna, 1994; van Netten and Kros, 2000; van Netten *et al.*, 2003). The operational deflection range (about 90% transduction current modulation) of a hair cell bundle is directly related to the gating force and the thermal energy, kT, ($\sim 4.2 \times 10^{-21}$ J, @ room temperature) and is given by

$$\Lambda_{90} = \frac{6kT}{Z}, \tag{2}$$

which, in line with the values of Z, is usually on the order of 100 nm (van Netten *et al.*, 2003; cf. Markin and Hudspeth, 1995).

The nonlinear component of the ciliary bundle stiffness can be understood in terms of the gating spring mechanism proposed by Hudspeth and colleagues. That is, if a force applied to the bundle is mechanically linked to the transduction gate by an elastic spring, the opening and closing of that gate should reciprocally affect the applied force through the same mechanical link. Indeed, the measured stiffness of ciliary bundles declines with increasing levels of imposed deflection, reaching a maximum reduction in stiffness if approximately half of the channels are open. Upon stronger deflection, more channels open while the stiffness increases again to an asymptotic value. The reduction in the nonlinear stiffness component is termed gating compliance, Δ_N, and for N transducer channels has a maximum reduction that amounts to

$$\Delta_N = N\frac{Z^2}{4kT}, \tag{3}$$

and that depends on the molecular gating force, Z, and the thermal energy (kT). The nonlinear gating compliance may be of the same order of magnitude as the linear passive component, especially in vestibular and lateral line hair cells (van Netten and Kros, 2000). The effects of this molecularly induced gating compliance in hair bundle mechanics have been observed in the nonlinear dynamics of single hair bundles (Howard and Hudspeth, 1988; Russell *et al.*, 1992; Geleoc *et al.*, 1997; van Netten and Kros, 2000; Ricci *et al.*, 2002) and lateral line cupular responses (van Netten and Khanna, 1994; Ćurčić-Blake and van Netten, 2005).

Recently, a fundamental lower displacement detection threshold, σ_{min}, related to the stochastic nature of the hair cell's transduction channel gating, has been characterized (van Netten *et al.*, 2003). The value of σ_{min} appears to be proportional to the unavoidable thermal energy kT and is inversely proportional to the gating force, Z, according to

$$\sigma_{min} = \frac{1}{\sqrt{N}}\frac{2kT}{Z} = \sqrt{\frac{kT}{\Delta_N}}. \tag{4}$$

For a typical hair cell with 60 identical transducer channels, each having a gating force of 300 fN, the related detection threshold is about 3.5 nm. Integrating the signal over several hair cells lowers the stochastically imposed threshold, σ_{min}, with the square root of N, that is, the total number of transducer channels involved. The last equal sign in Eq. (4) shows a fundamental property of mechano-electrical transduction, since it demonstrates directly that the exquisite sensitivity of hair cells (σ_{min}) depends on the nonlinear dynamics of the gating mechanism—that is, on the gating compliance, Δ_N, and is degraded by the stochastic nature (i.e., thermal energy, kT) of the transduction process. This relationship will be used again to describe the sensitivity of the lateral line system (Section IV.A).

In addition to changes in the gating spring tension caused by hair bundle deflections to transient stimuli, gating spring tension also appears to be partially reset after prolonged exposure to an ongoing stimulus. The resetting of spring tension is reflected by the phenomenon of adaptation, in which the hair cell's response to a sustained stimulus declines with time. It has been reported that the tension resetting is controlled by molecular (actin-myosin) motors (Holt *et al.*, 2002; Kros *et al.*, 2002) that may cause the upper insertion points of tip links to slide up or down the adjacent (longer) stereocilium; downward sliding in response to an excitatory stimulus relaxes the gating tension, whereas upward sliding to an inhibitory stimulus increases the gating tension. Adaptation is dependent on Ca^{2+} entry through the channel (Howard and Hudspeth, 1988; Hacohen *et al.*, 1989; Crawford *et al.*, 1991). This means that there is a Ca^{2+}-dependent active feedback mechanism, which regulates the transduction current during the adaptation process (Eatock *et al.*, 1987). Functionally, this has been suggested to enhance a hair cell's sensitivity (Hudspeth and Gillespie, 1994) and to adjust or possibly optimize the signal-to-noise ratio of signal transduction by hair cells (Dinklo *et al.*, 2003). The mechanical consequences of motor-driven adaptation, in combination with the gating compliance, have been related to increased mechano-sensitivity in bullfrog saccular hair cells (Hudspeth *et al.*, 2000) and amplification processes in the mammalian cochlea (Chan and Hudspeth, 2005).

IV. LATERAL LINE MECHANICS AND HYDRODYNAMICS

In Section III, we described the micromechanical processes that give rise to individual hair cell responses. In this section, we describe the gross biomechanical and hydrodynamic processes that give rise to the responses of many (<50 to >1000) hair cells packaged as individual sense organs, the superficial and canal neuromasts of the lateral line system. For each type of neuromast, the motion of the surrounding water is coupled to that of the underlying hair cell cilia by a gelatinous cupula, which has a distinct infrastructure or columnar organization that presumably enhances the mechanical coupling between it and the ciliary bundles of the underlying hair cells (Kelly and van Netten, 1991). The cupula, being of nearly the same density as the surrounding fluid (e.g., Jielof *et al.*, 1952), is thought to be driven primarily by viscous forces (Harris and Milne 1966; Flock, 1971; van Netten and Kroese, 1987, 1989; Kalmijn, 1988, 1989; Denton and Gray, 1989). This viscous drag hypothesis of cupular excitation means that the cupular displacement response is largely proportional to the velocity of water flowing past it. For this reason, superficial neuromasts are often described as being sensitive to water velocity (see Section IV.A for further detail). Indeed, electrophysiological measurements of the phase and amplitude responses of superficial neuromast fibers to sinusoidal signals confirm that superficial neuromasts function predominantly as flow velocity detectors in several species of fish (Coombs and Janssen, 1990; Kroese and Schellart, 1992; Montgomery and Coombs, 1992) and amphibians (Görner, 1963; Kroese *et al.*, 1978). When responsiveness, as measured in terms of the amplitude of afferent fiber activity, is plotted as a function of frequency for a given flow velocity amplitude, the form of the function describes a low-pass filter. That is, responsiveness is best at low frequencies and begins to decline at some high-frequency cutoff. High-frequency cutoffs for superficial neuromast responses are found in the range of 10 to 60 Hz (Münz, 1989).

Similarly, if cupulae are displaced in proportion to velocity inside the canal, and canal fluid velocity is proportional to net accelerations outside the canal (see Section IV.B for further detail), it follows that canal neuromasts will respond in proportion to net outside accelerations. For sinusoidal signals of different frequencies but of constant acceleration amplitude, evoked spike activity from canal neuromast fibers does indeed show a relatively flat, low-pass behavior (e.g., Coombs and Janssen, 1990; Kroese and Schellart, 1992; Montgomery and Coombs, 1992; Engelmann *et al.*, 2000, 2002). The high-frequency cutoffs are usually found in the range of 60 to 150 Hz, thus exceeding those of the velocity-sensitive superficial neuromasts (e.g., Münz, 1989). It should be remembered that this "view" of the frequency response depends upon the dimensions in which fluid motion is expressed (e.g., m/s or m/s^2). Thus, both superficial and canal neuromasts have been described previously as low-pass systems, but the former with respect to fluid velocity and the latter with respect to fluid acceleration. Choosing different frames of reference in this case emphasizes the dependence of each on the different dimensions of velocity and acceleration. A later comparison of the two in the same frame of reference (flow velocity) illustrates their functions in passing or rejecting common types of low-frequency noise (Section IV.B). The functional implications of these distinctions are that canal neuromasts are better suited for detecting high-frequency, transient, or rapidly changing events, whereas superficial neuromasts are better suited for processing low-frequency, sustained (constant velocity), or slow events. The mechanics and hydrodynamics of cupula and canal fluid motion and their functional consequences for sensory processing and encoding are described in greater detail in the following sections.

A. Cupular Mechanics and Hydrodynamic Excitation

Direct laser-interferometric measurement of the submicrometer displacement of canal cupulae in the supraorbital canal of the ruffe (*Gymnocephalus cenuus*) (van Netten and Kroese, 1987) has confirmed earlier assumptions that cupular displacements were proportional to the velocity of fluid motion inside the canal (Denton and Gray, 1983b) at low frequencies, where viscous drag forces are predicted to dominate (Figure 4.5). These measurements also revealed an unexpected cupular resonance that could be explained only by the dominance of inertial fluid forces acting on the cupula at higher frequencies, but still below the electrophysiologically measured cutoff frequency (e.g., van Netten, 1991). The presence and relevance of inertial fluid forces acting on the cupula at higher frequencies can be understood from the hydrodynamic and mechanical properties of cupulae, as described below.

On the basis of comparisons between measured cupular motion in the lateral line canal of ruffe and African knife fish (*Xenomystus nigri*) and a hydrodynamic model of cupular motion (van Netten, 1991), four physical parameters have been identified as important to cupular motion: the viscosity, μ, and density, ρ, of the fluid that excites the cupula and the cupula size (radius), a, and sliding stiffness, K. Cupular excitation in this model is based on periodic Stokes flow around a (hemi)sphere, which is representative of cupulae for which direct measurements of motion have been made (van Netten and Kroese, 1987; Wiersinga-Post and van Netten, 2000). A vibratory fluid flow past the cupula results in a frequency-dependent boundary layer with thickness

$$\delta = \sqrt{\mu / \rho \pi f}, \tag{5}$$

where f is the frequency of the vibratory fluid stimulus and therefore that of the resulting vibration of the cupula. The displacement amplitude of a cupula, X_0, can then be derived in response to an excitatory fluid velocity with constant amplitude, V_0, and frequency f (van Netten, 1991, 2005). The results show that the four physical parameters reduce to only two independent parameters that completely describe the frequency response of a cupula. A direct implication of this is that cupulae with different morphological properties and dimensions may still share the same mechanical frequency response. This may be a contributing factor to the observation that canal neuromasts with different peripheral morphology may yet possess similar, low-pass properties (e.g., Coombs and Montgomery, 1992).

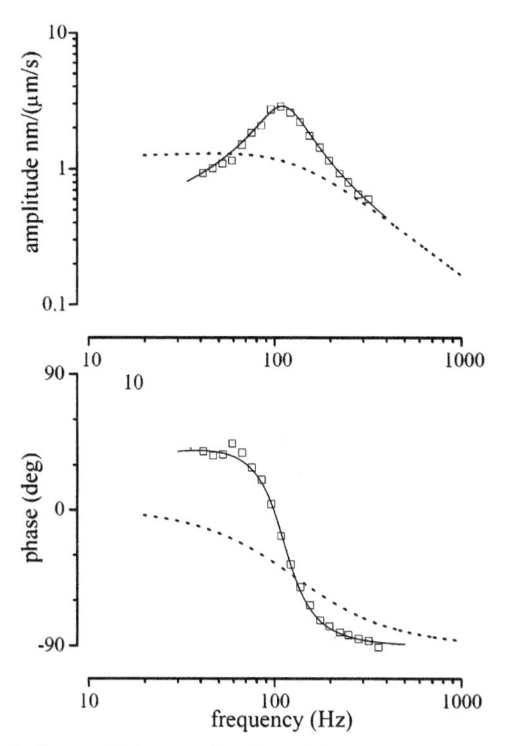

Fig. 4.5. Cupular velocity sensitivity as a function of frequency. Data points show measured amplitude (A) and phase (B) of cupular velocity sensitivity of a supra-orbital canal neuromast of ruffe. Solid lines are fits to the data and correspond to $f_t = 10$ Hz and $N_r = 70$ [Eqs. (6) and (7)]. Using Eq. (8), these data result in a low-frequency sensitivity of 0.23 nm/(μm/s). The resonance at about 110 Hz is in line with $N_r \gg 1$. At resonance, cupular sensitivity (S_r) is about 3 nm/(μm/s), yielding a gain of about 22 dB, ($S_r/S_v \approx 13$), as compared to the low-frequency velocity sensitivity (S_v). The dashed lines predict the sensitivity of a cupula with 50 times fewer underlying hair cells and a 10-fold smaller radius. Such a neuromast can be expected to have a constant velocity sensitivity of about 1 nm/(μm/s) for frequencies up to 60 Hz. (Data from Wiersinga-Post and van Netten, 2000.)

The two independent parameters of modeled cupular motion are a transition frequency,

$$f_t = \mu/2\pi\rho a^2, \tag{6}$$

at which inertial fluid forces become more important than viscous forces, and a dimensionless resonance number, defined by van Netten (1991, 2005):

$$N_r = (Ka\rho)/(6\pi\mu^2). \tag{7}$$

It appears that a cupula has constant low-frequency velocity sensitivity, S_v, in terms of cupular displacement, X_0, per unit fluid velocity, V_0:

$$S_v = X_0/V_0 = (2\pi f_t N_r)^{-1} \tag{8}$$

in a frequency range from DC to the transition frequency f_t. In the case that the resonance number considerably exceeds one ($N_r \gg 1$), deviation from this effective velocity detection starts approximately at frequencies exceeding the transition frequency, f_t, at which inertial fluid forces take over, and results in a maximum sensitivity, S_r, at a resonance frequency, f_r. This resonance frequency can be approximated by $f_r \cong f_t\sqrt{3N_r}$. The extra gain obtained from cupular resonance as compared to the low-frequency sensitivity (S_r/S_v) increases with increasing resonance number and may amount to more than one order of magnitude (see Figure 4.5 and Ćurčić-Blake and van Netten, 2005).

The cupulae investigated so far have all been located in the supra-orbital canal and classified as resonating, with resonance numbers thus largely exceeding 1 (ruffe: $N_r \cong 60$, African knife fish:$N_r \cong 20$, Wiersinga-Post and van Netten, 2000; Clown knife fish: $N_r \cong 170$, van Netten and Khanna, unpublished). The resonance behavior of these cupulae can be explained by their relatively large sizes and numbers of underlying hair cells (>1000), which all impart their hair bundle stiffness to the cupular base. This results in a relatively high product of cupular sliding stiffness and radius, Ka, with a concomitantly large resonance number, $N_r = (Ka\rho)/(6\pi\mu^2)$.

Using the threshold of a single hair cell of a few nanometers [e.g., Eq. (4)], the overall detection limit of a ruffe's canal neuromast can be predicted to be tenths of a nanometer, assuming independent averaging across the several thousands of hair cells underlying a cupula. Together with the range of velocity sensitivities obtained for the ruffe ($S_v = 0.23$–3 nm/(μm/s)), this predicts an optimum velocity sensitivity on the order of 1 μm/s (e.g., van Netten, 2005).

It is interesting to consider the responses that the cupular excitation model predicts if the resonance number is smaller than 1 ($N_r \ll 1$). In that case, the cupula exhibits very little, if any, resonance and is predominantly viscously driven; in essence, it becomes a pure velocity-sensitive detector as implied by the viscous drag hypothesis of cupular excitation. Its velocity sensitivity, $S_v = (2\pi f_t N_r)^{-1}$, is expressed by the same equation as the low-frequency sensitivity of a resonating cupula but is constant over the entire bandwidth extending from DC to a cutoff frequency, f_c, given by $f_c = f_t N_r$. The cupula thus acts as a linear first-order filter (Figure 4.5) with a fixed sensitivity bandwidth product, $S_v f_c = 1/(2\pi)$ (van Netten, 2005). This means that increased sensitivity can be gained at the expense of bandwidth, a trade-off principle that could be used to adapt peripheral lateral line processing to different environmental constraints or demands.

So far, theoretically predicted curves with $N_r \ll 1$ have not been confirmed by direct experimental data on cupular mechanics, but it seems likely that this description may apply to cupulae with relatively small dimensions and/or small numbers of underlying hair cells supplying a small sliding stiffness. From this point of view, this description seems appropriate for superficial neuromasts, since they usually possess smaller numbers of hair cells (e.g., Münz, 1989).

B. Canal Mechanics and Hydrodynamics

In a series of pivotal studies on lateral line canal hydrodynamics, Denton and Gray (1983, 1988, 1989) measured and modeled fluid motions inside actual lateral line canals and canal-like structures (e.g., capillary tubes) as a function of the frequency of sinuosoidal water motions outside the canal. Their model essentially computes the flow impedance outside the canal relative to the impedance inside the canal and is described by the following equation relating water displacements inside the canal (X_{in}) to water displacements of frequency f outside the canal (X_{out}):

$$\frac{X_{in}}{X_{out}} = \frac{j\omega I_{out}}{j\omega I_{in} + R_{in}}, \tag{9}$$

where $\omega = 2\pi f$ and j are the complex operators. I (inertance) represents the inertial component of the acoustic impedance, whereas R (resistance) represents the frictional (viscous) component calculated for steady (Hagen-Poiseuille) flow conditions [see also Eqs. (11)–(13)].

For relatively narrow canals, these investigators demonstrated that flow velocity inside the canal is essentially proportional to the net acceleration between the surrounding water and the fish over a relatively wide range of frequencies (Figure 4.6A; low-frequency flat parts of the curves). Although different (velocity and acceleration) frames of reference show clearly that canal neuromasts respond best to changes in fluid velocity, whereas superficial neuromasts respond best to constant velocity, they tend to obscure the biomechanical filtering action of canals (Montgomery *et al.*, 1994). When the responsiveness of canal neuromast fibers (or canal fluid motions) is plotted as a function of flow velocity outside the canal, keeping velocity amplitude constant at all frequencies, the low-pass nature of this system with respect to outside flow acceleration (Figure 4.6A) is transformed into a high-pass filter with respect to outside flow velocity (Figure 4.6B). The smaller the canal diameter, the more effective the filter is in reducing responsiveness to outside water motions at low frequencies. The high-pass nature of the canal arises from a viscous-dominated resistance to internal flow around the circumference of the canal, as evidenced by low (<10) Reynolds numbers associated with small-diameter canals (see Sections IV.B and IV.C for a computational

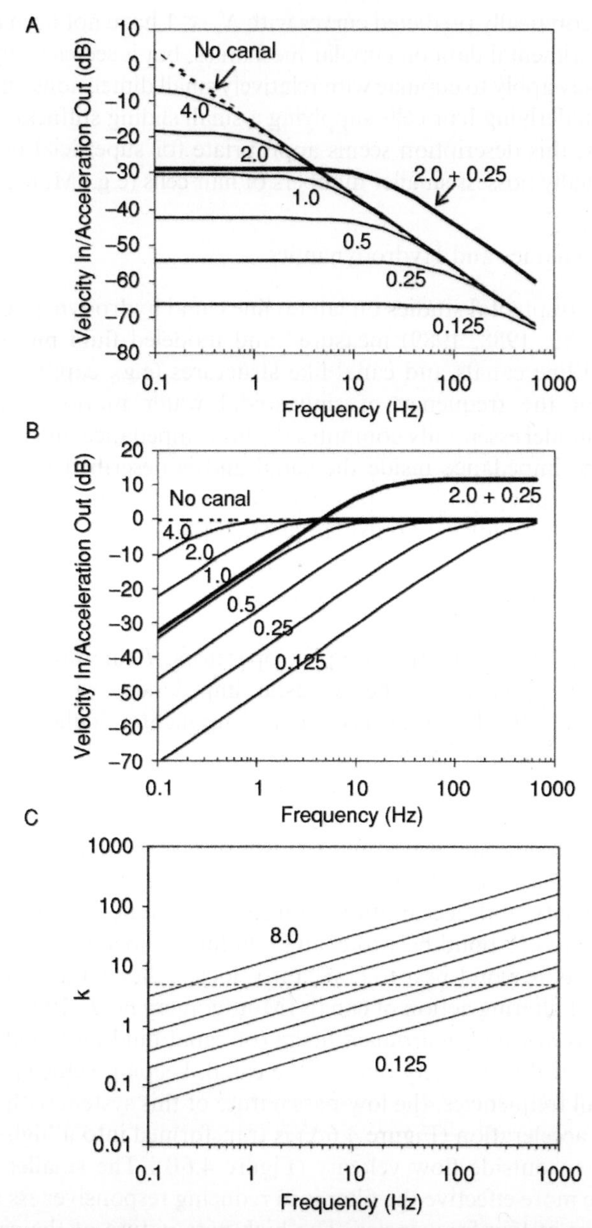

Fig. 4.6. The effects of stimulus frequency and canal diameter on (1) the ratio of flow velocity inside the canal relative to flow acceleration (A) or velocity (B) outside the canal [defined by Eqs. (9), (11), (12), and (13), Sections IV.B and IV.C] and (2) the spatial distribution of flow velocities inside the canal, as reflected by the dimensionless quantity, k (C) [defined in Eq. (10), Section IV.B]. Each

model of this effect). One way to look at this is that smaller canals have increased internal surface area/volume ratio, resulting in boundary layer formation at low frequencies around the entire internal circumference of the canal extending into the center. A neuromast cupula submerged in a thick boundary layer means that the flow velocity along the cupula is significantly reduced relative to that of the excitatory free stream outside the canal. As the frequency increases, however, and viscous forces are dominated by inertial fluid forces in the canal, the boundary layer becomes thinner, so that the cupula is exposed to a less impeded velocity and responsiveness improves.

As Denton and Gray (1989) noted, the assumption for steady (Hagen-Poiseuille) flow conditions is reasonable at low frequencies for the small-diameter (~0.1–0.2 mm) canals they investigated in sprat and herring. At higher frequencies and in wider canals, however, the flow patterns can be more complicated. At these frequencies, their model assumes that the mass-related inertial term [I_{in} in Eq. (9)] dominates. The dimensionless quality

$$k = \sqrt{\frac{2\pi f \rho}{\mu}} r \tag{10}$$

used by Schlichting (1979) to describe oscillating flows in pipes gives a good approximation of whether the parabolic (laminar) flow conditions are met. The value of k depends on the radius, r, of the canal, the frequency of water motion, f, and the fluid viscosity and density, μ and ρ.

When $k \ll 5$, there is a thick boundary layer and the distribution of velocity across the canal is parabolic, since viscous forces dominate. When $k \gg 5$, there is still a thin boundary layer along the canal wall, but a large mass of water in the center of the canal is outside the boundary layer and moves at the same velocity as freestream water outside the canal. In essence, inertial forces have taken over. Therefore, k can be considered the canal-related Reynolds number, demonstrating the relative importance of viscous versus inertial forces. When $k \cong 5$, flow at a given point along the width of the canal is out of phase with that at another, and the maximum velocity is not in the center of the canal. Experiments on fluid flow profiles in lateral line canals have shown transitions from low to high frequencies, with

---

successive function (thin lines) depicts the effects of doubling the canal diameter, which ranges from 0.125 to 8 mm. Dashed lines represent the case in which no canal is present (A, B) or $k = 5$ (C). The heavy solid line in (A) and (B) represents the computed fluid motions inside a constricted portion of the canal where the widest diameter (D_w) = 2 mm, the narrow to wide diameter ratio (D_n/D_w) = 0.25 and the narrow to wide length ratio (L_n/L_w) = 1 (see Section IV.C).

accompanying shifts of the maximum flow from the center to the canal wall (Tsang and van Netten, 1997). As can be seen in Figure 4.6C, k varies from $\ll 5$ to $\gg 5$ for the range of canal diameters and frequencies important to the lateral line system. For canal diameters <0.5 mm and frequencies <100 Hz, k is generally <5, but for canal diameters >1 mm and frequencies >10 Hz, k is generally >5. When Figure 4.6C is compared to Figure 4.6A, it can be seen that k is generally <5 over the range of frequencies at which fluid velocity inside the canal is proportional to fluid acceleration outside the canal. From a signal processing point of view, the canal system is therefore governed by parabolic flow distributions over this range, uncomplicated by the annular flow profile effects associated with $k > 5$ (e.g., van Netten, 2005).

When the frequency responsiveness of both superficial and canal neuromasts is plotted in the same (velocity) frame of reference, the signal-to-noise processing capabilities of these two submodalites can be directly compared (Montgomery *et al.*, 1994). If fish were to sense exclusively through superficial neuromasts, it can be seen that their ability to detect weak, high-frequency signals (e.g., from prey) would be compromised in the presence of high levels of low-frequency noise (e.g., ambient water motions or self-generated breathing or swimming motions) because more of the low-frequency noise than high-frequency signal is passed by the system. Conversely, this ability would be enhanced by canal neuromasts, because most of the low-frequency noise will be rejected, whereas the high-frequency signal will be passed. Recent behavioral experiments indicate that Lake Michigan mottled sculpin (*Cottus bairdi*) can indeed detect relatively weak ($\sim$1–10 μm/s estimated at the location of the fish), high-frequency (50 Hz), prey-like signals in the presence of strong (2–8 cm/s), DC ambient flows (Kanter and Coombs, 2003). This ability is consistent with the high-pass filtering actions of lateral line canals and neurophysiological findings showing that the responsiveness of superficial but not canal neuromast fibers to the same signal is compromised when low-frequency noise (DC ambient flow) is added (Engelmann *et al.*, 2000, 2002).

C. Modeling the Functional Consequences of Morphological Variations

So far, the biomechanical and hydrodynamic features of lateral line canal and superficial neuromasts have been discussed largely as if the structures comprising each of these submodalities exhibit no intra- or interspecific variability. Although the generic preceding descriptions probably give a good first approximation of response properties for many superficial and canal neuromasts in the majority of fishes, there are multiple dimensions along which lateral line structures vary (both within and between species), as reviewed in Coombs *et al.* (1988), Denton and Gray (1988), and Webb

(1989b). The size and shape of a single neuromast and its overlying cupula, as well as the total number of hair cells, can vary; the consequence of this for cupula sliding stiffness and resonance is discussed in Section IV.A. Likewise, lateral line canals can vary in diameter from ~0.1 to 7 mm. Canal walls can be rigid or compliant and with or without pores; canal openings can be singular pores or branched tubules with multiple pores. Thus, the question arises as to whether these structural variations have functional consequences and if so, whether these represent sensory adaptations for, e.g., a particular habitat or lifestyle (e.g., Dijkgraaf, 1963; Coombs *et al.*, 1991).

In this regard, mathematical models of biomechanical and hydrodynamic performance can be extremely informative—if only as a means of generating testable hypotheses. In addition to the mathematical models developed by van Netten and colleagues for describing the frequency response of cupulae of different sizes and sliding stiffnesses (see Section IV.A), canal filter models developed by Denton and Gray (1988) provide sufficient components to simulate some of the morphological variations that have been observed in lateral line canals, including a narrowing of the canal diameter in the vicinity of the neuromast. In this case, the lateral line canal is modeled as a tube of two cross-sectional areas: one wide, a_w, for a given length, l_w, and the other narrow, a_n, for the length, l_n. The inertance (I) and resistance (R) terms in Eq. (9) can thus be computed as follows:

$$I_{out} = \frac{\rho(l_w + l_n)}{a_n} \tag{11}$$

$$I_{in} = \rho\left(\frac{l_w}{a_w} + \frac{l_n}{a_n}\right) \tag{12}$$

$$R_{in} = 8\pi\mu\left(\frac{l_w}{a_w^2} + \frac{l_n}{a_n^2}\right), \tag{13}$$

where density of the surrounding fluid, ρ, is 1000 kg/m^3 and dynamic viscosity, μ, is 0.001 PaS.

For constant diameter canals (the simplified case in which $a_n = a_w$), a decrease in canal diameter leads to a progressive attenuation of responsiveness at low frequencies and an upward shift in the cutoff frequency (CF) when viewed in the velocity frame of reference (Figure 4.6B). Decreasing the diameter of the canal to a narrow section near the neuromast ($D_n = 0.5$ mm) relative to the wider section ($D_w = 2$ mm) causes (1) a further (*extra*) attenuation of response at low frequencies, (2) an upward shift in the CF, and (3) response gain at higher frequencies. These effects can be seen in Figures 4.6A and 4.6B by comparing thick line (constriction present;

$D_n/D_w = 0.25$) and thin line (constriction absent; $D_n/D_w = 1$) functions for the same value of D_w (2 mm).

When the effects of varying the relative lengths (L_n/L_w) in the narrow and wide canal sections are also considered for a given D_w (2 mm) (Figure 4.7), it can be seen that CF is virtually independent of L_n/L_w (Figure 4.7A) and inversely proportional to the square of D_w. (Figure 4.7A). Furthermore, there is a tradeoff between the amount of *extra* low-frequency (LF) attenuation (Figure 4.7B) and high-frequency (HF) amplification (Figure 4.7C) with respect to L_n/L_w. That is, HF gain is maximized when L_n/L_w is minimized (Figure 4.7C), whereas *extra* LF attenuation is maximized when L_n/L_w is maximized (Figure 4.7b). Note also that as D_n/D_w decreases, the amount of HF amplification for any given L_n/L_w approaches asymptotically the value $(L_n/L_w) + 1$, whereas the amount of LF attenuation continues to increase steeply. In general, this means that lateral line canals, whatever their internal geometry, are more efficient in reducing responses to low frequencies than in enhancing responses to high frequencies. Thus, even though levels of *extra* LF attenuation may be very low or even negative for a restricted range of L_n/L_w and D_n/D_w combinations (e.g., $L_n/L_w \leq 0.25$ and $\sim 0.5 > D_n/D_w < 1$; Figure 4.6B) the level of *total* LF attenuation exceeds that of HF gain over the entire range of L_n/L_w and D_n/D_w combinations. It should be noted, however, that resonant structures, such as flexible membranes covering the walls or pores of lateral line canals, may amplify signals in certain frequency ranges (e.g., see Figure 4.23.11 in Denton and Gray, 1988) and thus improve the overall gain, albeit at the sacrifice of temporal resolution.

D. Comparing Models of Lateral Line Biomechanics with Neural Responses

Do morphological variations in canal dimensions reflect diverse functional adaptations in species that show wide variations? This question was addressed in a series of comparative studies on lateral line function in several different species of notothenioid (antarctic) fishes (Coombs and Montgomery, 1992; Montgomery *et al.*, 1994). This is a monophyletic, perciform suborder that exhibits considerable interspecific variation in lateral line morphology, but for which variation due to different phylogenetic origins can be ruled out. Furthermore, environmental conditions during the evolutionary history of this group (e.g., extreme seasonal changes in food availability and light intensity, 4 months of total darkness in winter) argue favorably for functional adaptations in nonvisual sensory systems.

In these studies, lateral line function was assessed at the level of individual nerve fibers by using neurophysiological techniques to measure nerve fiber responsiveness to different frequencies of a sinusoidally vibrating sphere,

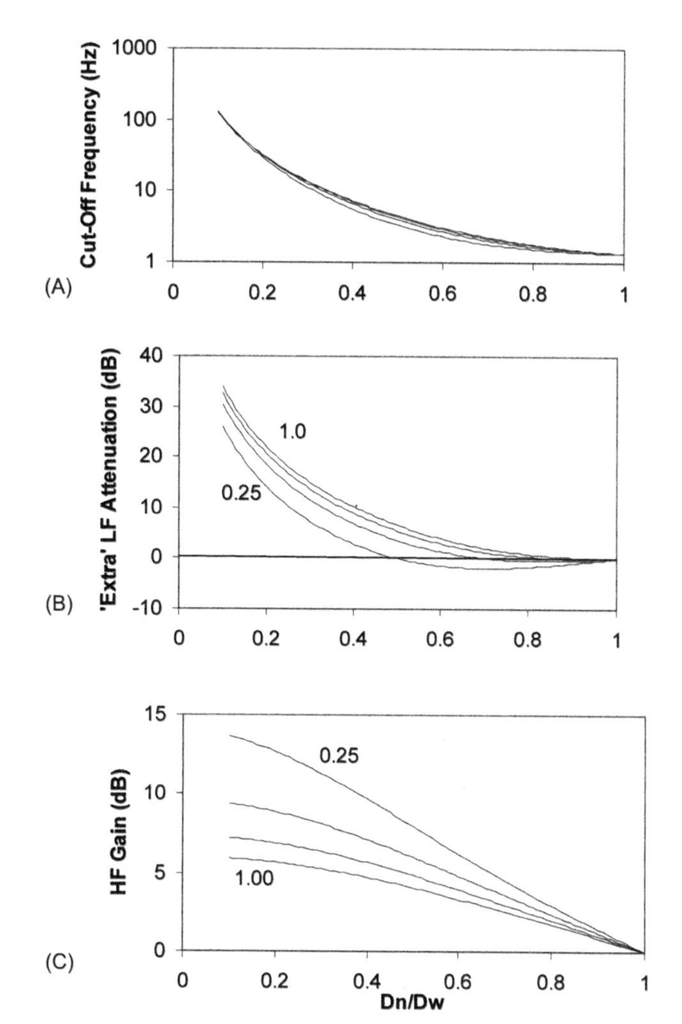

Fig. 4.7. The effects of canal constrictions on the frequency responsiveness of canal fluid motions, as characterized by cutoff frequency, CF (A), extra LF attenuation (B), and HF gain (C). Effects are plotted for the case in which the widest part of the canal (D_w) = 2 mm and the narrow to wide diameter ratios (D_n/D_w) vary from 0.1 to 1 for narrow to wide length ratios (L_n/L_w) of 0.25, 0.5, 0.75, and 1, as computed from Eqs. (11) and (12) in Section IV.C. Note that LF attenuation and HF gain are expressed with respect to values expected from a canal of constant diameter of D_w so as to produce frequency-independent values. Thus, negative LF attenuation values in (B) when $L_n/L_w = 0.25$ and $D_n/D_w \cong 0.5 - 0.9$ mean that the level of *extra* LF attenuation is less than that expected for a canal of the same constant diameter and not that there is overall LF gain relative to outside water motions.

holding acceleration amplitudes constant. In addition, a sufficient number of anatomical dimensions (e.g., canal diameter, neuromast area, hair cell number) were also measured for incorporation into the parametric models of cupular excitation (van Netten, 1991) and canal fluid (Denton and Gray, 1988) motion, as summarized in Sections IV.A and IV.D. Essentially, the Denton and Gray models of canal fluid motion were used as the input to the van Netten model of cupular motion, and the resultant cupular motion was then compared to the measured neural responses at different frequencies (for further modeling details see Coombs and Montgomery, 1992; Montgomery *et al.*, 1994). Because acceleration amplitudes were held constant at suprathreshold levels, the overall shape of the frequency response, rather than sensitivity at each frequency, was measured. Under these circumstances, then, a perfect match between the modeled and measured response would suggest that biomechanical processes at the periphery were sufficient for shaping the neural response. Conversely, a significant mismatch between the two would likely mean that there was additional shaping of the response by intervening neural processes.

Despite marked differences in canal widths, neuromast size, and hair cell number, the frequency response characteristics of nerve fibers innervating mandibular canal neuromasts were remarkably similar across six species belonging to two different families (Montgomery *et al.*, 1994), and even within a single species for neuromasts located in different-sized canals on the head and trunk (Coombs and Montgomery, 1992). That is, lateral line fibers in all cases exhibited responses that were largely proportional to outside fluid acceleration over the low-frequency end (less than ~40 Hz) of the response function. In this respect, the models did a relatively good job of predicting neural responses due to the high-pass (relative to outside velocity) filter action of canals (compare Figures 4.6A and 4.6B). In the case of the giant (~1.5 m SL) Antarctic cod, *Dissostichus mawsoni*, which has a relatively wide-bore mandibular canal (2 mm in diameter) relative to other species examined (<0.5 mm), reduced responsiveness at low frequencies was accomplished with a low D_n/D_w ratio (~0.25) relative to that in other species (~0.6) (see Figures 4.6 and 4.7).

In contrast, the models consistently overestimated by a wide margin the overall bandwidths and high-frequency CFs of neural response functions (plotted in an acceleration frame of reference). Taken together, these findings illustrate two important points. First, signal processing in the lateral line system, as in other octavolateralis systems, involves a cascade of different biomechanical and electrical/neural filters located at different levels in the nervous system. As Figure 4.4 illustrates, there is at least one important, intervening level of neural processing between cupula motion and afferent nerve fiber response that was neither measured nor modeled in these studies:

the hair cell response. This includes (1) a change in the membrane potential at the "incoming" or "receiving" side of the cell and (2) the release of chemical transmitter at the synapse between the hair cell and the innervating fiber at the "outgoing" or "transmission" side of the cell. In this case, the mismatch between measured results (spike activity in afferent nerve fibers) and modeled predictions for canal fluid plus copular motion (the proximal stimulus to canal neuromasts) is most likely due to intervening processes at the level of individual hair cells and may be related to the temperature-dependent kinetics of hair cell tuning (e.g., Wiersinga-Post and van Netten, 2000) in these cold-adapted species (Coombs and Montgomery, 1992). In this regard, it is interesting to note that the resonance behaviors of cupulae modeled and measured by van Netten and colleagues in the supra-orbital canal of ruffe (Section IV.A; Figure 4.5) were likewise not reflected in the response properties of their innervating nerve fibers. Instead, electrophysiological measures of responsiveness revealed an acceleration-sensitive system without tuning (Wubbels, 1992). In this case, the mismatch between cupula motion (modeled and measured) and nerve fiber responsiveness can be understood from the compensatory effects of cupular mechanics and canal filtering (van Netten, 2005). In other words, the overall neural response did not show evidence of any resonance, but instead exhibited a low-pass filter shape (regarding stimulus amplitudes of equal acceleration) with a high-frequency CF that was located between the CF of the canal (Sections IV.B and IV.C) and the resonance frequency of the cupula (see Section IV.A).

The second important point is that reduced responsiveness at low frequencies (with respect to outside velocity amplitudes) appears to be a highly conserved and dominant function of lateral line canals that persists despite considerable variation in canal or cupular morphology. For this reason, it has been argued that at least some of the lateral line canal diversity observed in notothenioid fishes may, in fact, be functionally homogenous and, in that sense, nonadaptive (Coombs and Montgomery, 1994). This view is consistent with computational models showing that a wide range of L_n/L_w and D_n/D_w combinations is effective in reducing responses to low-frequency signals (more so than it is at providing gain at high frequencies) and, hence, in maintaining the basic function of canals as high-pass filters (Figures 4.6B and 4.7). The flip side of this coin is that information about accelerating or higher frequency motions is preserved.

E. Lateral Line Canal Neuromasts as Spatial Filters and Pattern Encoders

Another way of thinking about generic canal biomechanics is that there has to be an external pressure difference across canal pores in order for fluid to flow inside the canal. Since there is typically one neuromast between every two canal pores, the response of each neuromast will be proportional to the pressure difference across the two adjacent pores, which in turn is proportional to the net acceleration between fish and surrounding water at that location (Denton and Gray, 1983a,b, 1988; Kalmijn, 1988, 1989). Conveniently, there appears to be very little, if any, mechanical coupling between fluid motions in adjacent canal segments (Sand, 1981; Denton and Gray, 1983b), meaning that each canal neuromast functions as an independent sensor to sample water motion in a restricted region of space. This idea is further supported by direct fluid flow measurements in a lateral line canal from which the overlying skin was partly removed, showing that fluid flow past a given cupula is insignificantly affected by fluid flow in adjacent canal segments (Tsang, 1997). Each canal neuromast also communicates to the brain with a dedicated set of nerve fibers that do not contact neuromasts in adjacent canal segments (Münz, 1979). Given that interpore distances tend to be a fixed fraction (~0.01–0.02) of standard body length (Coombs, unpublished data), the spatial sampling period of the lateral line canal system is scaled to fish size.

For external sources of stimulation, the pattern of water motions along the body of the receiving fish can vary significantly with the position of the source, as first documented by Denton and Gray (1983a) for an oscillating (dipole) source and by Hassan (1985) for a stationary source immersed in the flow field of a moving fish. In more recent years, Coombs and colleagues have systematically modeled and measured stimulation (pressure difference) patterns to canal neuromast arrays in response to a small dipole source (50 Hz oscillating sphere) at different locations, distances, and orientations relative to the receiving fish (Coombs *et al.*, 1996, 2000). These patterns show a fairly good match with those measured electrophysiologically for canal neuromast nerve fibers when modeling parameters are set to values used under experimental conditions (Coombs *et al.*, 1996). The pattern of stimulation varies not only in terms of pressure difference amplitude but also in terms of pressure difference polarity (positive or negative), which translates into fluid motion inside the canal in one of two opposing directions along the long axis of the canal. Given that canal neuromasts contain two populations of separately innervated, oppositely oriented hair cells, each "poised" to respond best to one of these two directions (Figures 4.2 and 4.3; Section II. B), the system is designed to encode these directional differences.

The pattern of both pressure difference directions and amplitudes along canal neuromast arrays conveys information about different stimulus dimensions. An increase in source distance is encoded by a decrease in the peak amplitude of stimulation and an increase in the overall width of stimulation (Figure 4.8A). Source location is likewise encoded by the relative

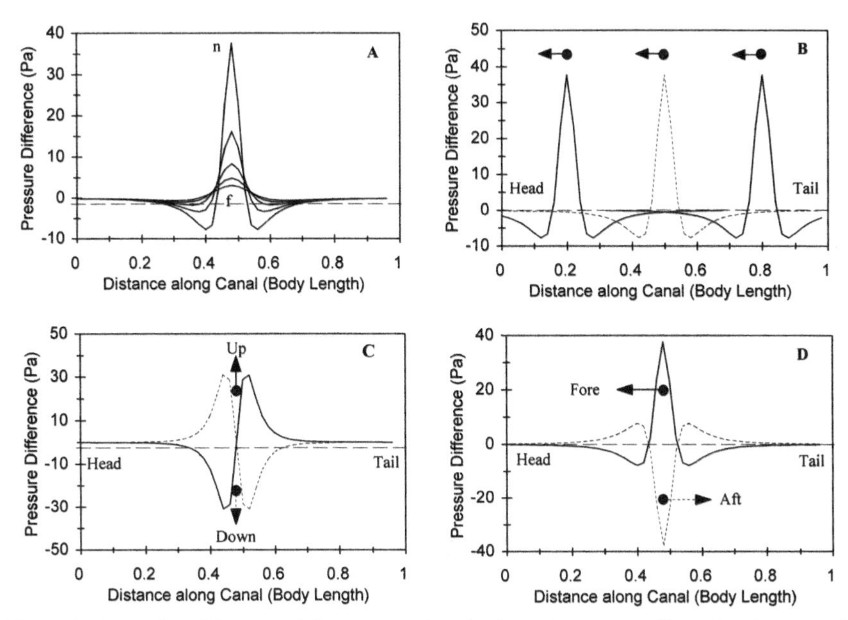

Fig. 4.8. Examples of how spatial patterns of activation along a two-dimensional array of sensors (e.g., canal neuromasts along the trunk canal of fish) can convey different types of information about a vibrating source, including *distance*, from near (n) to far (f) (A), *location*, from head to tail (B), *polarity of movement* (dashed versus solid lines in C, D) and *orientation* (axis of vibration) (head/tail axis in C versus up/down axis in D). Level of activation is expressed as the pressure difference across the two canal pores surrounding each neuromast (see Coombs *et al.*, 1996 for modeling details). (Reprinted with kind permission of Springer-Science and Business Media. From Coombs and Montgomery, 2005.)

location of peak amplitude along the length of the fish (Figure 4.8B). Finally, the axis of source vibration (e.g., parallel or orthogonal to the long axis of the fish) is encoded by the shape of the overall amplitude pattern (compare bimodal patterns in Figure 4.8C to trimodal patterns in Figure 4.8D), whereas the polarity of source vibration (e.g., up/down or fore/aft) is encoded by the pattern of pressure difference directions (compare dashed with solid lines in Figure 4.8C or 4.8D). Recent behavioral experiments have demonstrated that, in fact, canal neuromasts, but not superficial neuromasts, are required for the unconditioned and naturally occurring prey-orienting response of mottled sculpin, *Cottus bairdi* (Coombs *et al.*, 2001a). Thus, for locating small, punctate sources, it would seem that the pattern of activation along lateral line canals is important.

F. Superficial Neuromasts as Spatial Integrators

In contrast to the features of the lateral line canal system that make it ideally suited as a series of high-resolution spatial filters (i.e., decoupling of fluid motions in adjacent canal segments, interpore distances that are a small fraction of the fish's body length, and a dedicated set of nerve fibers for each canal neuromast), superficial neuromasts may be better suited for the spatial integration of surrounding fluid motions for several reasons. First, it is less likely that fluid motions over closely spaced neuromasts on the skin surface are decoupled, although clearly structures like pits or papillae surrounding each neuromast could function in this way. Second, superficial neuromasts are often aligned in rows with several, if not all, of the neuromasts in a given row being innervated by the same nerve fiber (Münz, 1979). Thus, the activation of each nerve fiber depends on the integration of information from hair cells in several adjacent superficial neuromasts. Third, there is good behavioral evidence that superficial, but not canal, neuromasts underlie the rheotactic ability of fish to determine the overall direction of a relatively uniform current (Montgomery *et al.*, 1997). This could best be accomplished by integrating information over a wide range of directionally sensitive sensors to, in effect, average out the small perturbations and deviations from the general direction likely to be present in natural settings.

V. CONCLUDING REMARKS

Biophysical principles play a dominant role in appreciating how the proximal stimulus is conveyed to lateral line neuromasts, and hence what kind of information is extracted by different submodalities (superficial versus canal neuromasts) of this system. In particular, the biophysical properties (e.g., mass, stiffness) of grouped hair cells and their overlying structures (cupula, canal), as well as the relative importance of viscous and inertial forces at different frequencies of surrounding fluid motions, heavily influence the response properties of different modalities and submodalities. The tuning characteristics of the lateral line system, as well as other response properties, arise at several different levels or stages of the mechano-electrical transduction process (Figure 4.4) and involve both macro- and micromolecular events. These include (1) the high-pass filtering action of canals, arising from viscous resistance to canal fluid motion at low frequencies, (2) the low-pass filtering action at the cupula/fluid interface due to viscous drag forces, (3) resonant (band-pass) characteristics of cupula motion determined by the number of underlying hair cell bundles and their combined stiffness, (4) molecular gating of ion channels by changes in stereociliary tip link tensions, and (5) resetting of stereociliary tip link tension by molecular motors during adaptation to a sustained stimulus.

Biomechanical constraints also shape the directional response properties of hair cells. At the level of individual hair cells, the stepwise arrangement of stereocilia and their filamentous links dictates the preferred bending direction. Likewise, directionally selective changes in tip link tensions determine both the direction (on or off) and amplitude (number of open or closed gates) of molecular gates controlling the mechano-electrical transduction process.

In addition to the biomechanical action of lateral line canals as spectral filters to reduce responsiveness to low-frequency signals (and in some cases, to enhance responses to higher frequency signals), the pore-neuromast-pore configurations of canals act as spatial filters to sample different regions of the stimulus field. Whereas spatial stimulation patterns along an array of lateral line canal organs can instantaneously encode information about the location, distance, and polarity of movement of discrete sources, those along superficial neuromasts are more likely to encode low-frequency information about, for example, the general direction of uniform currents.

There are also a multitude of structures associated with the lateral line system that have gone largely uninvestigated from a mechanical and hydrodynamic point of view. These range from large structures, including the overall form of the fish's body and body parts (e.g., fins) and the impact that they will have on flow patterns in the vicinity of the lateral line (e.g., Coombs *et al.*, 2001b), to smaller structures more intimately associated with the lateral line, including the complexly branched tubules associated with lateral line canal pores and the small pits and papillae associated with superficial neuromasts on the skin surface (reviewed in Coombs *et al.*, 1988). In addition, while the mechanical responses of low-lying, rather large canal neuromast cupulae have been measured in several species, those of superficial neuromasts have not. Superficial cupulae in many species tend to have an elongate, flag-like shape, which may therefore be more flexible than canal cupulae and more likely to differentially respond to flow in the boundary layer on the skin surface of the animal. Mechano-physiological experiments on cupulae like these are required to resolve their dynamics into more detail.

Although current views and models of cupula/canal fluid motion and spatial stimulation patterns have greatly increased our understanding of lateral line function, they will no doubt undergo further revisions and refinements as new information becomes available or as more complex and realistic conditions are studied. For example, there is very little information on the composition and viscosity of canal fluids. Likewise, there is a relative paucity of data and models dealing with how the lateral line system encodes and extracts information from vortex structures and in the presence of turbulent flow conditions, despite their pervasiveness in nature. More creative and widespread use of computational fluid dynamics and digital imaging techniques (e.g., digital pratical imaging velocimetry [DPIV]) to describe more complex and biologically realistic flow patterns will likely generate new insights on lateral line function in the future.

ACKNOWLEDGMENTS

Work done by the authors and their colleagues has been supported by the National Science Foundation (NSF), the National Institute of Deafness and Communicative Disorders (NIDCD), and the Office of Naval Research (ONR) (S.C.) and the Netherlands Organization for Scientific Research (NWO) and the School of Behavioral and Cognitive Neurosciences (S.V. N.). We would also like to pay special tribute to Sir Eric Denton and Sir John Gray for their entire body of work and their many insightful contributions to our understanding of the biomechanics and hydrodynamics of the lateral line system.

REFERENCES

Blaxter, J. H. S., Denton, E. J., and Gray, J. A. B. (1981). Acousticolateralis system in clupeid fishes. *In* "Hearing and Sound Communication in Fishes" (Tavolga, W. N., Popper, A. N., and Fay, R. R., Eds.), pp. 39–60. Springer-Verlag, New York.

Braun, C. B., Coombs, S., and Fay, R. R. (2002). What is the nature of multisensory interaction between octavolateralis sub-systems? *Brain Behav. Evol.* **59,** 162–176.

Chan, D. K., and Hudspeth, A. J. (2005). Ca(2+) current-driven nonlinear amplification by the mammalian cochlea in vitro. *Nat. Neurosci.* **8,** 149–155.

Coombs, S., and Janssen, J. (1990). Behavioral and neurophysiological assessment of lateral line sensitivity in the mottled sculpin, *Cottus bairdi. J. Comp. Physiol. A* **167,** 557–567.

Coombs, S., and Montgomery, J. C. (1992). Fibers innervating different parts of the lateral line system of the Antarctic fish, *Trematomus bernacchii,* have similar neural responses despite large variations in peripheral morphology. *Brain Behav. Evol.* **40,** 217–233.

Coombs, S., and Montgomery, J. (1994). Structural diversity in the lateral line system of Antarctic fish: Adaptive or non-adaptive? *Sens. Syst.* **8,** 150–155.

Coombs, S., and Montgomery, J. C. (2005). Comparing octavolateralis sensory systems: What can we learn? *In* "Comparative Hearing: Electroreception" (Popper, A. N., and Fay, R. R., Eds.), pp. 318–359. *Springer Handbook of Auditory Research,* Springer-Verlag, New York.

Coombs, S., Braun, C. B., and Donovan, B. (2001a). Orienting response of Lake Michigan mottled sculpin is mediated by canal neuromasts. *J. Exp. Biol.* **204,** 337–348.

Coombs, S., Anderson, E. J., Braun, C. B., and Grosenbaugh, M. A. (2001b). How fish body parts alter local hydrodynamic stimuli to the lateral line. *SICB Abstracts* **149.**

Coombs, S., Janssen, J., and Webb, J. F. (1988). Diversity of lateral line systems: Evolutionary and functional considerations. *In* "Sensory Biology of Aquatic Animals" (Atema, J., Fay, R. R., Popper, A. N., and Tavolga, W. N., Eds.), pp. 553–593. Springer-Verlag, New York.

Coombs, S., Janssen, J., and Montgomery, J. (1991). Functional and evolutionary implications of peripheral diversity in lateral line systems. *In* "The Evolutionary Biology of Hearing" (Webster, D., Popper, A. N., and Fay, R. R., Eds.), pp. 267–294. Springer-Verlag, New York.

Coombs, S., Hastings, M., and Finneran, J. (1996). Modeling and measuring lateral line excitation patterns to changing dipole source locations. *J. Comp. Physiol.* **176,** 359–371.

Coombs, S., Finneran, J., and Conley, R. A. (2000). Hydrodynamic imaging by the lateral line system of the Lake Michigan mottled sculpin. *Phil. Trans. Roy. Soc. Lond.* **355,** 1111–1114.

Corey, D. P., and Hudspeth, A. J. (1983a). Analysis of the microphonic potential of the bullfrog's sacculus. *J. Neurosci.* **3,** 942–961.

Corey, D. P., and Hudspeth, A. J. (1983b). Kinetics of the receptor current in bullfrog saccular hair cells. *J. Neurosci.* **3,** 962–976.

Corey, D. P., Garcia-Anoveros, J., Holt, J. R., Kwan, K. Y., Lin, S. Y., Vollrath, M. A., Amalfitano, A., Cheung, E. L., Derfler, B. H., Duggan, A., Geleoc, G. S., Gray, P. A., Hoffman, M. P., Rehm, H. L., Tamasauskas, D., and Zhang, D. S. (2004). TRPA1 is a candidate for the mechanosensitive transduction channel of vertebrate hair cells. *Nature* **432,** 723–730.

Crawford, A. C., and Fettiplace, R. (1985). The mechanical properties of ciliary bundles of turtle cochlear hair cells. *J. Physiol.* **364,** 359–379.

Crawford, A. C., Evans, M. G., and Fettiplace, R. (1991). The actions of calcium on the mechano-electrical transducer current of turtle hair cells. *J. Physiol.* **434,** 369–398.

Ćurčić-Blake, B., and van Netten, S. M. (2005). Rapid responses of the cupula in the lateral line of ruffe (Gymnocephalus cernuus). *J. Comp. Physiol. A,* **191,** 393–401.

Denk, W., Holt, J. R., Shepherd, G. M., and Corey, D. P. (1995). Calcium imaging of single stereocilia in hair cells: Localization of transduction channels at both ends of tip links. *Neuron* **15,** 1311–1321.

Denton, E. J., and Gray, J. A. B. (1983a). The rigidity of fish and patterns of lateral line stimulation. *Nature* **297,** 679–681.

Denton, E. J., and Gray, J. A. B. (1983b). Mechanical factors in the excitation of clupeid lateral lines. *Proc. Roy. Soc. Lond. B* **218,** 1–26.

Denton, E. J., and Gray, J. A. B. (1988). Mechanical factors in the excitation of the lateral line of fishes. *In* "Sensory Biology of Aquatic Animals" (Atema, J., Fay, R. R., Popper, A. N., and Tavolga, W. N., Eds.), pp. 595–617. Springer-Verlag, New York.

Denton, E. J., and Gray, J. A. B. (1989). Some observations on the forces acting on neuromasts in fish lateral line canals. *In* "The Mechanosensory Lateral Line: Neurobiology and Evolution" (Coombs, S., Görner, P., and Münz, H., Eds.), pp. 229–246. Springer-Verlag, New York.

De Rosier, D. J., Tilney, L. G., and Egelman, E. (1980). Actin in the inner ear: The remarkable structure of the stereocilium. *Nature* **287,** 291–296.

Dijkgraaf, S. (1963). The functioning and significance of the lateral-line organs. *Biol. Rev.* **38,** 51–105.

Dinklo, T., van Netten, S. M., Marcotti, W., and Kros, C. J. (2003). Signal processing by transducer channels in mammalian outer hair cells. *In* "Biophysics of the Cochlea: From Molecule to Model" (Gummer, A. W., Ed.), pp. 73–79. World Scientific, Singapore.

Eatock, R. A., Corey, D. P., and Hudspeth, A. J. (1987). Adaptation of mechanoelectrical transduction in hair cells of the bullfrog's sacculus. *J. Neurosci.* **7,** 2821–2836.

Engelmann, J., Hanke, W., Mogdans, J., and Bleckmann, H. (2000). Hydrodynamic stimuli and the fish lateral line. *Nature* **40,** 51–52.

Engelmann, J., Hanke, W., and Bleckmann, H. (2002). Lateral line reception in still- and running water. *J. Comp. Physiol. A* **188,** 513–526.

Flock, Å. (1965a). Transducing mechanisms in the lateral line canal organ receptors. *Cold Spring Harbor Symp. Quant. Biol.* **30,** 133–145.

Flock, Å. (1965b). Electron microscopic and electro-physiological studies on the lateral line organ. *Acta Oto*-Laryngol. Suppl. **199,** 1–90.

Flock, Å. (1971). Sensory transduction in hair cells. *In* "Handbook of Sensory Physiology, Vol. 1: Principles of Receptor Physiology" (Lowenstein, W. R., Ed.), pp. 396–441. Springer-Verlag, Berlin.

Flock, Å, and Orman, S. (1983). Micromechanical properties of sensory hairs on receptor cells of the inner ear. *Hearing Res.* **11,** 249–260.

Geisler, C. D. (1993). A model of stereociliary tip-link stretches. *Hearing Res.* **65,** 79–82.

Géléoc, G. S. G., Lennan, G. W. T., Richardson, G. P., and Kros, C. J. (1997). A quantitative comparison of mechanoelectrical transduction in vestibular and auditory hair cells of neonatal mice. *Proc. Roy. Soc. Lond. B Biol. Sci.* **264,** 611–621.

Goodyear, R. J., and Richardson, G. P. (2003). A novel antigen sensitive to calcium chelation that is associated with the tip links and kinocilial links of sensory hair bundles. *J. Neurosci.* **23,** 4878–4887.

Görner, P. (1963). Untersuchungen zur Morphologie und Electrophysiologie des Seitenlinienorgans vom Krallenfrosch (Xenopus laevis Daudin). *Z. Vergl. Physiol.* **47,** 316–338.

Hacohen, N., Assad, J. A., Smith, W. J., and Corey, D. P. (1989). Regulation of tension on hair-cell transduction channels: Displacement and calcium dependence. *J. Neurosci.* **9,** 3988–3997.

Harris, G. G., and Milne, D. C. (1966). Input-output characteristics of the lateral-line sense organs of *Xenopus laevis*. *J. Acoust. Soc. Am.* **40,** 32–42.

Harris, G. G., and van Bergeijk, W. A. (1962). Evidence that the lateral-line organ responds to nearfield displacements of sound sources in water. *J. Acoust. Soc. Am.* **34,** 1831–1841.

Hassan, E. S. (1985). Mathematical analysis of the stimulus for the lateral line organ. *Biol. Cybern.* **52,** 23–36.

Hassan, E. S. (1986). On the discrimination of spatial intervals by the blind cave fish (*Anoptichthys jordani*). *J. Comp. Physiol.* **159A,** 701–710.

Hoekstra, D., and Janssen, J. (1985). Non-visual feeding behavior of the mottled sculpin, *Cottus bairdi*, in Lake Michigan. *Environ. Biol. Fish.* **12,** 111–117.

Hofer, B. (1908). Studien uber die hautsinnesorgane der fische. I. Die funktion der seitenorgane bei den fischen. *Ber. Kgl. Bayer Biol. Versuchsstation Munchen* **1,** 115–164.

Holt, J. R., Gillespie, S. K., Provance, D. W., Shah, K., Shokat, K. M., Corey, D. P., Mercer, J. A., and Gillespie, P. G. (2002). A chemical-genetic strategy implicates myosin-1c in adaptation by hair cells. *Cell* **108,** 371–381.

Howard, J., and Ashmore, J. F. (1986). Stiffness of sensory hair bundles in the sacculus of the frog. *Hearing Res.* **23,** 93–104.

Howard, J., and Hudspeth, A. J. (1988). Compliance of the hair bundle associated with gating of mechanoelectrical transduction channels in the bullfrog's saccular hair cell. *Neuron* **1,** 189–199.

Hudspeth, A. J. (1982). Extracellular current flow and the site of transduction by vertebrate hair cells. *J. Neurosci.* **2,** 1–10.

Hudspeth, A. J., and Gillespie, P. G. (1994). Pulling springs to tune transduction: Adaptation by hair cells. *Neuron* **12,** 1–9.

Hudspeth, A. J., Choe, Y., Mehta, A. D., and Martin, P. (2000). Putting ion channels to work: Mechanoelectrical transduction, adaptation, and amplification by hair cells. *Proc. Natl. Acad. Sci. USA* **97,** 11765–11772.

Jielof, R., Spoor, A., and de Vries, H. (1952). The microphonic activity of the lateral line. *J. Physiol.* **116,** 137–157.

Kachar, B., Parakkal, M., Kurc, M., Zhao, Y, and Gillespie, P. G. (2000). High-resolution structure of hair-cell tip links. *Proc. Natl. Acad. Sci. USA* **97,** 13336–13341.

Kalmijn, A. J. (1988). Hydrodynamic and acoustic field detection. *In* "Sensory Biology of Aquatic Animals" (Atema, J., Fay, R. R., Popper, A. N., and Tavolga, W. N., Eds.), pp. 83–130. Springer-Verlag, New York.

Kalmijn, A. J. (1989). Functional evolution of lateral line and inner-ear sensory systems. *In* "The Mechanosensory Lateral Line: Neurobiology and Evolution" (Coombs, S., Görner, P., and Münz, H., Eds.), pp. 187–215. Springer-Verlag, New York.

Kanter, M., and Coombs, S. (2003). Rheotaxis and prey detection in uniform currents by Lake Michigan mottled sculpin (*Cottus bairdi*). *J. Exp. Biol.* **206,** 59–60.

Kelly, J. P., and van Netten, S. M. (1991). Topography and mechanics of the cupula in the fish lateral line. Variations of cupular structure and composition in three dimensions. *J. Morphol.* **207,** 23–36.

Kroese, A. B. A., and Schellart, N. (1992). Velocity- and acceleration-sensitive units in the trunk lateral line of the trout. *J. Neurophysiol.* **68,** 2212–2221.

Kroese, A. B. A., van der Zalm, J. M., and Van den Bercken, J. (1978). Frequency response of the lateral-line organ of *Xenopus laevis*. *Pflueg. Arch.* **375,** 167–175.

Kros, C. J., Rüsch, A., and Richardson, G. P. (1992). Mechano-electrical transducer currents in hair cells of the cultured neonatal mouse cochlea. *Proc. Roy. Soc. Lond. B Biol. Sci.* **249,** 185–193.

Kros, C. J., Marcotti, W., van Netten, S. M., Self, T. J., Libby, R. T., Brown, S. D., Richardson, G. P., and Steel, K. P. (2002). Reduced climbing and increased slipping adaptation in cochlear hair cells of mice with Myo7a mutations. *Nat. Neurosci.* **5,** 41–47.

Markin, V. S., and Hudspeth, A. J. (1995). Gating-spring models of mechanoelectrical transduction by hair cells of the internal ear. *Annu. Rev. Biophys. Biomol. Struct.* **24,** 59–83.

Maruska, K. P., and Tricas, T. C. (2004). The mechanotactile hypothesis: Neuromast morphology and response dynamics of mechanosensory lateral line primary afferents in stingray. *J. Exp. Biol.* **207,** 3463–3476.

Montgomery, J. C., and Coombs, S. (1992). Physiological characterization of lateral line function in the Antarctic fish (*Trematomus bernacchii*). *Brain Behav. Evol.* **40,** 209–216.

Montgomery, J. C., Coombs, S., and Janssen, J. (1994). Form and function relationships in lateral line systems: Comparative data from six species of Antarctic notothenioid fish. *Brain Behav. Evol.* **44,** 299–306.

Montgomery, J., Baker, C., and Carton, A. (1997). The lateral line can mediate rheotaxis in fish. *Nature* **389,** 960–963.

Münz, H. (1979). Morphology and innervation of the lateral line system in *Sarotherodon niloticus* L. (Cichlidae, Teleostei). *Zoomorpholgie* **93,** 73–86.

Münz, H. (1989). Functional organization of the lateral line periphery. *In* "The Mechanosensory Lateral Line: Neurobiology And Evolution" (Coombs, S., Görner, P., and Münz, H., Eds.), pp. 285–297. Springer-Verlag, New York.

Murray, R. W. (1955). The lateralis organs and their innervation in *Xenopus laevis*. *Quart. J. Micr. Sc.* **96,** 351–361.

Nicolson, T., Rusch, A., Friedrich, R. W., Granato, M., Ruppersberg, J. P., and Nusslein-Volhard, C. (1998). Genetic analysis of vertebrate sensory hair cell mechanosensation: the zebrafish circler mutants. *Neuron* **20,** 271–283.

Northcutt, R. G. (1989). The phylogenetic distribution and innervation of craniate mechanoreceptive lateral lines. *In* "The Mechanosensory Lateral Line: Neurobiology and Evolution" (Coombs, S., Görner, P., and Münz, H., Eds.), pp. 17–78. Springer-Verlag, New York.

Osborne, M. P., Comis, S. D., and Pickles, J. O. (1984). Morphology and cross-linkage of stereocilia in the guinea pig labyrinth examined without the use of osmium as a fixative. *Cell Tissue Res.* **237,** 43–48.

Partridge, B., and Pitcher, T. J. (1980). The sensory basis of fish schools: Relative roles of lateral line and vision. *J. Comp. Physiol.* **135,** 315–325.
Pickles, J. O. (1993). A model for the mechanics of the stereociliar bundle on acousticolateral hair cells. *Hearing Res.* **68,** 15972.
Pickles, J. O., Comis, S. D., and Osborne, M. P. (1984). Cross-links between stereocilia in the guinea pig organ of Corti, and their possible relation to sensory transduction. *Hearing Res.* **15,** 103–112.
Pickles, J. O., Brix, J., Comis, S. D., Gleich, O., Koppl, C., Manley, G. A., and Osborne, M. P. (1989). The organization of tip links and stereocilia on hair cells of bird and lizard basilar papillae. *Hearing Res.* **41,** 31–41.
Pickles, J. O., Rouse, G. W., and von Perger, M. (1991). Morphological correlates of mechanotransduction in acousticolateral hair cells. *Scanning Microsc.* **5,** 1115–1128.
Ricci, A. J., Crawford, A. C., and Fettiplace, R. (2002). Mechanisms of active hair bundle motion in auditory hair cells. *J. Neurosci.* **22,** 44–52.
Roberts, B. L., and Meredith, G. E. (1989). The efferent system. *In* "The Mechanosensory Lateral Line: Neurobiology and Evolution," (Coombs, S., Görner, P., and Münz, H., Eds.), pp. 445–460. Springer-Verlag, New York.
Rouse, G. W., and Pickles, J. O. (1991). Paired development of hair cells in neuromasts of the teleost lateral line. *Proc. Roy. Soc. Lond. B Biol. Sci.* **246,** 123–128.
Russell, I. J., Kossl, M., and Richardson, G. P. (1992). Nonlinear mechanical responses of mouse cochlear hair bundles. *Proc. Roy. Soc. Lond. B Biol. Sci.* **250,** 217–227.
Sand, O. (1981). The lateral line and sound reception. *In* "Hearing and Sound Communication in Fishes" (Tavolga, W. N., Popper, A. N., and Fay, R. R., Eds.), pp. 459–480. Springer-Verlag, New York.
Satou, M., Takeuchi, H. A., Tanabe, M., Kitamura, S., Okumoto, N., Iwata, M., and Nishii, J. (1994). Behavioral and electrophysiological evidences that the lateral line is involved in the inter-sexual vibrational communication of the hime salmon (landlocked red salmon, *Oncorhynchus nerka). J. Comp. Physiol. A—Sens. Neur. Behav. Physiol.* **174,** 539–549.
Saunders, J. C., and Szymko, Y. M. (1989). The design, calibration, and use of a water microjet for stimulating hair cell sensory hair bundles. *J. Acoust. Soc. Am.* **86,** 1797–1804.
Schlichting, H. (1979). "Boundary Layer Theory" (7th edn.), McGraw-Hill, New York.
Sidi, S., Friedrich, R. W., and Nicolson, T. (2003). NompC TRP channel required for vertebrate sensory hair cell mechanotransduction. *Science* **301,** 96–99.
Siemens, J., Lillo, C., Dumont, R. A., Reynolds, A., Williams, D. S., Gillespie, P. G., and Muller, U. (2004). Cadherin 23 is a component of the tip link in hair-cell stereocilia. *Nature* **428,** 950–955.
Söllner, C., Rauch, G. J., Siemens, J., Geisler, R., Schuster, S. C., the Tübingen 2000 Screen Consortium, Müller, U., and Nicolson, T. (2004). Mutations in cadherin 23 affect tip links in zebrafish sensory hair cells. *Nature* **428,** 955–959.
Strelioff, D., and Flock, Å. (1984). Stiffness of sensory-cell hair bundles in the isolated guinea pig cochlea. *Hearing Res.* **15,** 19–28.
Sutterlin, A. M., and Waddy, S. (1975). Possible role of the posterior lateral line in obstacle entrainment by brook trout (*Salvelinus fontinalis*). *J. Fish. Res. Bd. Can.* **32,** 2441–2446.
Tsang, P. T. S. K. (1997). Laser interferometric flow measurements in the lateral line organ. Ph.D. thesis, University of Groningen, Netherlands.
Tsang, P. T. S. K., and van Netten, S. M. (1997). Fluid flow profiles measured in the supraorbital lateral line canal of the ruff *In* "Diversity in Auditory Mechanics" (Lewis, E. R., Long, G. R., Lyon, R. F., Narris, P. M., Steele, C. R., and Hecht-Poinar, E., Eds.), pp. 25–31. World Scientific, Singapore.

van Netten, S. M. (1991). Hydrodynamics of the excitation of the cupula in the fish canal lateral line. *J. Acoust. Soc. Am.* **89,** 310–319.
van Netten, S. M. (1997). Hair cell mechano-transduction: Its influence on the gross mechanical characteristics of a hair cell organ. *Biophys. Chem.* **68,** 43–52.
van Netten, S. M. (2005). Hydrodynamic detection by cupulae in a lateral line canal: Functional relations between physics and physiology. *Biol. Cybern.* in press.
van Netten, S. M., and Khanna, S. M. (1994). Stiffness changes of the cupula associated with the mechanics of hair cells in the fish lateral line. *Proc. Natl. Acad. Sci. USA* **91,** 1549–1553.
van Netten, S. M., and Kroese, A. B. A. (1987). Laser interferometric measurements on the dynamic behavior of the cupula in the fish lateral line. *Hearing Res.* **29,** 55–61.
van Netten, S. M., and Kroese, A. B. A. (1989). Dynamic behavior and micromechanical properties of the cupula. *In* "The Mechanosensory Lateral Line: Neurobiology and Evolution" (Coombs, S., Görner, P., and Münz, H., Eds.), pp. 247–263. Springer-Verlag, New York.
van Netten, S. M., and Kros, C. J. (2000). Gating energies and forces of the mammalian hair cell transducer channel and related hair bundle mechanics. *Proc. Roy. Soc. Lond. B Biol. Sci.* **267,** 1915–1923.
van Netten, S. M., Dinklo, T., Marcotti, W., and Kros, C. J. (2003). Channel gating forces govern accuracy of mechano-electrical transduction in hair cells. *Proc. Natl. Acad. Sci. USA* **100,** 15510–15515.
Voigt, R., Carton, A. G., and Montgomery, J. C. (2000). Responses of lateral line afferent neurons to water flow. *J. Exp. Biol.* **203,** 2495–2502.
Webb, J. F. (1989a). Developmental constraints and evolution of the lateral line system in teleost fishes. *In* "The Mechanosensory Lateral Line: Neurobiology and Evolution" (Coombs, S., Görner, P., and Münz, H., Eds.), pp. 79–99. Springer-Verlag, New York.
Webb, J. F. (1989b). Gross morphology and evolution of the mechanoreceptive lateral line system in teleost fish. *Brain Behav. Evol.* **33,** 34–53.
Webb, J. F. (2000). Mechanosensory lateral line: Microscopic anatomy and development. *In* "The Handbook of Experimental Animals—The Laboratory Fish" (Ostrander, G. K., Ed.), pp. 463–470. Academic Press, San Diego.
Wiersinga-Post, J. E. C., and van Netten, S. M. (1998). Amiloride causes changes in the mechanical properties of hair cell bundles in the fish lateral line similar to those induced by dihydrostreptomycin. *Proc. Roy. Soc. Lond. B* **265,** 615–623.
Wiersinga-Post, J. E. C., and van Netten, S. M. (2000). Temperature dependency of cupular mechanics and hair cell frequency selectivity in the fish canal lateral line organ. *J. Comp. Physiol.* **186,** 949–956.
Wubbels, R. J. (1992). Afferent response of a head canal neuromast of the Ruff (Acerina cernua) lateral line. *Comp. Biochem. Physiol.* **102A,** 19–26.
Zhao, Y., Yamoah, E. N., and Gillespie, P. G. (1996). Regeneration of broken tip links and restoration of mechanical transduction in hair cells. *Proc. Natl. Acad. Sci. USA* **93,** 15469–15474.

5

SKIN AND BONES, SINEW AND GRISTLE: THE MECHANICAL BEHAVIOR OF FISH SKELETAL TISSUES

ADAM P. SUMMERS
JOHN H. LONG, JR.

I. INTRODUCTION

The mechanical workings of fishes have been approached in a two-pronged framework, with (1) muscle as the engine of motion and force and (2) water as the external source of resistance and purchase. The interaction of muscle and water certainly lays the foundation for behaviors as diverse as swimming, breathing, and feeding, but the interaction between them is only part of the picture. Understanding fishes as mechanical actors requires study of a third factor: the skeleton. Here we define skeleton broadly to include connective tissues such as tendon, ligament, cartilage, and bone that have a large component of extracellular collagen fibers. These connective tissues form skeletal structures as diverse as joints, myosepta, skin, scales, and bones. How these structures behave mechanically—how they reconfigure

Fish Biomechanics: Volume 23
FISH PHYSIOLOGY

DOI: 10.1016/S1546-5098(05)23005-4

in response to the physical loads applied by muscle and water—is the focus of this chapter.

As implied by the use of the terms "tissue" and "structure," we study mechanical behavior by analyzing a skeletal system's material (the tissue) and its shape (the structure). While the contributions of material and shape can be examined separately, it is important to note that only in conjunction do they predict mechanical behavior and, hence, function. This can be shown in a simplified way. Engineers have developed a theory that predicts the downward deflection (y, expressed in meters, m) in a cantilevered beam of length l (in meters), when a weight, F (in Newtons, N), is placed on its end (Figure 5.1). This beam will bend as the load is applied and will eventually come to rest in a new shape predicted by

$$y = \frac{Fl^3}{3EI} \tag{1}$$

What should strike you immediately is the intuitively satisfying relation between deflection and the applied weight. Also intuitive is the prediction that deflection will increase if the length of the beam increases; less intuitive is the fact that this will happen in proportion to the cube of the beam's length. The deflection is predicted to be inversely proportional to two other features of the beam, E, its material stiffness or Young's modulus (in Pa or Nm^{-2}) and, I, its second moment of area (m^4). Together, E and I form EI, a composite variable known as flexural stiffness (in Nm^2). The deflection of the beam can thus be said to be inversely proportional to the beam's flexural stiffness.

But what is this flexural stiffness? In an engineering beam, such as a steel girder, the material stiffness refers to the shape-independent properties of

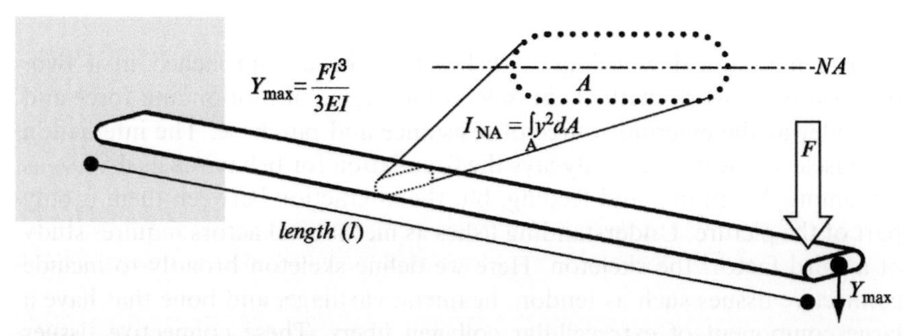

Fig. 5.1. Bending in a beam supported at one end (cantilevered), a situation common in fish skeletal elements. The deflection Y_{max} is dependant on the length of the element (l), the applied force (F), the stiffness of the material from which it is made (E), and the second moment of cross-sectional area of the material (I). The second moment relative to the neutral axis (I_{NA}) is a measure of the area weighted by the distance from the neutral axis (NA).

steel. In contrast, the second moment of area refers to the material-independent properties of the beam's shape in cross-section. Taken together, E and I quantify the combined impact of material and cross-sectional shape, of tissue and structure. To complete the structural view, we need to add the spatial dimension perpendicular to the cross-section, namely, length. To sum up, in this simplified situation the beam's mechanical behavior, deflection, is dependent on the beam's material and shape.

While engineering theory *is* applicable to biological systems, one must take care to recognize that fishes, compared to steel beams, have skeletal structures, tissues, and force environments that are more complicated. Even supposedly simple structures like fin rays—which appear at first glance to be candidate cantilevers—have shapes of tapering thickness and bilateral branches (Lundberg and Baskin, 1969; Videler, 1974; Madden and Lauder, 2003; Thorsen and Westneat, 2005). Fin rays are bent by forces that arise from antagonistic muscles, adjacent rays, and the surrounding water (Videler, 1974; Lundberg and Marsh, 1976; Blake, 1979; Geerlink and Videler, 1987; Daniel, 1988). Moreover, fin rays are in nearly constant motion, reconfiguring and never achieving the kind of "at rest" equilibrium required in Eq. (1) (Webb, 1973; Blake, 1983; Geerlink, 1983; see also Chapter 10 in this volume).

Even though the skeletal systems of fishes are complicated, analysis is helped by the often clear connection between skeletal structure and mechanical function, particularly when that correlation has convergently evolved. A clear causal connection is easily seen between the teeth and prey processing in heterodontid sharks and in sparid fish: their robust molariform teeth permit the crushing of hard prey such as molluscs and echinoderms (Smith, 1942; Hernandez and Motta, 1997). Scombriods (tunas and marlins) and lamnid sharks have independently evolved a lunate tail with a narrow peduncle, a structural combination that functions to generate high-speed swimming through lift-based thrust production (Bernal *et al.*, 2001; Donely *et al.*, 2004).

However, without guidance from clear causal connections and convergent evolution, it is more difficult to infer mechanical function. For example, what is the mechanical function of the vertical septum (Videler, 1993)? Or intermuscular bones (Gemballa and Britz, 1998)? Do amiid fish retain a gular plate in their lower jaw for some mechanical function (Grande and Bemis, 1998)? Why have living lungfishes lost bony vertebral centra (Bemis, 1984)? To circumscribe the range of possible functions these skeletal structures might convey to the organism, the analysis of mechanical behavior has been used to specify the structure's functional capacities. In this chapter, we review the functional capacities of fish skeletal structures and tissues, as those capacities have been determined by the measurement of mechanical properties under life-like loading conditions. An important note about what we mean when we use the term "fishes": "fishes" is broadly construed as including two lineages of jawless fishes (Myxiniformes and Petromyzontiformes), the cartilaginous fishes (Chondrichthyes), and two lineages of bony fishes (Actinopterygians and Sarcopterygians), but not tetrapods (Liem *et al.*, 2001). No phylogenetic monophyly is implied. Given this wide evolutionary territory, fishes provide a fertile ground for comparisons among disparate groups that have had the opportunity for large timescale-independent character acquisition. Since the tetrapod lineage is nested within the Sarcopterygians, the most well-understood tissues (those of rodents and humans) provide both a useful comparison to other fish lineages and a powerful motivation to understand the response of tissues to load in other animals.

II. A PRIMER ON MECHANICAL BEHAVIOR

This section is intended to introduce to the novice some of the basic concepts that are needed to understand the mechanical properties and behavior of skeletal systems. If you know the difference between stress and strain, force and moment, stiffness and compliance, then we suggest that you jump straight to Section III. If the brief treatment offered here is insufficient, we recommend several well-written texts on material properties and skeletons that deal with the derivation and interdependence of material and structural properties, including Martin *et al.* (1998), Vincent (1992), and Gordon (1984).

Consider a blue marlin doing its spectacular, rattling tail walk as it tries to throw a hook (Rockwell *et al.*, 1938). Its lunate tail, as it slashes back and forth through the water, deforms, and the extent of that deformation is critical to the marlin's performance. Too wimpy a tail, and its lobes will collapse and fail to hold the marlin up out of the water. Too rigid a tail, and the lobes might break if they allow the marlin to generate too much force. A tail just right for that mechanical situation will be stiff enough to avoid collapse, yet flexible enough to change shape without breaking. As illustrated in Eq. (1), we can talk formally about the tail's flexural stiffness, EI. If we are interested in how EI might determine the mechanical behavior of the tail, and we judge that the tail's flexibility is the behavior we are after, then we can rearrange Eq. (1) as follows:

$$\frac{y}{F} = \frac{l^3}{3EI} \tag{2}$$

In words, we can say that we have defined the mechanical behavior of flexibility as the ratio of the spanwise displacement of the tail tip, y, to the force loading the tail in that direction, F (recall that this equation is for a force acting at the tip of a cantilever; appropriate adjustments for a distributed load on the tail can be found in most engineering texts on static mechanics). Equation (2) allows us to predict that flexibility will decrease as the flexural stiffness, EI, of the tail increases. Also, small increases in the tail's spanwise length, l, will have a disproportionately large impact on flexibility because of l's cubic exponent. For the tail-walking marlin, this means that to have a lunate, lift-generating tail with its extremely long span length, the tail will need to have a very high flexural stiffness to avoid collapse.

Flexural stiffness, as mentioned in Section I, is a composite variable with a shape-independent factor, E, the Young's modulus or material stiffness, and a material-independent factor, I, the second moment of area. To understand Young's modulus, we need to consider the relation between stress and strain in a simplified situation. Consider a cylinder of a uniform solid that is being pulled on at either end (Figure 5.2). Stress, σ (in Pascals or Nm^{-2}), is the force, F, per unit cross-sectional area, A (in m^2) of the cylinder. Strain, ϵ (no units), is the deformation or reconfiguration of the cylinder relative to its resting length. An experiment in which the strain is gradually increased while the stress is recorded would yield a stress–strain curve similar to Figure 5.2. Notice that over the linear portion of the curve the material's response to load can be described with the equation $\sigma = E\epsilon$, in which E, which has units of Pascals, is referred to as Young's modulus, or, in the case of this pulling experiment, tensile material stiffness. If you alter the experiment by pushing on the cylinder, you would be measuring a compressive material stiffness. When the curve is no longer linear, we can still find E at any point by recognizing that E is the instantaneous slope:

$$E = \frac{d\sigma}{d\epsilon} \tag{3}$$

Because the E in tension can differ from the E in compression, it is important in our case to measure an E that is appropriate to bending. To do so, we can use the setup suggested in Eq. (1). Keep in mind that unlike the linear force that creates tension and compression, bending is created by a force couple, a torque, that creates a bending moment, M (in Nm), from the combination of a force acting to rotate through the length of a moment or lever arm. The bending moment, M, causes the cylinder to bend, that is, to acquire curvature, κ (in m^{-1}) in proportion to the bending material stiffness:

$$M = E\kappa \tag{4}$$

Again, to recognize that we will most often see a nonlinear relation between moment and curvature, the bending material stiffness is generally represented as the instantaneous slope:

$$E = \frac{dM}{d\kappa} \tag{5}$$

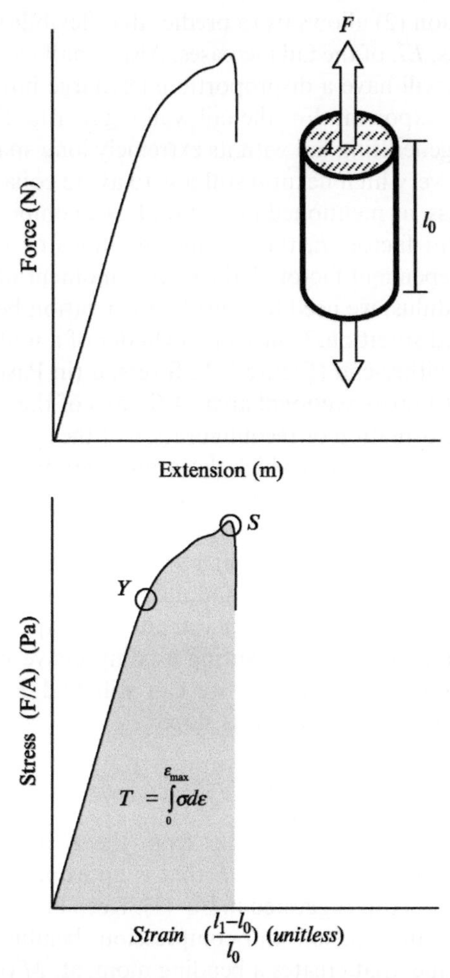

Fig. 5.2. A cylinder of a homogeneous solid is subjected to tensile force causing it to elongate from l0 to l1. The force extension curve is the output of an LVDT and force transducer. To remove sample dimensions from consideration, the stress (force per unit area) is plotted against the strain (% elongation). In this example, the material will return to its original length as long as strain does not exceed Y, the yield point. The slope of the line to Y is the elastic, or Young's modulus, and the stress at S is the ultimate strength. The stress at Y is called the yield strength. The toughness (T) is the area under the curve or the integral of stress.

Keep in mind that in most biological structures and tissues, E, of any of the three kinds just mentioned, will likely vary with strain (or curvature), the rate of strain, and temperature (Vincent, 1990).

Material stiffness is not the only mechanical property that can be gleaned from Figure 5.2. In the linear region of the curve, the cylinder will return to its original shape if the load is removed; this is called the elastic region. Beyond that region the cylinder will deform plastically, remaining deformed once the stress is released. The stress at the elastic/plastic transition point is called the yield strength. Yield strength is biologically relevant for skeletal tissues because in the plastic region damage is occurring that must be later repaired. Beyond this plastic region (and in brittle skeletal tissues this region is insignificantly short), the material breaks at a stress called the ultimate strength. The area under the stress–strain curve up to the point of fracture is called the toughness, T (in Joules) ($T = \int_0^{\epsilon_{max}} \sigma d\epsilon$), and it represents the mechanical work that must be done to break the material. Toughness is a property of pressing interest for those fishes that crush hard prey; it is a measure of how much energy they will have to invest (at a minimum) to get their prey out of its shell.

From a technical perspective, the characteristics of biological materials are often difficult to measure. All biological materials are composites, made up of a heterogeneous mixture of disparate materials, each playing some role in the response of the tissue to load. The particulars of the composite nature vary widely with the location and orientation of a selected sample—a square of skin from the flank of a fish will be quite different from a similar patch taken from the head. Heterogeneity in the ultra- and microstructure of tissues leads to anisotropy: variation in material properties dependent on sample orientation. This greatly complicates the process of determining material properties and also muddies the interpretation of results, as they depend heavily on understanding the magnitude and direction of biologically relevant loads. Fortunately, as biologists we are usually concerned with responses to narrowly defined loading regimes rather than the general responses that interest engineers. In other words, we seek to describe a structure's or tissue's mechanical behavior that is physiologically relevant.

Another difficulty is posed by the nature of biological materials. Though fishes live in a fluid medium and appear to anyone who has handled them to be quite solid, there are few tissues in the fish (or any animal) that nicely fit into the category of solid or fluid. Teeth and tears are very nearly the only examples of a biological solid and fluid. The rest of the tissues, and indeed the whole fish, exhibit a mixture of solid and fluid properties and are therefore "viscoelastic" materials.

Solids and fluids respond to loads in radically different fashion. Any shearing load that does not irreversibly damage a solid causes it to deform elastically—when the load is removed the solid will regain its original shape. In contrast, a fluid continuously deforms under any shear load, no matter how slight. In solids, the amount of strain determines stress, but for fluids,

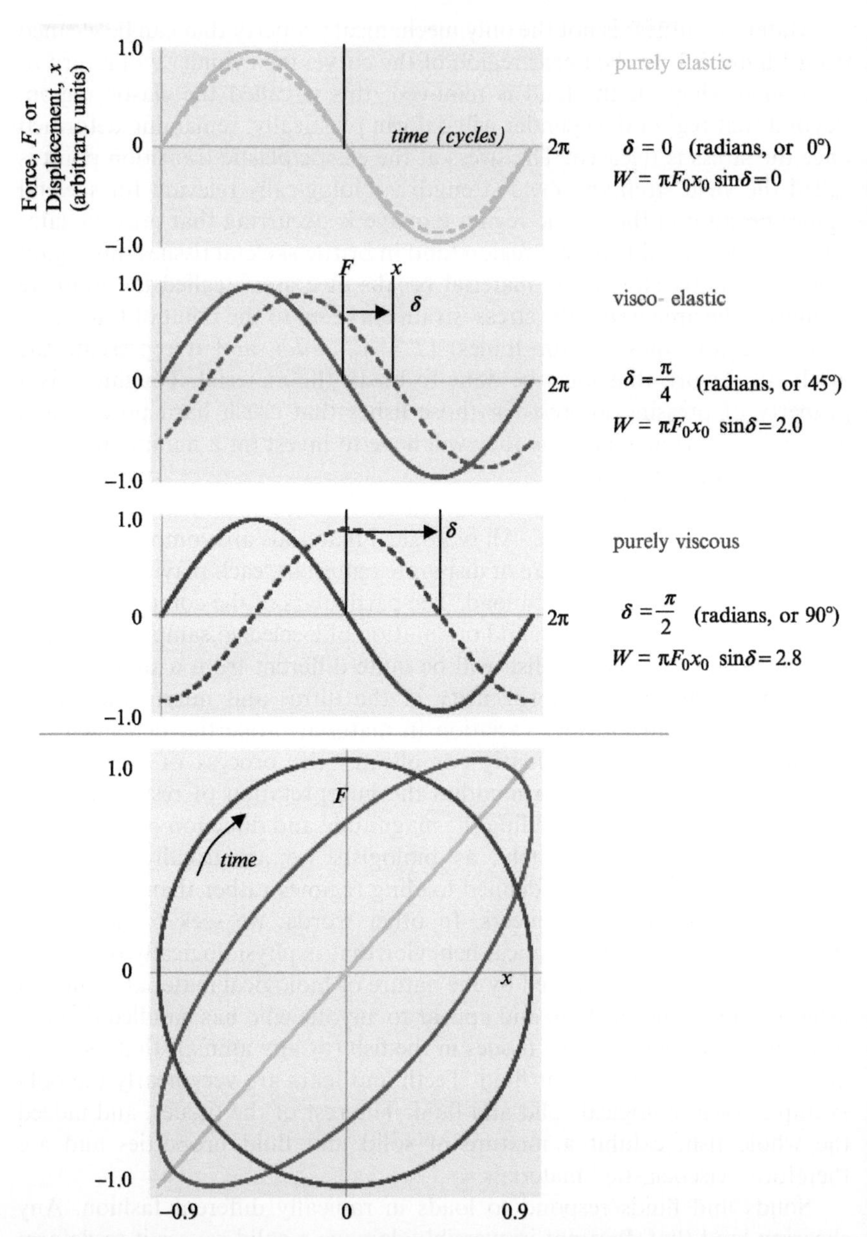

Fig. 5.3. Dynamic, time-dependent mechanical behavior can be defined by the phase angle, δ, between the sinusoidal driving force, F, that causes a tissue's or structure's sinusoidal reconfiguration (measured as displacement), x . In a purely elastic system, $\delta = 0$ radians and x and F occur at the same time, as measured by the zero crossings. Over a full cycle of 2π radians, no net

the rate of strain ($\frac{du}{dy}$) determines stress such that $\sigma = \eta \frac{du}{dy}$, where η is the coefficient of dynamic viscosity in Poise (0.1 Pa s). The key to this contrast between solids and fluids is the element of time: it does not matter in solids and is central for fluids. As we will see, for viscoelastic materials time matters, but at biologically relevant scales in some cases, it may not.

Viscoelastic materials exhibit both an elastic and a fluid response, behaving in a time-dependent way with load. As an example, consider the cartilage in your knees. For short-term loads, such as footfalls, the cartilage acts as a stiff spring. However, when a static load is applied over a long time, fluid squeezes out of the cartilage and the deformation occurs over the entire time course. Eventually the cartilage reaches an equilibrium with the applied load and stops deforming. There are several conceptual models used to predict the behavior of a viscoelastic material, including the Maxwell (spring-dashpot in series), the Kelvin-Voigt (spring-dashpot in parallel), and the standard linear model, which combines the two. The equations relating strain to stress are simply those associated with the physical realities of springs and dashpots. These models are useful for understanding transient loading phenomena, including creep, stress relaxation, and equilibrium modulus; we refer the reader to Lakes (1999). We now move on to a biologically relevant effect of viscoelasticity—the behavior when cyclically loaded.

Imagine an elastic solid, the marlin's caudal fin ray, for example, subjected to a sinusoidal cycle of strain (Figure 5.3). The stress or internal force developed within the element would be perfectly in phase with the strain such that if $\epsilon = \epsilon_0 \sin(\omega t)$ then $\sigma = \sigma_0 \sin(\omega t)$, where ω is the circular or angular frequency (in radians s^{-1}) and t is time (in s). If the same sinusoidal strain were applied to a viscoelastic material, the intervertebral joint, for example, the stress waveform would be offset from the strain by some phase lag, δ (in radians), such that $\sigma = \sigma_0 \sin(\omega t + \delta)$. The relation between stress

---

work, W, is done by F; that is, F is in phase with x and all the mechanical work stored in the tissue or structure is recovered during the return to resting shape. When $\delta = \pi/4$, F occurs before x, and, as a result, net work, W, is done by the component of the F that is in phase with the tissue's or structure's reconfiguration velocity. In a purely viscous system, $\delta = \pi/2$, F is entirely in phase with the velocity of the tissue's or structure's reconfiguration, net work, W, is maximal, and no elastic storage or recovery occurs. "Time" is in radians, where 2π is one complete cycle of arbitrary duration. In practice, δ in radians is calculated as follows: the difference in time (in seconds) between zero-crossings of the two curves is divided by the cycle period (in seconds) and multiplied by 2π radians. In the bottom diagram, the mechanical behaviors in three top figures are represented in work loops, with x as the abscissa, F as the ordinate, and time parameterized within the trajectory of the curve. Colors code for same δ as above. As δ increases, the area circumscribed by the ellipse increases, indicating increased cost of reconfiguring, W (or net work; same as above).

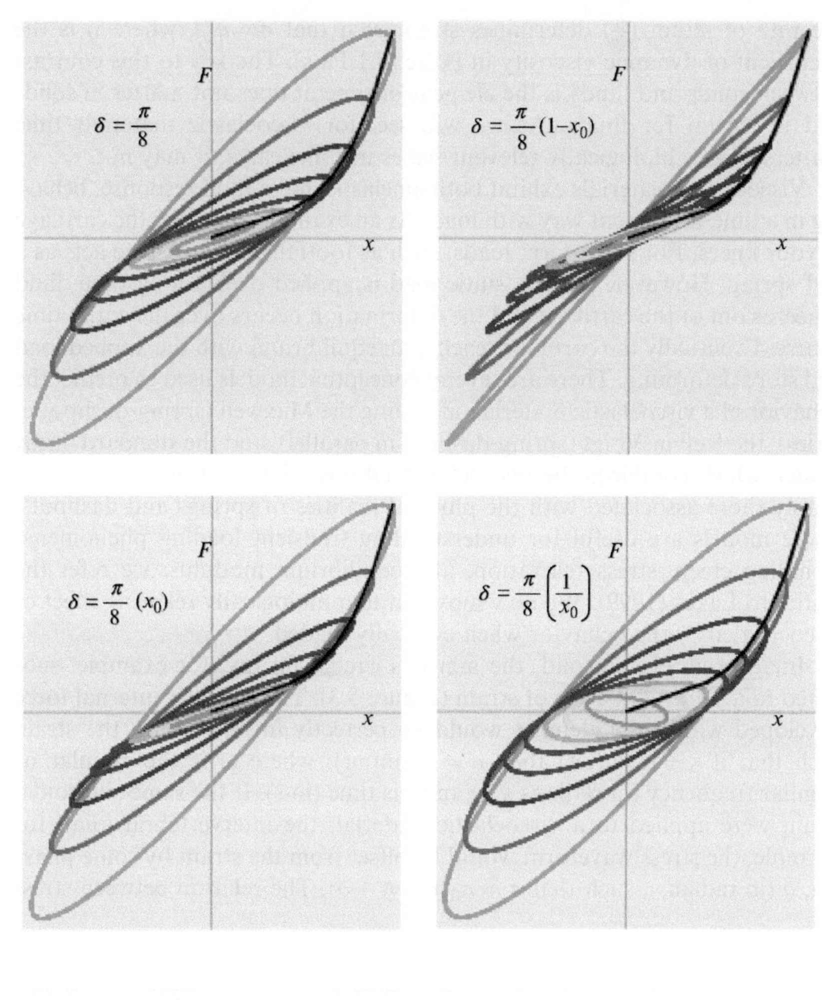
F
$\delta = \frac{\pi}{8}$
x
F
$\delta = \frac{\pi}{8}(1-x_0)$
x
F
$\delta = \frac{\pi}{8}(x_0)$
x
F
$\delta = \frac{\pi}{8}\left(\frac{1}{x_0}\right)$
x

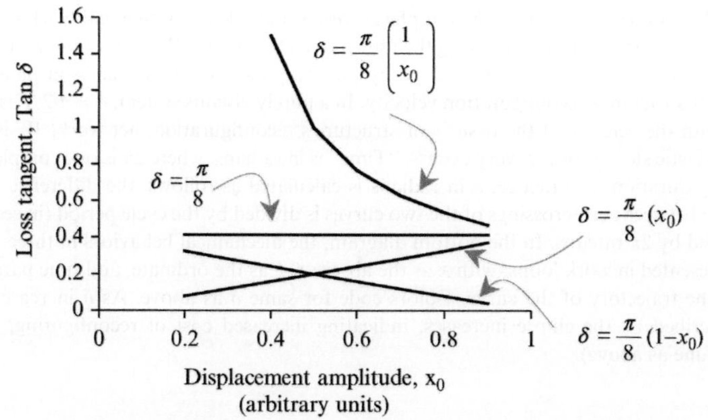
Loss tangent, Tan δ
1.6
1.4
1.2
1
0.8
0.6
0.4
0.2
0
0
0.2
0.4
0.6
0.8
1
$\delta = \frac{\pi}{8}\left[\frac{1}{x_0}\right]$
$\delta = \frac{\pi}{8}$
$\delta = \frac{\pi}{8}(x_0)$
$\delta = \frac{\pi}{8}(1-x_0)$
Displacement amplitude, x_0
(arbitrary units)

and strain in this time-dependent or dynamic situation generates a complex type of stiffness. The complex modulus, $E = E' + iE''$ (in Pa), has a component in phase with the strain, E', called the storage or elastic modulus and a component in phase with strain velocity, E'', called the loss or damping modulus, which is 90° out of phase with strain (a sine versus cosine function, since one differentiates sine to move from displacement to velocity and, in so doing, creates a cosine). The ratio of the loss and storage moduli is sometimes referred to as the system's relative damping or loss tangent, and works out trigonometrically (recall the preceding sine and cosine relations) as

$$\tan\delta = \frac{E''}{E'} \tag{6}$$

Since most fish structures and tissues are loaded cyclically, this dynamic approach to understanding the response to load seems most appropriate (for caveats, see Figure 5.4). Finding out tan δ at a relevant frequency is a guide to whether simple quasistatic testing, such as that described for Eq. (1), is appropriate. The closer a material is to an elastic solid, the lower tan δ. Quasistatic testing should capture the important aspects of response to load if tan δ is less than 0.1. Bone, for example, has a tan δ of about 0.01 and hence is usually tested quasistatically.

We are now in a position to return to our tail-walking marlin. We had defined flexibility such that it was inversely proportional to flexural stiffness, EI [see Eq. (2)]. Next we defined E, the shape-independent material property known as stiffness. Now we return to I, the material-independent shape factor known as second moment of area. The second moment of area accounts for the cross-sectional distribution of material; the farther a piece

Fig. 5.4. Why mechanical behavior cannot always be captured by a quasistatic loading curve (black lines). The four different behaviors (top four figures) are determined solely by the relation of phase lag, δ (see Figure 5.3) to the displacement amplitude, x_0 (see equations on figure) for a hypothetical system. Constant among the behaviors are the eight pairs of amplitudes (demarcated by color) of the driving force, F_0, and the resulting displacement, x_0, as indicated by the quasistatic loading curve. The work loops (see Figure 5.3) show the dynamic mechanical behavior as measured using sinusoidal cyclic loading tests at a frequency of 1 Hz (angular frequency of 2π radians per second). Since δ determines the energy loss (W, net work per cycle; see Figure 5.3), that aspect of mechanical behavior is ignored in the quasistatic loading domain. While the quasistatic curve is identical among the figures, the work loops are not. (Bottom) The mechanical behavior of each of the four dynamic systems can be summarized by the relation of the tangent of δ, Tan δ, to the displacement amplitude, x_0. Tan δ, also called the loss tangent, is useful because it is equivalent to the ratio of the loss modulus, E′, to the storage modulus E″ [see text and Eq. (6)].

of area is from the axis of bending, the greater is its impact on the resistance to bending. As an example, for our solid cylinder,

$$I = \frac{\pi r^4}{4}, \tag{7}$$

where r is the radius of the cylinder's cross-section. Please note that shape is nearly always treated as invariant over biological loads with motion coming at defined joints (e.g., Alfaro *et al.*, 2004; Westneat, 2004; Schaefer and Summers, 2005). In reality, it is clear that many structures deform considerably during normal use, and these deformations are the product of both structure, I, and material, E. A spectacular example from birds is the deflection and reshaping of the hummingbird beak during insect capture (Yanega and Rubega, 2003). Studies on the deformation of pectoral fin rays are emblematic of a trend toward considering the complications of deformable structures in fishes rather than ignoring them (Gibb *et al.*, 1994; Wilga and Lauder, 2000; Madden and Lauder, 2003).

III. BONE

Bone, a crystalline composite shot through with collagen fibers, is the most familiar rigid skeletal material, providing support and protection and serving as a reservoir of calcium in tetrapods (Figure 5.5) It is the least viscoelastic of skeletal materials, with such a large, well-organized mineral fraction that the response of bone to load is very nearly completely elastic (Currey, 1999). There is an extensive literature on the mechanical properties of bone from tetrapods, with stiffness ranging from 1 to 30 GPa, and ultimate strengths from 90 to 400 MPa (Currey, 2002).

Among recent fishes, only the Osteichthians have significant bone, though dermal bone was ancestrally present in many lineages, and endochondral bone has been found in extinct sharks (Smith and Hall, 1990; Coates *et al.*, 1998). Though the material of fish bone (calcium phosphate hydroxyapatite) is the same as tetrapod bone, the organization is quite different in that it is often acellular (Moss, 1961a). The lack of Haversian systems, trabeculae, and the remodeling that accompanies cellularity implies significant differences in the ultrastructural organization of fish bone. Indeed, Lee and Glimcher (1991) noted that the distinctive "twisted plywood" arrangement of collagen fibers and mineral crystals found in mammalian bone is far simpler in fish bones.

Rho *et al.* (2001) took advantage of the more linear arrangement of mineral crystal in fish bone to assess the effects of crystal orientation on material properties. A nanoindenter, a material testing machine capable of

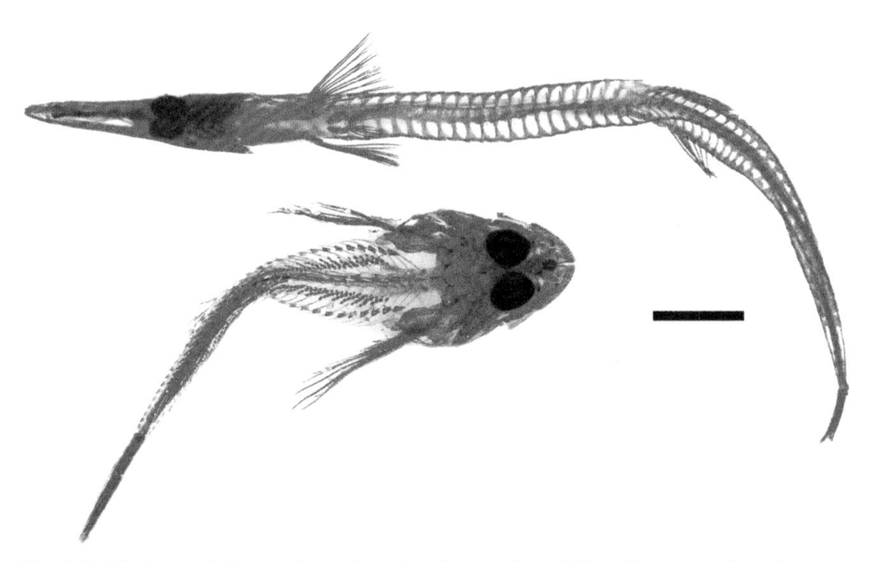

Fig. 5.5. The bony skeleton, shown in red, of two teleost fishes. On top is the tubesnout, *Aulorhynchus flavidus,* the sister taxon to the group containing seahorses and pipefishes, demonstrating a rudimentary precursor of the complete dermal armor found in the latter groups. On the bottom, the northern sculpin, *Icelinus borealis*, shows well-mineralized dermal scales in addition to endoskeletal and dermal bone. Both specimens have small but functionally important cartilages shown in blue. In particular, the ethmoidal cartilage sits between the ascending process of the premaxilla and the ethmoidal region of the skull to lubricate the sliding joint that allows upper jaw protrusion. Scale bar is 1 cm.

measuring the microNewton loads needed to press a sharp diamond tip just 700 nm into herring (*Clupea harengus*) intramuscular bone, was used to assess indentation modulus (a measure of stiffness). Along the length of the intramuscular bone, from the poorly mineralized distal tip to the fully mineralized proximal end, the indentation modulus increased from 3.5 GPa to 19 GPa. By comparing the modulus of sections cut longitudinally to those cut transversely it was possible to assess the anisotropy of the bone. In the least mineralized region the bone was nearly isotropic—that is, the response to load was the same regardless of the angle of load application. As mineralization increased, so too did anisotropy until at the most mineralized region the indentation modulus in the longitudinal direction was twice that in the transverse direction.

Using similar methodology, Roy *et al.* (2001) took advantage of the high osteocalcin (bone GLA-protein [BGP]) content of carp (*Cyprinus carpio*) rib bone to determine the relationship between BGP and indentation modulus. Because BGP has a strong affinity for hydroxyapatite it is an important organizing molecule in mineralizing bone. Nanoindentation revealed a significant relationship between stiffness and BGP content and between stiffness and crystal orientation. Surprisingly, there was no relationship between BGP and crystal orientation. The stiffness and anisotropy of carp rib were similar to herring intramuscular bone: 15.9 GPa in the transverse axis and 8.86 GPa along the longitudinal axis.

These two studies illustrate the power of using the fish as a model system for understanding basic skeletal biology, but shed little light on the evolutionary context of bone as a skeletal material. Erickson *et al.* (2002) used a three-point bending test to measure stiffness, ultimate strength, and yield strain of the pelvic metapterygium of *Polypterus* (Table 5.1). The data were compared to data from their own and literature studies of femora (the homolog of the pelvic metapterygium) across the vertebrate phylogeny. In spite of the simpler micro-architecture of fish bone, the mechanical properties are invariant across the phylogeny. This constancy implies a basic constraint on material properties of bone. The nature of this constraint is not known; however, mineralization, stiffness, and anisotropy increase in concert to the phylogenetic mean, implying that the relationship among the oblong crystals of hydroxyapatite and the surrounding matrix may be the key.

Table 5.1
The Mechanical Properties of Various Tissues from Fishes

Tissue	Taxa	Stiffness	Strength	Yield strain
Bone	Osteichthyes	3.5–19.4 GPa[a]	155 MPa[b]	8807 ($\mu\epsilon$)[c]
Cartilage	Lamprey	0.71–4.85 MPa[d]		
	Elasmobranch (vertebrae)[e]	0.53 GPa	23 MPa	
	Elasmobranch (jaws)	29 MPa	41 MPa	
Tendon	Hagfish[f]	290 MPa	48 MPa	22%
	Osteichthyes[g]	1.2–1.4 MPa	~30 MPa	
Slime fibers	Hagfish[h]	6.4 MPa	180 MPa	34%
Skin	Osteichthyes[i]	6.6–10 MPa		>8–>48%

[a]Rho *et al.*, 2001; Erickson *et al.*, 2002.
[b]Erickson *et al.*, 2002.
[c]Erickson *et al.*, 2002.
[d]Courtland *et al.*, 2003.
[e]Porter and Summers, 2004.
[f]Summers and Koob, 2002.
[g]Shadwick *et al.*, 2002.
[h]Fudge *et al.*, 2003.
[i]Brainerd, 1994b.

Separating those mechanical properties due to the shape of a skeletal element from those due purely to the material is not always simple (Currey, 1998). The long, thin shape of the femur and the homologous pelvic metapterygium make them ideal elements for determining bone strength and stiffness through bending tests. The close agreement between the bending and the nanoindenter data suggest that though an entire skeletal element was being tested, the changes in mechanical properties were due to the changes in material rather than changes in shape.

In spite of the acellular nature of most teleost bone, it would appear from these limited data that it has much the same properties as mammalian compact bone. The only indication that this might not be the case is a study of *in vivo* bone strain during suction feeding. Lauder and Lanyon (1980) attached rosette strain gages to the opercular bone of bluegills and found that while raw strains were similar to those reported for load-bearing bones in mammals ($\sim$1800 $\mu\epsilon$ in tension and 2700 $\mu\epsilon$ in compression), the rate of strain was not. During the explosive expansion phase of suction feeding, the bluegill operculum strained maximally at a rate of 615,000 $\mu\epsilon$/s, over 10 times the rate recorded from footfall impact during locomotion in mammals (Lanyon and Baggott, 1976). Since strain rate has a pronounced effect on fatigue, and unlike cellular bone acellular bone cannot repair microfractures, these data suggest that there may be unappreciated effects of acellularity (Moss, 1961b; Currey, 2002).

IV. CARTILAGE

Cartilage is a viscoelastic composite formed by condrocytes that secrete an extracellular matrix (ECM) rich in proteoglycans and collagen. The proteglycans ensure a large swelling pressure, or ability to absorb water, while the collagen serves as a reinforcing tensile element. In tetrapods hyaline cartilage serves as a model for endochondral bone, as an articular surface in joints, and as contour filler as in noses and ears. Fibrocartilage, distinguished by a high percentage of well-organized collagen fibers, provides a cushion against shearing loads in joints. In addition to these roles, cartilage is an important skeletal material in fishes, and in some taxa is the skeletal material throughout life (Figure 5.6). In bony fishes, cartilage is the primary skeletal material of larval life. Though there is a growing literature on the performance of larval fishes, including swimming and feeding kinematics, there are no data on mechanical properties of larval cartilage, loading regimes, or *in vivo* deformation of the skeleton in larval bony fishes (Hernandez, 2000; Mueller and van Leeuwen, 2004).

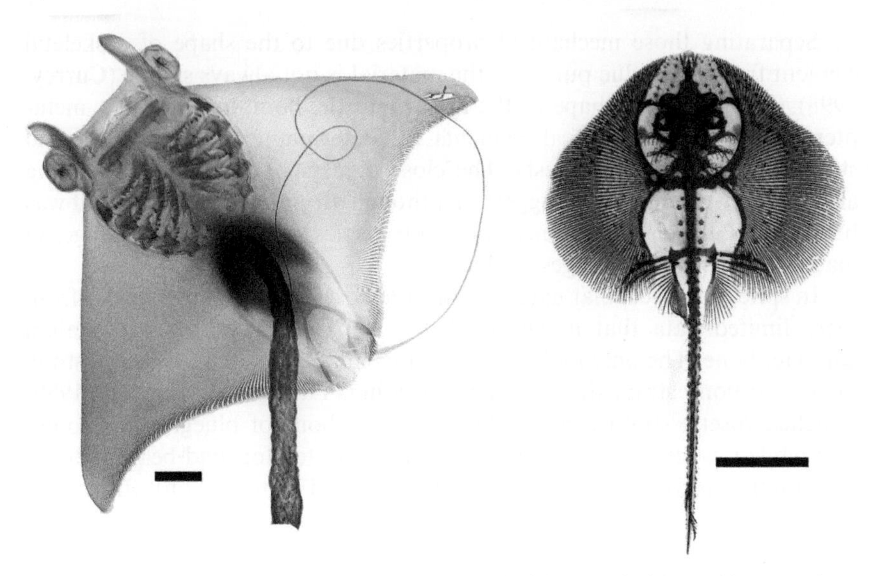

Fig. 5.6. The wholly cartilaginous skeleton of two batoid fishes. On the left, a mid-term embryo of the smooth-tailed mobula, *Mobula thurstoni,* shows no mineralized elements (red). The cephalic wings are fully dormed and are supported by wholly cartilaginous segmented radials. On the right is a hatchling hedgehog skate, *Leucoraja erinacea,* in which the dermal denticles have mineralized, while the rest of the cartilaginous skeleton remains unmineralized. In spite of being made of very different materials, there is no obvious difference in structural complexity between these skeletons and those of the bony fishes in Figure 5.1. Scale bars are 1 cm.

A wide variety of cartilages in bony fishes appears to serve a variety of functions (Symmons, 1979; Benjamin, 1989, 1990; Benjamin *et al.*, 1992; Schmitz, 1995). These putative functions include acting as a low friction-bearing surface, a flexible joint with little range of motion, and a stiffener in thin structures. While these cartilages are variable in cellularity and ECM, there have been no studies of the implications of this variation for response to loads.

The entire skeleton of agnathans and chondrichthian fishes is composed of cartilage that is mineralized to varying degrees. It is not clear that these cartilages are all of the same basic type, nor that they are homologous to tetrapod cartilage. The gross morphology appears similar to hyaline cartilage of tetrapods, but there are significant differences, including surface calcification and regions of acellularity in chondrichthians and noncollagenous fibers in the ECM of hagfishes and lampreys (Ørvig, 1951; Moss, 1977; Junquiera *et al.*, 1983; Wright *et al.*, 1983b, 1984; Robson *et al.*, 1997; Robson *et al.*, 2000). This variation makes the cartilage in these groups of fishes a good system for examining the relationships among structure, mechanical properties, and biochemical constituents.

The viscoelastic behavior of lamprey annular and pericardial cartilage has been examined and compared with the supporting cartilage from cow ear (Courtland *et al.*, 2003). The equilibrium modulus rather than Young's modulus was measured for these tissues; this is one of the transient properties that can be used to characterize viscoelastic materials. In this type of test, an instantaneous strain is applied to a viscoelastic material. The stress in the sample is initially high but decreases as the specimen takes on or loses water (stress relaxation), until an equilibrium length is reached. Several different initial strains are applied to a sample, each yielding a datum on equilibrium stress. The slope of the line through these successive tests is the equilibrium modulus. This modulus is similar in concept and identical in units to Young's modulus.

Both annular and pericardial cartilage became more than 50% stiffer as the lamprey senesced. Samples taken immediately from upstream migrating animals were less stiff than those taken from animals kept for 4 months in tanks postcapture. The comparison between bovine auricular cartilage and presenescence lamprey cartilage revealed no difference in equilibrium modulus, but there was an interesting 4-fold increase in the time to reach equilibrium (Table 5.1). The increased time to equilibrium in the lamprey is assumed to be due to decreased permeability of the tissue to water. The authors hypothesize that the branching nature of lamprin, the noncollagenous fibrous protein of the lamprey ECM, is responsible for the low permeability (Wright *et al.*, 1983a; Courtland *et al.*, 2003).

Neither lampreys nor hagfishes have heavily mineralized cartilage, nor are they high performance fishes. In contrast, the skeleton of ratfishes, sharks, and rays is sometimes heavily mineralized, and many species perform at functional extremes. There are no mechanical tests of the skeleton of ratfishes (holocephalans), though they eat hard prey and presumably have quite stiff jaws. The cartilage of their jaws is mineralized in a different fashion than the elasmobranch fishes, being a more diffuse mineralization running through the skeleton rather than a thin surface shell (Lund and Grogan, 1997).

The cartilage of sharks and rays (elasmobranches) has two forms, a thin layer of heavily mineralized tiles laying on the surface of a "hyaline" core and discrete internal mineralization, termed "aereolar," that can form complex solid shapes within an element (Ridewood, 1921; Ørvig, 1951; Halstead, 1974). The latter is found in the vertebral centra of sharks and rays, while the former is characteristic of the cranial and appendicular skeleton. The mechanical properties of both these forms of mineralized cartilage have widespread implications for the biomechanics of these fishes. The vertebrae and intervertebral disks determine the properties of the vertebral column, which in part determines the efficiency of swimming at different speeds (Long and Nipper, 1996). The stiffness and strength of the remainder of the skeleton are determinants of functional niche in that the mechanical properties must serve to transmit loads with appropriate deformation characteristics (Summers, 2000).

The vertebrae of six species of sharks and rays had mineral fractions similar to trabecular bone (40–50%) (Porter and Summers, 2004). When these vertebrae were subjected to compressive tests to failure between flat platens, they exhibited the strength of bone, the extensibility of tetrapod cartilage, and an intermediate stiffness (Table 5.1). There was no relationship between proteoglycan or collagen content and mechanical properties, though there was a weak correlation with mineral content. Tantalizing data indicate that sharks pressurize as they swim faster (Wainwright *et al.*, 1978). If this is so, and the fiber angle in the skin is less than 54°, then there should be a compressive load on the vertebral column during swimming and this loading should increase with increasing speed. However, the estimated swimming speed of the species tested by Porter and Summers did not predict mechanical properties, with the exception that the vertebrae of the only nonundulator, the torpedo ray (*Torpedo californica*), were neither as stiff nor as strong as those of sharks.

Jaw cartilage from three species of shark and a stingray, including a deep-sea form, the sleeper shark (*Somniosus pacificus*) and a hard prey crusher, the cownose ray (*Rhinoptera bonasus*), were compressed to failure. The least stiff species were not even as stiff as tetrapod articular cartilage, while the stiffest were as stiff as trabecular bone (Table 5.1). Not surprisingly, the trabecular cartilage of the cownose ray was the stiffest measured, though even when the mineralized struts were chemically removed (via ethylenediaminetetraacetic acid [EDTA]) the remaining unmineralized cartilage was stiffer than the shark species (Figure 5.7).

The response of a skeletal element to load is determined by both the material properties and the structure of the element (Vogel, 2003). A formula similar to that used earlier [Eq. (1)] to predict deflection in a caudal ray moving through the water can be used to predict deflection in a jaw crushing a snail:

$$y\text{max} = \frac{FL^3}{3EI}.$$

The denominator has the product of E (Young's modulus) and I, the second moment of area. To review, the former is a descriptor of the material stiffness and the latter a metric for the ability of a particular cross-section to resist flexion about a particular "neutral" axis. Second moment of area,

Fig. 5.7. A three-dimensional reconstruction from a micro CT scan (0.067 mm slices) of the upper jaw of an adult cownose ray (*Rhinoptera bonasus*) showing the complicated mineralized trabeculae that support the tooth plates. The mineralization takes the form of small (0.1–0.5 mm) blocks called tesserae. Prey is crushed at the asterisk (*), as is evident from the thinning of the tooth plates in this high wear region. The gap between the tooth plates and the jaw is filled with an elastin-rich dental ligament. The jaw illustrates the difficulty in understanding the mechanical behavior of whole structures, as there are certainly interactions between the enamel and dentine teeth, the dental ligament, and both the mineralized and unmineralized jaw. Scale bar is 1 cm.

sometimes called the area moment of inertia, is $I_{NA} = \int_A y^2 dA$, where y is the distance from the neutral axis (NA) and A is the area of the cross-section. By visualizing the mineralization of cartilaginous elements with CT scans, the I_{NA}, or structural contribution to flexural stiffness, can be calculated (Figure 5.8).

The hard prey-crushing cartilaginous fishes provide an interesting phylogenetic context for testing the notion that material properties and structural properties can vary independently. A completely durophagous diet has evolved at least five times in cartilaginous fishes. Comparing the second moment of area of the jaws among different lineages makes it clear that there are several different solutions to stiffening the jaws. The myliobatid stingrays have far more mineralization than horn sharks, both internally and

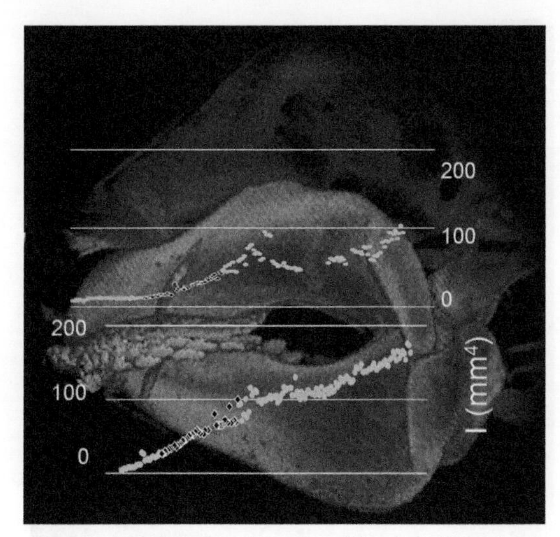

Fig. 5.8. A three-dimensional reconstruction of a CT scan of the head of a horn shark, *Heterodontus franscisi.* The second moment of area (I_{NA}) has been computed for 180 slices taken along the length of the upper and lower jaw. Symbols have been outlined in white in areas where the teeth are molariform, indicating high crushing forces. Figure 5.2 shows the importance of I_{NA} on the deflection of the jaw during feeding. (From Summers *et al.*, 2004.)

externally, but the arrangement of the mineral in the horn sharks will resist flexural deformation better (Summers *et al.*, 2004). The same material, arranged in different ways, can lead to alternative behaviors of the whole mechanical system.

V. TENDON

Tendon is most familiar as the bright white, linearly arrayed tissue connecting muscle to bone, as exemplified by the Achilles tendon of mammals. Alternatively, tendon exists as a flat sheet of organized arrays of collagen fibers that serves as muscle attachment sites or to distribute force or pressure (Vogel and Gemballa, 2000; Summers and Koob, 2002). Between the extremes of tendon morphology there are morphological intermediates—dense bands of connective tissue embedded within the flat sheet of a myoseptum (Hagen *et al.*, 2001; Donely *et al.*, 2004).

The flat sheet morphology arose first, presumably to transfer the force of myotomal muscle via myoseptal collagen fibers to the axis (Weitbrecht and Gemballa, 2001). The morphology of this form of tendon (well-organized layers of collagen fibers) implies both a high tensile stiffness and strength and a significant anisotropy. There are no mechanical tests of myoseptal tendon to support this supposition, nor are there data to suggest whether myosepta act primarily to resist pressure loads normal to the plane or in-plane tensile loads. Experimental data on myoseptal mechanics are sorely needed to test theories of myoseptal function gleaned from detailed and comprehensive studies of a range of fishes and other aquatic vertebrates (Alexander, 1969; Vogel and Gemballa, 2000; Hagen *et al.*, 2001; Azizi *et al.*, 2002; Gemballa and Vogel, 2002b) (see Chapter 7 of this volume for a description of mysoseptal architecture). While putative functions of the horizontal septum have been inferred from the arrangement of tendons in this septum in various fishes, it would be nice to see mechanical tests performed on excised septa and whole structures (Westneat *et al.*, 1993; Gemballa *et al.*, 2003). No doubt the inhomogeneity of the myosepta in terms of fiber direction will lead to complex mechanical behavior.

While the lack of mechanical data on myoseptal tendon is probably due largely to the difficulty of isolating samples and fixturing, the paucity of data on linearly arrayed tendon is explained by a lack of good examples in fishes. The vast majority of tendinous tissue in fishes is in flat sheets (often with embedded linear regions). Linearly arrayed tendon first appears in the basal craniates (hagfishes) as a tongue retractor and protractor tendon (Summers and Koob, 2002), and there are several other good examples of linear tendon, most notably associated with the horizontal septa of tunas and the fin inclinators of the ocean sunfish (*Mola mola*) (Raven, 1939; Westneat *et al.*, 1993).

The strength of hagfish tongue retractor tendon is similar to that of mammalian tendon, though the stiffness is comparatively low (Table 5.1) (Summers and Koob, 2002). This tendon is acting in a familiar role, transferring the force generated by the retractor muscle to the cartilaginous rasping tooth plate. Mammalian tendon can also serve to store mechanical energy during oscillatory locomotion, leading to significant efficiency gains (Alexander, 1984; Biewener *et al.*, 1998). Since undulatory locomotion in fishes is driven with an oscillatory mechanism, there is a potential for elastic energy storage as the caudal fin reverses direction.

Stored energy can be calculated from the area under the force/extension curve for the tendon *in vivo*. Because tendon typically has a very high resilience (>90%), all this stored energy is returned to the skeleton. Tunas, in addition to having prominent linearly arrayed tendons that transfer the myotomal muscle force to the tail, are also high-speed cruising specialists. Thus, they would seem to be an ideal place to look for significant storage of elastic strain energy. However, Shadwick *et al.* (2002) found that tuna tail tendons are as stiff as mammalian tendons, and because there appear to be several redundant systems for transmitting force, the net stress in the tail tendons is a small fraction of the muscular stress. This leads to very low *in vivo* strains, on the order of 0.5%, and consequently low storage of energy (~40 J/kg). This is less than 10% of the energy stored in tendon during high-speed running in mammals (Biewener *et al.*, 1988). A close examination of the mako shark (*Isurus oxyrinchus*), perhaps the most extreme cartilaginous high-speed burst swimmer, revealed force transmission between body musculature and tail through a system of tendons as in tuna (Donely *et al.*, 2004). This similarity makes it unlikely that significant elastic strain energy is stored in the tendons of chondrichthian fishes; instead, it appears that virtually all the muscle shortening results in tail movement directly.

VI. SKIN

Breder (1926) recognized that the mechanical properties of fish skin have an effect on swimming style and probably on performance. His assertions that the stiff skin of gars prevents anguiliform locomotion demonstrate the ease with which function can be imputed to a characteristic of material. Unfortunately, Breder's instincts were wrong, as shown in an analysis of the undulatory swimming of the longnose gar (Webb *et al.*, 1992; Long *et al.*, 1996). It is difficult to assess skin's contribution to locomotor performance without manipulating the stiffness *in vivo*, or measuring the whole body stiffness of fishes with and without skin. Excised skin is truly difficult to test mechanically in a biologically relevant way, so many assertions of function remain speculation.

The structure of fish skin is at the root of the difficulty in determining its mechanical properties: as a layered, pliable composite it is in a class of materials that resist easy characterization (Wainwright *et al.*, 1976; Vogel, 2003). The two principal layers are an outer, ectodermally derived, stratified, and occasionally cornified epidermal layer, and an inner, mesodermally derived, often quite thick dermal layer. These layers are further complicated by dermally derived bony scales, sometime with epidermally derived enamel sheaths or coatings.

A. Epidermis

The epidermis is quite thin relative to the dermis and likely plays little role in the stiffness and strength of the skin. However, the epidermally derived mucus-producing cells produce proteoglycan that certainly has large effects both on the drag coefficient of the fish and the friction coefficient (Shephard, 1994). While the former property is properly in the realm of the fluid dynamicist, the latter bears examination by biomaterials scientists, as variation in friction and adhesive properties can serve important antipredator and locomotor functions (Pawlicki *et al.*, 2004). This suggests a biologically relevant though somewhat unconventional test of the "slipperiness" imparted to fish skin *in vivo* by the normal secretion of mucus. Though no test of this sort has been performed, there have been tests on the epidermal secretions of the hagfish.

Hagfishes are well known for their ability to produce astonishing quantities of thick slime when disturbed. This slime is composed of a "slippery" proteoglycan-rich matrix in which long, strong fibers are suspended (Downing *et al.*, 1981; Fudge *et al.*, 2003). While the properties of the matrix have not been explored, the fibers have proven to be an interesting model material. Intermediate filaments (IFs) are one of three structurally important fibers supporting the cytoskeleton. Since intermediate filaments are the principal fiber in α-keratins, the material properties of keratins might yield insight into the contribution of IFs to the integrity of the cytoskeleton. Unfortunately, even in hard α-keratins the matrix surrounding the IFs contributes to the mechanical properties, making it difficult to extrapolate tests on wool, hair, and horn to the cell. The filaments in hagfish slime are also IFs, but the matrix proteins that confound other studies are not present. The results from hagfish slime threads are completely at odds with those from the data on hard α-keratins: IFs are far less stiff than F-actin and microtubules. They are also more extensible and presumably far tougher (Fudge *et al.*, 2003; Fudge and Gosline, 2004). Though these results have not been incorporated in models of cytoskeletal function, it is satisfying to see that there is functional separation, rather than complete redundancy, in the three structural fibers.

B. Dermis

The dermis, with its highly organized *stratum compactum*, is intuitively principally responsible for the mechanical response of the skin as a whole, though the effect of scales needs further attention (Figure 5.9) (Long *et al.*, 1996; Gemballa and Bartsch, 2002). The *stratum laxum* of most fishes is thinner, has fewer well organized collagen fibers, and is assumed to have little mechanical function. In contrast, the *stratum compactum* is composed of layers of well-organized collagen fibers arranged such that the principal direction of collagen fibers alternates from one layer to the next. These layers presumably move relative to one another, contributing to the skin's anti-wrinkling properties (Motta, 1977; Wainwright *et al.*, 1978; Hebrank, 1980). Certainly this "crossed-ply" structure contributes to a classic stress strain curve in which there is a toe region of relatively low force extension during which the collagen fibers are sliding into an orientation parallel with the

Fig. 5.9. The multilayered stratum compatum of shark skin. This polarized light micrograph shows three layers of fibers from the skin of the blacknose shark, *Carcharhinus limbatus*. The skin can have from dozens to hundreds of layers of this dense connective tissue, which has two dominant axes of orientation. Adapted from Wainwright *et al.*, 1978.

applied load. When the embedded fibers can no longer slide relative to one another, skin suddenly stiffens and behaves more like tendon (Brainerd, 1994b).

Relative to bone and tendon, skin is more viscoelastic, a factor that complicates making repeatable, or biologically relevant, mechanical tests. In the event that assessing the degree of viscoelasticity is impractical or irrelevant, the approach of Brainerd (1994b) is informative. Since strain rate affects mechanical properties of viscoelastic materials, a shortcut to characterizing the material is to test it only at biologically relevant strain rate(s). Brainerd determined that 5%/s was the *in vivo* strain rate for pufferfish skin and used this rate in uniaxial tensile tests. Unfortunately, no *in vivo* data were available for comparative tests on filefish and sunfish, so differences in properties among the three species are difficult to interpret. In any case, the extensibility of pufferfish skin is many times that of other fish at this strain rate, implying significant differences in elastic and/or viscous components of the skin (Figure 5.10).

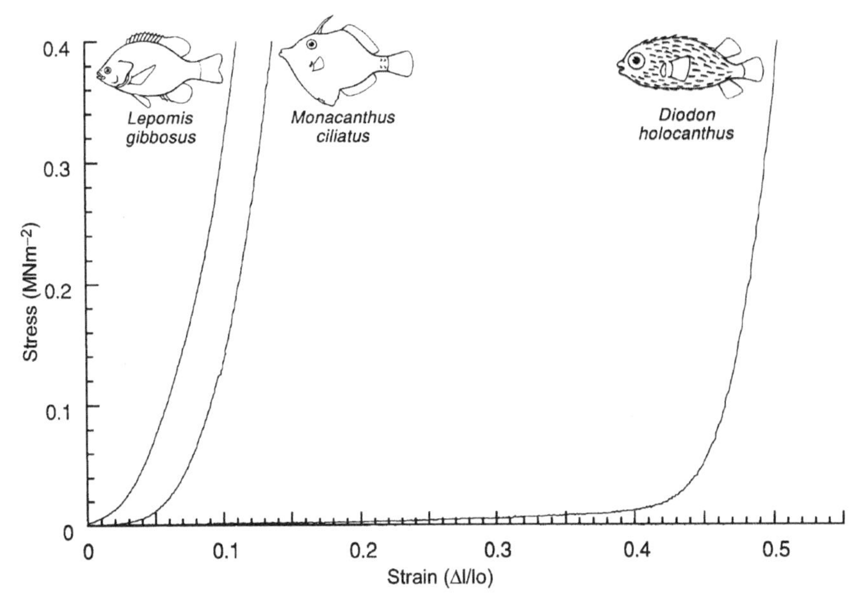

Fig. 5.10. The relationship between stress and strain for uniaxially loaded skin from three disparate teleost fishes. The tensile stiffness for the three species is similar once the fibers in the *stratum compactum* have been aligned with the direction of applied force. The major difference is in the length of the "toe" region of the curve, where the pufferfish skin elongates over 40% without significant increase in stress. (Adapted from Brainerd, 1994b.)

The tendinous myosepta insert on the *stratum compactum* in such a fashion that it seems inevitable that the skin plays a significant role in force transmission during locomotion (Gemballa and Treiber, 2003; Gemballa and Roder, 2004). In a provocative paper Wainwright *et al.* (1978) proposed that in sharks the skin does not just transmit force from the muscles; it also stiffens the shark, transmitting force like an external tendon. This hypothesis was based on measured variation in the intramuscular pressure at several swimming speeds combined with biaxial tensile tests of skin stiffness. The pressure measurements are not well explained, nor are raw pressure traces presented, but steady state increases in intramuscular pressure seem difficult to explain and are at odds with some experimental data (Martinez *et al.*, 2003; Horton *et al.*, 2004).

Further complicating the concept of skin as an exotendon are data from uniaxial and biaxial measurements of skin stiffness in teleost fishes. Eels share with sharks a thick, multi-ply *stratum compactum* that becomes stiffer along the longitudinal axis when strained in the hoop direction (circumferentially) (Wainwright *et al.*, 1978; Hebrank, 1980). In contrast, two bony fishes with radically different swimming styles, the sparid *Leiostomus xanthusrus* and the thunnid *Katsuwonus pelamis*, do not show this increase in longitudinal stiffness (Hebrank and Hebrank, 1986). It may be true that the thinner skin of the latter two fishes behaves completely differently than the former, though there is no morphological basis for supposing this. The biological significance of biaxial (and uniaxial) tests on excised skin samples is very difficult to interpret because the vital collagenous connections to the muscles, horizontal septum, and myosepta have been severed (Gemballa and Vogel, 2002a; Gemballa *et al.*, 2003; Gemballa and Roder, 2004). Furthermore, the very large energy losses during relaxation in the both the uni- and biaxial tests indicates that there is an irreversible reorganization of fibers in the excised sample that is likely to be highly dependent on sample size and shape (Hebrank, 1980; Hebrank and Hebrank, 1986).

Scales, from the imbricate armor of polypterids and lepisosteids to the more delicately connected ctenoid and cycloid scales of teleosts, also play a role in the mechanical properties of skin. Scales are attached to the dermis through collagenous fibers, with the degree of attachment varying widely among taxa. The scales of the heavily armored basal actipterygians cannot be separated from the surrounding dermis without extensive damage, while many teleost fishes (i.e., engraulids and salmonids) shed scales at the slightest disturbance and eventually regenerate them (Helfman *et al.*, 1997).

The mechanical role of the ganoid scales of *Polypterus* and *Lepisosteus* has been better studied than that of ctenoid or cycloid scales. The scales themselves are presumably as stiff and as strong as the bone and enamel of which they are made would suggest. That is, they are probably most like broad, flat teeth in their material properties. The interactions between these scales are mediated by collagen fiber bundles that render the skin highly anisotropic and permit counterintuitive deformation regimes. The polypterid fishes (ropefish and bichirs) power inhalation via elastic recoil of the fibers joining ganoid scales (Brainerd *et al.*, 1989). The cross-section of the fish is normally circular, but when air is expired from the lung, a flat spot appears on the ventral surface. Scales in this flat spot are extended relative to their rest position, which stretches the collagen fibers between the scales. When the fish relaxes the "exhalation" muscles, the scales snap back into position and air rushes into the lung (Brainerd, 1994a).

In both polypterids and lepisosteids (gars) the scale armor is continuous and overlapping. This implies that scales move relative to one another when the fish undulates. Gemballa and Bartsch (2002) showed that the deformation of the armor coat is not at all isotropic. The scales are tied into the dermis and to one another with collagen fibers. A peg and socket joint further restricts movements in some directions, but the skin seems relatively free to move in others.

One approach to understanding the role of skin and scales is to sequentially, surgically interrupt the function of various layers and assess the effect on function. These experiments can be carried out *in vivo* by comparing swimming kinematics from pre- and postsurgically altered fishes (Long *et al.*, 1996). In this study, the surgical interruption of the dermis of the longnose gar caused an increase in tail beat amplitude, consonant with a decrease in stiffness. Quasistatic tests on dead gars showed a decrease in stiffness with the removal of scales and the interruption of the dermis, but these tests were not performed dynamically (Figure 5.11). A similar approach was used in the sequential dissection and dynamic testing of the hagfish, *Myxine glutinosa* (Long *et al.*, 2002b). In this case, sections of the animal were oscillated at speeds and amplitudes seen in swimming fish. The resultant dynamic data showed that whole body stiffness is hardly affected by the presence of skin. Instead, the effect of the notochord made up over 70% of the stiffness, outweighing even the passive stiffness associated with the body musculature.

VII. WHOLE BODY MECHANICS

> "*As we analyze a thing into its parts or into its properties, we tend to magnify these, to exaggerate their apparent independence, and to hide from ourselves the essential integrity and individuality of the composite whole.*"
>
> —D'Arcy Thompson (1917: 1018).

Though obviously more complicated because of the large number of varied materials involved, the mechanical properties of the entire body are more closely tied to the performance of the whole organism (Aleev, 1969). Data addressing the mechanical properties of fish bodies have been gathered in the context of understanding passive and active stiffness because of the relationship between swimming efficiency at a particular frequency and body stiffness (Blight, 1977).

The initial investigation of the properties of the fish vertebral column might properly be attributed to Everard Home, who performed qualitative and quantitative tests on the character of the intervertebral ligaments and the viscous intervertebral fluid from the basking shark, *Cetorhinus maximus* (Home, 1809) (Figure 5.12); he suggested that the ligaments functioned as elastic springs. Spring-like mechanical behavior was also the hypothesis suggested by Rockwell *et al.* (1938) upon examination of the vertebral column of billfishes. Symmons (1979) came to a similar conclusion after manual manipulation of a wide range of fish vertebral columns. The first careful mechanical tests of vertebral columns were conducted by Hebrank

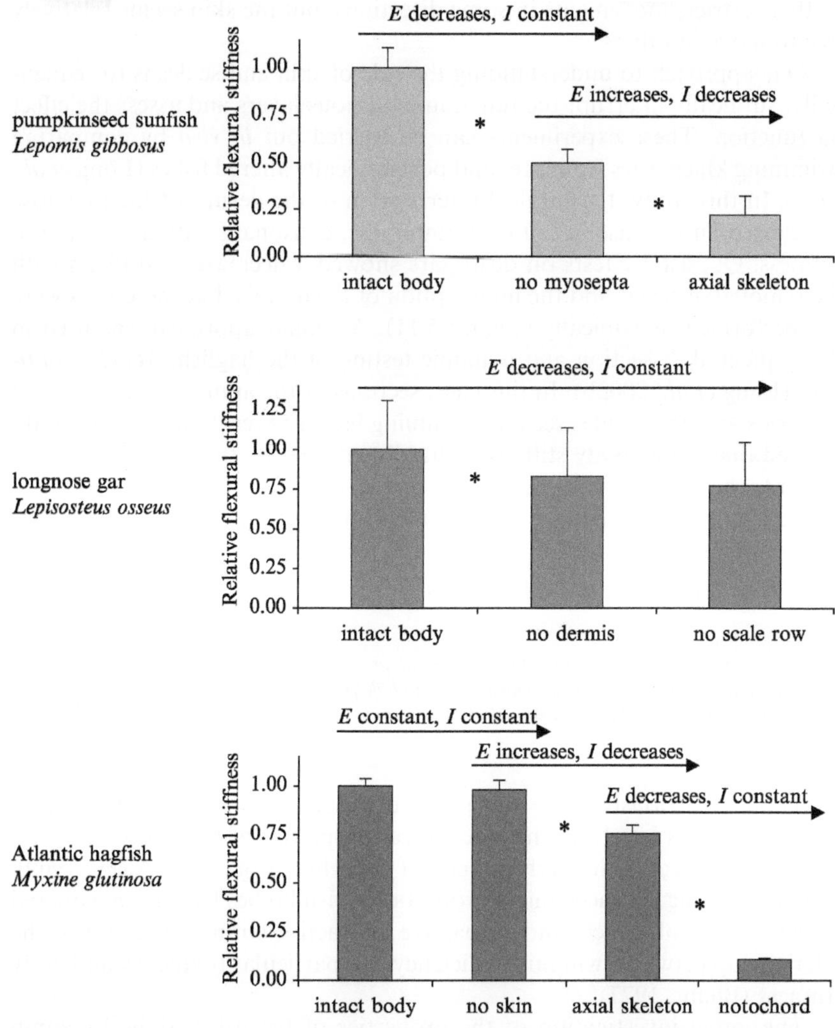

Fig. 5.11. Skeletal structures contribute in varying degrees to the mechanical behavior of intact fish bodies. As structures are sequentially removed from dead, intact fish bodies, the flexural stiffness, *EI*, decreases to varying degrees, depending on species and structure. Whether the change in *EI* is caused by a change in the apparent material stiffness, *E*, or by the composite second moment of area, *I*, is indicated. In sunfishes, maceration of the tendinous myosepta in the muscles of the caudal region reduce *EI* by 50%. Removal of the muscle and skin further reduces the body's *EI*. In gars, laceration of the dermis connecting two ganoid scale rows in the caudal region decreases the body's *EI*, but removal of the scale row itself does not. In hagfishes, the skin does not contribute to *EI* of the mid-precaudal region, while muscles and the outer fibrous sheath of the axial skeleton do. All experiments were cantilevered bending. Sunfish and gar experiments were conducted using quasistatic loading tests at frequencies under 1 Hz.

(1982), who found differences in quasistatic bending stiffness among species that appeared to be related to swimming performance. However, without measurement of viscous properties under physiological conditions (see also Hebrank *et al.*, 1990), quasistatic tests do not provide sufficient information to characterize the mechanical behavior of the vertebral column bending dynamically during swimming. The integrated mechanics of swimming were appreciated by Webb (1973), who, upon examination of swimming cephalochordates (*Branchiostoma lanceolatum*), postulated that the paramyosin fibers within the axial notochord could actively vary the stiffness of the body, facilitating forward and backward undulations by altering the gradient of stiffness. Building upon Webb's (1973) insight, Blight (1977) created the "hybrid oscillator" model, in which the mechanical properties of the whole body interact with the hydrodynamic forces of the surrounding water. While qualitative, Blight's hybrid oscillator model has proven to be highly influential, ushering in the modern era of closed-loop, force-coupled fish swimming models (e.g., Hess and Videler, 1984; Cheng and Blickhan, 1994; Carling *et al.*, 1998; Czuwala *et al.* 1999; Librizzi *et al.*, 1999; Long *et al.*, 2002a). The accuracy of these models, however, depends on knowing the dynamic mechanical properties of the whole body.

The realization that the dynamic bending during swimming could be mathematically modeled and used to compute bending moments led to the consideration of the whole fish as a resonating beam (Hess and Videler, 1984; Cheng and Blickhan, 1994; Long and Nipper, 1996). The power needed to drive a resonating beam is minimized at the natural frequency of the beam. The natural frequency of a beam depends directly on flexural stiffness, *EI*, that is, either a beam with a higher modulus of elasticity or one with a larger second moment of area will have a higher natural frequency. As long as the damping coefficient of the system is low relative to the critical damping coefficient, which appears to be the case in fishes (Long, 1992, 1995; Long *et al.*, 2002b), the dynamic stiffness of the beam is the primary determinant of natural frequency.

To alter swimming speed, fishes vary tail beat frequency over a continuous range, and the mechanical cost to flex the axial skeleton varies in proportion to its viscosity (Figure 5.13). To offset these operational costs,

---

Hagfish experiments were conducted using dynamic tests over a range of physiological frequencies and curvatures; means shown are pooled across frequency and curvature. In all three species, only a small axial section of the intact body was bent, including myomeres from no more than five body segments. Asterisk indicates significant difference ($p < 0.05$) between means ($\pm$ one standard error) as determined by *a priori* contrasts in ANOVA. (Data reanalyzed from Long *et al.*, 1994, Long *et al.*, 1996, and Long *et al.*, 2002b.)

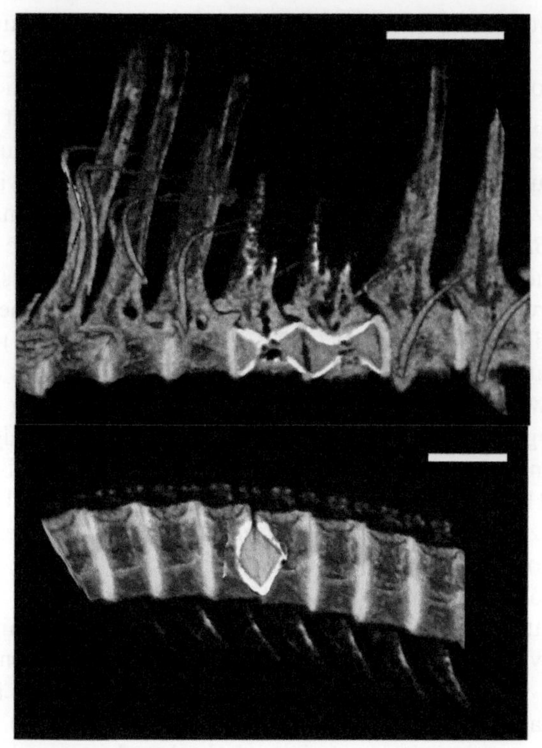

Fig. 5.12. An illustration of the mechanical complexity of the fish backbone. The top is a three-dimensional reconstruction from a CT of the spine from the rockhead poacher, *Bothragonus swanii,* and the bottom is from the spine of a sandbar shark, *Carcharhinus plumbeus*. In each, a central pair of vertebrae are sagitally sectioned to show the internal anatomy of the intervertebral joint. The circumferential and longitudinal fibers of the *annulus fibrosus* are shown in yellow, while the proteoglycan-rich gel of the *nucleus pulposus* is blue. Extensive bony zygopophyses can transfer load and serve to keep the vertebrae aligned in the bony fish. The shark has no such processes, and even with the slight ventroflexion in this scan, significant subluxation of the vertebrae relative to one another is obvious. The top scale bar is 1 mm; the bottom scale bar is 1 cm.

fishes may prefer to swim with a tail beat frequency at or near the body's natural frequency. To take advantage of the efficiencies of resonating, hagfishes (*Myxine glutinosus*) appear to have mechanically tuned bodies that would permit motion amplification without destructive high-amplitude bends during resonant bending (Long *et al.*, 2002b). Moreover, eels (*Anguilla rostrata*) have the capacity to alter body stiffness by a factor of three using their muscles, an ability that permits them to match the body's natural frequency to that of the beating tail over a range of swimming speeds (Long, 1998). In

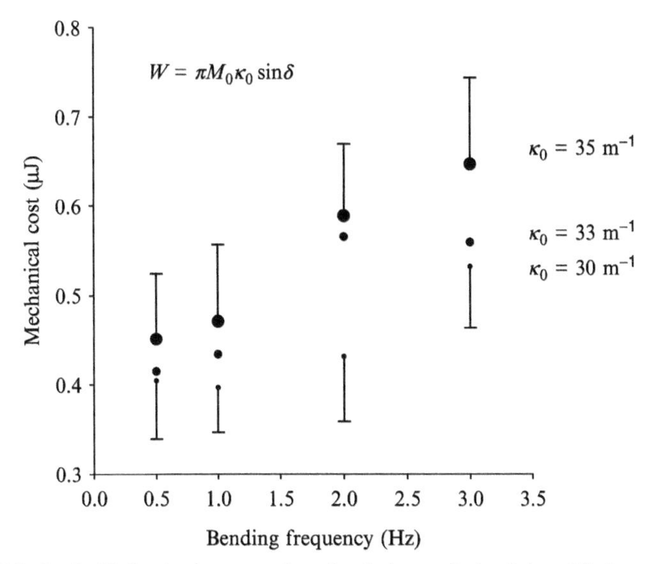

Fig. 5.13. Mechanical behavior in terms of mechanical cost. In hagfishes, W, the net work per cycle (stored–recovered elastic work) needed to bend a small section (4 mm) of the intact notochord measures the mechanical cost of operation (see also Figure 5.3). In these dynamic bending tests, W is proportional to the peak bending moment, M_0, the peak curvature, κ_0, and the phase angle, δ, between the two. Means are given for four individuals, with standard error bars for the lowest and highest κ_0. For a given κ_0, M_0 varies with changes in the structure's stiffness, or strain-proportional moment. (Data reanalyzed from Long *et al.*, 2002.)

swimming fishes, indirect evidence for passive and/or active tuning of the body's mechanical properties has been indirectly gleaned from harmonic analysis of the undulatory motion of the dynamically flexing body (Root *et al.*, 1999).

VIII. CONCLUSIONS

By measuring the mechanical properties of skeletal tissue and structures, we are able to begin modeling a few mechanical behaviors of a few species and to understand the integrated function of muscle, water, and skeleton. As this chapter details, however, much work remains to be done. Without more comparative data, physiologically relevant measurements, and accounting of intraspecific variability, we will not be able to understand the impact that the skeletal tissues and structures have had on the evolution of vertebrates. At the moment, because of the dearth of information, we are limited to examining correlations at the broadest phylogenetic levels (Koob and Long,

2000). But the evolution of vertebrates is a story driven by the evolutionary history of skeletons and skeletal tissues.

It would be fair to say that we have only the barest idea of the evolutionary history of the response of vertebrate tissues to load. The recent interest in tracing the evolutionary history of particular tissues, and the cell lines, makes it possible to conduct broad-scale comparisons across taxa. For example, there is consensus that some invertebrate cartilage is homologous to that of vertebrates (Cole and Hall, 2004). Since the familiar functions of vertebrate cartilage (bearing surface, contour filler) are so different from the role in invertebrates, the evolution of the mechanical properties is likely interesting.

ACKNOWLEDGMENTS

Tom Koob, Beth Brainerd, and Mason Dean provided input on the manuscript. The biomechanics group at the University of California, Irvine provided reviews and interesting papers we might otherwise have overlooked. This research has been supported by National Science Foundation grants to study cartilaginous fishes (IBN- 0317155 to A.P.S.) and the mechanics and evolution of vertebral columns and swimming (BCS-0320764 and DBI-0442269 to J.H.L.)

REFERENCES

Aleev, Y. G. (1969). "Function and Gross Morphology in Fish." Keter Press, Jerusalem.

Alexander, R. M. (1969). Orientation of muscle fibres in myomeres of fishes. *J. Mar. Biol. Assoc. UK* **49,** 263–290.

Alexander, R. M. (1984). "Elastic Mechanisms in Animal Movement." Cambridge University Press, Cambridge, UK.

Alfaro, M. E., Bolnick, D. I., and Wainwright, P. C. (2004). Evolutionary dynamics of complex biomechanical systems: An example using the four-bar mechanism. *Evolution* **58,** 495–503.

Azizi, E., Gillis, G. B., and Brainerd, E. L. (2002). Morphology and mechanics of myosepta in a swimming salamander (*Siren lacertina*). *Comp. Biochem. Physiol.* **133,** 967–978.

Bemis, W. E. (1984). Paedomorphosis and the evolution of the Dipnoi. *Paleobiology* **10,** 293–307.

Benjamin, M. (1989). Hyaline-cell cartilage (chondroid) in the heads of teleosts. *Anat. Embryol.* **179,** 285–303.

Benjamin, M. (1990). The cranial cartilages of teleosts and their classification. *J. Anat.* **169,** 153–172.

Benjamin, M., Ralphs, J. R., and Eberewariye, O. S. (1992). Cartilage and related tissues in the trunk and fins of teleosts. *J. Anat.* **181,** 113–118.

Bernal, D., Dickson, K. A., Shadwick, R. E., and Graham, J. B. (2001). Review: Analysis of the evolutionary convergence for high performance swimming in lamnid sharks and tunas. *Comp. Biochem. Physiol.* **129,** 695–726.

Biewener, A. A., Blickhan, R., Perry, A. K., Heglund, N. C., and Taylor, C. R. (1988). Muscle forces during locomotion in kangaroo rats: Force platform and tendon buckle measurements compared. *J. Exp. Biol.* **137,** 191–205.

Biewener, A. A., Konieczynski, D. D., and Baudinette, R. V. (1998). *In vivo* muscle force-length behavior during steady-speed hopping in tammar wallabies. *J. Exp. Biol.* **V201,** 1681–1694.

Blake, R. W. (1979). The mechanics of labriform locomotion in the angelfish (*Pterphyllum eimekei*): an analysis of the power stroke. *J. Exp. Biol.* **82,** 255–271.

Blake, R. W. (1983). "Fish Locomotion." Cambridge University Press, London.
Blight, A. R. (1977). The muscular control of vertebrate swimming movements. *Biol. Rev.* **52,** 181–218.
Brainerd, E. L. (1994a). Mechanical design of polypterid fish integument for energy storage during recoil aspiration. *J. Zool. (Lond.)* **232,** 7–19.
Brainerd, E. L. (1994b). Pufferfish inflation: functional morphology of postcranial structures in *Diodon holocanthus* (Tetraodontiformes). *J. Morphol.* **220,** 243–261.
Brainerd, E. L., Liem, K. F., and Samper, C. T. (1989). Air ventilation by recoil aspiration in polypterid fishes. *Science* **246,** 1593–1595.
Breder, C. M. J. (1926). The locomotion of fishes. *Zoologica* **4,** 159–297.
Carling, J. C., Williams, T. L., and Bowtell, G. (1998). Self-propelled anguilliform swimming: simultaneous solution of the two-dimensional Navier-Stokes equations and Newton's laws of motion. *J. Exp. Biol.* **201,** 3143–3166.
Cheng, J.-Y., and Blickhan, R. (1994). Bending moment distribution along swimming fish. *J. Theor. Biol.* **168,** 337–348.
Coates, M. I., Sequeira, S. E. K., Sansom, I. J., and Smith, M. M. (1998). Spines and tissues of ancient sharks. *Nature* **396,** 729–730.
Cole, A. G., and Hall, B. K. (2004). The nature and significance of invertebrate cartilages revisited: Distribution and histology of cartilage and cartilage-like tissues within the Metazoa. *Zoology* **107**(4), 261–274.
Courtland, H. W., Wright, G. M., Root, R. G., and DeMont, M. E. (2003). Comparative equilibrium mechanical properties of bovine and lamprey cartilaginous tissues. *J. Exp. Biol.* **206,** 1397–1408.
Currey, J. D. (1998). Mechanical properties of vertebrate hard tissues. *Proc. Inst. Mech. Eng.* **212,** 399–411.
Currey, J. D. (1999). The design of mineralised hard tissues for their mechanical functions. *J. Exp. Biol.* **202,** 3285–3294.
Currey, J. D. (2002). "Bones: Structure and Mechanics." Princeton University Press, Princeton, NJ.
Czuwala, P. J., Blanchette, C., Varga, S., Root, R. G., and Long, J. H., Jr. (1999). A mechanical model for the rapid body flexures of fast-starting fish. "Proceedings of the 11th International Symposium on Unmanned Untethered Submersible Technology,", pp. 415–426.
Daniel, T. L. (1988). Forward flapping flight from flexible fins. *Can. J. Zool.* **66,** 630–638.
Donely, J. M., Sepulveda, C. A., Konstantinidis, P., Gemballa, S., and Shadwick, R. E. (2004). Convergent evolution in mechanical design of lamnid sharks and tunas. *Nature* **429,** 61–65.
Downing, S. W., Salo, W. L., Spitzer, R. H., and Koch, E. A. (1981). The hagfish slime gland: A model system for studying the biology of mucus. *Science* **214,** 1143–1145.
Erickson, G. M., Catanese, J., and Keaveny, T. M. (2002). Evolution of the biomechanical material properties of the femur. *Anat. Rec.* **268,** 115–124.
Fudge, D. S., and Gosline, J. M. (2004). Molecular design of the alpha-keratin composite: Insights from a matrix-free model, hagfish slime threads. *Proc. Roy. Soc. Lond. B Biol. Sci.* **271,** 291–299.
Fudge, D. S., Gardner, K., Forsyth, V. T., Riekel, C., and Gosline, J. M. (2003). The mechanical properties of hydrated intermediate filaments: Insights from hagfish slime threads. *Biophys. J.* **85,** 2015–2027.
Geerlink, P. J. (1983). Pectoral fin kinematics of *Coris formosa* (Teleostei, Labridae). *Neth. J. Zool.* **33,** 515–531.
Geerlink, P. J., and Videler, J. J. (1987). The relation between structure and bending properties of teleost fin rays. *Neth. J. Zool.* **37**(1), 59–80.
Gemballa, S., and Bartsch, P. (2002). Architecture of the integument in lower teleostomes: Functional morphology and evolutionary implications. *J. Morphol.* **253,** 290–309.

Gemballa, S., and Britz, R. (1998). The homology of the intermuscular bones in acanthomorph fishes. *Am. Mus. Novitates* **3241,** 1–25.

Gemballa, S., and Roder, K. (2004). From head to tail: The myoseptal system in basal actinopterygians. *J. Morphol.* **259,** 155–171.

Gemballa, S., and Treiber, K. (2003). Cruising specialists and accelerators—Are different types of fish locomotion driven by differently structured myosepta? *Zoology* **106,** 203–222.

Gemballa, S., and Vogel, F. (2002a). Spatial arrangement of white muscle fibers and myoseptal tendons in fishes. *Comp. Biochem. Physiol. Mol. Integ. Physiol.* **133,** 1013–1037.

Gemballa, S., and Vogel, F. (2002b). The spatial arrangement of white muscle fibers and myoseptal tendons in fishes: How may muscle fibers and myoseptal tendons work together to induce bending? *Comp. Biochem. Physiol.* **133,** 1013–1037.

Gemballa, S., Hagen, K., Roder, K., Rolf, M., and Treiber, K. (2003). Structure and evolution of the horizontal septum in vertebrates. *J. Evol. Biol.* **16,** 966–975.

Gibb, A. C., Jayne, B. C., and Lauder, G. V. (1994). Kinematics of pectoral fin locomotion in the bluegill sunfish *Lepomis macrochirus*. *J. Exp. Biol.* **189,** 133–161.

Gordon, J. E. (1984). "The New Science of Strong Materials, or, Why You Don't Fall through the Floor." Princeton University Press, Princeton, NJ.

Grande, L., and Bemis, W. E. (1998). A comprehensive phylogenetic study of amiid fishes (Amiidae) based on comparative skeletal anatomy. An empirical search for interconnected patterns of natural history. *J. Vert. Paleontol. Sp. Mem.* **44**(suppl. to Vol. 18). 1–690.

Hagen, K., Gemballa, S., and Freiwald, A. (2001). Locomotory design and locomotory habits of *Chimaera monstrosa*. *J. Morphol.* **248,** 237.

Halstead, L. B. (1974). "Vertebrate Hard Tissues." Wykeham Publications distributed by Chapman & Hall, London.

Hebrank, J. H., Hebrank, M. R., Long, J. H., Block, B. A., and Wainwright, S. A. (1990). Backbone mechanics of the blue marlin *Makaira nigricans* (Pisces: Istiophoridae). *J. Exp. Biol.* **148,** 449–459.

Hebrank, M. R. (1980). Mechanical properties and locomotor functions of eel skin. *Biolog. Bull.* **158,** 58–68.

Hebrank, M. R. (1982). Mechanical properties of fish backbones in lateral bending and tension. *J. Biomech.* **15,** 85–89.

Hebrank, M. R., and Hebrank, J. H. (1986). The mechanics of fish skin: The lack of an external tendon in two teleosts. *Biolog. Bull.* **171,** 236–247.

Helfman, G. S., Collette, B. B., and Facey, D. E. (1997). "The Diversity of Fishes." Blackwell Science Malden, MA.

Hernandez, L. P. (2000). Intraspecific scaling of feeding mechanics in an ontogenetic series of zebrafish, *Danio rerio*. *J. Exp. Biol.* **203,** 3033–3043.

Hernandez, L. P., and Motta, P. J. (1997). Trophic consequences of differential performance: Ontogeny of oral jaw-crushing performance in the sheepshead, *Archosargus probatocephalus* (Teleostei, Sparidae). *J. Zool.* **243,** 737–756.

Hess, F., and Videler, J. J. (1984). Fast continuous swimming of Saithe (*Pollachius virens*): A dynamic analysis of bending moments and muscle power. *J. Exp. Biol.* **109,** 229–251.

Home, E. (1809). On the nature of the intervetebral substance in fish and quadrupeds. *Phil. Trans. Roy. Soc.* **99,** 177–187.

Horton, J. M., Drucker, E. G., and Summers, A. P. (2004). Swiftly swimming fish show evidence of stiff spines. *Integ. Comp. Zool.* **43,** 905.

Junquiera, L. C. U., Toledo, O. M. S., and Montes, G. S. (1983). Histochemical and morphological studies of a new type of acellular cartilage. *Basic Appl. Histochem.* **27,** 1–8.

Koob, T. J., and Long, J. H., Jr. (2000). The vertebrate body axis: Evolution and mechanical function. *Am. Zool.* **40**(1), 1–18.

Lakes, R. S. (1999). "Viscoelastic Solids." CRC Press, Boca Raton, FL.
Lanyon, L. E., and Baggott, D. G. (1976). Mechanical function as an influence on the structure and form of bone. *J. Bone Joint Surg.* **58,** 436–443.
Lauder, G. V., and Lanyon, L. E. (1980). Functional anatomy of feeding in the bluegill, *Lepomis macrochirus*: In vivo measurement of bone strain. *J. Exp. Biol.* **84,** 33–55.
Lee, D. D., and Glimcher, M. J. (1991). Three dimensional spacial relationship between collagen fibrils and the inorganic calcium phosphate crystals of pickeral and herring bone. *J. Mol. Biol.* **217,** 487–501.
Librizzi, N. N., Long, J. H., Jr., and Root, R. G. (1999). Modeling a swimming fish with an initial-boundary value problem: Unsteady maneuvers of an elastic plate with internal force generation. *Comp. Math. Model.* **30**(11/12), 77–93.
Liem, K. F., Bemis, W. E., Walker, W. F., Jr. and Grande, L. (2001). "Functional Anatomy of the Vertebrates: An Evolutionary Perspective, Third Ed." Harcourt, New York.
Long, J. H. (1992). Stiffness and damping forces in the intervertebral joints of blue marlin (*Makaira nigricans*). *J. Exp. Biol.* **162,** 131–155.
Long, J. H. Jr., McHenry, M. J. and Boetticher, N. C. (1994). Undulatory swimming: How traveling waves are produced and modulated in sunfish (*Lepomis gibbosus*). *J. Exp. Biol.* **192,** 129–145.
Long, J. H. (1995). Morphology, mechanics, and locomotion—The relation between the notochord and swimming motions in sturgeon. *Environ. Biol. Fishes* **44,** 199–211.
Long, J. H. (1998). Muscle, elastic energy, and dynamics of body stiffness in swimming eels. *Am. Zool.* **38**(4), 181–202.
Long, J. H., and Nipper, K. S. (1996). The importance of body stiffness in undulatory propulsion. *Am. Zool.* **36,** 678–694.
Long, J. H., Hale, M. E., McHenry, M. J., and Westneat, M. W. (1996). Functions of fish skin: Flexural stiffness and steady swimming of longnose gar *Lepisosteus osseus*. *J. Exp. Biol.* **199,** 2139–2151.
Long, J. H., Koob-Emunds, M., Sinwell, B., and Koob, T. J. (2002a). The notochord of hagfish *Myxine glutinosa*: Visco-elastic properties and mechanical functions during steady swimming. *J. Exp. Biol.* **205,** 3819–3831.
Long, J. H., Jr., Adcock, B., and Root, R. G. (2002b). Force transmission *via* axial tendons in undulating fish: A dynamic analysis. *Comp. Biochem. Physiol. A* **133,** 911–929.
Lund, R., and Grogan, E. D. (1997). Relationships of the chimaeriformes and the basal radiation of the Chondrichthyes. *Rev. Fish Biol. Fisheries* **7,** 65–123.
Lundberg, J. G., and Baskin, J. N. (1969). The caudal skeleton of the catfishes, order Siluriformes. *Am. Mus. Novitates* **2398,** 1–49.
Lundberg, J. G., and Marsh, E. (1976). The evolution and functional anatomy of the pectoral fin rays in cyprinoid fishes, with emphasis on the suckers (family Catostomidae). *Am. Midland Natur.* **96**(2), 332–349.
Madden, P. G., and Lauder, G. V. (2003). Pectoral fin locomotion in fishes: Fin ray deformation and material properties. *Integ. Comp. Biol.* **43,** 903.
Martin, R. B., Burr, D. B., and Sharkey, N. A. (1998). "Skeletal Tissue Mechanics." Springer, New York.
Martinez, G., Drucker, E. G., and Summers, A. P. (2003). Under pressure to swim fast. *Integ. Comp. Biol.* **42,** 1273–1274.
Moss, M. L. (1961a). Studies of the acellular bone of teleost fish. I. Morphological and systematic variation. *Acta Anatom.* **46,** 343–362.
Moss, M. L. (1961b). Studies of the acellular bone of teleost fish. II. Response to fracture under normal and acalcemic conditions. *Acta Anatom.* **48,** 46–60.
Moss, M. L. (1977). Skeletal tissue in sharks. *Am. Zool.* **17,** 335–342.

Motta, P. J. (1977). Anatomy and functional morphology of dermal collagen fibers in sharks. *Copeia* **1977,** 454–464.

Mueller, U. K., and van Leeuwen, J. L. (2004). Swimming of larval zebrafish: Ontogeny of body waves and implications for locomotory development. *J. Exp. Biol.* **207,** 853–868.

Ørvig, T. (1951). Histologic studies of Placoderm and fossil elasmobranchs. I. The endoskeleton, with remarks on the hard tissues of lower vertebrates in general. *Arkiv. Zoolog.* **2,** 321–454.

Pawlicki, J. M., Pease, L. B., Pierce, C. M., Startz, T. P., Zhang, Y., and Smith, A. M. (2004). The effect of molluscan glue proteins on gel mechanics. *J. Exp. Biol.* **207,** 1127–1135.

Porter, M., and Summers, A. P. (2004). Biochemical and structural properties of the cartilaginous vertebral column of sharks. *Integ. Comp. Biol.* **43,** 828.

Raven, H. C. (1939). On the anatomy and evolution of the locomotor apparatus of the nipple-tailed ocean sunfish (*Masturus lanceolatus*). *Bull. Am. Mus. Nat. His.* **76,** 143–150.

Rho, J. Y., Mishra, S. R., Chung, K., Bai, J., and Pharr, G. M. (2001). Relationship between ultrastructure and the nanoindentation properties of intramuscular herring bones. *Ann. Biomed. Eng.* **29,** 1082–1088.

Ridewood, W. G. (1921). On the calcification of the vertebral centra in sharks and rays. *Phil. Trans. Roy. Soc. Lond.* **210,** 311–407.

Robson, P., Wright, G. M., Youson, J. H., and Keeley, F. W. (1997). A family of non-collagen-based cartilages in the skeleton of the sea lamprey, *Petromyzon marinus*. *Comp. Biochem. Physiol. B* **118,** 71–78.

Robson, P., Wright, G. M., and Keeley, F. W. (2000). Distinct non-collagen based cartilages comprising the endoskeleton of the Atlantic hagfish, *Myxine glutinosa*. *Anat. Embryol.* **202,** 281–290.

Rockwell, H., Evans, F. G., and Pheasant, H. C. (1938). The comparative morphology of the vertebrate spinal column: its form as related to function. *J. Morphol.* **63,** 87–117.

Root, R. G., Courtland, H.-W., Pell, C. A., Hobson, B., Twohig, E. J., Suter, R. J., Shepherd, W. R., III, Boetticher, N., and Long, J. H., Jr. (1999). Swimming fish and fish-like models: The harmonic structure of undulatory waves suggests that fish actively tune their bodies. "Proceedings of the 11th International Symposium on Unmanned Untethered Submersible Technology," pp. 378–388.

Roy, M. E., Nishimoto, S. K., Rho, J. Y., Bhattacharya, S. K. A., Lin, J. S., and Pharr, G. M. (2001). Correlations between osteocalcin content, degree of mineralization, and mechanical properties of *C. carpio* rib bone. *J. Biomed. Mat. Res.* **54,** 547–553.

Schaefer, J. T., and Summers, A. P. (2005). Batoid wing skeletal structure: Novel morphologies, mechanical implications, and phylogenetic patterns. *J. Morphol.* **264,** 1–16.

Schmitz, R. J. (1995). Ultrastructure and function of cellular components of the intercentral joint in the percoid vertebral column. *J. Morphol.* **226,** 1–24.

Shadwick, R. E., Rapoport, H. S., and Fenger, J. M. (2002). Structure and function of the tuna tail tendons. *Comp. Biochem. Physiol.* **133A,** 1109–1126.

Shephard, K. L. (1994). Functions for fish mucus. *Rev. Fish Biol. Fisheries* **4,** 401–429.

Smith, B. G. (1942). The heterodontid sharks: their natural history, and the external development of *Heterodontus japonicus* based on notes and drawings by Bashford Dean. *In* "The Bashford Dean Memorial Volume—Archaic Fishes," (Gudger, E. W., Ed.), Vol. VIII, pp. 651–769. American Museum of Natural History, New York.

Smith, M. M., and Hall, B. K. (1990). Development and evolutionary origins of vertebrate skeletogenic and odontogenic tissues. *Biol. Rev.* **65,** 277–373.

Summers, A. P. (2000). Stiffening the stingray skeleton—An investigation of durophagy in myliobatid stingrays (Chondrichthyes, Batoidea, Myliobatoidea). *J. Morphol.* **243,** 113–126.

Summers, A. P., and Koob, T. J. (2002). The evolution of tendon. *Comp. Biochem. Physiol.* **133,** 1159–1170.

Summers, A. P., Ketcham, R., and Rowe, T. (2004). Structure and function of the horn shark (Heterodontus francsisi) cranium through ontogeny—The development of a hard prey crusher. *J. Morphol.* **260,** 1–12.
Symmons, S. (1979). Notochordal and elastic components of the axial skeleton of fishes and their functions in locomotion. *J. Zool. (Lond.)* **189,** 157–206.
Thompson, D. A. W. (1917). "On Growth and Form." Cambridge University Press, London.
Thorsen, D. H., and Westneat, M. W. (2005). Diversity of pectoral fin structure and function in fishes with labriform propulsion. *J. Morphol.* **263,** 133–150.
Videler, J. J. (1974). On the interrelationships between morphology and movement in the tail of the cichlid fish *Tilapia nilotica* (L). *Neth. J. Zool.* **25,** 143–194.
Videler, J. J. (1993). "Fish Swimming." Chapman and Hall, London.
Vincent, J. F. V. (1992). "Biomechanics—Materials: A Practical Approach." IRL Press at Oxford University Press, Oxford, UK.
Vincent, J. (1990). "Structural Biomaterials, Revised Ed." Princeton University Press, Princeton, NJ.
Vogel, F., and Gemballa, S. (2000). Locomotory design of 'cyclostome' fishes: Spatial arrangement and architecture of myosepta and lamellae. *Acta Zoolog. (Stockholm)* **81,** 267–283.
Vogel, S. (2003). "Comparative Biomechanics: Life's Physical World." Princeton University Press, Princeton, NJ.
Wainwright, S. A., Biggs, W. D., Currey, J. D., and Gosline, J. M. (1976). "Mechanical Design in Organisms." Princeton University Press, Princeton, NJ.
Wainwright, S. A., Vosburgh, F., and Hebrank, J. H. (1978). Shark skin: Function in locomotion. *Science* **202,** 747–749.
Webb, J. E. (1973). The role of the notochord in forward and reverse swimming and burrowing in the amphioxus *Branchiostoma lanceolatum. J. Zool. (Lond.)* **170,** 325–338.
Webb, P. W. (1973). Kinematics of pectoral fin propulsion in *Cymatogaster aggregata. J. Exp. Biol.* **59,** 697–710.
Webb, P. W., Hardy, D. H., and Mehl, V. L. (1992). The effect of armored skin on the swimming of longnose gar, *Lepisosseus osseus. Can. J. Zool.* **70,** 1173–1179.
Weitbrecht, G. W., and Gemballa, S. (2001). Arrangement and architecture of myosepta in the lancelet *Branchiostoma lanceolatum. J. Morphol.* **248,** 299.
Westneat, M. W. (2004). Evolution of levers and linkages in the feeding mechanisms of fishes. *Integ. Comp. Biol.* **44,** 378–389.
Westneat, M. W., Hoese, W., Pell, C. A., and Wainwright, S. A. (1993). The horizontal septum: Mechanisms of force transfer in locomotion of scombrid fishes (Scombridae, Perciformes). *J. Morphol.* **217,** 183–204.
Wilga, C. D., and Lauder, G. V. (2000). Three-dimensional kinematics and wake structure of the pectoral fins during locomotion in leopard sharks Triakis semifasciata. *J. Exp. Biol.* **203,** 2261–2278.
Wright, G. M., Keeley, F. W., and Youson, J. H. (1983a). Lamprin: A new vertebrate protein comprising the major structural protein of adult lamprey cartilage. *Experientia* **39,** 495–497.
Wright, G. M., Youson, J. H., and Keeley, F. W. (1983b). Lamprey cartilage: A new type of vertebrate cartilage. *Anat. Rec.* **205,** 221A.
Wright, G. M., Keeley, F. W., Youson, J. H., and Babineau, D. L. (1984). Cartilage in the Atlantic hagfish, *Myxinus glutinosa. Am. J. Anat.* **169,** 407–424.
Yanega, G., and Rubega, M. (2003). Placing intramandibular kinesis in hummingbirds in the context of tetrapod feeding evolution. *Integ. Comp. Biol.* **43,** 983–983.

6

FUNCTIONAL PROPERTIES OF SKELETAL MUSCLE

DOUGLAS A. SYME

I. INTRODUCTION

The majority of studies on the anatomy and contractile characteristics of fish skeletal muscle have focused on species that employ some degree of axial body undulation to power locomotion, typically anguilliform, carangiform, subcarangiform, and thunniform. This bending is generated by contraction of the myotomal skeletal muscles. A single, myotomal "muscle" may effectively act across dozens of vertebral joints and must act in concert with perhaps hundreds of other "muscles" both in parallel and in series, and with both antagonists and agonists. Swimming in such fishes is powered by a propagated wave of bending that progresses caudally from near the head. This requires spatial and temporal coordination of muscle contraction longitudinally (i.e., axially), perhaps radially, contra- and ipsilaterally, and in

Fish Biomechanics : Volume 23
FISH PHYSIOLOGY

DOI: 10.1016/S1546-5098(05)23006-6

many cases across different types of muscle. The muscles are charged with producing adequate power, using the same basic kinetic and kinematic strategy, over a wide range of swimming speeds and temperatures. Here I discuss muscle contraction in the context of such variation (particularly axial, fiber type, and thermal), both among species and within individuals; Chapter 7 focuses on integration of the muscles into a swimming animal.

Bone (1978) prefaced an earlier review of fiber types in fish muscle with the premise that two types of muscle with distinctly different characteristics are used to power body undulations during slow and fast swimming (see also Bone, 1966), but reiterated J. L. Austin's warning concerning the dangers of "tidy dichotomies." Almost three decades later, his premise remains bolstered by considerable experimental evidence, but, true to the warning, is now riddled with exceptions. Changes in muscle with development, training, and environment certainly complicate the practice of making definitive statements about a fiber's "type" (Sanger and Stoiber, 2001). Up to seven different fiber types have been described in some species based on histochemical evidence alone; the validity of these distinctions based on function is unknown and perhaps unlikely (Bone, 1978). While not the first to propose the idea, Lawrence C. Rome and his colleagues were among the first to take up the ambitious task of providing rigorous evidence that fast and slow swimming would require fast and slow muscles to power them, and in so doing proposed answers to two key questions: why animals (fishes) have different muscle fiber types (Rome *et al.*, 1988) and how fishes power swimming (Rome *et al.*, 1993). These studies and the original ideas themselves, that two fiber types power two forms of locomotion, provide a foundation on which to discuss new ideas about muscle design, contractile function of different muscles in fish, and ultimately how fishes power swimming.

The notion that two different fiber types are used by fishes to power fast and slow swimming can be captured in an analogy. Consider a bicycle (fish) with two riders (muscle types), each with their own gear (orientation) linked to a chain (tendon, skin, etc.) and the wheels (fins). The absolute range of speeds over which the bicycle can move and how long it can move are determined by its anatomy and the physiology of the riders. One rider is small and slow, yet possesses great stamina, while the other is large, fast, and powerful, yet rarely exercises. Using an appropriate gear, the small rider can propel the bicycle slowly over long distances; using a different but appropriate gear the large rider can power the bicycle rapidly over a short distance. Each rider excels at one extreme but not the other, and they contribute based on their abilities and the speed the bicycle is traveling. This two-muscle, two-gear model has simple and intuitive appeal, and has influenced greatly the way we think about how muscles are used in fish to power swimming. Accounts of new types of riders and new types of gears have broadened the binary confines of the original model, but the underlying idea, that different fiber types with different mechanical and metabolic capacities power different types of activities economically, effectively, or both, remains the central doctrine of how we perceive and study fish muscle. I will limit my discussions to the physiology of the muscle *per se*, but will hint at how their design and function appear to be rooted in how fish use them to move, which will be explored more fully in the following chapters.

In many respects, the mechanical performance of the axial skeletal muscles of fish is largely indistinguishable from that of other vertebrate skeletal muscles, qualitatively if not also quantitatively. I will relate the contractile characteristics of fish muscles most strictly within the confines of fishes, and so the reader will be reminded that many of these observations have been made previously on the more thoroughly studied muscles of amphibians and mammals. I will highlight characteristics that are particularly fishy, considered in the context of how they impinge on and have a basis in the ability to swim in these usually ectothermic, unusually diverse, and anatomically distinctive animals.

Despite the relatively recent arrival of fish muscle under the microscope of the experimental physiologist, progress in our understanding of how the muscles of fishes are designed, function, and are used to power swimming has been remarkable and now rivals and probably exceeds that of the terrestrial condition. This stems in large part from the oft-championed anatomical separation of the different fiber types in fishes, their apparent orderly recruitment patterns, the contrasting and relatively stereotypical steady "cruise" and unsteady "burst" swimming behaviors of most fishes, and no doubt from the inherent fascination with an animal that thrives over a staggering range of thermal environments. Others have summarized, in various contexts, many of the topics that I will touch upon (e.g., Johnston, 1980a, 1981, 1994; Rome, 1986, 1994; Altringham, 1994; Jayne and Lauder, 1994; van Leeuwen, 1995; Wardle *et al.*, 1995; Shadwick *et al.*, 1998; Altringham and Ellerby, 1999; Altringham and Shadwick, 2001; Sanger and Stoiber, 2001; Watabe, 2002; Coughlin, 2002a, 2003).

II. ULTRASTRUCTURE

The sarcomeric structure of fish myotomal muscle is fundamentally similar to that of the skeletal muscles of most other animals, consisting of overlapping actin and myosin filaments, the former anchored to dense Z-discs. Detailed accounts of the ultrastructure of locomotor muscle in fish can be found elsewhere (e.g., Bone, 1978; Johnston, 1980a, 1981; Sanger, 1992; Luther *et al.*, 1995; Sanger and Stoiber, 2001). Here I only highlight several intriguing characteristics potentially relevant to understanding how fish muscles may be used in swimming. The sarcolemmal transverse tubules in fish and lamprey trunk/myotomal muscles form triads with the terminal cisternae of the sarcoplasmic reticulum at the level of the Z-line, similar to what is observed in frogs. While in striated muscles of most other animals, including hagfish myotomal muscle and fish extraocular and swimbladder muscles, the triads are located at the junction of the A-I bands in the sarcomere (Smith, 1966; Johnston, 1981). The functional consequences of differing placements of the triads are not understood, but do not appear to bear in any important way on muscle speed or contraction kinetics. With regard to speed, fish muscles contain varied amounts of the soluble calcium-binding protein parvalbumin, which is suggested to play an important role in activation kinetics, particularly relaxation. The content and isoform vary by fiber type, muscle location, species, and ontogeny, perhaps belying a role in fine-tuning contraction kinetics (Sanger and Stoiber, 2001). High concentrations of parvalbumin may facilitate high-frequency contractions (reviewed in Syme and Josephson, 2002, and references therein). Indeed, parvalbumin is found in fast-twitch muscles, but its presence in slow red muscle is less certain (Gerday *et al.*, 1979; Zawadowska and Supikova, 1992). However, the speeds at which muscle must operate before the assistance of parvalbumin may be needed seem well beyond the realm of red and even white muscle in most cases. Perhaps it confers an energetic benefit (Rome and Klimov, 2000).

The stretch proprioceptors, muscle spindles, are conspicuous by their absence in the axial musculature of fishes. While central pattern generators can effectively drive rhythmic and relatively stereotypical body undulations in the absence of feedback from intrafusal proprioceptors, such feedback usually does modify the rhythm, and it would be surprising to find that there is no such feedback in fishes. Naked nerve endings in teleosts and limited proprioceptive structures in elasmobranchs have been described (Roberts, 1981) and may provide some sensory feedback. The lack of a strong acute thermal response in the timing of muscle activation during swimming, which results in a notable impairment of performance (discussed later), may be further evidence that proprioceptive tuning of swimming is absent or slight.

III. FIBER TYPES

A. Physiology

As in other animals, individual muscle cells (fibers) in the trunk musculature of fishes have signature metabolic and contractile profiles. These properties endow specific fibers with the capacity to effectively power specific types of activity. Any particular profile is somewhat exclusive of others and is thus largely to the disadvantage of certain types of activity. Current terminology defines three categories of fiber types in fishes based on color (red, pink, and white) and hence myoglobin content. Myoglobin content reflects oxidative capacity, and as metabolic and contractile characteristics tend to follow suit, color is also a crude but convenient surrogate to the mechanical characteristics of the fiber type. However, despite the preponderance of this three-fiber-type taxonomy, most fishes certainly have more than three fiber types (and perhaps some have fewer) depending on whether metabolic, histochemical, ultrastructural, contractile, protein isoform, or functional characteristics are assessed, excluding the influence of ontogeny (e.g., Sanger and Stoiber, 2001). Johnston (1985) stressed that it is not possible to reliably summarize the capacities of a muscle fiber using a single characteristic for classification, and Bone (1978) was no doubt correct in principle that we would be wise to abandon attempts to classify muscles based on a naive nomenclature of color, but the system has persisted and proven useful (Johnston, 1981).

Of all the cellular elements that determine the mechanical performance of a muscle cell, the particular myosin isoform present in the thick filaments is probably the most influential (it is normally identified using histochemical or immunochemical staining or gel electrophoresis). Goldspink *et al.* (2001) provided a review of myosin expression, regulation, and responses to growth, exercise, and temperature in fish muscle. Pink and white muscles share the same "fast" myosin light-chain complement, while red has a distinct "slow" isoform; the myosin heavy-chain complements of mature red, pink, white, as well as some other described fiber types are all unique (Sanger and Stoiber, 2001). The unique myosin heavy-chain isoforms in red, pink, and white fibers endow them with progressively increasing speed and power potential, and they appear to be recruited in this respective order with increased swimming speed (Figure 6.1) (e.g., Johnston *et al.*, 1977; Coughlin and Rome, 1999). There are gradual changes in the myosin isoform complement of muscle fibers during development conferring faster contraction kinetics in younger, smaller fish, as well as zones of heavy-chain transition between distinctly homogenous regions of fiber types (Scapolo *et al.*, 1988; Johnston, 1994; Coughlin *et al.*, 2001a; Goldspink *et al.*, 2001; Weaver *et al.*, 2001).

Red fibers are also commonly referred to as slow (based on their mechanical response), oxidative (based on their metabolic profile), or type I fibers (based on their myosin heavy-chain isoform). They form a relatively small component of the total muscle mass, about 10% in most fish but can be up to 30% in some and are entirely absent in a species of stickleback (Sanger and Stoiber, 2001). They are typically small in diameter ($\sim$30 μm, about half that of white fibers), have a high myoglobin and lipid content and high

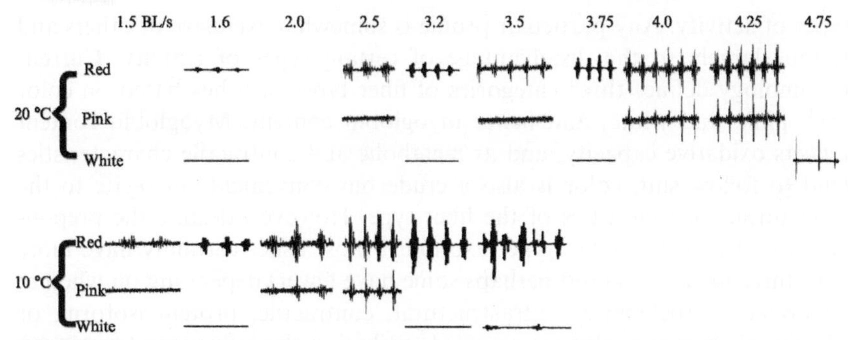

Fig. 6.1. EMG activity in red, pink, and white muscles of scup when swimming at different speeds (body lengths per second) and temperatures. Upper three traces are at 20 °C; lower three traces are at 10 °C. Each column represents a different swim speed as indicated at the top. Note the increased EMG activity with increased swim speed, the need to recruit faster fiber types (pink then white) at higher speeds, and the need to recruit faster fiber types at slower speeds in colder temperatures. (Adapted from Rome *et al.*, 1992a, and Coughlin and Rome, 1999, with permission of the Company of Biologists Ltd. and the Biological Bulletin.)

capillarization, have a surprisingly well-developed sarcoplasmic reticular system for a slow muscle, have a relatively low myofibrillar ATPase rate and low creatine phosphokinase activity, are remarkably 20–50% mitochondria by volume and thus have a relatively low myofibril volume density, have metabolic enzyme profiles that support high oxidative (very high succinate dehydrogenase [SDH] levels) and low anaerobic activity, and have fiber volume densities ranging from about 40 to perhaps 90% (reviewed in Johnston *et al.*, 1977; Bone, 1978; George and Stevens, 1978; Mosse, 1978; Johnston, 1981, 1985; Bone *et al.*, 1986; Rome and Sosnicki, 1990; Sanger and Stoiber, 2001). In the trunk, the fibers run roughly parallel with the long axis of the fish and are usually located superficially, approximately under the lateral line. A notable exception is in some scombrids, particularly tuna, and in lamnid sharks, where they either form a deep lateral wedge or are internalized and lie in close proximity to the spine (reviewed in Altringham and Shadwick, 2001); the functional consequences of this placement are discussed later. The fibers span the myosepta and thus their length depends on the size of the fish, the number of myotomes in series along its length, and the particular axial and radial location in which they are located; however, a length of about 1–10 mm is typical.

White (phasic or fast) fibers constitute much of the remainder of the trunk musculature, about 90%, and are clearly specialized for short duration bursts of high power output. They have a relatively large diameter (~75 μm and sometimes considerably larger); again, their length is variable depending on the fish and its anatomy, have a low myoglobin and lipid content and low capillarization (about one-sixth that of red fibers), have a highly developed sarcoplasmic reticulum and high concentrations of parvalbumin, have a relatively high myofibrillar ATPase rate and high creatine phosphokinase activity, containing from less than 1 to 4% mitochondria by volume, and have metabolic enzyme profiles that support low oxidative activity (no SDH activity) but with anaerobic activities that vary widely between species and are often not unlike those of red fibers (Johnston, 1980a, 1981; Bone *et al.*, 1986; Sanger and Stoiber, 2001). The myofibril volume density of white fibers is 75–95% (Sanger and Stoiber, 2001), substantially greater than that of red fibers due to the low mitochondrial volume in white fibers.

While less is known about pink (intermediate, fast-red/aerobic) fibers, they appear intermediate to red and white in most regards including distribution, oxidative capacity, fiber diameter, fatigue resistance, contraction kinetics, power output, optimal operating frequency, etc. (reviewed in Sanger and Stoiber, 2001). Like red fibers, they run parallel to the long axis of the body. They are sandwiched between the superficial red and proximal white muscles and can rival or slightly exceed red in mass (Johnston, 1980a, 1981; Coughlin *et al.*, 1996). The role of pink muscle in swimming has only recently been given careful consideration. They are recruited at intermediate swimming speeds (Figure 6.1), and their metabolic profile suggests that they help power fast but sustained swimming (Johnston *et al.*, 1977; Coughlin and Rome, 1999). Their relatively fast contraction kinetics and anatomical location suggest that they are employed when and where the slower contraction kinetics of red muscle are limiting (Coughlin and Rome, 1996, 1999; Coughlin *et al.*, 1996). They appear to play a critical role in powering swimming at cold temperatures where red muscle is no longer adequate due to its slow speed (Rome *et al.*, 2000).

Other fiber types have been described (e.g., red muscle rim fibers, transitional fibers, tonic fibers, superficial); see Sanger and Stoiber (2001) for a review. They form a relatively small component of the total fiber distribution and it is unlikely that they contribute significantly to powering swimming, thus they are not considered further here.

The myofibrillar ATPase rate of red muscle is about 2- to 5-fold slower than that of white muscle and about 2-fold slower than that of pink (reviewed in Johnston *et al.*, 1977; Johnston, 1980a, 1981; Rome *et al.*, 1999). Accordingly, the maximal velocity of shortening (V_{max}) of red fibers is slower than that of white (e.g., about half in the dogfish [Lou *et al.*, 2002]) and pink muscle. Likewise, the half width of the intracellular free calcium transient of red fibers is slower than that of white (about 5 fold longer in toadfish red than in white muscle; Rome *et al.*, 1996), resulting in slower twitch kinetics in red than in white muscle (see Figure 6.3). With these slow kinetics, red fibers are not capable of producing the rapid oscillations in force required to produce rapid tail beats for fast swimming but are well suited to power slow body undulations for cruise swimming (discussed later). Peake and Farrell (2004) noted a progressive switch from aerobic to aerobic/anaerobic to exclusively anaerobic metabolism with increased swim speed in smallmouth bass, suggesting a red to red/white to white sequence of muscle recruitment. Direct measures confirm that slow fibers are activated during steady, cruise swimming and pink fibers at intermediate speeds, while white fibers with their high myofibril ATPase rates and fast calcium transients/twitches power burst swimming activity such as kick-and-glide and the startle response (Figure 6.1) (e.g., Rayner and Keenan, 1967; Johnston *et al.*, 1977, 1993; Bone, 1978; Johnston, 1980b, 1981; Rome *et al.*, 1984, 1988, 1992a; Sisson and Sidell, 1987; Jayne and Lauder, 1994; Johnson *et al.*, 1994; Hammond *et al.*, 1998; Knower *et al.*, 1999; Shadwick *et al.*, 1999; Coughlin and Rome, 1999; Ellerby *et al.*, 2000; Ellerby and Altringham, 2001; Sanger and Stoiber, 2001).

While white fibers are the main source of power during high-speed swimming, continued recruitment of red fibers at high speeds may aid power production (Johnson *et al.*, 1994) or force transmission (Altringham and Ellerby, 1999, and references therein). Indeed, the sarcoplasmic reticulum and transverse tubule system in fish red muscle is extremely well developed, rivaling that of fast-twitch fibers in mammals (see previous discussion), which may bestow red muscle with the ability to contribute power at relatively high swimming speeds. Additionally, measurements in tuna indicate that their red muscles have ATPase rates conspicuously higher (perhaps double) than red muscles of most other fish, suggesting that red muscles in tuna may be well suited to power relatively high speeds of swimming (Johnston and Tota, 1974). Conversely, while red fibers are certainly the major source of power during sustained cruise swimming, white and pink fibers also appear to be recruited at these swim speeds in many teleosts, although Sanger and Stoiber (2001) argued that the limited aerobic capacity of white fibers will severely restrict a sustained contribution. Notably, in elasmobranchs and more primitive teleosts the division of tasks into red for cruise and white for burst appears more discrete and absolute (Bone, 1978; Johnston, 1980a, 1981).

B. Distribution

The axial and radial distribution of the different fiber types in the body have important consequences for their contributions to powering movement; Ellerby *et al.* (2000) provided a review and discussion of the functional consequences of muscle distribution. In general, red muscle is restricted to a thin, superficial, lateral wedge in the vicinity of the lateral line; pink muscle, if present, is located medial to the red; and white muscle makes up the remainder and typically the bulk of the muscle mass. Some scombrids (particularly tuna), and lamnid and alopiid sharks are notable exceptions, with their red muscle positioned more medially and often opposed to the spine (Fierstine and Walters, 1968; Westneat *et al.*, 1993; Ellerby *et al.*, 2000), and with the potential to elevate muscle temperature above ambient. The implications of this, in the context of warmer temperatures and mechanical advantage, are discussed below and in Chapter 7.

There is remarkable variability in red muscle distribution along the length of the body of fishes (Ellerby *et al.*, 2000), almost certainly related to variability in swimming styles, which has formed the basis of much debate about the function of red muscle: why its distribution may vary along the length of the fish, and how fishes actually power swimming using red muscle. In eels, the slow muscle is relatively evenly distributed along the body length, suggesting an even axial distribution of power in this anguilliform swimmer (Ellerby *et al.*, 2000). In scup, the relative proportion of the body cross-section occupied by red fibers increases from head to tail, but with the greatest absolute area occurring just caudal to mid-body (Zhang *et al.*, 1996). Likewise, in rainbow trout and brook char, the relative proportion of red fibers is maximal at about 65% fork length from the snout (Ellerby *et al.*, 2000; McGlinchey *et al.*, 2001). Initially, at least, this would suggest that the majority of power comes from more caudal myotomes in these fishes, but as discussed later, information about strain and activation patterns is required before this can be confirmed. In mackerel and bonito, the relative area of red muscle peaks about mid-body but is biased toward the caudal end; in tuna, it is also most prevalent at mid-body but is sometimes skewed toward the anterior of the fish, likely related to their highly derived mode of swimming (Graham *et al.*, 1983; Ellerby *et al.*, 2000; Altringham and Shadwick, 2001). Interestingly, the molecular mass of titin also increases moving from anterior to posterior in both red and white muscle, and it has been suggested that increasing muscle strains in posterior myotomes may modulate titin expression (Spierts *et al.*, 1997). The marine scup is the only species for which the distribution of pink fibers has been examined in great detail; their relative distribution does not change along the length of the fish (Zhang *et al.*, 1996). Whether this is a common or unusual condition is not known. Pink fibers are not found in European eels (Egginton and Johnston, 1982).

IV. PATTERNS OF INNERVATION

Myotomal red fibers of fishes receive polyneuronal (usually at least two neurons) and polyterminal innervation. They are cholinergic and typically tonic, producing graded membrane (junction) potentials but also action potentials (Stanfield, 1972; reviewed in Bone, 1978; Johnston, 1980a, 1981, 1985; Granzier *et al.*, 1983; Akster *et al.*, 1985). Some appear to produce only action potentials in response to a single stimulus and thus may not be tonic (Altringham and Johnston, 1988a; Curtin and Woledge, 1993a). The nerve terminals are dispersed along the length of the fibers, with the average spacing appearing related to the space constant of the membrane potential such that a safety factor for activation is maintained (Bone, 1978). However, there exists considerable variability in the spacing along individual fibers, between species, and with growth, and it is not known if the spacing is maintained for the terminals of a single axon as would be expected if a safety factor was the objective (Archer *et al.*, 1990).

The pattern of innervation of white fibers differs between species (Bone, 1978; Johnston, 1980a, 1981, 1983; Ono, 1983). In agnathans, elasmobranchs, dipnoans, chondrosteans, holosteans, and some primitive teleosts (e.g., eels, herring), the fibers are polyneurally innervated but with motor end plates clustered (focal) at one and rarely both ends of the fiber; the muscle cells respond to stimulation with action potentials. The more advanced teleosts as well as some primitive species (e.g., salmonidae) also receive polyneuronal innervation, however; much like with red fibers, the nerve terminals are spread more diffusely along the fiber length (Bone, 1964; Altringham and Johnston, 1981). The functional significance of the different patterns of innervation in fast fibers awaits clarification. The tetanic fusion frequency of focally innervated fast fibers (from skate and eel) is about 3- to 10-fold slower than for diffusely innervated fast fibers (from cod or sculpin) (Johnston, 1980b; Altringham and Johnston, 1988b). Hence, focally innervated fibers show less potential for graded force production. A related and intriguing observation is that fishes with focally innervated white fibers tend to recruit them only during burst activity, while fish with a more diffuse distribution of terminals appear to recruit their fast fibers even during sustained swimming. Further, a higher mitochondrial and capillary density in fast fibers with diffuse terminal distributions suggests the design may facilitate graded contributions of fast fibers during slower, sustained swimming (discussed in Johnston, 1981; Altringham and Johnston, 1988a; Sanger and Stoiber, 2001).

From the investigator's perspective, early work suggests that low-intensity stimulation of axons innervating white fibers results in junction potentials and graded contractions, while supramaximal stimulation produces propagating action potentials in response to as few as one junction potential in some muscles (reviewed in Bone, 1978). Yet other observations show that white fibers respond to stimulation from a single neuron with all-or-none action potentials, and there do not appear to be notable differences in the mechanical performance of fibers with focal or diffuse innervation patterns

(Altringham and Johnston, 1988a,b; see Archer *et al.*, 1990 for a review and discussion). Sculpin white and red fibers (Altringham and Johnston, 1988a,b) and dogfish white fibers (Curtin and Woledge, 1988) are capable of producing action potentials and thus respond well to direct electrical stimulation of the muscle. Some preparations appear to be activated primarily through the motor end plates (Johnson *et al.*, 1991a; Rome *et al.*, 1992b), hence the use of curare or acetylcholinesterase blockers on such preparations has marked effects on performance. For example, Rome and Sosnicki (1990) and Granzier *et al.* (1983) observed that it was not possible to attain maximal contractions in carp red muscle using direct electrical stimulation, and based on experiments using esserine and curare suggested that these red fibers may not produce action potentials but rather may rely on local, synaptic depolarization. The same may be true for the marine scup (Rome *et al.*, 1992b). Red fibers of cod also have a high stimulus threshold for direct electrical activation, suggesting that they too may not produce action potentials (D.A.S., personal observation). Cold temperatures appear to reduce synaptic efficacy, so diffuse, polyterminal innervation may serve as a safeguard against activation failure in the cold (Adams, 1989; see also Archer *et al.*, 1990). It has also been suggested that the pattern of polyneuronal innervation along sequential myotomes may serve to modulate the number of stimuli and thus force and power output during swimming (Altringham and Johnston, 1988a).

V. MECHANICS OF CONTRACTION

Because of the diversity in species, their habits and habitats (particularly thermal), fiber types, anatomical considerations, and approaches used to study the mechanics of fish muscle, it would be impractical and misleading to attempt to argue what the power of fish muscle is, how fast or how strong it is, etc. Rather, I will emphasize this diversity and highlight factors that influence performance, for this is the reason that we find such fascination with the muscles of fishes. In so doing there will be many numerical examples, but the reader should bear in mind that these are intended only to illustrate an idea; the exact figures are specific to circumstances. I first describe some fundamental characteristics of muscle mechanics in fish, including the length–tension relationship, the ability to generate isometric force, and rates of activation and relaxation, then discuss in some detail what we know about the ability of fish muscles to produce work and power, and how measurements that we make on isolated muscle do and do not bear on what occurs in swimming fishes. The final sections, dealing with scaling, axial variation, and the effects of temperature, draw on these discussions and place measurements of mechanics in the context of the anatomy of a swimming fish and its thermal environment.

A. Length–Tension Characteristics

The myofilament lengths of vertebrate skeletal muscles are remarkably conservative, thus sarcomere dimensions and length–tension characteristics of fish muscle are, not surprisingly, frog-like (e.g., Sosnicki *et al.*, 1991). Where does fish muscle operate on its length–tension relationship during swimming? Muscle strain (length change) is relatively straightforward to determine in a swimming fish, but absolute length is considerably more difficult to measure. Sosnicki *et al.* (1991), in an exhaustive study of anatomy and bending kinematics in carp, determined sarcomere lengths at different degrees of body curvature and showed that both red and white fibers operate over or near the plateau of their length–tension relationships during swimming, rarely shortening to lengths where less than 90% of maximal force is produced. Although this finding remains to be verified in more species, it seems like a sensible design, and in an animal like a fish where the contralateral axial muscles are a symmetrical antagonistic pair it would be quite surprising to find that the muscles were not operating over the plateau of their length–tension relationship. As indirect evidence, the changes in muscle length (i.e., strain) during cruise and sprint swimming, even C-starts, are relatively modest, typically 3–15% (e.g., Rome, 1994; Franklin and Johnston, 1997). With such strains, if the muscles operate over the plateau of their length–tension relationship they would produce no less than 85% of maximum force and usually well over 90% (Rome, 1994). The location and orientation of the fibers in the fish also have important consequences for where they operate on their length–tension relationships during swimming. These details are beyond the scope of this chapter, but in general white fibers are oriented such that they can power high-speed and extreme body undulations associated with the escape response, while red fibers can power relatively smaller amplitude and slower undulations while still operating over favorable portions of their length–tension relationship (Rome, 1994).

B. Twitches and Tetani

Although there are considerable differences between fiber types in their metabolic and dynamic contractile characteristics, when corrected for the content of intercellular adipose tissue, connective tissue, and water, they typically produce similar levels of isometric, tetanic force per cross-sectional area of muscle fiber, both within and across species (e.g., Granzier *et al.*, 1983; Akster, 1985; Akster *et al.*, 1985; Rome and Sosnicki, 1990; Johnson and Johnston,

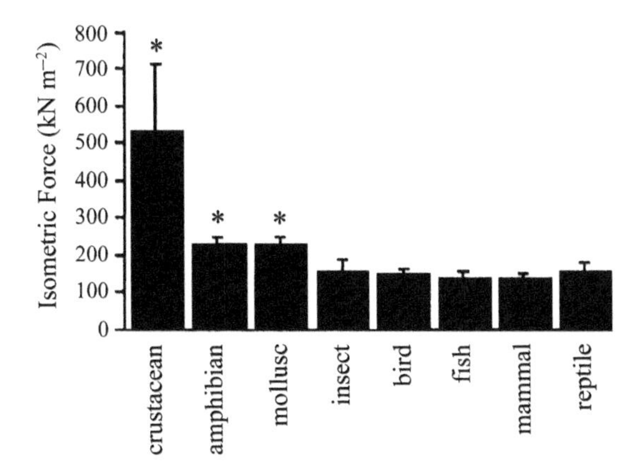

Fig. 6.2. Isometric tetanic force produced by some animals in different taxonomic groups. Asterisks indicate significantly greater forces in these groups than the rest ($p < 0.05$). (Adapted from Medler, 2002; used with permission.)

1991a; Rome and Swank, 1992; Coughlin *et al.*, 1996; reviewed in Medler, 2002), but not always (e.g., Lou *et al.*, 2002). However, reports of force per cross-section of muscle, which includes these intercellular components, vary widely in the literature, likely due in large part to variability in the intercellular content. Values can be as low as 65 kNm^{-2} for red muscle in yellow phase eels (Ellerby *et al.*, 2001a), typically fall between 100 and 250 kNm^{-2} in living bundles of muscles and skinned fibers (e.g., Curtin and Woledge, 1988; Luiker and Stevens, 1992; Altringham *et al.*, 1993; Wakeling and Johnston, 1998; Lou *et al.*, 2002; Medler, 2002), and can range upward of 300 kNm^{-2} in sculpin fast muscle (Langfeld *et al.*, 1989). While reported values of isometric force from fish muscle (both slow and fast) tend to fall below values typically reported for frogs, crustaceans, and molluscs, they are not atypical in the context of most other vertebrates (Figure 6.2). High-frequency sound-producing (swimbladder) muscles in toadfish appear to be an exception, producing only about one-tenth the force of the axial muscles due to extreme modifications in cross-bridge kinetics for high-speed operation (Rome *et al.*, 1999).

Quite unlike the stability of isometric force, the speed of twitch contractions increases from red to pink to white muscle (Figure 6.3) (e.g., Granzier *et al.*, 1983; Akster *et al.*, 1984; Akster, 1985; Coughlin *et al.*, 1996; Coughlin and Rome, 1996; Rome *et al.*, 1996; Coughlin, 2003). Twitch relaxation times in red muscle are about double those of white muscle (e.g., Coughlin *et al.*, 1996; Ellerby *et al.*, 2001a, 2001b), reflecting predominantly the large differences in the duration of the intracellular free calcium transient

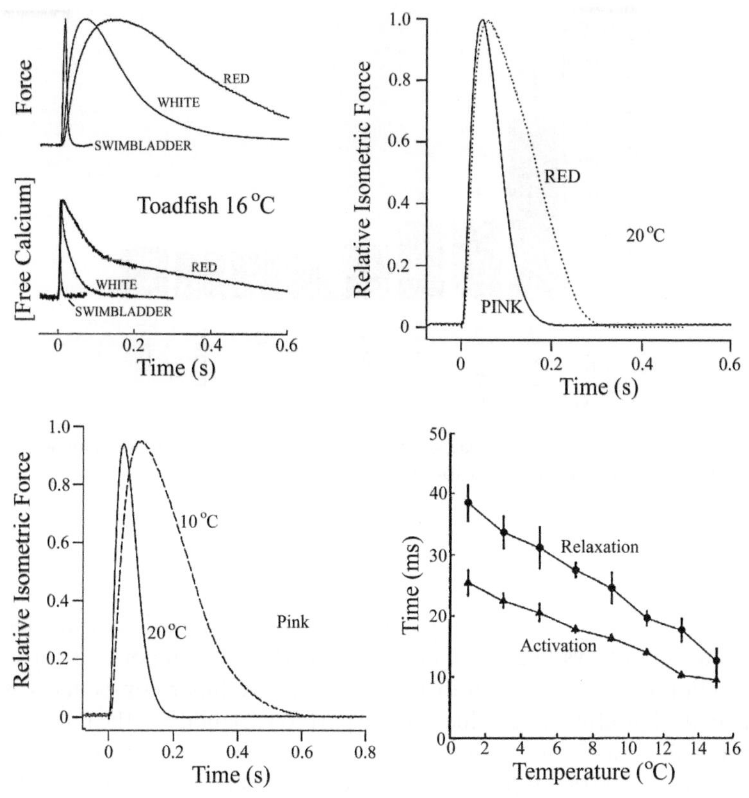

Fig. 6.3. Effects of fiber type and temperature on isometric twitches. Upper left are twitch force and the intracellular free calcium transients from red, white, and high-speed swimbladder muscle of toadfish. Upper right is twitch force from red and pink muscles of scup at 20 °C. Lower left is twitch force from pink muscle of scup at 10 and 20 °C. Lower right are twitch durations measured as the time for force to rise to half-maximal (activation) or to relax to half-maximal (relaxation) at different temperatures in sculpin fast fibers. (Adapted from Rome *et al.,* 1996, Coughlin *et al.,* 1996, and Langfeld *et al.,* 1989, with permission of the Company of Biologists Ltd. and copyright National Academy of Sciences USA.)

(Figure 6.3), among other potential factors (Rome *et al.*, 1996). The slower contraction kinetics of red muscle versus pink, and pink versus white, result in force becoming fused at slower stimulus frequencies in slower muscles. This has important implications for the ability of the different fiber types to power movements at different tail-beat frequencies, as twitch kinetics will critically limit the maximum cycle frequency at which a muscle can generate power (e.g., Wardle, 1975; Marsh, 1990; Rome *et al.*, 1996). These limitations also apply to a given fiber type when compared across species. For example,

Coughlin (2003) noted that the relaxation and contraction times of red muscles from rainbow trout are faster than for scup when compared at the same temperature, and Ellerby *et al.* (2001a) noted the relatively short relaxation times and high operating frequencies in red muscle of scup versus the long relaxation times and slow operating frequencies in red muscle of eels.

C. Work, Power, and Force-Velocity Characteristics

In the context of muscles used to power swimming, work is manifested as a muscle shortens against a load (a concentric contraction); it is this work that makes animals move. Muscles also absorb work when they are lengthened (eccentric contraction), producing heat or perhaps storing energy in compliant elements to be used later. Work has units of energy, typically Joules (Nm), and is calculated as the integral of force with respect to muscle length change (i.e., strain), assuming the force and strain vectors are parallel. Hence, factors that influence force (N) or muscle shortening (m) will influence the work done by muscles. The amount of work done by a muscle is normally quantified relative to some defined strain, such as work done during a tail-beat, or during the shortening or lengthening portion thereof. This gives rise to the commonly cited quantities "shortening work" (sometimes called positive work), which is the work done by the muscle while it shortens, and "lengthening work" (sometimes called negative work), which is the work required to lengthen the muscle and is work/energy absorbed by the muscle. Another quantity, "net work," is a derived term that falls out of the work loop technique (described later) and is simply the difference between shortening and lengthening work during a complete cycle of shortening and stretch; it is the "net" energy produced (or in some cases absorbed) by the muscle during one complete cycle (i.e., tail-beat). Muscles have passive viscoelastic properties, thus some of the work done while shortening is from elastic recoil of structures that were stretched during muscle lengthening, and some work is required to overcome viscous resistance during both stretch and shortening (Syme, 1990). Net work accounts crudely for these elements and reflects the net contribution the contractile elements in muscle make toward powering movement.

Power is the rate of doing work and is typically expressed in Watts (Js^{-1}). Power can be extracted from force-velocity data as the product of force and associated shortening velocity (Figure 6.4); in such cases it describes the instantaneous ability of a (usually) fully activated muscle to generate power. Force changes inversely with shortening velocity, and so with increasing shortening velocity power rises to a maximum and then falls (Figure 6.4). As is typical of most skeletal muscles, fully activated myotomal muscles of fishes produce maximal power when shortening at about one-third of their

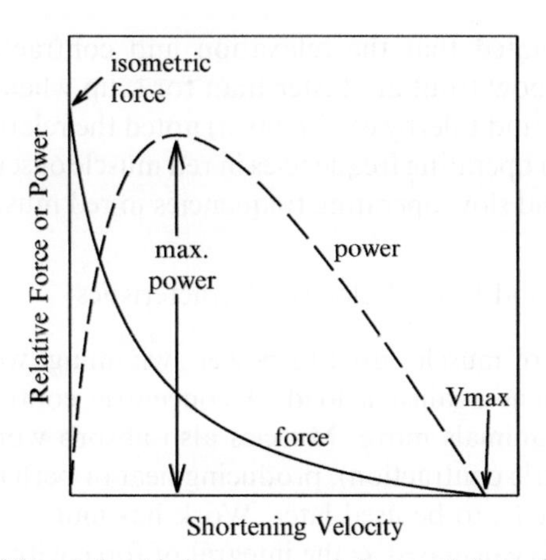

Fig. 6.4. Illustration of the relationship between muscle force and shortening velocity (solid line) and between power output and shortening velocity (broken line) in fully activated muscle during constant velocity shortening. Power is the product of force and shortening velocity and is maximal at an intermediate shortening velocity, typically about one-third of the maximal velocity of shortening (V_{max}).

maximal velocity of shortening (V_{max}); that is, when shortening at a velocity (V) such that the ratio V:V_{max} is about 1:3 (e.g., Curtin and Woledge, 1988; Rome *et al.*, 1988; Rome and Sosnicki, 1990; Coughlin *et al.*, 1996; James and Johnston, 1998; Lou *et al.*, 2002; Medler, 2002).

It might be expected that during swimming muscles would operate in a manner that produces maximal or near-maximal power. Comparison of the velocity of muscle shortening during swimming with V_{max} suggests that this is true during fast starts (Johnston *et al.*, 1995; James and Johnston, 1998) and in slow muscle of carp during steady swimming (Rome *et al.*, 1988). Other evidence suggests that some fishes do not use their myotomal muscles during swimming in a manner that maximizes power output. Muscles shorten at differing fractions of V_{max} along the length of the fish during the initial C-bend of an escape response, suggesting that power can not truly be maximized everywhere along the length of the body (James and Johnston, 1998). Further, studies using strain and activation parameters measured from swimming fishes show that their muscles do not always operate in a fashion that maximizes power (see "Section V.D"). James and Johnston (1998) cautioned that extrapolation from observations made using steady-state force-velocity relationships to those from swimming fish may be deceptive; as noted later, enhanced force due to stretch of muscle during swimming can increase power locally as well as in more posterior myotomes but does not occur during typical force-velocity measurements (see also Stevens, 1993).

V_{max} is the velocity of shortening at which external force produced by the muscle is zero (Figure 6.4) and is the fastest that a muscle can shorten; it is limited by and also an index of the inherent rate kinetics of the cross bridges (Hill, 1938; Woledge *et al.*, 1985). V_{max} reported from fish myotomal muscles varies by well over an order of magnitude (e.g., Medler, 2002) and depends heavily on fiber type, species, and temperature. Clearly muscles could not operate at V_{max} in nature, and so V_{max} is not in itself a functionally important parameter in this context, yet it is generally a useful gauge of the inherent speed- and power-producing ability of a muscle, although not always (e.g., Rome *et al.*, 1999; Syme and Josephson, 2002). Notably, the shapes of the scaled force-velocity relationships of red and white muscle in fishes (dogfishes) are virtually identical, which is in particular contrast to the muscles of mammals, in which slower fibers have force-velocity relationships that are relatively much more curved than faster fibers (Lou *et al.*, 2002). A greater curvature signifies a less powerful muscle. Thus, relative to fast fibers, slow fibers in fish appear better able to produce power than slow fibers in mammals. This, in combination with the unusually well-developed t-tubule system and high density of sarcoplasmic reticulum in fish red muscle (see the preceding), suggests that the slow muscles of fishes are designed for operation at speeds relatively higher than the slow fibers of mammals. The seeming lack of proprioception in fish muscle (see Section II) and the relatively superior ability of red muscle to produce power (compared with mammals, at least) may have a basis in the buoyant lifestyle of fishes. Mammals tend to be terrestrial and thus rely heavily on red muscles for postural control, which requires heightened proprioception but very little power, while fishes tend to be aquatic and near neutrally buoyant and thus do not require their red muscles to provide fine postural control but do use them to power aerobic swimming.

Power and work extracted from isokinetic and isotonic contractions, such as those used in force-velocity measurements, provide information about the inherent capacity of the contractile elements. They tell us much less about the ability of the muscle as an integrated machine to power movement in an animal undergoing alternate cycles of flexion and extension. To this end, Josephson (1985) modified the "work loop" technique, originally developed by Machin and Pringle (1959) for measuring work and power from asynchronous insect flight muscle, to allow measurements from synchronous muscle and for the imposition of controlled strain trajectories as would occur in moving animals. This innovation opened a new era in the study of muscle that continues today. It allows investigators to study muscle function under conditions that closely mimic those in an animal, such as a muscle in a swimming fish undergoing alternate cycles of extension and flexion, and to ask how different aspects of muscle design impinge on its ability to power movement and how variables internal and external to the muscle influence its use in a physiologically relevant context.

In the work loop technique, a muscle, a small bundle of fibers, or a single muscle fiber is normally isolated from the animal and bathed in physiological saline to maintain viability. One end of the muscle is attached to the arm of a servomotor that is used to control muscle length. The other end is attached to a force transducer, or in cases in which the servomotor also possesses galvanometer circuitry to measure force, it is attached to a rigid link. The muscle can be activated through an intact nerve or directly by passing current between plates adjacent to the muscle. If the length of the muscle is oscillated cyclically by the servomotor and it is activated at an appropriate point and for an appropriate duration during the length change cycle, the muscle can be made to contract and do work much like it would in a moving animal. Knowing the length changes and the force produced by the muscle, the work done by the muscle can be quantified accurately (Figure 6.5). The technique gets its name from the loop formed when force produced by the muscle is plotted versus muscle length during a stretch/shorten cycle. The area underneath the lengthening arm of the force-length plot represents the lengthening or negative work. The area underneath the shortening arm of the plot represents the shortening or positive work. The difference between shortening and lengthening work is the net work, which is represented by the area within the loop. Typically the loops are traversed in a counterclockwise direction, such that force during shortening is greater than force during lengthening, and thus net work has a positive value (i.e., the muscle contributes net positive mechanical energy during a cycle). However it is not uncommon for some animals to use their muscles in eccentric contractions such that they are activated while being lengthened; then force during lengthening is greater than force during shortening, the loop is traversed in a clockwise direction, and net work has a negative value (i.e., the muscle absorbs net mechanical energy and acts as a brake). Whether the muscle contributes or absorbs net energy depends largely on the stimulus phase and duration (when and for how long during the cycle the muscle is activated), discussed later. Work and power are a function of both the amplitude of movement (i.e., strain), which is directly related to muscle length, and the force produced, which is directly related to the cross-sectional area of the muscle. Thus work and power are normally expressed relative to the mass of the muscle, which is directly related to the product of length and cross-sectional area (i.e., volume) through the density of the tissue. Work is expressed as Joules per kilogram of muscle (Jkg^{-1}) and power as Watts per kilogram (Wkg^{-1}).

In work loop experiments on fish muscle, the imposed strain trajectory is usually sinusoidal, which has been shown to be a good approximation of the

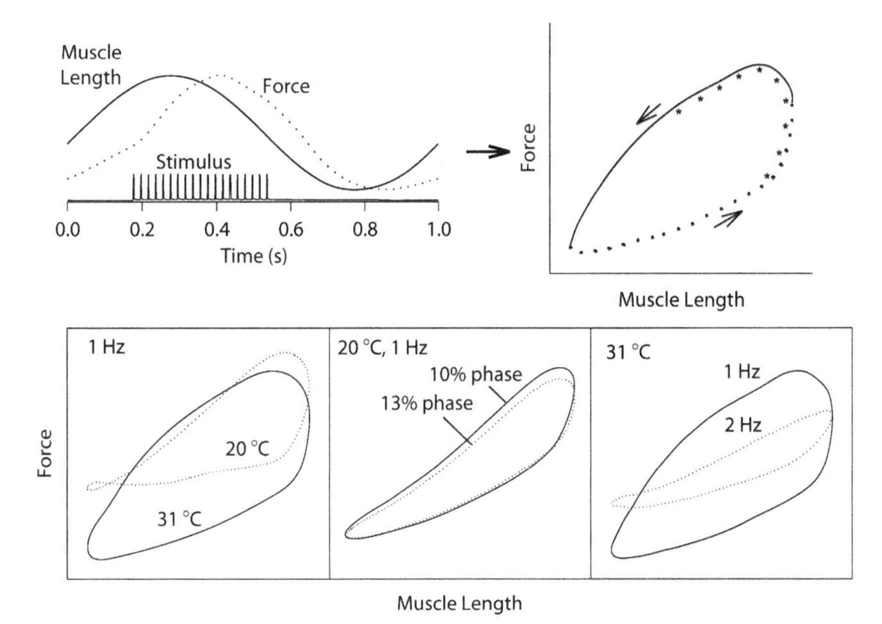

Fig. 6.5. Work loops from red muscle of salmon sharks (*Lamna ditropis*). Muscles were subjected to sinusoidal oscillations in length while being stimulated at the phase and duration resulting in maximal net work output. Strain was 12% peak-peak. Left upper panel shows time traces of muscle length, force, and the stimulus during one cycle (1 Hz cycle frequency). Right upper panel shows the loop that is formed when muscle force is plotted versus muscle length for one complete cycle; asterisks indicate the segment of the cycle over which the muscle was stimulated. The solid line shows the shortening portion of the cycle, and the broken line shows the lengthening portion, thus this loop is traversed in a counterclockwise direction and net work is positive. The area under the shortening portion of the loop is the work done while shortening, the area under the lengthening portion is the work required to lengthen the muscle, and the area inside the loop is the net work done. Lower panels: (Left) The effects of altering the temperature, where cooler temperatures reduced the extent of relaxation, resulting in increased lengthening work and thus reduced net work. (Middle) The effects of a small change in stimulus phase, where an earlier phase resulted in a 21% increase in net work done. (Right) The effects of cycle frequency, where at the higher cycle frequency the muscle cannot generate as much force during shortening and is unable to relax fully before lengthening, resulting in a near 75% reduction in net work done. (From D. A. Syme, R. E. Shadwick, D. Bernal, and J. Donley, unpublished data.)

strain experienced in some swimming fishes (e.g., Shadwick *et al.*, 1999; Ellerby *et al.*, 2000); triangular trajectories with rounded transitions have also been recommended (Swank and Rome, 2000). Symmetrical, sinusoidal strain oscillations do not necessarily result in maximal work from the muscle (e.g., Marsh, 1999), and linear strain patterns can increase work output but only by about 20% (Josephson, 1989). The rate of the length oscillation (cycle frequency) is analogous to the tail-beat frequency in a swimming fish.

The amplitude of the imposed length oscillation (strain amplitude) is related to the tail-beat amplitude. The onset of stimulation relative to the strain cycle (phase) and the duration of the stimulation reflect the electromyogram (EMG) onset phase and duration, respectively, in a swimming fish. The effects of altering any of these parameters on the work done by the muscle can then be measured directly (Figure 6.5). These parameters are often manipulated systematically (optimized) until work output is maximal, which provides information about the inherent ability of the muscle to do work and produce power when undergoing cyclic contractions. However, such results are usually not indicative of how the muscle is used in a swimming fish (see Section V.D). Alternatively, parameters of strain, cycle frequency, stimulus phase, and duration can be first measured from a swimming fish and then imposed on the muscles in a work loop experiment, providing information about how the muscle is actually performing in the fish. Comparison of maximal work with that produced in simulated swimming is a useful tool to provide insight into the function of muscle in fishes (see Section V.D). See Altringham and Ellerby (1999), Altringham and Johnston (1990a, 1990b), Johnson and Johnston (1991b), and Josephson (1993) for further discussion on application and interpretation of the work loop technique.

The net work done during a cycle tends to decrease with increasing cycle frequency (Figure 6.5) (e.g., Altringham and Johnston, 1990a,b; Rome and Swank, 1992; Hammond *et al.,* 1998). This is largely a consequence of the force-velocity characteristics of muscle, in which faster cycling is associated with faster shortening and thus less force and work, but also with the reduced time available for the muscle to be activated, resulting in less force, or with insufficient time to allow full relaxation before the muscle is lengthened (Josephson, 1993). An exception to the pattern of increased work with decreased cycle frequency sometimes occurs at very slow cycling frequencies (likely unphysiologically slow), at which work also decreases, perhaps caused by fatigue due to prolonged activation (Altringham and Johnston, 1990a; Johnson and Johnston, 1991b; Rome and Swank, 1992; Coughlin *et al.*, 1996; Hammond *et al.*, 1998; Syme and Shadwick, 2002).

The stimulus phase is a critical determinant of work done by the muscle during shortening and also the ability of a muscle to resist lengthening. Work loop studies on isolated fish muscle have shown that work and power are maximized when the stimulus phase precedes slightly the maximal muscle length; that is, activation of the muscle begins during the latter stages of muscle lengthening, just before the onset of shortening (Figure 6.5) (e.g., Johnson and Johnston, 1991b). Stimulating the muscle just prior to shortening allows adequate time for the processes of activation to occur before the shortening phase has begun, and, very importantly, results in a degree of force enhancement due to stretch that causes relatively high forces to be produced at the onset of shortening and hence increased work output. The muscle must then remain activated for as much of the shortening period as possible to maximize the force and thus work done, but must not be activated so long as to prevent it from relaxing before the lengthening portion of the cycle commences, which would result in increased work required to lengthen the muscle. To maximize work at faster cycle frequencies (shorter cycle periods), the onset of activation must precede the onset of shortening by a greater proportion of the cycle period (i.e., the muscle must be stimulated earlier in the cycle), and the duration of the stimulus must be reduced accordingly (e.g., Josephson, 1985, 1993; Altringham and Johnston, 1990a,b; Johnson and Johnston, 1991b; Rome and Swank, 1992). Whether fish employ such strategies while swimming is not well documented. Ellerby and Altringham (2001) noted that during fast swimming in rainbow trout the onset of EMG activity does occur during the latter stages of muscle lengthening, thus enhancing the work done during subsequent shortening. They also noted that phase is relatively earlier toward the caudal portion of the fish, but suggested that rather than maximizing work, this might result in the caudal muscles acting as rigid elements that transmit power from anterior myotomes to the tail. Without direct measures of force and work, or simulations using work loop analysis, it is not possible to state definitively what the effects of the axial phase shift are. Other investigators have made measurements of activation patterns in swimming fish and applied them using work loop analyses (e.g., Johnston *et al.*, 1993; Rome *et al.*, 2000; Swank and Rome, 2000) and while most fish appear to activate their muscles in a manner that results in net positive work output, they do not always maximize work output (see Section V.D). The implications of activation phase on muscle function in swimming fishes are discussed more fully in Chapter 7 and elsewhere (e.g., Wardle *et al.*, 1995; vanLeeuwen, 1995; Shadwick *et al.*, 1998; Altringham and Ellerby, 1999; Coughlin, 2002a).

The strain that muscles experience in swimming fishes has a marked influence on the work they can do. In general, muscle work or power increases with increasing strain amplitude up to a maximum, after which it declines (e.g., Altringham and Johnston 1990a,b; Johnson and Johnston, 1991b; Johnston *et al.*, 1993). The particular strain at which work is maximized is muscle and temperature specific, with white fibers tending to operate at smaller strains than red fibers, and typically falling in the range of 5–15% peak-to-peak, which agrees well with the range of strains observed in the muscles of swimming fishes (e.g., Rome and Sosnicki, 1991; Rome and Swank, 1992; Hammond *et al.*, 1998; Shadwick *et al.*, 1999; Coughlin, 2000; Ellerby *et al.*, 2000, and references therein). Somewhat surprisingly, the strain giving maximal power is not highly sensitive to the cycle frequency (Altringham and Johnston, 1990b), so a single strain could be employed to

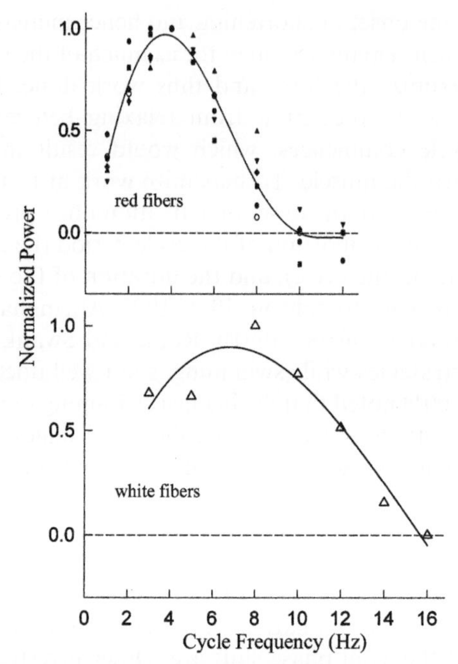

Fig. 6.6. Normalized power output from work loop analysis of red and white muscle from yellowfin tuna at different cycle frequencies. Temperature was 25 °C, strain was 10% peak-peak, and stimulus phase and duration were adjusted to maximize power output. Note that white fibers produce maximal power at a higher cycle frequency than red and can produce power at frequencies at which red muscle is unable to produce positive power. However, both red and white muscles produce substantial amounts of power over a broad and overlapping range of cycle frequencies. (From D. A. Syme and R. E. Shadwick, unpublished data.)

swim over a wide range of speeds without compromising power output. As might be predicted from this observation, muscle strain does not appear to change with tail-beat frequency in swimming fish (e.g., Webb *et al.*, 1984; Hammond *et al.*, 1998; Ellerby *et al.*, 2000).

Power is normally derived from work loop analyses as the product of the net work done per cycle and the cycle frequency, and much like power calculated from force-velocity data tends to increase with increasing cycle frequency up to some maximum, after which it declines (Figure 6.6) (e.g., Altringham and Johnston, 1990a,b; Johnson and Johnston, 1991b; Rome and Swank, 1992; Johnston *et al.*, 1993; Ellerby *et al.*, 2001b). Josephson (1993) presented a thorough review of factors that influence power from skeletal muscles and its interpretation, and provided many useful references to the performance of fish muscle. Power from work loop analysis is an average over the complete cycle (net power) and accounts for activation and relaxation, usually variable rates of shortening/lengthening, and the time and energy required to lengthen the muscle. As such, net power from work loop analysis is generally a more accurate reflection of the power available from muscle for movement in a swimming fish than is peak power from force-velocity measurements. Further, because power calculated from work loops is an average over a complete cycle that includes the time required to lengthen the muscle and because it is limited by activation, relaxation, etc., it will be considerably less than power measured from force-velocity data (e.g., Josephson, 1993; Rome, 1994; James *et al.*, 1996; Coughlin *et al.*, 1996; Caiozzo and Baldwin, 1997). For example, at 20 °C the maximal power from work loop analysis in scup pink muscle is 44 Wkg^{-1}, while the peak power during steady shortening and maximal activation is 133 Wkg^{-1} (Coughlin *et al.*, 1996). Likewise, the peak power measured during shortening in a C-start (about 220 Wkg^{-1} for sculpin fast muscle; James and Johnston, 1998) is much higher than the average power measured from work loops and is comparable to the conceptually similar force-velocity values. Interestingly, the peak power during contraction of contralateral muscles after they have been stretched by the initial C-bend can be much higher again than during the initial C-bend (about 307 Wkg^{-1}; James and Johnston 1998). These relationships appear to be body position dependent, as fast muscles from the anterior region of Antarctic rock cod display a near-doubling of peak power from the initial to contralateral C-bend, but caudal muscles actually produce about 15% less power during the contralateral C-bend (Franklin and Johnston, 1997). Mean power output during the initial and contralateral C-bends, however, is not markedly different from one another in either fish.

Does power change with increased tail-beat frequencies in fishes as it does with increased cycle frequency in work loop studies when the variables of stimulus phase and duration are manipulated to maximize power? Few measurements have been made. Johnson and Johnston (1991b) note that with fast fibers from short horned sculpin the cycle frequencies required to maximize power in work loop studies coincide with the maximum tail-beat frequencies in fishes of similar size. Whether this fiber type is recruited under such conditions in swimming fishes or if it is recruited in a fashion that maximizes power is not known. In red muscle of bass, scup, and rainbow trout, there does appear to be an increase in power from the muscle during swimming with increased tail-beat frequency (Coughlin, 2000; Rome *et al.*, 2000), which in itself would contribute to satisfying the increased demand for power with increased swim speed. To the contrary, when the red muscles of brook char are activated as occurs in a swimming fish there does not appear to be a change in power with changes in tail-beat frequency

(McGlinchey *et al.*, 2001). This suggests that with changes in tail-beat frequency the fishes are not changing activation patterns appropriately to maximize power output, and the increased power for faster swimming must be a result of increased muscle recruitment.

Work and power output are critically dependent on the inherent speed of the muscle, which will have a major impact on how a muscle is used to power swimming (Rome *et al.*, 1988). That a particular fiber type is "faster" than another can be defined at several levels, including force-velocity characteristics (e.g., higher maximal velocity of shortening or a force-velocity relationship that is less curved), twitch duration (faster rates of contraction and relaxation or briefer period of the twitch), and ability to produce power during cyclic contractions (a combination of factors, including rapid activation/deactivation kinetics and an ability to produce substantial force during rapid shortening). Fast muscles produce more power than slow, in part due to their ability to maintain higher forces during shortening, which will enhance the work done, and in part due to their ability to turn on and off more rapidly, which will enhance the work done but also allow the muscle to operate at higher tail-beat frequencies and thus do more cycles of work in a given time period. The relationship between these parameters, the ability to power cyclic activities, and their cellular bases have been discussed in detail elsewhere (e.g., Marsh, 1990; Altringham and Johnston, 1990a; Josephson, 1993; Caiozzo and Baldwin, 1997; Rome and Lindstedt, 1998; James *et al.*, 1998; Syme and Josephson, 2002). As with twitch characteristics, a given fiber type when compared across species can have substantially different contractile and power-producing capacities. Lou *et al.* (2002) reviewed results from red muscles compared across species, noting a nearly 2-fold range of maximal velocities of shortening and maximal forces, marked differences in the curvatures of the force-velocity relationships, and an almost 3-fold range for power outputs.

Faster fiber types not only are able to produce more power than slow types, but also can produce work and power at higher cycling frequencies (Figure 6.6). For example, white muscle of the sculpin produces about 30 Wkg^{-1} during a tail-beat cycle and can do so at a cycle frequency of 6 Hz (six length oscillations or tail-beats per second), while red muscle produces only 7 Wkg^{-1} and at a frequency of only 2 Hz (Altringham and Johnston, 1990a). White muscle from eels produces about 16 Wkg^{-1} at 2 Hz (Ellerby *et al.*, 2001b), while red muscle produces 1.2 Wkg^{-1} at 0.5 Hz (Ellerby *et al.*, 2001a). Red muscle of scup produces about 28 Wkg^{-1} at 5 Hz, while pink muscle produces 44 Wkg^{-1} at about 7 Hz (Coughlin *et al.*, 1996). White muscle in dogfish produces maximal power at a cycle frequency of 3.5 Hz, while red muscle does so at only 1 Hz (Curtin and Woledge, 1993a,b). Similar relative differences in maximal power and optimal operating speeds between red and white fibers are also noted based on force-velocity measurements (Rome *et al.*, 1988); for example, peak power in dogfish red muscle is 29 $W kg^{-1}$ at a shortening velocity of 0.54 muscle lengths per second, while in white muscle it is 122 $W kg^{-1}$ at 1.2 lengths per second (Lou *et al.*, 2002).

Despite faster muscles producing maximal power at higher cycle frequencies, there does tend to be a considerable range of cycle frequencies over which both faster and slower fiber types can simultaneously produce substantial amounts of power (Figure 6.6); hence, it is not always clear from a mechanical viewpoint which fiber type would be better suited to power certain movements. Still, these large differences in inherent power output have important implications for the ability of slow and fast muscle to power swimming and the speeds and temperatures at which they can contribute (Rome *et al.*, 1988). The power produced by pink muscle of scup is double to quadruple that of the slower red muscle, the differences tending to be greater in the anterior regions of the fish where strain is small and thus the inherent ability of the muscle to relax is most critical and most limiting (Coughlin *et al.*, 1996) (see also Section VII). Also, because pink muscle relaxes faster than red, has faster force-velocity characteristics, and produces maximal power at a higher cycle frequency, it is recruited at faster swimming speeds and at colder temperatures to compensate for the inadequacies of red (Figure 6.1) (Coughlin *et al.*, 1996; Coughlin and Rome, 1999; Rome *et al.*, 2000).

D. *In Vivo* and *In Vitro*

Is muscle power maximized in swimming fishes, or are there other demands or constraints placed on it that reduce power? The power produced by muscle depends on many factors, including the strain, cycling frequency, phase of stimulus onset, duration, and frequency. When applying the work loop technique, these parameters can be varied and optimized to maximize the power output of the muscle. The specific combination of parameters that maximizes power will change if any one of the others is changed. Do fishes exercise the same luxury of freedom to maximize power? Rome and Swank (1992) were among the first to attempt to obtain realistic estimates of cyclic performance from fish muscle by applying strains and activation patterns measured in swimming fishes to muscle using the work loop technique. With foresight, they cautioned that animals in life may not simply maximize work and power while swimming, that other factors may affect or perhaps dominate how muscle is used, and thus that work and power obtained by simply optimizing work loop parameters may prove misleading. In their studies (e.g., Rome and Swank, 1992; Swank and Rome, 2000; Rome *et al.*, 2000) they found that at 20 °C, scup use their red muscles in a way that produces 87–98% of maximum power. But at 10 °C, the result is dramatically different,

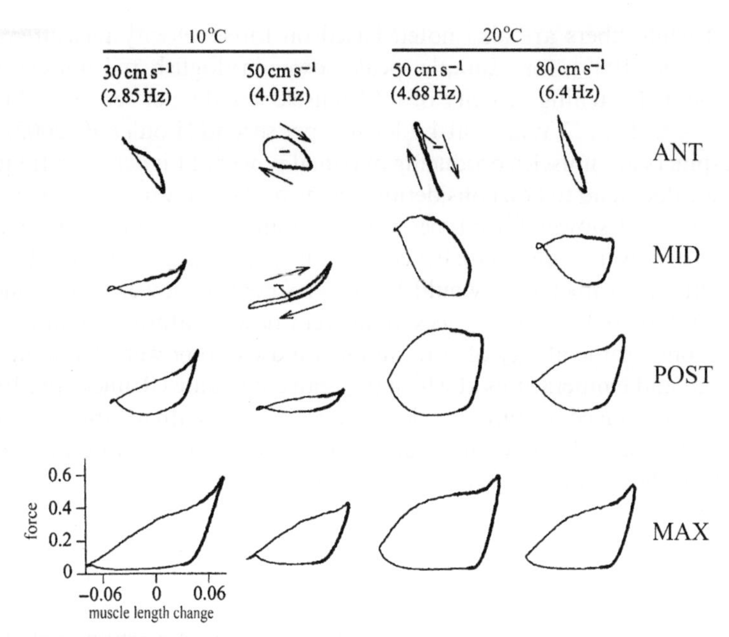

Fig. 6.7. Work loops from scup red muscle at 10 °C (left pair of columns) and 20 °C (right pair of columns); each column represents a different swim speed (associated tail-beat frequency in parentheses). The upper three rows of data are from muscles at anterior (ANT), middle (MID), and posterior (POST) locations, respectively, on the fish, and have been subjected to the same strain and stimulus patterns measured in swimming fish at the associated speeds and temperatures; they thus represent what the muscles are doing in a swimming fish. The bottom row (MAX) is a muscle from the anterior position, but subjected to strains and stimulus conditions that maximize work output at the associated cycle frequency and temperature; they thus represent what the muscle is capable of doing. It is clear that despite the poor performance of the muscles in the fish under certain conditions, and the apparent poor performance of ANT muscle under all swimming conditions, ANT muscle is quite capable of doing substantial amounts of work at all temperatures and cycle frequencies experienced. Note also that when swimming at 50 cms^{-1} the muscle performs poorly at all axial locations at 10 °C, but does relatively well at 20 °C. Negative signs and clockwise loops indicate the muscle absorbed rather than produced net work. (Adapted from Rome *et al.*, 2000, with permission of the Company of Biologists Ltd.)

with the muscles producing only about 20% of the power they are capable of producing (Figure 6.7), and producing maximum power at a cycle frequency about half of the tail-beat frequency observed in fishes at this temperature. Similarly, Ellerby *et al.* (2001a,b) found in red and white muscle of eels that maximum power from work loop analysis was 2- to 4-fold greater than what the muscles appear to produce in swimming fishes (although their estimates for activation and strain in swimming fishes came from a closely related but different species of eel). Hammond *et al.* (1998) also noted that while red muscles from anterior through posterior locations in rainbow trout produced net positive power at most cycle frequencies when activated as they are during swimming, at higher cycle frequencies and in all cases in muscle from anterior myotomes, they produced substantially less power during swimming than they are capable of. The lengthening work (negative) component increased in muscle toward the tail of the fish so that the muscles there absorbed almost as much work as they produced, particularly at higher cycle frequencies. Similarly, Coughlin (2000) noted that red muscle in rainbow trout produced only about 30% of maximal power when working as it does during cruise swimming, at both anterior and posterior locations. Many other investigators have used a similar approach to further our understanding of how fishes use their muscle during swimming (e.g., Altringham *et al.*, 1993; Rome *et al.*, 1993; Johnson *et al.*, 1994; Coughlin and Rome, 1996; Franklin and Johnston, 1997; James and Johnston, 1998; Wakeling and Johnston, 1998; McGlinchey *et al.*, 2001; Syme and Shadwick, 2002). Patterns of muscle function are arising (Wardle *et al.*, 1995; Altringham and Ellerby, 1999), but further investigation is required.

As was foreshadowed by Rome, a major pitfall to assumptions that fishes should maximize work or power output during cruise swimming is that fishes would rarely require maximum power for routine cruise swimming (although it is reasonable to expect that individual fibers may be used in a fashion that maximizes power, and it is simply the number of fibers being recruited that determines the overall power output of the muscle). Further, other factors such as economy, stability, or maneuverability may dominate muscle function. However, by studying systems in which it is fully expected that power should be maximal, more robust conclusions may be drawn. Franklin and Johnston (1997) studied C-starts in Antarctic rock cod, in which it can be more safely assumed that muscle power should be maximized for escape maneuvers. They noted that during these extreme behaviors the white fibers were activated and underwent strains that resulted in the muscle producing 90–100% of maximal power. Likewise, tuna cruise at relatively high speeds using only their aerobic red muscle, and it might be expected that they also recruit the muscle in a fashion that maximizes power. Syme and Shadwick (2002) imposed strain and activation patterns measured from swimming fishes (Shadwick *et al.*, 1999) on red muscle bundles from skipjack tuna and noted that the muscles produced about 90% of the maximum power they were capable of producing. Further, Donley (2004) noted that the phases/durations of EMG activity measured in swimming shortfin mako sharks are very similar to those that maximize work output by the red muscles. Thus, there is evidence that at times fishes use their muscles, both red and white, in a fashion that maximizes power. However, to date there are considerably more examples that suggest they usually do not (also see following discussions of axial variation and effects of temperature).

While the work loop technique provides important information about the physiologically relevant potential of muscle to produce work and power, it should be applied with caution, and it should not be assumed that fish routinely employ a strategy that maximizes power output. A particularly striking example: the Q_{10} for maximal power output in scup red muscle is about 2.3 when all parameters are optimized in work loop analysis. But when power is measured under the conditions observed in swimming fishes, the Q_{10} is highly variable and sensitive to the oscillation frequency, ranging from 1 at slow frequencies to upward of 14 at higher frequencies, and in some cases is indeterminate when the muscles actually absorb rather than produce power (Swank and Rome, 2000).

In addition to the possibility that fishes may use their muscles in a way that maximizes power output, perhaps they use them in a way that maximizes efficiency or the economy of swimming, particularly during sustained swimming. The economy of swimming is a function of both the effectiveness of the transfer of work done by muscle to forward movement and the efficiency of muscle contraction. We know little about either. The efficiency of working fish muscle does not appear to be notably different from that of other skeletal muscles working at similar temperatures and frequencies. Reports of efficiency for fish muscle include 33 and 41% for dogfish white muscle during constant velocity shortening and cyclic sinusoidal movements, respectively (Curtin and Woledge, 1991, 1993b), 51% for dogfish red muscle during sinusoidal movements (Curtin and Woledge, 1993a), about 24% at 4 °C and 8% at 15 °C for sculpin white fibers (Johnson *et al.*, 1991b), and 12–21% at 4 °C for cod fast muscle during sinusoidal contractions (Moon *et al.*, 1991). Efficiency appears to be temperature dependent, being higher at cooler temperatures with a Q_{10} of about 2.5 (Johnson *et al.*, 1991b); it is noteworthy that in this study energy consumption per work cycle was temperature independent while work per cycle was greater at the colder temperature and slower cycle frequency, resulting in greater efficiency in the cold. An explanation for this observation was not offered and would be enlightening to pursue.

It is often assumed, although not critically justified, that a sensible design would be one in which the operating conditions that maximize power are closely associated with those that maximize efficiency; in this way, animals can maximize mechanical performance and economy of movement simultaneously. The limited evidence we have from fish muscle suggests that this is not a simple doctrine. Curtin and Woledge (1991) found during isovelocity shortening in fully activated dogfish white muscle that maximal efficiency and maximal power occur at V/V_{max} ratios of 0.14 and 0.28, respectively. While the two quantities are not maximized in chorus, efficiency does remain greater than 90% of maximal over a broad range of velocities and is probably not significantly less than maximal when power is also maximal. This might suggest that power and efficiency can be maximized simultaneously. However, Curtin and Woledge (1993b) found in dogfish white muscle, this time undergoing cyclic, sinusoidal movements, that the cycle frequency giving maximal efficiency was only about two-thirds of that giving maximal power. Under these circumstances, efficiency was about three-fourths maximal when power was maximal, and power was only about two-thirds maximal when efficiency was maximal. Likewise, using cyclic contractions in dogfish red muscle, efficiency was maximal at a cycle frequency of 0.74 Hz while power was maximal at a cycle frequency of 1.02 Hz (Curtin and Woledge, 1993a); power and efficiency remained about 80% maximal under the operating conditions that maximized the other. Thus far, it may be said that power output tends to be maximized at a higher velocity of shortening (or cycling frequency) than does efficiency, but whether they can be maximized simultaneously depends on the muscle, the mode of measurement, and the rigor with which "maximal" is defined.

In these preceding studies, the duration of stimulation was not systematically varied to truly maximize power or efficiency, making conclusions tentative. In a subsequent study on dogfish white muscle, Curtin and Woledge (1996) described relationships between power and efficiency when stimulus phase, duration, cycle frequency, and strain amplitude were varied. For a given stimulus duty cycle (duration of stimulus/duration of cycle), phase can be adjusted so that power and efficiency are maximized simultaneously over a wide of cycle frequencies. The question remains as to whether fishes employ such fine-tuning during swimming. However, altering the stimulus duration appears to result in a tradeoff between power and efficiency; power tends to increase while efficiency decreases as the stimulus duration is increased. Thus, brief activation promotes high efficiency (32% greater than if power is maximized) while prolonged activation augments power (82% greater than if efficiency is maximized). Curtin and Woledge (1996) attributed the reduced efficiency with longer activation primarily to a continued high turnover of energy during the lengthening portion of the cycle when work is being absorbed instead of produced. The existence of such a tradeoff between power and efficiency has been alluded to previously by Johnson *et al.* (1991b). It is clear that the cycle frequencies that maximize power and efficiency in white muscle are considerably faster than those for red, and hence red muscle is not well suited to power fast movements either mechanically or energetically, while white muscle is not well suited to power slow movements. Yet relationships between efficiency of muscle contraction, work, power, and the economy of swimming continue to be poorly understood and remain a fruitful area of research, particularly in the context of how fishes actually use their muscles while swimming.

VI. SCALING

Larger fishes tend to swim with slower kinematics than do smaller fishes, thus aspects of muscle contraction that bear on dynamic performance (twitch kinetics, V_{max}, factors influencing work and power output) might be expected to scale with body size as well. With increased body size there is a decrease in the tail-beat frequency used by swimming fishes (e.g., Bainbridge, 1958; Webb, 1976; Webb *et al.*, 1984). Likewise, the cycle frequency in work loop studies that results in maximal power output slows with increased body length (-0.5 exponent) and mass (−0.17 exponent) (Figure 6.8), and does so in concert with prolonged relaxation kinetics (0.29 exponent) and to a lesser extent slowed activation kinetics (Wardle, 1975; Archer *et al.*, 1990; Altringham and Johnston, 1990b; Videler and Wardle, 1991; Anderson and Johnston, 1992; James *et al.*, 1998). Wardle (1975) suggested that the slowed swimming kinematics in larger fishes are directly limited by the slowed twitch kinetics of the muscle, and presented compelling evidence to this effect. Twitch kinetics do not appear to scale with body mass in yellowfin tuna (Altringham and Block, 1997).

Like twitches, the maximal velocity of muscle shortening (V_{max}) is another measure of the inherent speed of a muscle, and the two typically scale in concert with fiber type (i.e., fast fibers have both a fast V_{max} and fast twitch kinetics). But as V_{max} reflects a cross-bridge rate function while twitch kinetics reflect calcium kinetics, troponin rate kinetics, cross-bridge rate functions, and characteristics of the series compliance, the two measures do not inevitably scale in unison. Thus, it is readily understood how different muscles with a similar V_{max} could have markedly different twitch speeds or vice versa. For example, in scup red muscle there is axial variation in twitch relaxation rates but not in V_{max} (e.g., Swank *et al.*, 1997). Does V_{max} scale with body mass? V_{max} scales inversely with body mass almost universally over a broad range of species in both terrestrial and flying animals (about mass$^{-0.2}$), but quite remarkably in swimming animals (fishes) it does not appear to scale at all (Figure 6.9) (Curtin and Woledge, 1988; Medler, 2002). Medler (2002) hinted at the intriguing possibility that the lack of a scaling relationship in fishes may hearken to their buoyant lifestyle, but also warns that the seeming lack of a mass scaling relationship for V_{max} in fish muscle remains uncertain. Further, within a given species of fish there are reports of negative body mass scaling coefficients for both V_{max} and myofibrillar ATPase rates (Witthames and Greer-Walker, 1982; James *et al.*, 1998; Coughlin *et al.*, 2001a). Also, the maximal swimming velocity (body lengths/s) and tail-beat frequency of fish scale inversely with body mass while absolute velocity (m/s) scales positively (Domenici and Blake, 1997),

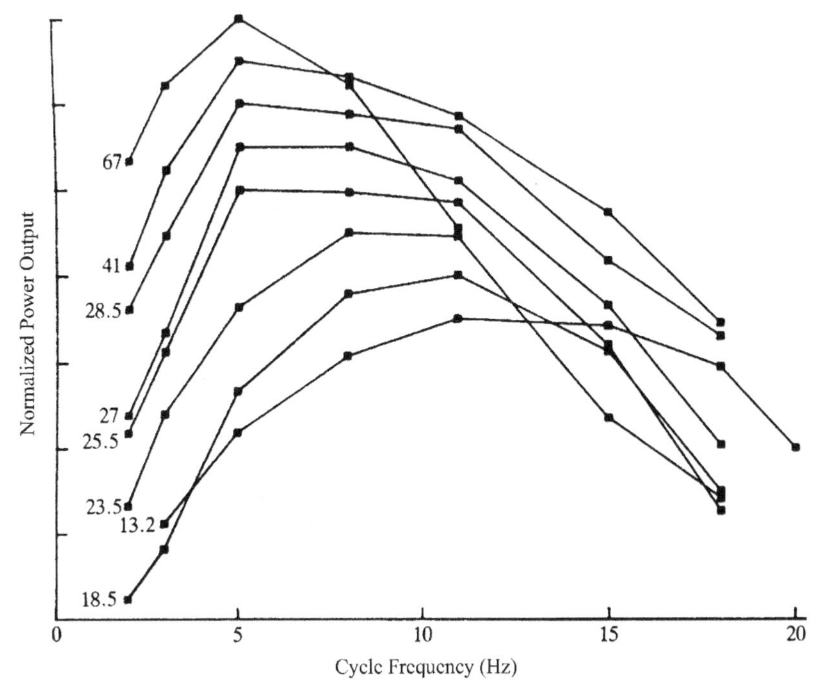

Fig. 6.8. Relative power output of fast muscle from different-sized cod during work loop analysis at different cycle frequencies. Each curve shows the results of muscle from a different sized fish; body length (cm) is shown next to the curve. Power is normalized for each fish, and for clarity the curves have been vertically shifted on the y axis. Note the rightward shift in the frequency at which power is maximal with decreasing fish length. The frequency at which power is maximal is described by the equation freq $= 46.8L^{-0.52}$, where L is fish length in cm, $r^2 = 0.97$. Strain was 10% peak-peak, the stimulus conditions were adjusted to maximize power output, and temperature was 4 °C. (Adapted from Altringham and Johnston, 1990b, with permission of the Company of Biologists Ltd.)

as one might expect if muscle shortening velocity scaled with a negative exponent. It should also be noted that a regression against mass alone (Figure 6.9) could prove misleading, as it does not consider animal length or the viscosity of the medium through which the animal moves, which would bear on the demands placed on the muscles and thus their design, nor does the swimming cohort in this analysis include any small invertebrates.

Certainly during developmental growth there is a marked slowing of both twitch kinetics and V_{max} of red muscle, which are associated with slowed tail-beat frequencies during swimming (e.g., Coughlin *et al.*, 2001a). Developmental changes do not appear to be entirely dependent on body size per se, but more so on the parr-smolt transition and the associated changes in distribution of

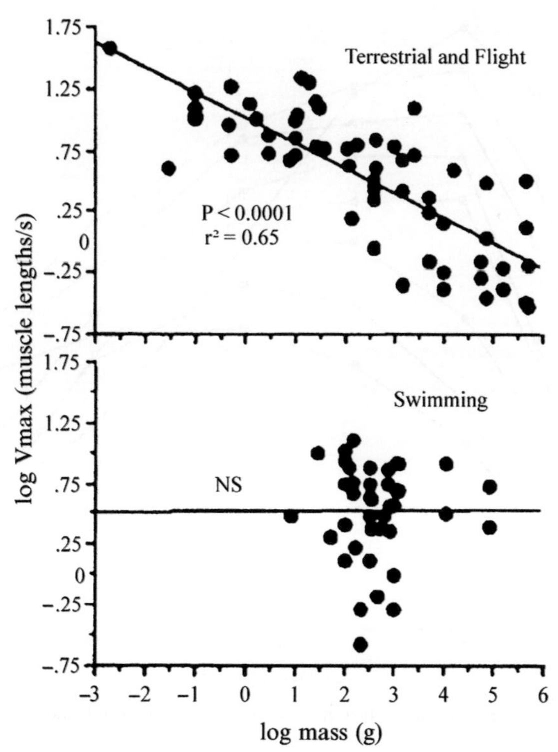

Fig. 6.9. Scaling relationships between maximal velocity of muscle shortening (V_{max}) and body mass. (Upper panel) Muscles used for terrestrial locomotion and flight (primarily insects, birds, amphibians, reptiles, and mammals) show a significant decline in V_{max} with increasing body mass. (Lower panel) Muscles used for swimming (primarily fish) do not show a significant relationship. (Adapted from Medler, 2002; used with permission.)

myosin heavy-chain isoforms, as demonstrated by induction with thyroid hormone treatment (e.g., Coughlin *et al.*, 2001b; Coughlin, 2002b). The mechanism responsible for the change, or lack thereof, in V_{max} with mass in fishes is as yet poorly understood. There are reasonably well described changes in myosin heavy- and light-chain isoforms with growth in terrestrial mammals; such changes are less well documented in fishes (James *et al.*, 1998), but recent advances are being made (reviewed in Coughlin, 2002b).

As mass specific power output reflects the ability to generate force (which is typically mass independent) and muscle shortening velocity (which is questionably mass independent), perhaps it is not surprising to find some observations that muscle power is not mass dependent [James and Johnston (1998) in fast muscle during escape responses] and other observations that it is [Anderson and Johnston (1992) in fast muscle during cyclic contractions,

$M^{-0.10}$]. This discrepancy may have a basis in the different types of contractions studied. In the one-shot shortening contractions typical of escape responses, the kinetics of activation/relaxation (which are mass dependent) are not particularly pertinent to power output, and thus power in this type of contraction may not be mass dependent. However, during cyclic contractions, rates of relaxation and activation can critically limit power output, and hence cyclic power may appear mass dependent. For example, with increasing body size and the accompanying slowing of twitch kinetics, muscles must be both activated and deactivated earlier in the tail-beat cycle to maximize work output (earlier phase and shorter stimulus duration, respectively) (Altringham and Johnston, 1990b). Slowed activation kinetics in larger fishes would be expected to reduce cyclic work and power output, particularly at higher operating frequencies, in the same way that they may limit maximal tail-beat frequencies.

Interestingly, during work loop studies the muscle strain that results in maximal power output appears highly dependent on body mass (Anderson and Johnston, 1992). Fast muscle from large Atlantic cod has optimal strains of 10% peak to peak, and the power/strain relationship is very steep, whereas not only does muscle from small fishes show a broad range of strains over which power is maximized, but also their optimal strain is about double that of muscle from larger fishes. The authors briefly discussed how this corresponds with the proportionately larger tail-beat amplitudes observed in smaller fishes (e.g., Webb *et al.*, 1984) and potential morphometric foundations.

Characteristics that are less important to swimming kinematics do not appear to scale with body mass. The maximum, isometric, tetanic force produced by fish muscle is, as with most muscles in other animals, scale independent (James *et al.*, 1998; Medler, 2002). Isometric tetanic force is a steady-state measure that reflects the density of cross-bridges and their rate constants within the cross-bridge cycle. These rate constants are typically such that the same proportion of the available cross-bridge population is attached at any given moment in a fully active muscle, hence the scale independence from tetanic force (e.g., Rome *et al.*, 1999). However, this is not to say that the rate constants are the same in fast and slow muscle, and thus the potential still exists for differences in V_{max} between fiber types and for scaling of V_{max} with body size.

VII. AXIAL VARIATION

A wave of contraction propagates down the length of many fishes during swimming. This wave is rarely uniform in amplitude or velocity (see Chapter 7), and so we might expect differences in the inherent properties of the muscles at different axial locations and differences in the manner in which they are recruited. What, exactly, these differences might be are difficult to predict, barring a simple assumption that fishes always use their muscles in a manner that maximizes power output, which seems to be more the exception than the rule. In this section I summarize observations on axial variation in contraction kinetics in fish muscle, and how this may bear on their ability to power swimming.

Isometric force is not noted to be body position dependent, but see Coughlin *et al.* (2001a) for a unique exception, in which posterior red muscle in parr and smolt rainbow trout produces about 20% more force than anterior muscle. However, dynamic aspects of muscle contraction that impact cyclic activity often do show axial variation. James *et al.* (1998) noted that in sculpin fast muscle the unloaded velocity of muscle shortening (closely related to V_{max}) is about 38% faster in anterior muscles. In contrast, Swank *et al.* (1997) did not observe axial variation in V_{max} of scup red muscle, nor did Coughlin *et al.* (2001a) in rainbow trout red muscle. Whether V_{max} varies axially or not, there do not appear to be any substantial axial differences in the inherent ability of muscle to generate mechanical power, although anterior muscles with their faster contraction kinetics (see later) sometimes excel marginally (e.g., Coughlin, 2000). For example, despite variations in V_{max} there are not any axial differences in power (based on force-velocity characteristics) in sculpin fast muscle (James *et al.*, 1998). Likewise, under conditions in which power was maximized in work loop analysis there is not a significant difference in mass-specific power production between anterior versus posterior locations in sculpin fast muscle (Johnston *et al.*, 1993), eel red and white muscle (D'Aout *et al.*, 2001; Ellerby *et al.*, 2001a,b), yellowfin tuna red muscle (Altringham and Block, 1997), scup red muscle (Rome *et al.*, 2000), or brook char red muscle (McGlinchey *et al.*, 2001). But despite this lack of axial differences in the inherent ability of muscle to generate power, in swimming fishes there are often substantial axial differences in contraction kinetics, strain, and activation kinetics, leading to often large axial differences in mass-specific power from the muscles *when used as during swimming*. I first touch upon some of these differences, and then return to their implications for generating power and muscle function.

Axial variation in muscle strain, strain rate, and twitch kinetics in fishes is particularly well documented. The amplitude of muscle strain in the myotomes of fishes that employ (sub)carangiform and even thunniform swimming generally increases posteriorly and falls in the range of 4–18% peak to peak (i.e., ±2–9%); see Figure 3 in Coughlin (2002a) and Figure 7.11 in Chapter 7 of this volume. The rostro-caudal increase in strain amplitude is modest in some fishes, but is typically a doubling or more. Assuming the tissues in the body of a fish deform as a homogenous beam, which is not

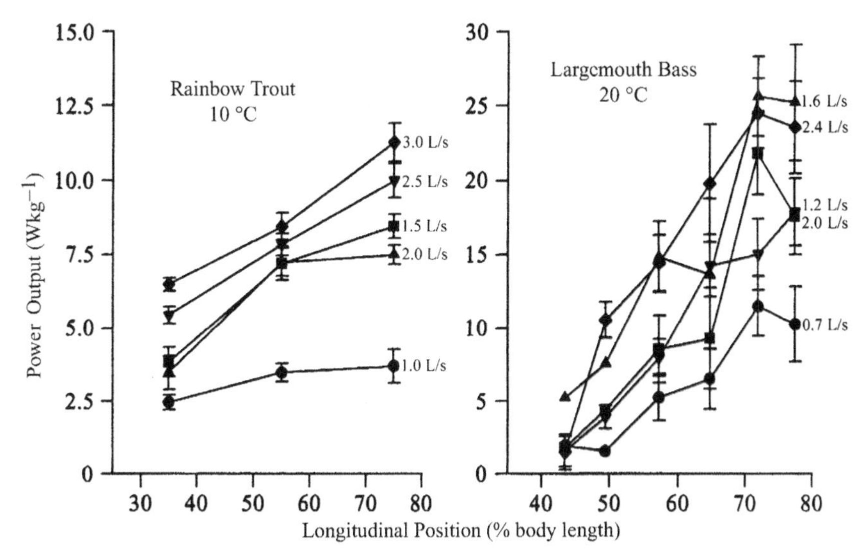

Fig. 6.10. Power output of red muscle from rainbow trout at 10 °C (left panel) and largemouth bass at 20 °C (right panel) at different axial (longitudinal) locations during work loop analysis. The muscles were exposed to the same strains and stimulus patterns as experienced in swimming fishes. Each line represents a different swim speed, as indicated. Note the pronounced increase in power output of the muscles moving from anterior toward posterior locations, and a general increase in power with increased swim speed. (Adapted from Coughlin, 2000, with permission of the Company of Biologists Ltd.)

true in tuna and some sharks (Katz *et al.*, 2001; Donley *et al.*, 2005; also see Chapter 7 of this volume), the small strain in anterior myotomes will translate directly into less mass-specific work done by the anterior muscle (Figures 6.7 and 6.10). For example, when working at their respective strains, anterior red muscles of skipjack tuna produce about two-thirds the work per unit mass of posterior muscle (Syme and Shadwick, 2002), anterior red muscles of scup produce only about 20% as much power as posterior muscle (Rome *et al.*, 1993, 2000), anterior pink muscles of scup produce about one-third the power of posterior muscles at 20 °C and only 3–10% at 10 °C (Coughlin *et al.*, 1996), anterior red muscles of rainbow trout produce 50–100% the power of posterior muscles (Coughlin, 2000), and anterior red muscles of bass produce 10- to 15-fold less power than posterior muscles (Coughlin, 2000). Brook char appear to be an exception, having larger strains in posterior red muscle but no axial variation in mass-specific power production (McGlinchey *et al.*, 2001). Given the lack of axial variation in the inherent ability to generate power (discussed previously), these axial differences in power output of muscle in swimming fishes are simply a consequence of the way that the muscle is used by the fish (e.g., Rome *et al.*, 1993). Notably, Ellerby *et al.* (2001a,b) also reported this marked influence of strain on power output of muscle from eels (anguilliform swimmers); however, in muscle from yellow phase eels the anterior (but not posterior) muscles produce less power with increased strain. The anterior muscles are not slower by any obvious measure, and thus it is not clear why their performance is so different from posterior muscle under the same conditions of strain and cycling frequency.

L. C. Rome has postulated that the limited strain in the anterior muscle of fishes will limit the extent of shortening-induced deactivation in the muscles there, which will in turn limit their ability to produce cyclic work and power. To compensate, muscles in the anterior of the fish should possess more rapid twitch kinetics to allow them to contract and relax at the same frequency as posterior muscles. In support of this hypothesis, anterior muscles typically have notably faster twitch kinetics in both red muscle (Rome *et al.*, 1993; Coughlin and Rome, 1996; Altringham and Block, 1997; Swank *et al.*, 1997; Altringham and Ellerby, 1999; Coughlin, 2000; Rome *et al.*, 2000; Altringham and Shadwick, 2001; Ellerby *et al.*, 2001a) and white muscle (Wardle *et al.*, 1989; Altringham *et al.*, 1993; Davies *et al.*, 1995; Ellerby *et al.*, 2001b; Thys *et al.*, 2001; also reviewed in Coughlin, 2002a). In general, rates of relaxation show axial variation, while rates of contraction are not markedly different. Yet despite a seemingly convincing literature documenting axial variation in twitch kinetics, the apparent differences are not always statistically significant (e.g., Johnston *et al.*, 1993; Hammond *et al.*, 1998; James *et al.*, 1998; Syme and Shadwick, 2002). Even suggestive trends are sometimes absent. Activation and relaxation times do not vary along the length of short-horn sculpin (Johnston *et al.*, 1995). Red muscles of brook char do not show any axial variation in twitch kinetics (McGlinchey *et al.*, 2001). The eel shows no (D'Août *et al.*, 2001) or little (Ellerby *et al.*, 2001a,b) axial variation in twitch kinetics in either red or white muscle; this may not be surprising for an anguilliform swimmer with relatively large and uniform strains and contraction kinetics along the body length (D'Août *et al.*, 2001; Ellerby *et al.*, 2001b). Red muscles of shortfin mako sharks do not show axial variation in twitch kinetics (Donley, 2004). Skipjack tuna may be an exception in having anterior muscles with slower rates of activation than posterior, although the differences are small and relaxation rates are not different axially (Syme and Shadwick, 2002). Yellowfin tuna appear to be like many other fishes with posterior muscles having slower rates of activation (R. E. Shadwick and D. A. Syme, unpublished observation).

The distribution of red and pink muscle in scup also supports the contention that faster fibers are required in the anterior region. Coughlin *et al.* (1996) noted that pink muscle is concentrated anteriorly in scup, where its faster contractions allow it to compensate for the reduced abilities of slower red muscle, and showed that the disparity between the ability of red and pink muscle to produce work becomes progressively greater at smaller strains (i.e., pink does relatively better). Similarly, Rome *et al.* (2000) showed that in all anterior and some posterior locations scup red muscles produce net negative work (absorb energy rather than produce it) at high swimming speeds and cold temperatures, while the faster pink muscles continue to produce positive work. Further, the area occupied by slow red muscle in the anterior of the fish is only about half that in the posterior, whereas there is about twice as much pink muscle in the anterior versus posterior (Zhang *et al.*, 1996); thus, the amount of faster pink muscle greatly exceeds red toward the head, where strain is most limited, but slower red muscle greatly exceeds pink toward the tail, where strain is least limited. Likewise, in fishes that bend as a homogenous beam, faster fibers might be needed with increasing proximity to the backbone (the neutral axis of bending), where strain becomes progressively smaller (e.g., Ellerby and Altringham, 2001). Indeed, faster pink muscles tend to be located more proximal to the slower red (Zhang *et al.*, 1996). In yellowfin tuna there is a considerable component of depth to the red muscle, and a speeding of the twitch is also observed with increasing depth (Altringham and Block, 1997). However, tuna do not appear to bend as a homogenous beam (Katz *et al.*, 2001), and the strain experienced by deep red muscle in tuna is considerably higher than in other fishes, about 12% in skipjack tuna and 10.6% in yellowfin tuna versus 5.6% at mid-body in bonito (Ellerby *et al.*, 2000; Katz *et al.*, 2001; Shadwick *et al.*, 1999). Thus, the limitation of reduced strain in more proximal muscle may not be a functional concern in tuna, and the need for faster muscle near the spine may be likewise muted. Why tuna also show a radial gradient in the speed of red muscle is therefore not clear.

There are several physiological mechanisms by which twitch speed could be altered (Rome *et al.*, 1996). In anterior white muscle of cod, both V_{max} and contraction/relaxation rates increase in concert, suggesting faster cross-bridge kinetics and thus a different myosin isoform as at least one modification (Altringham *et al.*, 1993; Davies *et al.*, 1995). In addition, Thys *et al.* (1998, 2001) found axial variation in the ratio of troponin T isoforms 1 and 2 (more type 2 in anterior muscles), increased levels of parvalbumin, and the presence of two novel soluble calcium-binding proteins in anterior white muscle of Atlantic cod and bass, any of which could result in faster twitch kinetics. Likewise, anterior red muscles of rainbow trout exhibit faster activation kinetics and relatively greater expression of the S2 isoform of troponin T, while red muscles of brook char do not show consistent axial variation in either activation kinetics or expression of the S2 and S1 isoforms of troponin T (Coughlin *et al.*, 2005), although the authors noted that within individual brook char there is a suggestive but not universal trend for axial variation in activation kinetics and troponin T isoform expression. Swank *et al.* (1997), in seeking to explain axial variation in the twitch speed of scup red muscle, found no differences in V_{max} (i.e., myosin isoforms) nor in the series compliance between anterior and posterior muscles. However, inhibiting the calcium pumps on the sarcoplasmic reticulum was effective at slowing the relaxation rate of anterior muscles to equal that of posterior muscle, suggesting that differences in the rates of calcium sequestration into the sarcoplasmic reticulum are responsible for the different rates of relaxation. There does not appear to be a higher pump density in the anterior muscles, so perhaps differences in the pump isoform may be responsible for the different relaxation rates.

Based on observations from work loop studies, slower activation and relaxation kinetics in posterior muscles would require that they be activated and deactivated sooner in the strain cycle in order to maximize work output. Such qualitative patterns are observed in swimming fishes (Altringham and Ellerby, 1999). Whether the quantitative changes are adequate to yield maximal power or whether substantial eccentric (energy absorbing) contractions occur in posterior muscles is currently debated. In fishes that swim with the anterior region of their body held relatively rigid (e.g., scup, largemouth bass, rainbow trout), the majority of power for cruise swimming appears to come from the posterior red muscles (Rome *et al.*, 1993, 2000; Johnson *et al.*, 1994; Coughlin and Rome, 1996, 1999; Coughlin, 2000, 2003). In these fishes, the axial variation in strain amplitude is large relative to fishes that employ more carangiform or anguilliform movements, so that anterior muscles experience very small strains and thus produce small amounts of work and power. Application of work loop analysis to isolated segments of muscle from various regions of such fishes, using strain and activation parameters measured in swimming fishes, confirms that anterior muscles do produce little mass-specific power while middle and posterior muscles produce the most (reviewed in Coughlin, 2002a) (Figures 6.7 and 6.10). The bias toward higher power in posterior regions is also apparent in the absolute power output of the muscles when considering their axial distribution (e.g., Coughlin, 2002a, 2003). Coughlin (2002a, 2003) commented that the approximately 2-fold axial variation of power from red muscle of rainbow trout is actually quite modest in comparison with fishes such as scup and bass, and suggested that perhaps rainbow trout should not be included in the list of fishes that preferentially power swimming with caudal muscles. However, in comparison with brook char, which show virtually no axial variation, and in conjunction with reports of preferential recruitment of posterior muscles at slow swim speeds in rainbow trout but not in brook char (Coughlin *et al.*, 2004), rainbow trout do appear to stand apart from brook char in where and when the power for cruise swimming is generated.

Perhaps due to these rostro-caudal patterns of power output, at slow swimming speeds only the more posterior red muscles are recruited, and as swim speed increases the anterior red muscles are recruited along with faster pink muscles to supplement the red (Johnston *et al.*, 1977; Coughlin and Rome, 1999; Coughlin *et al.*, 2004). Gillis (1998) observed a similar caudal-to-rostral pattern of initial recruitment speed in red muscle of eels, which use a considerably different pattern of body movement than (sub)carangiform swimmers. This pattern of recruiting red only at slow speeds then pink also at faster speeds is also expected based on the relatively high power output of red muscle at low oscillation frequencies but low power at high frequencies, and the relatively low power output of pink muscle at low oscillation frequencies but high power at high frequencies (Coughlin *et al.*, 1996).

In contrast, several studies indicate that caudal muscles in some fishes may serve a dual role in swimming, undergoing eccentric contractions during the early phase of the tail-beat cycle and thus acting to transmit power from anterior muscles to the tail, but then producing positive power for propulsion during the latter phase of the cycle (although the amount of positive power they produce may be very limited in some instances) (e.g., Hess and Videler, 1984; Williams *et al.*, 1989; van Leeuwen *et al.*, 1990; Altringham *et al.*, 1993; van Leeuwen, 1995; Hammond *et al.*, 1998; reviewed in Altringham and Ellerby,1999; Wardle *et al.*, 1995). While Hammond *et al.* (1998) suggested that the posterior red muscles of rainbow trout serve this dual function, Coughlin (2000) questioned this interpretation, arguing that recruitment patterns measured from larger trout with slower muscles were applied to work loop analysis on muscles from smaller trout with faster muscles, which may have resulted in an artifactual eccentric phase. Such arguments aside, in addition to Rome's posit that anterior muscles may be fast to accommodate small strains, it is intriguing to consider that posterior muscles may be slow to facilitate economical eccentric contractions and a dual role in powering swimming (e.g., Wardle *et al.*, 1995). Again, we are cautioned against assuming that muscles are designed and used simply to produce maximal power. Further, we may well be advised to study more carefully the eccentric characteristics of contractions in fish muscle, which to date are largely if not completely ignored.

In thunniform swimmers (e.g., tuna, mackerel) and anguilliform swimmers (e.g., eels) the mass-specific power output of muscle when used as during swimming appears more evenly distributed along the length of the fish (Shadwick *et al.*, 1998; Altringham and Ellerby, 1999; D'Août *et al.*, 2001; Coughlin, 2002a). In tuna in particular, all of the red muscle appears to be recruited in a fashion that maximizes power output, perhaps belying a unique role in simply serving to power forward swimming in this fast, pelagic predator. The relationship between EMG phase/duration and muscle strain at all locations in swimming tuna is consistent with those relationships that result in maximal work (Shadwick *et al.*, 1999), and work loop studies using isolated segments of red muscle from anterior and posterior locations confirm that the muscles are being used in a fashion that maximizes power output (Katz *et al.*, 2001; Syme and Shadwick, 2002). In tuna, this power is transmitted to the tail economically using tendons; thus, body undulations are minimal in thunniform swimming, and the muscles are freed from the potential constraints of creating a wave of body bending or undergoing eccentric contractions in the caudal region so that they can simply act as power producers for propulsion. Likewise, estimates based on EMG and strain patterns suggest that red muscles in mackerel may behave similarly (Shadwick *et al.*, 1998). Further, using activation phases recorded from swimming eels and results from work loop analysis, D'Août *et al.* (2001) concluded that both red and white muscles produce net positive power at all axial locations, although they did not have enough data to conclude whether power output was maximized. It is interesting to note that studies on brook char (McGlinchey *et al.*, 2001) suggested that they too show little axial variation in mass-specific power (or strain or contraction kinetics), although red muscle is still concentrated toward the caudal end of both brook char and rainbow trout, resulting in substantially higher absolute power in this region (Coughlin, 2002a). This in conjunction with the observations that brook char and rainbow trout may not show substantial axial variation in mass-specific power leaves us uncertain whether the large axial variation in power in carangiform swimmers such as scup and bass is widespread.

VIII. EFFECTS OF TEMPERATURE

Studying the effects of temperature and thermal acclimation on fish muscle is an enormous and active field, and easily the focus of an entire volume. In this section, I restrict the discussion to several observations particularly relevant to the ability of fish muscle to produce power for swimming: specifically, how temperature affects force production, twitch kinetics, and the ability to do work. Fishes are largely ectothermic poikilotherms and live in an environment where acute and seasonal temperature fluctuations are common. They are thus prone to the effects of temperature on muscle performance. In turn, they may be expected to have muscles that readily adapt to changes in temperature or that are limited by temperature; this is certainly one of the more fascinating aspects of the study of muscle biology in fishes. Johnston (1980a) and Guderley (1990) (and many references therein) provided accounts of the effects of temperature and thermal acclimation on cell physiology and metabolic function of fish muscle, including marked shifts in the catalytic activity of myosin, calcium sensitivity of the regulatory proteins, protein stability, catalytic efficiency, membrane fluidity, and enzyme activity in metabolism. There is also a large literature on the effects of temperature on muscle performance in ectotherms, with considerable emphasis on fishes (e.g., Bennett, 1984, 1985; Johnston *et al.*, 1990; Rall and Woledge, 1990; Rome, 1990; Rome and Swank, 1992; Johnson and Bennett, 1995; Altringham and Block, 1997; Rome *et al.*, 2000; Swank and Rome, 2000; Sanger and Stoiber, 2001; Johnston and Temple, 2002; Katz, 2002). The acute affects of cooling tend to limit performance, while acclimation can compensate to some extent. Conversely, thermal acclimation to warm temperatures results in improved performance at warm temperatures and impaired performance in the cold. Some of these improvements include or result from greater tetanic force production, muscle hypertrophy, altered recruitment strategies, faster twitches, faster V_{max}/higher myofibrillar ATPase activity, and increased mitochondrial volume density or enzyme activity (reviewed in Gerlach *et al.*, 1990; Johnston *et al.*, 1990; Guderley and St-Pierre, 2002; Johnston and Temple, 2002). Temperature acclimation does not appear to affect calcium sensitivity of force production, but calcium sensitivity does show a pronounced decrease at colder temperatures (Johnston *et al.*, 1990). Some acclimation effects are dependent on developmental stage and fish species and some extend the range of temperatures over which an animal can operate, but none appear to actually improve whole animal performance over the control condition; they are compensatory, at best (Wakeling *et al.*, 2000; Johnston and Temple, 2002).

Isometric, tetanic force in vertebrate skeletal muscle is commonly held to be largely independent of temperature (Bennett, 1984; Rall and Woledge, 1990). In fishes, isometric tetanic force in living muscle and in skinned fibers has a relatively small R_{10} of about 1.0–1.1 when measured around physiological temperatures (e.g., Johnston and Brill, 1984; Rome, 1990; Rome and Sosnicki, 1990; Johnson and Johnston, 1991a; Rome *et al.*, 1992b, 2000; Wakeling and Johnston, 1999; Katz, 2002). However, extremes of temperature, particularly cold, do impair force production (Figure 6.11), and there are several examples of relatively discrete temperature optima for tetanic force with progressive failure toward extremes and eventually muscle damage or death in fish and other ectotherms (Johnston and Altringham, 1985; Luiker and Stevens, 1994; see Langfeld *et al.*, 1989; Johnson and Johnston, 1991a for a discussion and further examples). Yet when observed across species from Antarctic to more tropical climates, where normal body temperatures range tremendously from below 0 °C to well over 20 °C, isometric tetanic force measured at the organism's preferred temperature is

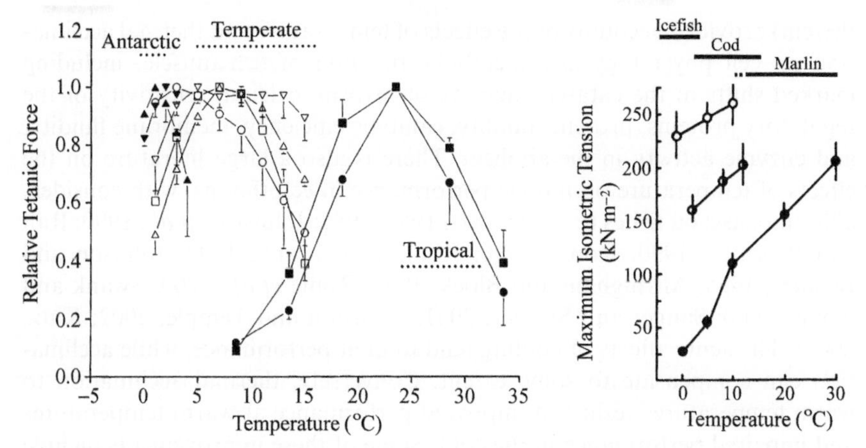

Fig. 6.11. Maximal isometric tetanic tension versus temperature. (Left) (live muscle fibers) Filled triangles are muscle from Antarctic species; open symbols are from temperate species; filled circles and squares are from tropical species. Horizontal dotted lines represent the normal environmental temperature ranges of the three groups of fishes. (Right) (skinned fibers): Open circles are icefish (Antarctic), half filled circles are cod (temperate), and filled circles are marlin (tropical). Horizontal bars represent the normal environmental temperature ranges. Note that isometric force is maximal in the normal environmental temperature range but falls at the extremes of and outside this range. Also note that isometric force is similar between the three groups when compared in the normal environmental temperature range. (Adapted from Johnston and Altringham, 1985, and Johnson and Johnston, 1991a, used with permission, copyright Springer.)

remarkably invariable (Figure 6.11) (Johnson and Johnston, 1991a). Thermal acclimation, a more acute response to temperature exposure, sometimes does not have a marked effect on isometric force (e.g., Swank and Rome, 2001) but sometime does (e.g., Langfeld *et al.*, 1989; Fleming *et al.*, 1990), particularly in skinned muscle fibers (reviewed in Johnston and Temple, 2002).

Quite unlike tetanic contractions, twitches are rate dependent and are very temperature sensitive (Bennett, 1984), reflecting in part the rate that a muscle can turn on and turn off. The primary effects are a slowing of the rate of activation and a marked slowing of the rate of twitch relaxation at cooler temperatures (Figure 6.3). In fish, Q_{10}'s are typically 1.5–3, both within species and across species at preferred body temperatures (e.g., Langfeld *et al.*, 1989; Johnson and Johnston, 1991a; Rome *et al.*, 1992b, 2000; Luiker and Stevens, 1994; Altringham and Block, 1997; Wakeling and Johnston, 1999). In scup, the Q_{10} for twitch relaxation rate is smaller in pink (~3) than in red (~4) muscle (Rome and Swank, 1992; Coughlin *et al.*, 1996), which may assist pink muscle in compensating for the reduced power of red muscle at cold temperatures. For example, while red muscle dominates power production at 20 °C in scup, at 10 °C it contributes virtually nothing, so that pink muscle becomes the sole power producer despite having only one-third the mass of red (Rome *et al.*, 2000). Notably, the rate of both activation and relaxation in muscle is less temperature sensitive during shortening than during isometric contractions, hence the impact of temperature on working muscles via an influence on twitch kinetics is less pronounced than might be predicted based on isometric properties (Luiker and Stevens, 1994). This will have important consequences in attempting to extrapolate results obtained from isometric contractions to swimming fishes.

Comparing across species with different thermal niches, there is evidence of only limited temperature compensation in twitch speed, such that the muscle twitches from fishes that live in the cold are considerably slower than those from fishes that are warm (reviewed in Johnson and Johnston, 1991a). There does appear to be some compensation due to acclimation; when measured at a cool temperature, twitches in cold acclimated fishes are faster than those from warm acclimated fishes. The compensation can be quite marked, but again not to the extent of allowing cold fishes to behave as if warm (e.g., Fleming *et al.*, 1990; Johnson and Bennett, 1995) and in some cases it is absent (Altringham and Block, 1997; Wakeling *et al.*, 2000). Some compensation in twitch speed may be due to a reduction in fiber diameter (shorter diffusion distances) and changes in the density or type of calcium pumps in cold fishes, but seemingly not to changes in the parvalbumin isoform or concentration (Fleming *et al.*, 1990; Rodnick and Sidell, 1995; see also Johnson and Johnston, 1991a; Johnston, 1990 for discussions). Since the ability to perform cyclic work is limited in part by twitch kinetics, fishes often swim with slower tail-beat frequencies in the cold (e.g., Stevens, 1979; Johnston *et al.*, 1990). There is also evidence that, like twitch duration itself, the Q_{10} for twitch relaxation may be body position dependent; Rome *et al.* (2000) noted that in scup red muscle the Q_{10} for twitch relaxation is about 2.8 in the slower posterior muscles and 3.6 in the faster anterior muscles, although the difference only approaches statistical significance ($p = 0.13$).

The maximal velocity of muscle shortening (V_{max}) is highly temperature dependent, being faster at warmer temperatures (Figure 6.12). Bennett (1984) reported a Q_{10} for V_{max} of about 2 for mixed skeletal muscle of many species. Similarly, Rome and Sosnicki (1990) and Rome *et al.* (1992b) reported a Q_{10} of 1.6 between 10 and 20 °C for slow red muscle of carp and scup, while Johnston *et al.* (1985) reported 2.1 between 7 and 23 °C in slow skinned carp fibers. The Q_{10} for V_{max} of scup pink muscle is 1.6 (Coughlin *et al.*, 1996), similar to that of red muscle. Q_{10}'s for V_{max} of fast white fibers are 1.8–2.5 (Wakeling and Johnston, 1998, 1999; Langfeld *et al.*, 1989). There is evidence for acclimation of V_{max} in some species, so that

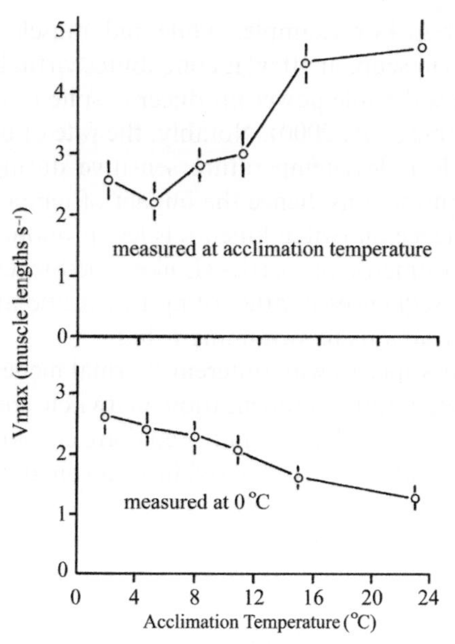

Fig. 6.12. Maximal velocity of muscle shortening (V_{max}) of skinned, fast fibers from carp at different temperatures. (Upper) Fishes were acclimated to the temperature noted on the x axis and V_{max} was measured at this same temperature. In this case, there is an increase in V_{max} with increasing temperature. (Lower) Fishes were acclimated to the temperature noted on the x axis but V_{max} was always measured at 0 °C. In this case, V_{max} is highest in muscle from fishes acclimated to cool temperatures and declines with increasing acclimation temperature. (Adapted from Crockford and Johnston, 1990, used with permission, copyright Springer.)

when measured at a cold temperature the V_{max} of muscle from fishes acclimated to cold is greater than from fishes acclimated to warm (Figure 6.12) (Johnston *et al.*, 1985; Crockford and Johnston, 1990; Fleming *et al.*, 1990; Langfeld *et al.*, 1991). Johnston and Temple (2002) reviewed several reports of increased myofibrillar ATPase activity in muscle from cold-acclimated fishes relative to warm-acclimated at all temperatures. Alternatively, Swank and Rome (2001) found no acclimation effect on V_{max} in scup red muscle (i.e., when tested at the same temperature, muscles from warm- and cold-acclimated scup had the same V_{max}). Perhaps the relative thermal stability of the habitat of scup may account for its lack of an acclimation response. Gerlach *et al.* (1990) noted that the acclimation effect on myofibrillar ATPase and activation energy is more pronounced in white than red muscle, and suggested that this may serve to preserve burst swim performance in the cold. Interestingly, the acclimation effect in white muscle is prevented by starvation, suggesting that substantial protein synthesis is required for the change (Heap *et al.*, 1986).

Surprisingly, there does not appear to be any notable temperature compensation of V_{max} across species from warm to cold climates. Velocity simply scales with temperature no matter what the animal's thermal niche, so that the V_{max} of muscles from fishes living in cold climates is not different than would be expected based on the Q_{10} scaling of V_{max} from fishes in warm climates (e.g., Johnston and Altringham, 1985). For example, V_{max} of fast fibers in Antarctic species living at −1 °C is only 0.9 Ls^{-1} and at 0 °C is about 2 Ls^{-1}; in coldwater species living at 1–8 °C it is about 1–4 Ls^{-1}, but it is up to a remarkable 20 Ls^{-1} in warm water species living at 25 °C (Johnston and Brill, 1984; Johnston and Altringham, 1985; Langfeld *et al.*, 1989; Johnson and Johnston, 1991a).

The drop in V_{max} with cold temperatures in conjunction with little or no change in maximal force (see the preceding) results in a reduction in muscle power output in the cold (Figure 6.13). The Q_{10}' for maximal power calculated from force-velocity data is about 2.3 (Bennett, 1984), not surprisingly similar to the Q_{10} for V_{max}. Likewise, acute changes in temperature of fish muscle affect maximal cyclic power output and the cycling frequency at which it is maximized (Figure 6.13) (e.g., Rome, 1990; Johnson *et al.*, 1991b; Johnson and Johnston, 1991b; Rome and Swank, 1992; Altringham and Block, 1997; Rome *et al.*, 2000). Independent of any particular cycle frequency and when optimized for stimulus and strain, the Q_{10} for cyclic power of red muscle is about 2 (Rome and Swank, 1992; Altringham and Block, 1997; Katz, 2002). For sculpin white muscle it is about 1.2 in summer-caught fish, but in winter-caught fish it is 0.5 (Johnson and Johnston, 1991b). In red muscle from yellowfin tuna not only is there a notable increase in power output with increasing muscle temperature, but also the deeper muscles (which tend to be warmer and more homeothermic) both are more temperature sensitive and have a higher temperature optimum than the more superficial fibers (Figure 6.13) (Altringham and Block, 1997). Thus, the internalization and subsequent chronic warming of the red muscle in tuna may lead to increased power production and thus swimming speed. Katz (2002) questioned an adaptive role for the warming of muscles in tuna, noting that when compared at similar temperatures, these tuna muscles actually underachieve in comparison with muscles from fishes that do not maintain elevated temperatures. In fairness to both sides, it should be recognized that estimating viable muscle mass is notoriously difficult in fish muscle, particularly red, with its fragility and high connective tissue content, and thus comparisons of absolute muscle power between preparations are fraught with uncertainty

Of interest, and likely considerable significance to performance in swimming fishes, the Q_{10} for power is highly dependent on the cycle or tail-beat frequency (Figure 6.13). In scup red muscle Q_{10} ranges from 1 at a frequency

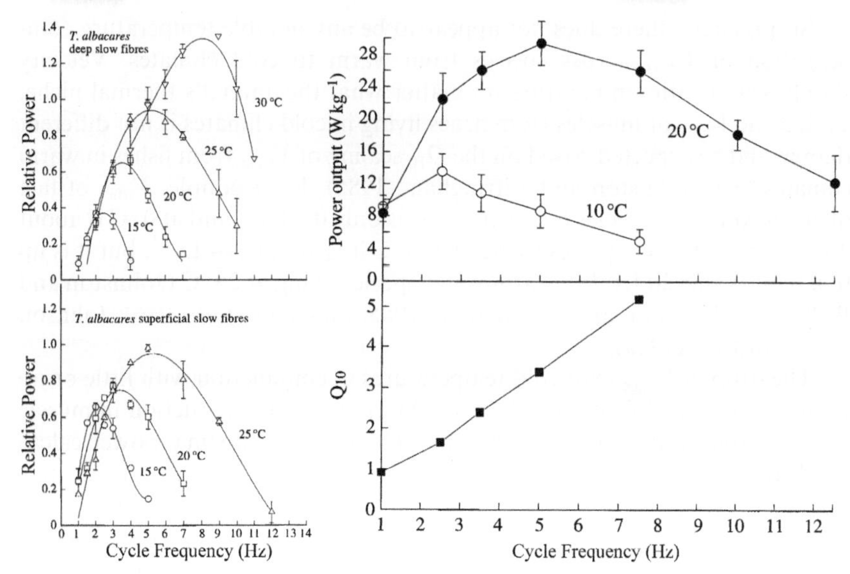

Fig. 6.13. Effects of temperature on maximal muscle power output. (Left) Relative power output from red muscle of yellowfin tuna measured over a range of cycle frequencies and temperatures (upper panel is deep red muscle and lower panel is superficial red muscle). Note that the deep red muscle has a higher thermal sensitivity than the superficial, and that power output is not temperature sensitive at slow cycle frequencies in muscle from both locations. (Upper right) Power output from red muscle of scup at 10 and 20 °C. (Lower right) The Q_{10} for power output is highly dependent on the cycle frequency, being 1 at slow cycle frequencies as in the tuna muscle, but increasing with increasing cycle frequency. (Adapted from Altringham and Block, 1997, and Rome and Swank, 1992, with permission of the Company of Biologists Ltd.)

of 1 Hz to about 5 at 7.5 Hz (Rome and Swank, 1992). In sculpin white muscle Q_{10} ranges from near 1 at 3 Hz to about 2 at 15 Hz (Johnson and Johnston, 1991b). In red muscle of yellowfin tuna Q_{10} would remain near 1 at slow cycle frequencies over appropriate temperature ranges, but would approach 9 at a cycle frequency of 7 Hz when measured over a similar temperature range (Figure 6.13). As a consequence, fishes that swam with a relatively slow tail-beat frequency would be less temperature sensitive in terms of mechanical performance.

In addition to the Q_{10} for power being highly dependent on the cycle frequency, the Q_{10} for power can be much different when the muscle is recruited as occurs in swimming fishes, than when power is simply maximized in work loop analysis (Rome and Swank, 1992; Rome *et al.*, 2000; Swank and Rome, 2000). In work loop analysis the experimenter can fine-tune the phase and duration of muscle activation at different temperatures to account for changing contraction kinetics in the muscle; under such circumstances the Q_{10} for maximal power output is about 2. But when swimming, fishes do not appear to adjust muscle activation to fully compensate for the changes in contraction kinetics of their muscles with changes in temperature, and power is thus compromised in the cold (Swank and Rome, 2000). Rome and Swank (1992) showed that power output of red muscle at some axial locations is near maximal in scup swimming at 20 °C, yet at 10 °C the muscles produce much less power than they are capable of, in some cases being negative (Figure 6.7). Further, power is maximal at a cycle frequency of only 2.5 Hz in the cold, much slower than the tail-beat frequencies in swimming scup. Thus, the Q_{10} for power in swimming fishes can be extremely large; values up to 14 have been recorded, and at some temperatures Q_{10} is indeterminate or negative (reviewed in Rome *et al.*, 2000). Extrapolating the effects of temperature on power output from work loop analysis to swimming performance must be done with due caution.

Like V_{max}, power output does not appear to be well maintained, if at all, across species that live in different thermal environments, ranging from about 25 Wkg^{-1} at 0 °C in cold water species to about 150 Wkg^{-1} at 20 °C in fish inhabiting warm water (Wakeling and Johnston, 1998, and references therein). Since the power required to swim and the velocity of muscle shortening at a given swim speed are both relatively temperature independent (Rome, 1994), a change in the power available from muscle with a change in temperature will affect the ability to power swimming in fishes that both are and are not strict poikilotherms, and will impact strategies for muscle recruitment (e.g., Rome *et al.*, 1984, 2000; Sisson and Sidell, 1987; Altringham and Shadwick, 2001). Without other forms of compensation, cold fishes must swim more slowly than warm fishes. Yet with acclimation to cold temperatures there are a variety of mechanisms that appear to partially compensate for the loss of power from muscle, including muscle hyperplasia/hypertrophy, conversion to different (i.e., faster) myosin isoforms (both heavy and light chains), reduced activation energy, a more flexible but less thermostable myosin molecule, increased rates of calcium pumping by the sarcoplasmic reticulum via faster pumps, and altered recruitment strategies both between and within fiber types (reviewed in Johnson and Bennett, 1995; Johnston and Temple, 2002; Watabe, 2002). One well-documented strategy reveals that with an acute decrease in temperature, the loss of power available from red muscle may be mitigated by recruiting more muscle or by recruiting faster (i.e., pink) fiber types at slower swimming speeds, coined compression of recruitment order (Figure 6.1) (e.g., Stevens, 1979, Rome *et al.*, 1984, 1985, 2000; Rome, 1986; Sisson and Sidell, 1987). Differences between species in the inherent contractile properties of specific fiber types will influence the utility of recruiting those fiber types at any given temperature and swimming speed. For example, Coughlin (2003) noted that at 10 °C, red muscles of rainbow trout are faster than red muscles of scup, so much so that the power of scup red muscle is near zero and the mass specific power output of trout red muscle rivals that of pink muscle in scup. Thus, at 10 °C, trout can power most of their swimming by recruiting only red muscle, while scup must recruit mostly pink muscle. Even within a given fiber type there can be differences in temperature sensitivity of power output. Power is more temperature sensitive in deep than superficial red muscles in yellowfin tuna (Figure 6.13), suggesting that the deeper (and warmer) red muscle of tuna has become specialized to tolerate warm but relatively stable temperatures, sustaining high power at the expense of fairing relatively poorly at cooler temperatures (Altringham and Block, 1997).

While not fully compensatory, the changes in recruitment patterns and contraction kinetics with thermal acclimation to the cold allow greater power output from red muscle and thus less reliance on faster pink muscle (e.g., Rome and Swank, 2001; Swank and Rome, 2001). Red muscles from cold-acclimated scup have significantly faster rates of activation (20–40% depending on axial location) than those from warm-acclimated fishes when operating at cold temperatures (Swank and Rome, 2001), and the duration of muscle activation during a tail-beat is reduced in cold-acclimated fishes (Rome and Swank, 2001). These changes in the muscle, along with alterations in the timing of recruitment (phase and duration), result in the red muscles of cold-acclimated fishes producing 3–9 times more power than the muscles of warm-acclimated fish when working in cold temperatures and 2.5-fold more power when power was simply maximized in work loop analysis. Thus, cold-acclimated fishes can swim at higher speeds in the cold before having to recruit the faster pink muscles. Likewise, Johnston *et al.* (1990) and Johnson and Bennett (1995) found that fast muscles in cold-acclimated fishes have faster rates of relaxation compared with muscle from warm-acclimated fishes when tested at cold temperatures. But there is a compromise. The fast muscles of winter-caught fishes (cold-acclimated) produce only about half the power of muscles from summer-caught fishes (warm-acclimated) when tested at warm temperatures, both when power is maximized (Johnson and Johnston, 1991b) and when using activation parameters measured from swimming fishes (Temple *et al.*, 2000). Cold acclimation thus appears to improve performance in the cold but impairs it when warm. Rainbow trout show relatively small acclimation effects, in either contractile kinetics or myosin isoform expression (Johnson *et al.*, 1996). Hence the specific response to cold exposure is highly variable between species and even life stages, and the resulting mechanical performance is highly dependent on acclimation and test temperatures. Although changes in muscle itself appear quite sensitive to changes in temperature, they do not

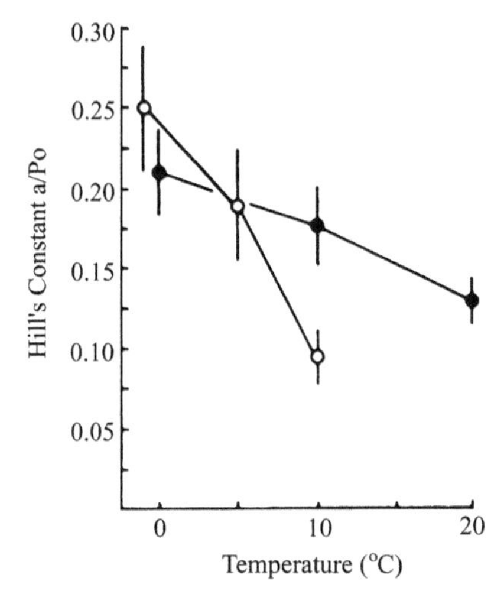

Fig. 6.14. Effects of temperature on the curvature of the force-velocity relationship of skinned, fast muscle fibers from Antarctic icefish living at −1 °C (open circles) and tropical marlin living at 20 °C (filled circles). The ratio of the Hill's constant a/Po describes the curvature of the force-velocity relationship; higher ratios indicate a straighter relationship and a relatively more powerful muscle. (Adapted from Johnston and Altringham, 1985, used with permission, copyright Springer.)

always translate into observable improvements in swimming performance (reviewed in Johnston and Temple, 2002).

The curvature of the force-velocity relationship is associated with the relative power of the muscle, with more powerful muscles having straighter relationships. When working at warm temperatures, muscles from fishes that live in warm water tend to maintain straighter relationships than muscles from fishes that live in the cold, and when working at cold temperatures, fishes that live in the cold maintain straighter relationships than fishes that live in the warm (i.e., the muscles appear designed to maintain higher powers at their normal body temperatures) (Figure 6.14). Interestingly, the force-velocity relation of skinned and both living red and white fibers from fishes is often relatively straighter at colder temperatures (Figure 6.14) (Johnston and Altringham, 1985; Langfeld *et al.*, 1989; Rome and Sosnicki, 1990), but not so in most other ectotherms (see Langfeld *et al.*, 1989), nor universally in all fish muscle (e.g., Johnston *et al.*,1985, in skinned carp fibers; Johnston and Salamonski, 1984, in skinned red and white blue marlin fibers; Rome *et al.*, 1992b, in living scup red fibers). While absolute power is still lessened in the cold, this straightening of the force-velocity relationship at cooler temperatures allows the muscles to produce more power than they could otherwise;

Langfeld *et al.* (1989) estimated a 15% improvement in fast fibers of the sculpin. This would partially offset the loss in power associated with cooler temperatures and allow muscles to power higher speeds of swimming in the cold or to maintain power output at cooler temperatures (Johnston and Altringham, 1985; Langfeld *et al.*, 1989; Rome and Sosnicki, 1990; Rome, 1990). For example, based on measurements of muscle power it has been estimated that as temperature drops from 20 to 10 °C, carp would have to recruit approximately 50% more red fibers to maintain the power required to swim at a given speed assuming similar levels of activation (Rome and Sosnicki, 1990), while scup would have to almost double the amount of muscle recruited (Rome *et al.*, 1992b). The carp's apparent advantage is that its red muscles have a straighter force-velocity relationship at cooler temperatures, while scup do not. Perhaps carp have adopted this strategy to remain eurythermic, while scup have sacrificed thermal tolerance for higher power and swimming speeds in warm water; the red muscles of scup can produce about 50% more power than those of carp (Rome *et al.* 1992b).

IX. SUMMARY

The axial muscles of fishes remain classified based largely on their color (red, pink, and white), which is a metaphor for their metabolic and mechanical characteristics and is associated with their location in the body. From a mechanical viewpoint, red fibers are relatively slow both at shortening and in their rates of activation and relaxation, pink fibers are intermediate, and white fibers are fast. Red fibers in fishes appear relatively faster and more powerful than their mammalian counterparts. The mechanical attributes of each fiber type are in turn associated primarily with the type of myosin heavy chain and to some extent light chains present in the fibers, and with the quantity and quality of the calcium handling machinery. Color and speed are further associated with a host of metabolic, structural, and contractile characteristics. The increased "speed" of contraction from red to pink to white is responsible for the increased power output and the ability to power swimming at higher tail-beat frequencies. Red fibers tend to power slow, sustained swimming, white fibers power burst activity, and pink fibers contribute at intermediate speeds and perhaps at cold temperatures, at which red muscle performs poorly. Direct measures of muscle recruitment in swimming fishes have confirmed these recruitment patterns. While there are differences across species in the contractile abilities within a given fiber type, these differences are usually overshadowed by interfiber-type variability in performance such that differences in the speed and power between a red and white fiber are immediately recognizable.

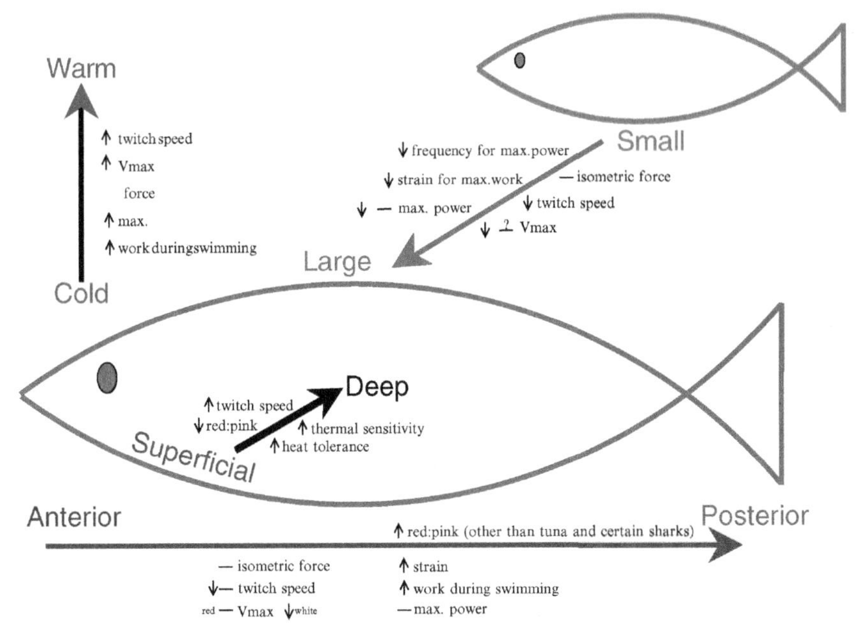

Fig. 6.15. The effects of body size, axial and radial location of the muscle sample, and temperature on mechanics of muscle in fishes. Upward arrows signify an increase along the indicated axis, downward arrows a decrease, and horizontal bars no change. Note that the symbols indicate the predominant trend and that in many cases there are exceptions reported (see text). More than one symbol indicates repeated disparate observations in the literature. "Max. power" is the maximal power the muscle is capable of producing during work loop analysis when all variables are systematically optimized. "Work during swimming" is the work produced by the muscle in work loop analysis when subjected to the strain and stimulation conditions measured in swimming fishes. The "superficial to deep" axis refers to red muscle in tuna.

Fish size, the location of the muscle in the fish, and temperature have pronounced effects on many aspects of muscle contraction (Figure 6.15). In many species, but not all, there is notable axial variation in the speed of red muscle within an individual, with the more-anterior muscles tending to be faster than the posterior. It is suggested that this may serve to compensate for the smaller strains experienced in the anterior myotomes of many fishes during swimming, allowing the muscles there to produce relatively more power than if they had the slower kinetics of more-posterior muscles. There is also often notable axial variation in power output of red muscle in swimming fishes, but this is associated more with the manner in which the muscle is used than with its inherent abilities. Radial variation in muscle speed has also been noted in tuna, in which the deeper red fibers are faster and perhaps better equipped to work when warm. This variability in the contractile characteristics of a given fiber type within an individual appears unique to the fish, and may have a basis in compensating for axial variation in strain and bending kinematics, or in promoting axial variation in muscle function during swimming. As with most other animals, there is a trend toward slowing of the muscles with increased body size in fishes, although V_{max} may be uniquely independent of body mass in some fishes.

Temperature has a major impact on many aspects of contraction, with measures such as V_{max}, twitch kinetics, and power output changing with a Q_{10} of 2–3. The muscles of cold-acclimated fishes show improvements over those from warm fishes when operating at cool temperatures, but cold fishes still tend to have muscles that are slower and less powerful than their warm counterparts. Perhaps surprisingly, fishes that live in cold climates do not show particularly impressive compensation for the depressive effects of cold on muscle performance. The reduced power from cold muscle is mitigated in part by hypertrophy, altered myosin isoform expression or posttranslational modification, a flattening of the force-velocity relationship at colder temperatures in some species, and altered recruitment strategies. Notably, limited evidence suggests that fishes swimming in acutely cooled environments do not make many of the adjustments necessary to compensate for the physiological changes in their muscle associated with cooling, and hence produce substantially less power than they might otherwise. Thus, extrapolating from the literature on work and power measured from isolated muscles under experimentally optimized conditions to the performance of these muscles in swimming fishes must be done with great caution.

Integrative studies combining measures of muscle strain and activation patterns in swimming fishes with measures of the contractile performance of isolated muscle have become popular as a tool to understand how fishes use their muscles to power swimming. The work loop technique is widely employed in such studies and has provided considerable insight into the function of the different fiber types and the effects of swim speed, body location, and temperature on muscle performance. Results from such studies suggest that a number of strategies to power swimming are employed, including fishes that use their red muscles at all axial locations exclusively to produce power (e.g., tuna, perhaps some sharks, and eels), fishes that generate most of the power for swimming from the more-posterior muscles, which undergo large strains and hence produce large amounts of work (e.g., scup and bass), and fishes that use more anterior muscles to produce power while posterior muscles both produce power and undergo an eccentric phase argued to transmit power to the tail (e.g., rainbow trout). We have also learned that we must use great prudence and avoid overlooking details when applying such techniques to understand how fishes use their muscles when they swim.

X. FUTURE DIRECTIONS

It is widely accepted that altered recruitment strategies are an important tool that fishes employ to regulate power for swimming. Yet it is entirely unknown how certain aspects of altering recruitment may affect mechanical performance and the energetics of swimming. EMG activity in fish muscle waxes and wanes in various patterns during each tail-beat cycle. However, during work loop experiments the muscle is typically activated at a constant frequency, which elicits maximal isometric force; this frequency likely exceeds that occurring in fishes at all but perhaps the highest level of exertion, and supposes an unrealistic neural command of fully ON or OFF with no intermediate states. More force and power can be attained by increasing the rate of activation of a given muscle fiber, by recruiting more fibers of a given type, or by recruiting faster fibers. When do fishes employ these various mechanisms, and what are the mechanical and energetic consequences? Such knowledge will add a new dimension to the utility of work loop analysis, and almost certainly new insights into muscle function.

Despite a growing literature about how fishes use their muscles during swimming, and despite several recent and competent reviews of this topic, there is as yet a lack of consensus about what exactly fishes are doing with their muscles when they swim and why. Additional careful studies of muscle recruitment and muscle strain in swimming fishes and then analysis of the mechanical characteristics of these muscles in more species are needed to better understand how muscles are used and to validate previous measurements in this challenging field. Analysis of the energy consumed by muscles when used as in swimming will certainly add insight, as we currently do not know if fishes are attempting to be economical when they swim; to date we have focused almost exclusively on the mechanical aspects of muscle contraction. Such studies will simultaneously add to our understanding of comparative mechanics in fish muscle. Further, there is good evidence for sizable variability in the phase and duration of muscle activation on a tail-beat to tail-beat basis, particularly when fishes are not swimming at high speeds. What is the foundation of this variability? A closer look here may go a long way toward understanding axial variation in activation patterns.

Finally, while we know a good deal about the thermal biology of fish muscle, the remarkable variability in the thermal environments of fishes makes this a still-fruitful area of research. We are only beginning to understand the effects of thermal change on muscle function, and we know much less than we should about muscle recruitment and performance in swimming fishes under acute and chronic thermal stress. What we have learned has been fascinating and should motivate us to learn more about 'how fishes power swimming.'

ACKNOWLEDGMENTS

This chapter was written while under the support of an NSERC Discovery grant. I thank an anonymous reviewer and the editors for critical comments and thoughts about the functional basis of proprioceptive feedback and relative power of red muscle in fishes.

REFERENCES

Adams, B. A. (1989). Temperature and synaptic efficacy in frog skeletal muscle. *J. Physiol.* **408,** 443–455.

Akster, H. A. (1985). Morphometry of muscle fibre types in the carp (Cyprinus carpio L.), relationships between structural and contractile characteristics. *Cell Tissue Res.* **241,** 193–201.

Akster, H. A., Granzier, H. L. M., and ter Keurs, H. E. D. J. (1985). A comparison of quantitative ultrastructural and contractile characteristics of muscle fibre types of the perch, *Perca fluviatilis L. J. Comp. Physiol. B* **155,** 685–691.

Altringham, J. D. (1994). How do fish use their myotomal muscle to swim? *In vitro* simulations of *in vivo* activity patterns. *In* "Mechanics and Physiology of Animal Swimming" (Maddock, L., Bone, Q., and Rayner, J. M. V., Eds.), pp. 99–110. Cambridge University Press, Cambridge, UK.

Altringham, J. D., and Block, B. A. (1997). Why do tuna maintain elevated slow muscle temperatures? Power output of muscle isolated from endothermic and ectothermic fish. *J. Exp. Biol.* **200,** 2617–2627.

Altringham, J. D., and Ellerby, D. J. (1999). Fish swimming: Patterns in muscle function. *J. Exp. Biol.* **202,** 3397–3403.

Altringham, J. D., and Johnston, I. A. (1981). Quantitative histochemical studies of the peripheral innervation of cod (*Gadus morhua*) fast myotomal muscle fibres. *J. Comp. Physiol.* **143A,** 123–127.

Altringham, J. D., and Johnston, I. A. (1988a). Activation of multiply innervated fast and slow myotomal muscle fibres of the teleost *Myoxocephalus scorpius*. *J. Exp. Biol.* **140,** 313–324.

Altringham, J. D., and Johnston, I. A. (1988b). The mechanical properties of polyneuronally innervated, myotomal muscle fibres isolated from a teleost fish (*Myoxocephalus scorpius*). *Pflugers Arch.* **412,** 524–529.

Altringham, J. D., and Johnston, I. A. (1990a). Modelling muscle power output in a swimming fish. *J. Exp. Biol.* **148,** 395–402.

Altringham, J. D., and Johnston, I. A. (1990b). Scaling effects on muscle function: Power output of isolated fish muscle fibres performing oscillatory work. *J. Exp. Biol.* **151,** 453–467.

Altringham, J. D., and Shadwick, R. E. (2001). Swimming and muscle function. *In* "Tuna: Physiology, Ecology, and Evolution" (Block, B. A., and Stevens, E. D., Eds.), pp. 313–341. Academic Press, London.

Altringham, J. D., Wardle, C. S., and Smith, C. I. (1993). Myotomal muscle function at different locations in the body of a swimming fish. *J. Exp. Biol.* **182,** 191–206.

Anderson, M. E., and Johnston, I. A. (1992). Scaling of power output in fast muscle fibres of the Atlantic cod during cyclical contractions. *J. Exp. Biol.* **170,** 143–154.

Archer, S. D., Altringham, J. D., and Johnston, I. A. (1990). Scaling effects on the neuromuscular system, twitch kinetics and morphometrics of the cod, *Gadus morhua*. *Mar. Behav. Physiol.* **17,** 137–146.

Bainbridge, R. (1958). The speed of swimming of fish as related to size and to the frequency and amplitude of the tail-beat. *J. Exp. Biol.* **35,** 109–113.

Bennett, A. F. (1984). Thermal dependence of muscle function. *Am. J. Physiol.* **247,** R217–229.

Bennett, A. F. (1985). Temperature and muscle. *J. Exp. Biol.* **115,** 333–344.

Bone, Q. (1964). Patterns of muscular innervation in the lower chordates. *Int. Rev. Neurobiol.* **6,** 99–147.

Bone, Q. (1966). On the function of the two types of myotomal muscle fiber in elasmobranch fish. *J. Mar. Biol. Assoc. UK* **46,** 321–349.

Bone, Q. (1978). Locomotor muscle. *In* "Fish Physiology, Locomotion" (Hoar, W. S., and Randall, D. J., Eds.), Vol. 7, pp. 361–424. Academic Press, New York.

Bone, Q., Johnston, I. A., Pulsford, A., and Ryan, K. P. (1986). Contractile properties and ultrastructure of three types of muscle fibre in the dogfish myotome. *J. Muscle Res. Cell Motil.* **7,** 47–56.

Caiozzo, V. J., and Baldwin, K. M. (1997). Determinants of work produced by skeletal muscle: Potential limitations of activation and relaxation. *Am. J. Physiol.* **273**(42), C1049–1056.

Coughlin, D. J. (2000). Power production during steady swimming in largemouth bass and rainbow trout. *J. Exp. Biol.* **203,** 617–629.

Coughlin, D. J. (2002a). Aerobic muscle function during steady swimming in fish. *Fish Fisheries* **3,** 63–78.

Coughlin, D. J. (2002b). A molecular mechanism for variations in muscle function in rainbow trout. *Integ. Comp. Biol.* **42,** 190–198.

Coughlin, D. J. (2003). Steady swimming by fishes: Kinetic properties and power production by the aerobic musculature. *In* "Fish Adaptations" (Val, A. L., and Kapoor, B. G., Eds.), pp. 55–72. Science Publishers Inc., Enfield N. H.

Coughlin, D. J., and Rome, L. C. (1996). The roles of pink and red muscle in powering steady swimming in scup, *Stenotomus chrysops. Amer. Zool.* **36,** 666–677.

Coughlin, D. J., and Rome, L. C. (1999). Muscle activity in steady swimming scup, *Stenotomus chrysops*, varies with fiber type and body position. *Biol. Bull.* **196**(2), 145–152.

Coughlin, D. J., Zhang, G., and Rome, L. C. (1996). Contraction dynamics and power production of pink muscle of the scup (*Stenotomus chrysops*). *J. Exp. Biol.* **199,** 2703–2712.

Coughlin, D. J., Burdick, J., Stauffer, K. A., and Weaver, F. E. (2001a). The parr-smolt transformation in rainbow trout (*Oncorhynchus mykiss*) involves a transition in red muscle kinetics, swimming kinematics and myosin heavy chain isoform. *J. Fish Biol.* **58,** 701–715.

Coughlin, D. J., Forry, J. A., McGlinchey, S. M., Mitchell, J., Saporetti, K. A., and Stauffer, K. A. (2001b). Thyroxine induces transitions in red muscle kinetics and steady swimming kinematics in rainbow trout (*Oncorhynchus mykiss*). *J. Exp. Zool.* **290**(2), 115–124.

Coughlin, D. J., Spiecker, A., and Schiavi, J. M. (2004). Red muscle recruitment during steady swimming correlates with rostral-caudal patterns of power production in trout. *Comp. Biochem. Physiol. A* **137,** 151–160.

Coughlin, D. J., Caputo, N. D., Bohnert, K. L., and Weaver, F. E. (2005). Troponin T expression in trout red muscle correlates with muscle activation. *J. Exp. Biol.* **208,** 409–417.

Crockford, T., and Johnston, I. A. (1990). Temperature acclimation and the expression of contractile protein isoforms in the skeletal muscles of the common carp (*Cyprinus carpio* L.). *J. Comp. Physiol. B* **160,** 23–30.

Curtin, N. A., and Woledge, R. C. (1988). Power output and force-velocity relationship of live fibres from white myotomal muscle of the dogfish, *Scyliorhinus canicula. J. Exp. Biol.* **140,** 187–197.

Curtin, N. A., and Woledge, R. C. (1991). Efficiency of energy conversion during shortening of muscle fibres from the dogfish *Scyliorhinus canicula. J. Exp. Biol.* **158,** 343–353.

Curtin, N. A., and Woledge, R. C. (1993a). Efficiency of energy conversion during sinusoidal movement of red muscle fibres from the dogfish *Scyliorhinus canicula*. *J. Exp. Biol.* **185,** 195–206.

Curtin, N. A., and Woledge, R. C. (1993b). Efficiency of energy conversion during sinusoidal movement of white muscle fibres from the dogfish *Scyliorhinus canicula*. *J. Exp. Biol.* **183,** 127–147.

Curtin, N. A., and Woledge, R. C. (1996). Power at the expense of efficiency in contraction of white muscle fibres from the dogfish *Scyliorhinus canicula*. *J. Exp. Biol.* **199,** 593–601.

D'Août, K., Curtin, N. A., Williams, T. L., and Aerts, P. (2001). Mechanical properties of red and white swimming muscles as a function of the position along the body of the eel *Anguilla anguilla*. *J. Exp. Biol.* **204,** 2221–2230.

Davies, M. L. F., Johnston, I. A., and van de Wal, J. (1995). Muscle fibres in rostral and caudal myotomes of the Atlantic cod (*Gadus morhua*) have different mechanical properties. *Physiol. Zool.* **68,** 673–697.

Domenici, P., and Blake, R. W. (1997). The kinematics and performance of fish fast-start swimming. *J. Exp. Biol.* **200,** 1165–1178.

Donley, J. (2004). Mechanics of steady swimming and contractile properties of muscle in elasmobranch fishes. Ph.D. dissertation, University of California, San Diego, CA.

Donley, J. M., Shadwick, R. E., Sepulveda, C. A., Konstantinidis, P., and Gemballa, S. (2005). Patterns of red muscle strain/activation and body kinematics during steady swimming in a lamnid shark, the shortfin mako (*Isurus oxyrinchus*). *J. Exp. Biol.* **208,** 2377–2387.

Egginton, S., and Johnston, I. A. (1982). A morphometric analysis of regional differences in myotomal muscle ultrastructure in the juvenile eel (*Anguilla anguilla* L.). *Cell Tissue Res.* **222,** 579–596.

Ellerby, D. J., and Altringham, J. D. (2001). Spatial variation in fast muscle function of the rainbow trout *Oncorhynchus mykiss* during fast-starts and sprinting. *J. Exp. Biol.* **204,** 2239–2250.

Ellerby, D. J., Altringham, J. D., Williams, T., and Block, B. A. (2000). Slow muscle function of Pacific bonito (*Sarda chiliensis*) during steady swimming. *J. Exp. Biol.* **203(13),** 2001–2013.

Ellerby, D. J., Spierts, I. L. Y., and Altringham, J. D. (2001a). Slow muscle power output of yellow- and silver-phase European eels (*Anguilla anguilla* L.): Changes in muscle performance prior to migration. *J. Exp. Biol.* **204,** 1369–1379.

Ellerby, D. J., Spierts, I. L. Y., and Altringham, J. D. (2001b). Fast muscle function in the European eel (*Anguilla anguilla* L.) during aquatic and terrestrial locomotion. *J. Exp. Biol.* **204,** 2231–2238.

Fierstine, H. L., and Walters, V. (1968). Studies in locomotion and anatomy of scombroid fishes. *Mem. S. Calif. Acad. Sci.* **6,** 1–31.

Fleming, J. R., Crockford, T., Altringham, J. D., and Johnston, I. A. (1990). Effects of temperature acclimation on muscle relaxation in the carp: A mechanical, biochemical, and ultrastructural study. *J. Exp. Zool.* **255,** 286–295.

Franklin, C. E., and Johnston, I. A. (1997). Muscle power output during escape responses in an Antarctic fish. *J. Exp. Biol.* **200,** 703–712.

George, J. C., and Stevens, E. D. (1978). Fine structure and metabolic adaptation of red and white muscles in tuna. *Env. Biol. Fish* **3**(2), 185–191.

Gerday, C., Joris, B., Gerardin-Otthiers, N., Collin, S., and Hamoir, G. (1979). Parvalbumins from the lungfish (*Protopterus dolloi*). *Biochimie* **61,** 589–599.

Gerlach, G., Turay, L., Malik, K. T. A., Lida, J., Scutt, A., and Goldspink, G. (1990). Mechanisms of temperature acclimation in the carp: A molecular biology approach. *Am. J. Physiol.* **259,** R237–244.

Gillis, G. B. (1998). Neuromuscular control of anguilliform locomotion: Patterns of red and white muscle activity during swimming in the American eel (*Anguilla rostrata*). *J. Exp. Biol.* **201,** 3245–3256.

Goldspink, G., Wilkes, D., and Ennion, S. (2001). Myosin expression during ontogeny, post-hatching growth, and adaptation. *In* "Fish Physiology: Muscle Development and Growth" (Johnston, I. A., Ed.), Vol. 18, pp. 43–72. Academic Press, London.

Graham, J. B., Koehrn, F. J., and Dickson, K. A. (1983). Distribution and relative proportions of red muscle in scombrid fishes: Consequences of body size and relationships to locomotion and endothermy. *Can. J. Zool.* **61,** 2087–2096.

Granzier, H. L. M., Wiersma, J., Akster, H. A., and Osse, J. W. M. (1983). Contractile properties of a white- and a red-fibre type of the m. Hyohyoideus of the carp (*Cyprinus carpio* L.). *J. Comp. Physiol. B* **149,** 441–449.

Guderley, H. (1990). Functional significance of metabolic responses to thermal acclimation in fish muscle. *Am. J. Physiol.* **259,** R245–252.

Guderley, H., and St-Pierre, J. (2002). Going with the flow or life in the fast lane: Contrasting mitochondrial responses to thermal change. *J. Exp. Biol.* **205,** 2237–2249.

Hammond, L., Altringham, J. D., and Wardle, C. S. (1998). Myotomal slow muscle function of rainbow trout *Oncorhynchus mykiss* during steady swimming. *J. Exp. Biol.* **201,** 1659–1671.

Heap, S. P., Watt, P. W., and Goldspink, G. (1986). Myofibrillar ATP-ase activity in the carp *Cyprinus carpio*: Interactions between starvation and environmental temperature. *J. Exp. Biol.* **123,** 373–382.

Hess, F., and Videler, J. J. (1984). Fast continuous swimming of saithe (*Pollachius virens*): A dynamic analysis of bending moments and muscle power. *J. Exp. Biol.* **109,** 229–251.

Hill, A. V. (1938). The heat of shortening and the dynamic constants of muscle. *Proc. Roy. Soc. Lond. B* **126,** 136–195.

James, R. S., and Johnston, I. A. (1998). Scaling of muscle performance during escape responses in the fish *Myoxocephalus scorpius* L. *J. Exp. Biol.* **201,** 913–923.

James, R. S., Young, I. S., Cox, V. M., Goldspink, D. F., and Altringham, J. D. (1996). Isometric and isotonic muscle properties as determinants of work loop muscle power output. *Pflugers Arch.* **432,** 767–774.

James, R. S., Cole, N. J., Davies, M. L. F., and Johnston, I. A. (1998). Scaling of intrinsic contractile properties and myofibrillar protein composition of fast muscle in the fish *Myoxocephalus scorpius* L. *J. Exp. Biol.* **201,** 901–912.

Jayne, B. C., and Lauder, G. V. (1994). How swimming fish use slow and fast muscle fibers: Implications for models of vertebrate muscle recruitment. *J. Comp. Physiol. A* **175,** 123–131.

Johnson, T. P., and Bennett, A. F. (1995). The thermal acclimation of burst escape performance in fish: An integrated study of molecular and cellular physiology and organismal performance. *J. Exp. Biol.* **198,** 2165–2175.

Johnson, T. P., and Johnston, I. A. (1991a). Temperature adaptation and the contractile properties of live muscle fibres from teleost fish. *J. Comp. Physiol. B* **161,** 27–36.

Johnson, T. P., and Johnston, I. A. (1991b). Power output of fish muscle fibres performing oscillatory work: Effects of acute and seasonal temperature change. *J. Exp. Biol.* **157,** 409–423.

Johnson, T. P., Altringham, J. D., and Johnston, I. A. (1991a). The effects of Ca^{2+} and neuromuscular blockers on the activation of fish muscle fibres. *J. Fish Biol.* **28,** 789–790.

Johnson, T. P., Johnston, I. A., and Moon, T. W. (1991b). Temperature and the energy cost of oscillatory work in teleost fast muscle fibres. *Pflugers Arch.* **419,** 177–183.

Johnson, T. P., Syme, D. A., Jayne, B. C., Lauder, G. V., and Bennett, A. F. (1994). Modeling red muscle power output during steady and unsteady swimming in largemouth bass. *Am. J. Physiol.* **267,** R481–488.

Johnson, T. P., Bennett, A. F., and McLister, J. D. (1996). Thermal dependence and acclimation of fast-start locomotion and its physiological basis in rainbow trout (*Oncorhynchus mykiss*). *Physiol. Zool.* **69,** 276–292.

Johnston, I. A. (1980a). Specialization of fish muscle. *In* "Development and Specialization of Skeletal Muscle" (Goldspink, D. F., Ed.), pp. 123–148. Cambridge University Press, Cambridge, UK.

Johnston, I. A. (1980b). Contractile properties of fish fast muscle fibres. *Mar. Biol. Lett.* **1,** 323–328.

Johnston, I. A. (1981). Structure and function of fish muscles. *In* "Vertebrate Locomotion" (Day, M., Ed.). *Symp. Zool. Soc. Lond.* **48,** 71–113.

Johnston, I. A. (1983). On the design of fish myotomal muscles. *Mar. Behav. Physiol.* **9,** 83–98.

Johnston, I. A. (1985). Sustained force development: Specializations and variation among the vertebrates. *J. Exp. Biol.* **115,** 239–251. Academic Press, London.

Johnston, I. A. (1994). Developmental aspects of temperature adaptation in fish muscle. *Bas. Appl. Myol.* **4,** 353–368.

Johnston, I. A., and Altringham, J. D. (1985). Evolutionary adaptation of muscle power output to environmental temperature: Force-velocity characteristics of skinned fibres isolated from Antarctic, temperate and tropical marine fish. *Pflugers Arch.* **405,** 136–140.

Johnston, I. A., and Brill, R. J. (1984). The thermal dependence of contractile properties of single skinned muscle fibres from Antarctic and various warm water marine fishes including skipjack tuna (*Katsuwonus pelamis*) and kawakawa (*Euthynnus affinis*). *J. Comp. Physiol.* **155,** 63–70.

Johnston, I. A., and Salamonski, J. (1984). Power output and force-velocity relationship of red and white muscle fibres from the Pacific blue marlin (*Makaira nigricans*). *J. Exp. Biol.* **111,** 171–177.

Johnston, I. A., and Temple, G. K. (2002). Thermal plasticity of skeletal muscle phenotype in ectothermic vertebrates and its significance for locomotory behavior. *J. Exp. Biol.* **205,** 2305–2322.

Johnston, I. A., and Tota, B. (1974). Myofibrillar ATPase in the various red and white trunk muscles in the tunny (*Thunnus thynnus* L.) and the tub gurnard (*Trigla lucerna* L.). *Comp. Biochem. Physiol. B* **49,** 367–373.

Johnston, I. A., Davison, W., and Goldspink, G. (1977). Energy metabolism of carp swimming muscles. *J. Comp. Physiol.* **114,** 203–216.

Johnston, I. A., Sidell, B. D., and Driedzic, W. R. (1985). Force-velocity characteristics and metabolism of carp muscle fibres following temperature acclimation. *J. Exp. Biol.* **119,** 239–249.

Johnston, I. A., Fleming, J. D., and Crockford, T. (1990). Thermal acclimation and muscle contractile properties in cyprinid fish. *Am. J. Physiol.* **259,** R231–236.

Johnston, I. A., Franklin, C. E., and Johnson, T. P. (1993). Recruitment patterns and contractile properties of fast muscle fibres isolated from rostral and caudal myotomes of the short-horned sculpin. *J. Exp. Biol.* **185,** 251–265.

Johnston, I. A., Van Leeuwen, J. L., Davies, M. L. F., and Beddow, T. (1995). How fish power predation fast starts. *J. Exp. Biol.* **198,** 1851–1861.

Josephson, R. K. (1985). Mechanical power output from striated muscle during cyclic contraction. *J. Exp. Biol.* **114,** 493–512.

Josephson, R. K. (1989). Power output from skeletal muscle during linear and sinusoidal shortening. *J. Exp. Biol.* **147,** 533–537.

Josephson, R. K. (1993). Contraction dynamics and power output of skeletal muscle. *Annu. Rev. Physiol.* **55,** 527–546.

Katz, S. L. (2002). Design of heterothermic muscle in fish. *J. Exp. Biol.* **205,** 2251–2266.

Katz, S. L., Syme, D. A., and Shadwick, R. E. (2001). Enhanced power in yellowfin tuna. *Nature* **410,** 770–771.

Knower, T., Shadwick, R. E., Katz, S. L., Graham, J. B., and Wardle, C. S. (1999). Red muscle activation patterns in yellowfin (*Thunnus albacares*) and skipjack (*Katsuwonus pelamis*) tunas during steady swimming. *J. Exp. Biol.* **202,** 2127–2138.

Langfeld, K. S., Altringham, J. D., and Johnston, I. A. (1989). Temperature and the force-velocity relationship of live muscle fibres from the teleost *Myoxocephalus scorpius*. *J. Exp. Biol.* **144,** 437–448.

Lou, F., Curtin, N. A., and Woledge, R. C. (2002). Isometric and isovelocity contractile performance of red muscle fibres from the dogfish *Scyliorhinus canicula*. *J. Exp. Biol.* **205,** 1585–1595.

Luiker, E. A., and Stevens, E. D. (1992). Effect of stimulus frequency and duty cycle on force and work in fish muscle. *Can. J. Zool.* **70,** 1135–1139.

Luiker, E. A., and Stevens, E. D. (1994). Effect of temperature and stimulus train duration on the departure from theoretical maximum work in fish muscle. *Can. J. Zool.* **72,** 965–969.

Luther, P. K., Munro, P. M. G., and Squire, J. M. (1995). Muscle ultrastructure in the teleost fish. *Micron* **26,** 431–459.

Machin, K. E., and Pringle, J. W. S. (1959). The physiology of insect fibrillar muscle. II. Mechanical properties of beetle flight muscle. *Proc. Roy. Soc. Lond. B* **151,** 204–225.

Marsh, R. L. (1990). Deactivation rate and shortening velocity as determinants of contractile frequency. *Am. J. Physiol.* **259,** R223–230.

Marsh, R. L. (1999). How muscles deal with real-world loads: The influence of length trajectory on muscle performance. *J. Exp. Biol.* **202,** 3377–3385.

McGlinchey, S. M., Saporetti, K. A., Forry, J., Pohronezny, J. A., and Coughlin, D. J. (2001). Red muscle function during steady swimming in brook trout, *Salvelinus fontinalis*. *Comp. Biochem. Physiol. A* **129,** 727–738.

Medler, S. (2002). Comparative trends in shortening velocity and force production in skeletal muscles. *Am J. Physiol.* **283,** R368–378.

Moon, T. W., Altringham, J. D., and Johnston, I. A. (1991). Energetics and power output of isolated fish fast muscle fibres performing oscillatory work. *J. Exp. Biol.* **158,** 261–273.

Mosse, P. R. L. (1978). The distribution of capillaries in the somatic musculature of two vertebrate types with particular reference to teleost fish. *Cell Tissue Res.* **187,** 281–303.

Ono, R. D. (1983). Dual motor innervation in the axial musculature of fishes. *J. Fish Biol.* **22,** 395–408.

Peake, S. J., and Farrell, A. P. (2004). Locomotory behavior and post-exercise physiology in relation to swimming speed, gait transition and metabolism in free-swimming smallmouth bass (*Micropterus dolomieu*). *J. Exp. Biol.* **207,** 1563–1575.

Rall, J. A., and Woledge, R. C. (1990). Influence of temperature on mechanics and energetics of muscle contraction. *Am. J. Physiol.* **259,** R197–203.

Rayner, M. D., and Keenan, M. J. (1967). Role of red and white muscles in the swimming of the skipjack tuna. *Nature* **214,** 392–393.

Roberts, B. L. (1981). The organization of the nervous system of fishes in relation to locomotion. *In* "Vertebrate Locomotion" (Day, M. H., Ed.), pp. 115–136. Academic Press, London.

Rodnick, K. J., and Sidell, B. D. (1995). Effects of body size and thermal acclimation on parvalbumin concentration in white muscle of striped bass. *J. Exp. Zool.* **272,** 266–274.

Rome, L. C. (1986). The influence of temperature on muscle and locomotory performance. *In* "Living in the Cold: Physiological and Biochemical Adaptations" (Heller, H. C., Musacchia, H. J., and Wang, L. C. H., Eds.), pp. 485–495. Elsevier, New York.

Rome, L. C. (1990). Influence of temperature on muscle recruitment and muscle function *in vivo*. *Am. J. Physiol.* **259,** R210–222.

Rome, L. C. (1994). The mechanical design of the muscular system. *Adv. Vet. Sci. Comp. Med.* **38A,** 125–178.

Rome, L. C., and Klimov, A. A. (2000). Superfast contractions without superfast energetics: ATP usage by SR-Ca^{2+} pumps and crossbridges in toadfish swimbladder muscle. *J. Physiol.* **526.2,** 279–298.

Rome, L. C., and Lindstedt, S. L. (1998). The quest for speed: Muscle built for high-frequency contractions. *News Physiol. Sci.* **13,** 261–268.

Rome, L. C., and Sosnicki, A. A. (1990). The influence of temperature on mechanics of red muscle in carp. *J. Physiol.* **427,** 151–169.

Rome, L. C., and Sosnicki, A. A. (1991). Myofilament overlap in swimming carp II. Sarcomere length changes during swimming. *Am. J. Physiol.* **260,** C289–C296.

Rome, L. C., and Swank, D. (1992). The influence of temperature on power output of scup red muscle during cyclical length changes. *J. Exp. Biol.* **171,** 261–281.

Rome, L. C., and Swank, D. M. (2001). The influence of thermal acclimation on power production during swimming. I. *In vivo* stimulation and length change pattern of scup red muscle. *J. Exp. Biol.* **204,** 409–418.

Rome, L. C., Loughna, P. T., and Goldspink, G. (1984). Muscle fiber activity in carp as a function of swimming speed and muscle temperature. *Am. J. Physiol.* **247,** R272–279.

Rome, L. C., Loughna, P. T., and Goldspink, G. (1985). Temperature acclimation improves sustained swimming performance at low temperatures in carp. *Science* **228,** 194–196.

Rome, L. C., Funke, R. P., Alexander, R. M., Lutz, G., Aldridge, H., Scott, F., and Freadman, M. (1988). Why animals have different muscle fibre types. *Nature* **335,** 824–827.

Rome, L. C., Choi, I., Lutz, G., and Sosnicki, A. (1992a). The influence of temperature on muscle function in the fast swimming scup. I. Shortening velocity and muscle recruitment during swimming. *J. Exp. Biol.* **163,** 259–279.

Rome, L. C., Sosnicki, A. A., and Choi, I. (1992b). The influence of temperature on muscle function in the fast swimming scup. II. The mechanics of red muscle. *J. Exp. Biol.* **163,** 281–295.

Rome, L. C., Swank, D., and Corda, D. (1993). How fish power swimming. *Science* **261,** 340–343.

Rome, L. C., Syme, D. A., Hollingworth, S., Lindstedt, S. L., and Baylor, S. M. (1996). The whistle and the rattle: The design of sound-producing muscles. *Proc. Natl. Acad. Sci. USA* **93,** 8095–8100.

Rome, L. C., Cook, C., Syme, D. A., Connaughton, M. A., Ashley-Ross, M., Klimov, A., Tikunov, B., and Goldman, Y. E. (1999). Trading force for speed: Why superfast crossbridge kinetics leads to superlow forces. *Proc. Natl. Acad. Sci. USA* **96,** 5826–5831.

Rome, L. C., Swank, D. M., and Coughlin, D. J. (2000). The influence of temperature on power production during swimming. II. Mechanics of red muscle fibres *in vivo*. *J. Exp. Biol.* **203,** 333–345.

Sanger, A. M. (1992). Quantitative fine structural diversification of red and white muscle fibres in cyprinids. *Environ. Biol. Fishes* **33,** 97–104.

Sanger, A. M., and Stoiber, W. (2001). Muscle fiber diversity and plasticity. *In* "Fish Physiology: Muscle Development and Growth" (Johnston, I. A., Ed.), Vol. 18, pp. 187–250. Academic Press, London.

Scapolo, P. A., Veggetti, A., Mascarello, F., and Romanello, M. G. (1988). Developmental transitions of myosin isoforms and organisation of the lateral muscle in the teleost *Dicentrarchus labrax*. *Anat. Embryol.* **178,** 287–295.

Shadwick, R. E., Steffensen, J. F., Katz, S. L., and Knower, T. (1998). Muscle dynamics in fish during steady swimming. *Am. Zool.* **38,** 755–770.

Shadwick, R. E., Katz, S. L., Korsmeyer, K. E., Knower, T., and Covell, J. W. (1999). Muscle dynamics in skipjack tuna: Timing of red muscle shortening in relation to activation and body curvature during steady swimming. *J. Exp. Biol.* **202,** 2139–2150.

Sisson, J. E., and Sidell, B. D. (1987). Effect of thermal acclimation on muscle fiber recruitment of swimming striped bass (*Morone saxatilis*). *Physiol. Zool.* **60,** 310–320.

Smith, D. S. (1966). The organization and function of the sarcoplasmic reticulum and T-system of muscle cells. *Prog. Biophys. Mol. Biol.* **16,** 107–142.

Sosnicki, A. A., Loesser, K., and Rome, L. C. (1991). Myofilament overlap in swimming carp. I. Myofilament lengths of red and white muscle. *Am. J. Physiol.* **260,** C283–288.

Spierts, I. L. Y., Akster, H. A., and Granzier, H. L. (1997). Expression of titin isoforms in red and white muscle fibers of carp (*Cyprinus carpio* L.) exposed to different sarcomere strains during swimming. *J. Comp. Physiol. B* **167,** 543–551.

Stanfield, P. R. (1972). Electrical properties of white and red fibres of the elasmobranch fish *Scyliorhinus canicula*. *J. Physiol.* **222,** 161–186.

Stevens, E. D. (1979). The effect of temperature on tail-beat frequency of fish swimming at constant velocity. *Can. J. Zool.* **57,** 1682–1635.

Stevens, E. D. (1993). Relation between work and power calculated from force-velocity curves to that done during oscillatory work. *J. Muscle Res. Cell Motil.* **14,** 518–526.

Swank, D. M., and Rome, L. C. (2000). The influence of temperature on power production during swimming. I. *In vivo* length change and stimulation pattern. *J. Exp. Biol.* **203,** 321–331.

Swank, D. M., and Rome, L. C. (2001). The influence of thermal acclimation on power production during swimming II. Mechanics of scup red muscle under *in vivo* conditions. *J. Exp. Biol.* **204,** 419–430.

Swank, D. M., Zhang, G., and Rome, L. C. (1997). Contraction kinetics of red muscle in scup: Mechanism for variation in relaxation rate along the length of the fish. *J. Exp. Biol.* **200,** 1297–1307.

Syme, D. A. (1990). Passive viscoelastic work of isolated rat, Rattus norvegicus, diaphragm muscle. *J. Physiol.* **424,** 301–315.

Syme, D. A., and Josephson, R. K. (2002). How to build fast muscles: Synchronous and asynchronous designs. *Integ. Comp. Biol.* **42,** 762–770.

Syme, D. A., and Shadwick, R. E. (2002). Effects of longitudinal body position and swimming speed on mechanical power of deep red muscle from skipjack tuna (*Katsuwonus pelamis*). *J. Exp. Biol.* **205,** 189–200.

Temple, G. K., Wakeling, J. M., and Johnston, I. A. (2000). Seasonal changes in fast-starts in the short-horn sculpin: Integration of swimming behavior and muscle performance. *J. Fish Biol.* **56,** 1435–1449.

Thys, T. M., Blank, J. M., and Schachat, F. H. (1998). Rostral-caudal variation in troponin T and parvalbumin correlates with differences in relaxation rates of cod axial muscle. *J. Exp. Biol.* **201,** 2993–3001.

Thys, T. M., Blank, J. M., Coughlin, D. J., and Schachat, F. (2001). Longitudinal variation in muscle protein expression and contraction kinetics of largemouth bass axial muscle. *J. Exp. Biol.* **204,** 4249–4257.

van Leeuwen, J. L. (1995). The action of muscles in swimming fish. *Exp. Physiol.* **80,** 177–191.

van Leeuwen, J. L., Lankheet, M. J. M., Akster, H. A., and Osse, J. W. M. (1990). Function of red axial muscles of carp (*Cyprinus carpio*): Recruitment and normalized power output during swimming in different modes. *J. Zool. (Lond.)* **220,** 123–145.

Videler, J. J., and Wardle, C. S. (1991). Fish swimming stride by stride: Speed limits and endurance. *Rev. Fish Biol. Fish* **1,** 23–40.

Wakeling, J. M., and Johnston, I. A. (1998). Muscle power output limits fast-start performance in fish. *J. Exp. Biol.* **201,** 1505–1526.

Wakeling, J. M., and Johnston, I. A. (1999). Predicting muscle force generation during fast-starts for the common carp *Cyprinus carpio*. *J. Comp. Physiol. B* **169,** 391–401.

Wakeling, J. M., Cole, N. J., Kemp, K. M., and Johnston, I. A. (2000). The biomechanics and evolutionary significance of thermal acclimation in the common carp *Cyprinus carpio*. *Am. J. Physiol.* **279,** R657–665.

Wardle, C. S. (1975). Limit of fish swimming speed. *Nature* **255,** 725–727.

Wardle, C. S., Videler, J. J., Arimoto, T., Franco, J. M., and He, P. (1989). The muscle twitch and the maximum swimming speed of giant bluefin tuna, *Thunnus thynnus* L. *J. Fish Biol.* **35,** 129–137.

Wardle, C. S., Videler, J. J., and Altringham, J. D. (1995). Tuning in to fish swimming waves: Body form, swimming mode and muscle function. *J. Exp. Biol.* **198,** 1629–1636.

Watabe, S. (2002). Temperature plasticity of contractile proteins in fish muscle. *J. Exp. Biol.* **205,** 2231–2236.

Weaver, F. E., Stauffer, K. A., and Coughlin, D. J. (2001). Myosin heavy chain expression in the red, white, and ventricular muscle of juvenile stages of rainbow trout. *J. Exp. Zool.* **290**(7), 751–758.

Webb, P. W. (1976). The effect of size on the fast-start performance of rainbow trout, *Salmo gairdneri* and a consideration of piscivorous predator-prey interactions. *J. Exp. Biol.* **65,** 157–177.

Webb, P. W., Kostecki, P. T., and Stevens, E. D. (1984). The effect of size and swimming speed on locomotor kinematics of rainbow trout. *J. Exp. Biol.* **109,** 77–95.

Westneat, M. W., Hoese, W., Pell, C. A., and Wainwright, S. (1993). The horizontal septum: Mechanisms of force transfer in locomotion of scombrid fishes (Scombridae, Perciformes). *J. Morph.* **217,** 183–204.

Williams, T. L., Grillner, S., Smoljaninov, V. V., Wallen, P., Kashin, S., and Rosignol, S. (1989). Locomotion in lamprey and trout: The relative timing of activation and movement. *J. Exp. Biol.* **143,** 559–566.

Witthames, P. R., and Greer-Walker, M. (1982). The activity of myofibrillar and actomyosin ATPase in the skeletal muscle of some marine teleosts in relation to their length and age. *J. Fish Biol.* **20,** 471–478.

Woledge, R. C., Curtin, N. A., and Homsher, E. (1985). Mechanics of contraction. *In* "Monographs of the Physiological Society No. 41, Energetic Aspects of Muscle Contraction," pp. 27–115. Academic Press, San Diego, CA.

Zawadowska, B., and Supikova, I. (1992). Parvalbumin in skeletal muscles of teleost (*Tinca tinca* and *Misgurnus fossilis*). Histochemical and immunohistochemical study. *Folia Histochem. Cytobiol.* **30,** 63–68.

Zhang, G., Swank, D. M., and Rome, L. C. (1996). Quantitative distribution of muscle fibre types in the scup *Stenotomus chrysops*. *J. Morphol.* **229,** 71–81.

FURTHER READING

Anderson, E. J., McGillis, W. R., and Grosenbaugh, M. A. (2001). The boundary layer of swimming fish. *J. Exp. Biol.* **204,** 81–102.

Jayne, B. C., and Lauder, G. V. (1995). Red muscle motor patterns during steady swimming in largemouth bass: Effects of speed and correlation with axial kinematics. *J. Exp. Biol.* **198,** 657–670.

7

STRUCTURE, KINEMATICS, AND MUSCLE DYNAMICS IN UNDULATORY SWIMMING

ROBERT E. SHADWICK
SVEN GEMBALLA

I. INTRODUCTION

Axial undulation is a common mechanism for powering slow and continuous movements in fishes, and, because it derives power from a musculature that may comprise 50% or more of the body mass, this propulsive system can also produce high thrust forces for fast swimming and high acceleration (Webb and Blake, 1985). Forward undulatory swimming depends on the coordinated action of lateral muscles to propagate a propulsive wave that travels with increasing amplitude from head to tail along the

Fish Biomechanics: Volume 23
FISH PHYSIOLOGY

DOI: 10.1016/S1546-5098(05)23007-8

body. Specific features of this wave vary somewhat among fishes because of differences in body morphology and undulatory mode, but the underlying biomechanics principles are similar (see Chapter 11 of this volume). In early kinematics studies using frame-by-frame analysis of high-speed ciné films, Gray (1933a,b,c) showed that during steady forward swimming, the wave of bending must travel backward along the body faster than the fish travels forward, and that thrust is generated continuously throughout the tail-beat cycle. In addition to the large-amplitude waves readily observed on eel-like swimmers, Gray also showed that other undulatory modes obeyed the same rules, although their propulsive waves were not as apparent in real time due to higher velocities, longer wavelengths, and lower amplitude profiles (see Figure 11.1, Chapter 11 of this volume for description of swimming modes). These initial and many subsequent studies produced a substantial collection of empirical data and theoretical predictions of the interactions between the fish body and the surrounding water, which are now being tested directly (see Chapters 10, 11).

More recently, internal mechanics have become an important area of research in fish locomotion. In order to understand the consequences of muscle action, researchers turned their attention to investigating the activation patterns and the contractile properties of the lateral muscles that generate the propulsive wave. In the past decade, the application of new methods in digital image analysis and techniques to measure muscle strain *in vivo* and to characterize muscle power output *in vitro* has provided the focus for experimental and modeling studies on the action of lateral muscle in powering undulatory swimming (reviewed in Altringham and Ellerby, 1999; Katz and Shadwick, 2000; Coughlin, 2002). At the same time, new morphological techniques have provided important insights into the arrangement of tendons in the segmented musculature, and their relationship with muscle fibers and axial skeleton (e.g., Gemballa and Vogel, 2002; Gemballa *et al.*, 2003a,b). The three-dimensional structure of the musculotendinous system provides the mechanical linkage that translates the muscle action into waves of body undulation, and is thus essential for a complete biomechanical analysis. This chapter discusses the structure of the musculotendinous system that provides the power for locomotion, and the relationship between muscle and body kinematics in steady swimming. Patterns of muscle activation and strain in different undulatory modes are summarized, as are recent studies on specializations related to high-performance swimming in tunas and lamnid sharks. For further discussion of unsteady swimming, readers should consult Chapters 8 and 9 of this volume.

II. MYOMERE STRUCTURE AND FORCE TRANSMISSION PATHWAYS

A. Three-Dimensional Morphology of Segmented Muscle Units and Their Myosepta

The simplest arrangement of lateral muscle segments in an undulating animal that would provide the ability to propagate a propulsive wave would be a series of discrete, axially arranged blocks. However, such simple arrangement of the major muscle units (= myomeres) does not occur in fishes. Even representatives of the basal notochordate taxa, such as lancelets, hagfishes, and lampreys, possess muscle units of complex three-dimensional shape (Nursall, 1956; Vogel and Gemballa, 2000; Gemballa *et al.*, 2003c). In the gnathostome stem lineage, myomeres have evolved to even more complex and three-dimensionally folded structures that bear anterior and posterior projecting cones (Greene and Greene, 1913; Nursall, 1956; Gemballa *et al.*, 2003b) (Figure 7.1). Although each myomere spans several vertebral segments along the body, the number of vertebral segments still equals the number of myomeres, that is, there is one muscle unit per vertebral segment, and their spatial accommodation is accomplished by serial nesting of the cones. The muscle unit receives innervation from the spinal nerve of its corresponding vertebral segment, and sometimes from adjacent segments (Westerfield *et al.*, 1986; Altringham and Johnston, 1989; Raso, 1991). The orientation of muscle fibers within the myomeres is also complex, with fibers spanning adjacent myosepta at variable angles, depending on location (Gemballa and Vogel, 2002). Three-dimensional reconstruction from serial sections (Alexander, 1969), and more recently from microdissection (Gemballa and Vogel, 2002), shows that the trajectory of fibers across sequential myomeres follows a helical path, around the axis of the cones. Modeling of this arrangement produced the hypothesis that fiber shortening should be relatively uniform regardless of lateral distance from the backbone, and that a given degree of fiber shortening would result in an amplified degree of body bending compared to the same shortening in superficial fibers (Alexander, 1969; Rome *et al.*, 1988).

Adjacent myomeres are separated by thin layers of collagenous connective tissue, the myosepta. In gnathostome fishes these myosepta are bisected by a collagenous horizontal septum into epaxial and hypaxial moieties (Figure 7.1). Because the muscle fibers insert into myosepta and the horizontal septum, muscular forces will be transferred along tensile collagenous fibers of these septa and effect lateral bending of the backbone. The possible interactions of this musculotendinous network have been investigated in recent years and lend further support to the idea that the collagenous septa

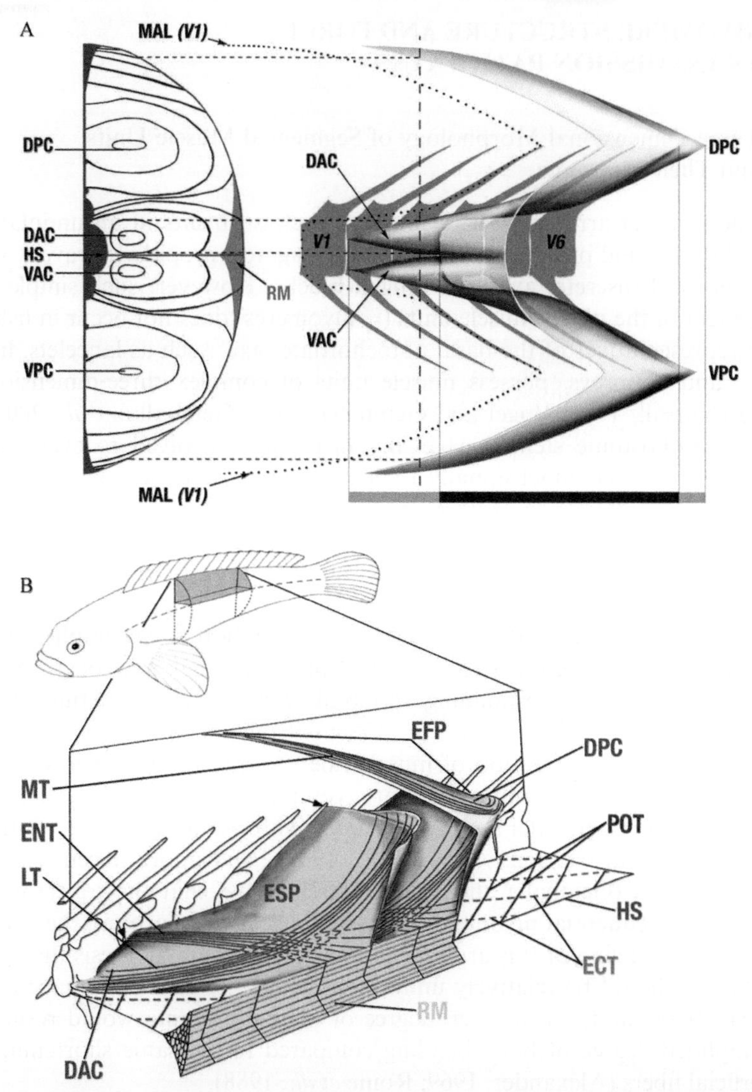

Fig. 7.1. Schematic illustration of gnathostome myosepta (MS) with associated red muscles (RM). (A) Lateral view (right) and corresponding transverse section (left) of a myoseptum. In the postanal region, a horizontal septum (HS, not shown in lateral view) bisects the MS into epaxial and hypaxial moieties. Lateral view: A region of six vertebral segments (V1 to V6) is shown. Only one out of the six MS of this region is shown. The anterior pointing main anterior cone (to the left) is subdivided into two subcones, the dorsal anterior cone (DAC) and ventral anterior cone (VAC). Two posterior pointing cones, the dorsal posterior and ventral posterior cones (DPC and VPC; to the right) are present. Medially, a MS attaches to the axial skeleton and vertical septum. This medial attachment line is shown for the MS that attaches to the

form the basis of a force transmission framework in the fish body (Westneat *et al.*, 1993; Gemballa and Vogel, 2002; Shadwick *et al.*, 2002; Gemballa *et al.*, 2003b,c; Gemballa and Treiber, 2003).

The biomechanical understanding of such a complex musculotendinous system is challenging and is best addressed by an integrative approach including kinematics and muscle dynamics as focal points. However, if the muscle action through tendinous structures and thus pathways of force transmission in swimming fishes is to be elucidated, the three-dimensional morphology of the musculotendinous system (e.g., myomeres and arrangement of tendons in myosepta and horizontal septa, insertion of muscle fibers on these tendons) becomes another relevant issue. With the application of new techniques over the past decade, important progress has been made regarding these issues.

Each myoseptum is a sheet of connective tissue that spans between its lateral line of attachment to the collagenous dermis and its medial line of attachment to the vertical septum and the connective tissue sheet of the axial skeleton. Both lines of attachment are formed by four legs that are oriented either caudoventrally or cranioventrally, thus giving the myoseptum a W shape turned by 90° (Figure 7.1). The four legs of the lateral attachment line are parallel to the orientations of the collagenous fibers of the dermis, whereas the four legs of the medial attachment are oriented more horizontally (Figure 7.1; Gemballa *et al.*, 2003b; Gemballa and Röder, 2004). Between its two lines of attachment a myoseptum does not form a plain sheet, but is drawn out into pointed cones. These cones, termed the dorsal posterior cone (DPC), ventral posterior cone (VPC), and main anterior cone (divided into dorsal and ventral subcones, DACs and VACs), project into the musculature at the level of the backward and forward flexure of the W,

anterior margin of V1 (the dotted line labeled as MAL [V1]). Notice that a MAL inserts into three subsequent vertebral segments. The black bar at the bottom indicates the rostrocaudal extension of MAL (V4). Myoseptal cones extend anteriorly and posteriorly beyond the MAL into the musculature (cone length, CL; grey bars at bottom). MAL and CL add up to the overall myoseptal length. Transverse section: Vertical dashed line in the lateral view indicates position of the transverse section; the lateral view is connected to corresponding points in the transverse section by horizontal dashed lines. A total of eight MS is present in the transverse section. Anterior cone length equals two segment lengths (see lateral view). Thus, two concentric rings are visible in the transverse section. (B) Graphic representation of three-dimensional shape and tendinous architecture of epaxial myosepta (MS; grey), horizontal septum (HS), and red muscles (RM) in an actinopterygian fish. Oblique dorsal and anterior view. HS bears posterior oblique tendons (POTs) and epicentral tendons (ECTs). Myoseptal tendons red: epineural tendon (ENT), lateral tendon (LT), myorhabdoid tendon (MT). EFP, epaxial flanking part; ESP, epaxial sloping part. Area between arrows marks attachment of MS to axial skeleton. (Combined after Gemballa *et al.*, 2003b,c; Donley *et al.*, 2004).

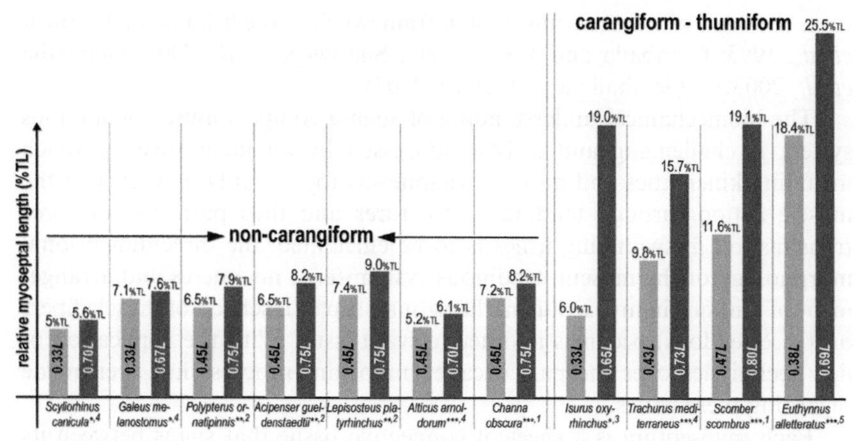

Fig. 7.2. Measurements of overall myoseptal length (ML) in 11 gnathostome fishes. Four carangifom and thunniform swimmers are shown on the right side, non-carangiform swimmers on the left side. ML equals the length of the lateral tendon (see text), and is given as percentage of total length of the fish (%TL). For each species a value at an anterior (light bars) and a posterior (dark bars) axial position is given at the top of the bars. Axial position is labeled on the bars. Data for widely separated taxonomic groups (*, sharks; **, basal actinopterygians; ***, percomorph teleosts) are combined from various sources: [1]Gemballa and Treiber, 2003; [2]Gemballa and Röder, 2004; [3]Donley *et al.*, 2004; [4]S. Gemballa and T. Hannich, unpublished; [5]S. Gemballa and P. Konstantinidis, unpublished.

and are evidenced in transverse sections by concentric rings (Figure 7.1A). A single myoseptum, reaching from the tip of its anterior to the tip of its posterior cone, spans across several vertebral segments. Within a series of myosepta this overall myoseptal length increases gradually from anterior to posterior myosepta. Interestingly, the degree of this elongation is related to the body morphology and swimming mode: in carangiform and thunniform swimmers myosepta are remarkably elongated (up to 25% of the body length) when compared to non-carangiform swimmers (Figure 7.2). Tunas have achieved the extreme in terms of myoseptal elongation, since single myosepta span between 10 and 17 vertebrae (Fierstine and Walters, 1968; Knower *et al.*, 1999).

The way that variations in the degree of elongation are achieved becomes clearer when viewing the overall myoseptal length as being composed of three parts: a middle part, in which the myoseptum is attached to the axial skeleton (medial attachment line), and an anterior and a posterior part, which are formed by the myoseptal cones (cone lengths; see black and grey bars in Figure 7.1A). The contribution of the medial attachment line to the overall myoseptal length is quite conservative in actinopterygians. Typically, a myoseptum is attached at the anterior margin of a vertebral centrum, N, and traverses posteriorly across three vertebral segments (the neural arches of N, N + 1, and N + 2, and the neural spine N + 2; Figure 7.1B; Gemballa *et al.*, 2003b; Gemballa and Treiber, 2003; Gemballa and Röder, 2004; Gemballa, 2004). In contrast, cone lengths differ remarkably between species. In non-carangiform swimmers, the maximum cone lengths in the posterior body do not exceed 2.5 segment lengths, whereas cone lengths reach up to 4 segment lengths in the carangiform mackerel. This tendency toward elongated cones in carangiform and thunniform swimmers is based on cone length measurements from cleared and stained specimens (Gemballa and Treiber, 2003; Gemballa and Röder, 2004), but this correlation is still weak because only a few species have been investigated in detail. However, though they are less accurate, estimations of cone lengths from transverse sections lend further support to this idea. In a transverse section, the number of concentric rings visible indicates the cone lengths. For example, one full concentric ring in a transverse section indicates a cone that spans one vertebral segment, whereas three full rings indicate a cone length of three vertebral segments. Obviously, myoseptal cones in scombrids are lengthened to at least 4–6 segment lengths; even more elongated cones were found in the thunniform lamnid sharks (e.g., Kafuku, 1950; Westneat *et al.*, 1993; Ellerby *et al.*, 2000; Westneat and Wainwright, 2001; Katz, 2002; Shadwick *et al.*, 2002; Donley *et al.*, 2004). These morphological comparisons not only reveal general differences in design among groups of fishes, but may also help to understand internal body mechanics of different locomotory modes, especially when myoseptal shape, length, and their collagen fiber architecture and the arrangement of tendons are considered.

B. The Collagen Fiber Architecture of Myosepta and Horizontal Septum: A Three-Dimensional Network of Specifically Arranged Tendons

Comparative studies on various vertebrates revealed that myosepta in gnathostome fishes are best considered as a set of specifically arranged tendons rather than homogenous sheets of connective tissue (Gemballa and Britz, 1998; Vogel and Gemballa, 2000; Azizi *et al.*, 2002; Gemballa *et al.*, 2003a,b,c; Gemballa and Ebmeyer, 2003; Gemballa and Treiber, 2003; Gemballa and Röder, 2004; Gemballa and Hagen, 2004). This specific arrangement of tendons occurred in gnathostome ancestors and is remarkably conserved within the group. This conservative anatomy of myoseptal tendons indicates functional significance and makes these tendons likely to form the basis of the force transmission framework in the fish body (Gemballa and Vogel, 2002; Gemballa *et al.*, 2003b).

Typically, two tendon-like structures are present in the epaxial region between anterior cone and dorsal posterior cone (the epaxial sloping part [ESP]). Only one of them, the epineural tendon, is connected to the axial skeleton. It inserts as a relatively distinct tendinous structure to the vertebral centrum, the neural arch, or the neural spine (arrows, Figure 7.1B). It runs caudolaterally within the dorsal part of the anterior myoseptal cone toward its insertion to the collagenous dermis (Figure 7.1B). Along this course, its collagen fibers usually diverge to form a broad insertion to the dermis. The second tendon-like structure of the ESP, the lateral tendon, is a broad and indistinct tendon that connects anterior and posterior cones (Figure 7.1B). It takes a curved pathway along the lateral part of the ESP. In its middle part, it intersects with the epineural tendon and is connected to the collagenous dermis. Values of myoseptal length (Figure 7.2) directly reflect the region of the body spanned by a lateral tendon. Thus, carangiform and thunniform swimmers appear to have a more heteronomous series of myosepta with lateral tendons being elongated toward the posterior body, whereas the non-carangiform swimmers appear to have a more homogenous series of myomeres. The medial part of the ESP does not show any tendon-like structures but consists of fewer and relatively thin collagenous fibers (Figure 7.1B) (Gemballa *et al.*, 2003b). Thus, the two tendons described contribute to the robustness of the ESP.

The epaxial flanking part (EFP) of a myoseptum bears one longitudinally oriented tendon, the myorhabdoid tendon. It runs along the lateral part of the EFP from its anterior tip toward the dorsal posterior cone, where it merges with the lateral tendon of the epaxial sloping part (Figure 7.1B). Here, both tendons, the myorhabdoid and the lateral tendon, fan out into horizontal fanlike tendons that project into the musculature. These projections are relatively weak between adjacent cones of the anterior body region, whereas in the posterior body region they form mechanical links between adjacent cones (Gemballa *et al.*, 2003b; Gemballa and Treiber, 2003; Figure 7.3B). They are present not only between posterior cones but also between anterior pointing cones. The epaxial set of three tendons, the epineural, lateral, and myorhabdoid tendon, has a mirror image in the hypaxial region, termed epipleural, lateral, and myorhabdoid tendon. In certain groups of teleost fishes (e.g., some osteoglossiforms, elopiforms, clupeiforms, myctophiforms; see Patterson and Johnson, 1995; Gemballa *et al.*, 2003a for ossification patterns) the whole set of tendons or part of it might ossify.

The horizontal septum of gnathostome fishes consists of an array of caudolaterally and craniolaterally oriented collagen fibers that connect vertebral column and skin (Figure 7.1B; Gemballa *et al.*, 2003a). In lower gnathostomes, such as chondrichthyans, fibers of both direction are evenly distributed in the horizontal septum. In actinopterygians, fibers of either one

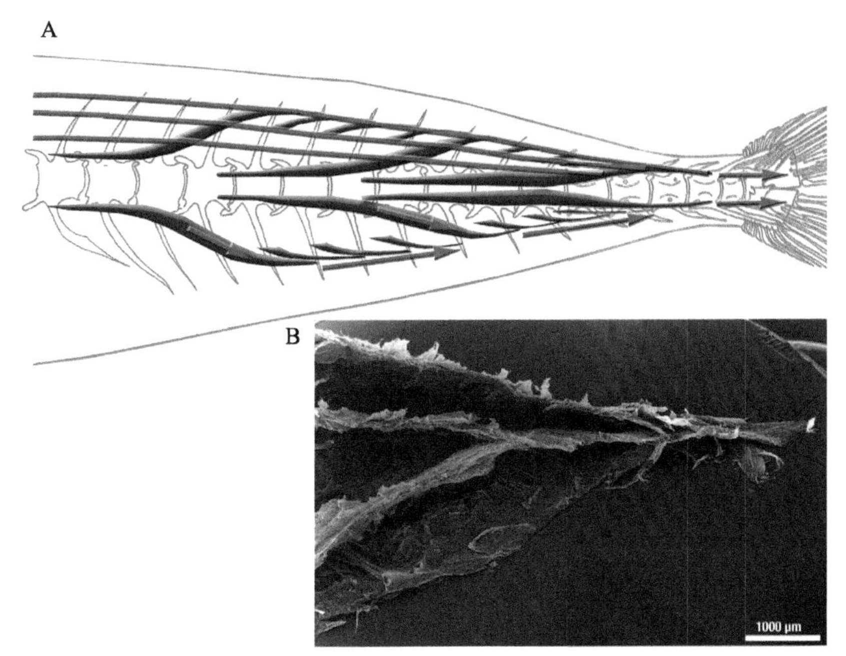

Fig. 7.3. (A) Possible pathways of transmission of red muscle forces to the caudal fin based on investigations of the mackerel *Scomber scombrus*. A series of nine epaxial and hypaxial lateral tendons (LTs) is shown (grey). Only three epaxial and hypaxial LTs are shown completely (from anterior to posterior cone, see Figure 7.2), six are cut close to their posterior ends at the posterior cones. Caudally, the series of LTs ends as caudal tendons (i.e., medial and great lateral tendons) that insert to the caudal fin. The association of three trajectories of red muscle fibers is shown in the epaxial part. Forces might be transmitted via LTs to the posterior cones and through subsequent cones to the caudal tendons and caudal fin (red arrows in hypaxial part; modified from Gemballa and Treiber, 2003). (B) Two subsequent dorsal posterior cones of the bichir *Polypterus ornatipinnis*. Anterior to left; medial view, cones are cut parasagitally to see their inner architecture. Myorhabdoid tendon (upper part) and lateral tendon (lower part of cone) merge at cone and fan out into a horizontal fanlike projection that mechanically links the two subsequent cones. (Original SEM micrograph, S. Gemballa, K. H. Hellmer, and J. Berger, Tübingen.)

or both directions are condensed to distinct tendons that are called epicentral tendons (ECTs = caudolateral orientation, anterior oblique tendon [AOT]) and posterior oblique tendons (POTs). These tendons are well pronounced in many teleosts, especially in scombrid fishes (Kafuku, 1950; Westneat *et al.*, 1993; Patterson and Johnson, 1995; Gemballa *et al.*, 2003a). However, some exceptions are known in which either one or both tendons were reduced (Gemballa *et al.*, 2003a).

The fact that myosepta and the horizontal septum are networks of specifically arranged tendons that are either embedded in the body musculature (lateral and myorhabdoid tendons) or connect vertebral axis and skin (epineural and epipleural tendons of myosepta, and epicentral tendons and POTs of horizontal septum) instead of being homogenous sheets of connective tissue has raised new questions in fish swimming mechanics. Tendons might either transmit muscular forces to generate bending (in-plane tensile loads on tendons) or resist radial expansion of contracting muscle fibers (normal to plane pressure loads). Both hypotheses have been inferred from anatomical studies (e.g., Westneat *et al.*, 1998; Gemballa and Vogel, 2002; Gemballa and Treiber, 2003; Gemballa and Hagen, 2004; Gemballa and Röder, 2004). A first step in understanding the complex arrangement of tendinous structures is to consider how muscle-tendon interactions may occur during swimming.

C. Muscle Recruitment and Force Transmission to the Axial Skeleton

Fishes generally have a distinct anatomical division between red (= slow, oxidative) and white (= fast, glycolytic) muscle fibers, the former generally occupying a lateral subcutaneous band and the latter making up the bulk of the underlying myomeres (Figures 7.1 and 7.2; see Chapter 6 of this volume for a detailed account of contractile properties). Notable differences are found in the tunas and Lamnid sharks, which have their red muscle located deep within the myotomal mass, where it is complexed with a circulatory heat exchanger to facilitate heat conservation (Bernal *et al.*, 2001a,b; see Section V). Although the relative proportions and positioning of the red fibers is variable among species, it has been generally observed that these fibers are primarily responsible for powering sustained slow swimming (Boddeke *et al.*, 1959; Bone, 1966; Rayner and Keenan, 1967; Bone *et al.*, 1978; Brill and Dizon, 1979; Rome *et al.*, 1984, 1988, 1992; Tsukamoto, 1984; Jayne and Lauder, 1993; Coughlin *et al.*, 2004). In most fishes, red muscle fibers are oriented longitudinally on either side of the horizontal septum, where epicentral tendons and POTs are present (Figure 7.1B). Based on an investigation of this arrangement in scombrids it was proposed that axial bending could be caused by red muscle forces that are transmitted via POTs to the backbone (Westneat *et al.*, 1993) and, more recently, that this mechanism may extend to many other groups (Gemballa *et al.*, 2003a). However, further analysis is needed. Since red muscle fibers do not insert directly to the POTs, but rather to the adjacent myosepta, it remains to be tested if these myoseptal fibers provide an adequate mechanical link between red muscles and POTs. Second, the series of POTs does not extend to the last vertebra; thus, it seems unlikely that they transmit red muscle forces directly to the caudal fin. Additionally, some fishes (e.g., gars, gadids, blenniids; see Gemballa *et al.*, 2003a) lack POTs.

A second mechanism for red muscle force transmission, the LT pathway, was proposed based on investigations of the musculotendinous system of the mackerel *Scomber scombrus* (Gemballa and Treiber, 2003). In this species, the red muscles fibers form a band of remarkable dorsoventral width and are directly connected to the lateral tendons of the myosepta. Thus, red muscle forces can be transmitted directly via lateral tendons to the dorsal posterior cones and via intermyoseptal linkages, the horizontal fanlike projections (Figure 7.3) toward the caudal fin. The same principal association of lateral red muscles with lateral tendons is present in sharks, suggesting that the LT pathway is widely distributed among gnathostomes (S. Gemballa *et al.*, submitted). Caudally, the series of posterior cones ends in flat sheets of connective tissue of the superficial caudal musculature or in distinct caudal tendons, both of which attach to the caudal fin rays (Gemballa and Treiber, 2003; Gemballa, 2004). Direct force measurements have proven that caudal tendons in two species of tunas transmit red muscle forces to the tail (Knower *et al.*, 1999; Shadwick *et al.*, 2002), but experimental data from any other species are lacking.

Recruitment of additional fibers, including white muscle, is required to power faster, unsteady movements (sprint and burst swimming, escape response and fast starts; Rome *et al.*, 1992; Jayne and Lauder, 1994, 1995a,b; Gillis, 1998; Coughlin and Rome, 1999; Ellerby and Altringham, 2001; Coughlin *et al.*, 2004; also see Figure 7.1, Chapter 6, and 9). The orientation of white muscle fibers within the myomeres is complex. So far, we have little understanding of muscle-tendon interactions, and what we do know is almost exclusively based on morphological descriptions. White muscle fibers do not attach to the skin or to the horizontal septum, but only to myosepta, the vertical septum and the connective tissue sheet that wraps around the axial skeleton. They insert into the myoseptal tendons at variable angles, depending on location (Gemballa and Vogel, 2002). Muscle fibers of one trajectory, the helical muscle fiber arrangement (HMFA; see Chapter 9 and Figure 7.6B) are at relatively acute angles (27–40° in lateral projections) with the lateral tendon and the distal part of the epineural tendon as well as with the vertebral axis and vertical septum. Deeper muscle fibers (forming the crossing muscle fibers; Chapter 9 and Figure 7.6B) are associated with the proximal part of the epineural tendon and the medial thin ESP. This arrangement does not seem to be adequate for force transmission, since muscle fibers are at obtuse angles with the myoseptum (58–63°). Fiber angles with myorhabdoid tendons have not yet been investigated.

According to the acute angels between HMFA and myoseptal tendons, this arrangement might be regarded as adequate for force transmission.

Likely pathways are via epineural tendons to the backbone or via lateral tendons to the posterior cones and via mechanical linkages between subsequent cones (Figure 7.3B) to the caudal fin (Gemballa and Vogel, 2002; Gemballa and Treiber, 2003; Gemballa, 2004). The latter way is similar to the LT pathway described previously for red muscle forces. In addition to steady swimming red muscle forces, caudal tendons of tunas also transfer much higher white muscle forces during bursts (Shadwick *et al.*, 2002). Furthermore, one model of axial-undulatory swimming has emphasized the role of longitudinally arranged tendons (Long *et al.*, 2002).

III. STEADY SWIMMING KINEMATICS

A. Waves on the Body

Undulatory swimming is controlled by motor patterns generated via the spinal cord. A propulsive body wave is initiated by coordinated, sequential contractions of muscle segments alternately along both sides of the fish, transmitting bending moments to the axial skeleton that result in characteristic waves of lateral displacement, which travel with increasing amplitude from head to tail. The increasing time delay of the muscle activation toward the posterior gives the impression, as recorded by electromyography (EMG), of a wave of activation progressing rostro-caudally along the body (e.g., Blight, 1976; Grillner and Kashin,1976; Jayne and Lauder, 1995b; Hammond *et al.*, 1998; Knower *et al.*, 1999; Ellerby *et al.*, 2000; Donley and Shadwick, 2003). The spatial and temporal coordination of muscle activation and contraction generates the propulsive wave of lateral bending that likewise travels along the body from head to tail (see Figures 7.4 and 7.5). The amplitude of this displacement wave typically increases posteriorly, with the shape of the amplitude envelope varying according to the degree of lateral undulation that characterizes different body forms and swimming modes (Altringham and Shadwick, 2001; Donley *et al.*, 2004; see Chapter 11). Regardless, the lateral peak-to-peak displacement at the tip of the caudal fin is approximately 0.2L and virtually independent of swimming speed in most fishes (e.g., goldfish, dace, rainbow trout [Bainbridge, 1958; Webb *et al.*, 1984]; jack, sardine, mackerel, anchovy [Hunter and Zweifel, 1971]; saithe and mackerel [Videler and Hess, 1984]; carp and scup [Rome *et al.*, 1990, 1992]; largemouth bass [Jayne and Lauder, 1995b]; and American eel [Gillis, 1998]). In contrast, sprinting bouts by 33 cm rainbow trout showed a significant increase in peak tail tip displacement from 0.15L to 0.2L when tail-beat frequency increased from 4 to 11 Hz (Ellerby and Altringham, 2001), while peak lateral motion of the caudal fin of a kawakawa tuna

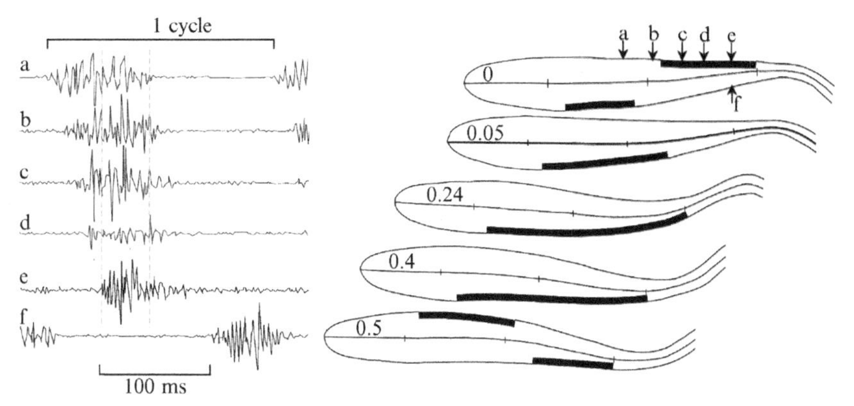

Fig. 7.4. EMG recordings from a largemouth bass (*Micropterus salmoides*) swimming at 2.4 Ls^{-1}, showing one tail-beat cycle. Axial locations of each electrode pair are shown in the top body outline in the left panel (right side, a = 0.43L, b =0.5L, c = 0.57L, d = 0.65L, e = 0.72L; left side, f = 0.72L). Vertical dashed lines in left panel indicate the time period when activation occurs simultaneously between sites a and e. Comparison of traces e and f shows that EMG is 180° out of phase on opposite sides of the body. Body outlines are taken from sequential digitized video images and shifted to the left to indicate the forward progression of the fish as a function of the tail-beat cycle (0 to 0.5T). Thick lines along the sides indicate muscle that is active, based on EMG recordings. (Derived from data in Jayne and Lauder, 1995b.)

(*Euthynnus affinis*) appeared to increase from 0.16L to 0.34L during sprints powered by tail-beat frequencies of 8–14 Hz (Fierstine and Walters, 1968). Such changes in amplitude at burst speeds may be transient and indicative of larger lateral excursions during acceleration not seen during steady, low-speed swimming (Gray, 1968). In a comprehensive study on trout, Webb *et al.* (1984) found that length-specific caudal amplitude in steady swimming was smaller in larger fish, scaling as $L^{-0.26}$.

Typically, in steady swimming, muscle activation (and presumably shortening) at any location is 180° out of phase with the opposite side and, since the activation duration (duty cycle) is normally <0.5 of a tail-beat period, there is rarely any temporal overlap observed. Thus, as each portion of the musculature shortens locally to create concave bending, the opposite side is relaxing and being stretched into local convex curvature. In most undulators (except anguilliform), the activation wave travels along the body at rates in excess of 1 length per cycle period (LT^{-1}), so that even with a rostral-caudal delay in onset, there is a substantial portion of each cycle when all the red muscle along one side will be active, although at different phases of shortening (see Section IV.C). For example, in a swimming bass (see Figure 7.4) the activation onset progresses at 1.3LT^{-1} and there is

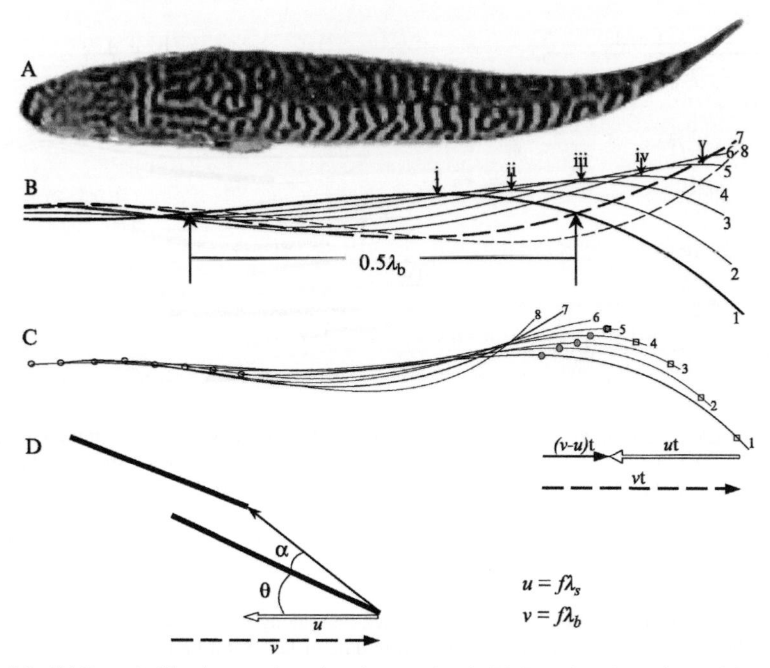

Fig. 7.5. (A) Dorsal video image of a swimming mackerel. (B) Fourth-order polynomial curves fit to body midlines from successive digitized video images of a mackerel swimming at $u = 3.25$ Ls^{-1}, with a tail-beat frequency of 4.5 Hz (T = 0.22 s), to show methods of determining v (the velocity of the propulsive wave) or λ_b (the propulsive wavelength). By aligning the midlines on the horizontal axis, the position of peaks in lateral deflection can be followed in time (arrows i–v, each separated by a time interval of 16.67 ms). In this example, the deflection peak travels 0.35L in 66.7 ms, yielding $v = 5.2$ Ls^{-1} and $\lambda_b = 1.15L$. Alternatively, the distance on the body between nodes of the body wave depicted by the intersection of midlines (large arrows) from the start of successive half tail beats yields a measurement of $0.5\lambda_b$; in this example $\lambda_b = 1.1L$, giving $v = 5.0$ Ls^{-1}, values very similar to the first method. (C) Midlines as in (B), here displaced horizontally to indicate the forward speed of the fish. The distance traveled forward through space by a point on the body (e.g., squares at the tail tip) is represented by ut, the distance traveled backward along the body by the crest of the propulsive wave is vt, while the difference in these quantities $(v\text{-}u)$t represents the relative progression of the wave crest (circles) backward in space. (D) The position of a tail segment at two successive times showing its forward and lateral movement. Its orientation relative to forward progression is the angle θ. The angle of attack of the segment relative to the fluid flow is α. The path angle of the segment is $(\alpha + \theta)$. When u approaches v, α decreases. If $u = v$, $\alpha = 0$ and no thrust is produced. λ_s is the wavelength of the path traced out by the body of the fish.

coincident activation of muscle between 0.43 and 0.72L along one side for about 22% of each tail-beat cycle (or 44% of the half-cycle when muscles on one side are shortening). This is similar to observations on other species:

activation wave velocities are 1.2–1.6LT^{-1} in trout, saithe, mackerel, carp, leopard shark (Wardle *et al.*, 1995; Hammond *et al.*, 1998; Donley and Shadwick, 2003), increasing to 1.7–2.1LT^{-1} in bonito, mako shark, scup, and yellowfin tuna (Wardle *et al.*, 1995; Knower *et al.*, 1999; Ellerby *et al.*, 2000; Donley *et al.*, 2005). Coincident activation from 0.3–0.8L on one side occurs for about 20–32% of the tail-beat period in subcarangiform and carangiform swimmers (saithe, mackerel, leopard shark, bass, carp, scup, and trout) and 14–19% in tunas and bonito (Altringham and Shadwick, 2001). In contrast, the activation waves in eels and lamprey travel at only about 0.6LT^{-1}, and there is virtually no coincident activation of the entire side of muscle (Grillner and Kashin, 1976; G. B. Gillis, personal communication).

B. Body Kinematics

The distance traveled forward in one tail-beat cycle is the stride length, λ_S. Treating the path traced through the water by the tail tip (or any other segment of the body) as a wave of forward progression, λ_s is the forward velocity u times the cycle period T or, u/f, where f is tail-beat frequency. Most measured values of λ_s fall in the range of 0.60–0.70L for steady swimming subcarangiform, carangiform, and thunniform fishes (see Videler and Wardle, 1991; Videler, 1993; Dewar and Graham, 1994; Donley and Dickson, 2000), although there is evidence that λ_s decreases at the lowest speeds in some species (Bainbridge, 1958, Hunter and Zweifel, 1971, Webb *et al.*, 1984; Dewar and Graham, 1994), and is typically only 0.35–0.5L in eels swimming at 1 Ls^{-1} or less. At the other extreme, λ_s may approach 1.0L in the fastest sprints by scombrids (Wardle and He, 1988). In order to generate forward thrust, the propulsive wave pushes against the water; because the water gives way the wave must travel backward along the body at a velocity, v, that is greater than the forward velocity of the fish (i.e., u/v 1.0; Figure 7.5). Consequently, the propulsive wavelength λ_b is greater than λ_S, as v is greater than u, and the path angle of a body segment is greater than the orientation of the segment relative to the direction of forward travel (see Figure 7.5C and D). This ensures that the segment has a lateral component of motion relative to the flow, that is, a positive angle of attack angle α (Figure 7.5D; Webb, 1975; Jayne and Lauder, 1995a). Experimental measurements of α during swimming are difficult to obtain and are limited; Bainbridge (1958) showed that in goldfish and dace α at the tail tip and caudal peduncle ranged up to 30° and was positive for about 75% of the tail-beat cycle. Jayne and Lauder (1995a) found lower values of α in swimming bass and significant periods of each cycle when $\alpha < 0$. In contrast, α was always positive in a swimming tuna (Fierstine and Walters, 1968), averaging 30° at mid-stroke when lateral velocity of the caudal fin was greatest.

The ratio u/v (= λ_s/λ_b) defines the slip of the body in the water. At slip values <1.0, the propulsive wave travels faster than the body and generates forward and lateral thrust. If u approaches v, α will decrease because the path of the segment coincides closer to the shape of the propulsive wave. If $u = v$, the propulsive and progression waves are coincident, $\alpha = 0$, there is no relative motion between the body and water, and no thrust is generated (Webb, 1975). In the case of slip >1.0, the body wave would travel forward relative to the water and decelerate the fish. The propulsive or Froude efficiency is the ratio (thrust power/total power output), which can be expressed as $(u + v)2v$ (Webb, 1975; Webb *et al.*, 1984). When u is close to v, the locomotor efficiency is optimized because the energy lost to lateral motion is minimized. For example, Froude efficiency would be 0.75 when $u/v = 0.5$, rising to 0.95 for $u/v = 0.9$. Empirical determination of v can be made directly from digitized video sequences by measuring the time taken for peaks in lateral deflection to travel a given distance posteriorly along the body (see Figure 7.5B); propulsive wavelength λ_b is then calculated as v/f. Equal spacing of the peaks in time in Figure 7.5B reveals that v is constant along the body, as is also found in other swimming fishes (Videler and Hess, 1984; Müller *et al.*, 1997; Katz and Shadwick, 1998; Katz *et al.*, 1999). Alternatively, λ_b can be measured graphically from internode distances of superimposed midlines (see Figure 7.5B; Webb *et al.*, 1984) and v calculated, but unlike the former, this method is highly sensitive to lateral drift of the fish during the sequence analyzed. Generally, λ_b is about $1.0L$, so u/v is about 0.7, although this decreases at lower speeds (i.e., $u < 1\ Ls^{-1}$). Consequently, the Froude efficiency is also decreased at low swim speeds. Propulsive waves are well above $1.0L$ in thunniform swimmers, but are much shorter in non-carangiform and anguilliform swimmers (Williams *et al.*, 1989; Videler, 1993; Dewar and Graham, 1994; Donley *et al.*, 2005), and this corresponds to the differences in the elongation of their myomeres and myoseptal tendons (Figure 7.2).

C. Speed Control

Undulatory swimming speed is primarily controlled by changes in tail-beat frequency, regulated by the rate at which the muscle activation and propulsive waves propagate along the body. A survey of numerous studies on fish swimming in water tunnels and pools revealed a surprisingly similar relationship between tail-beat frequency and swimming speed (Figure 7.6), regardless of differences in fish size, swimming mode (with the exception of eels and small larvae), apparatus, or measurement technique. By this measure, we can conclude that the basic kinematics of swimming by body

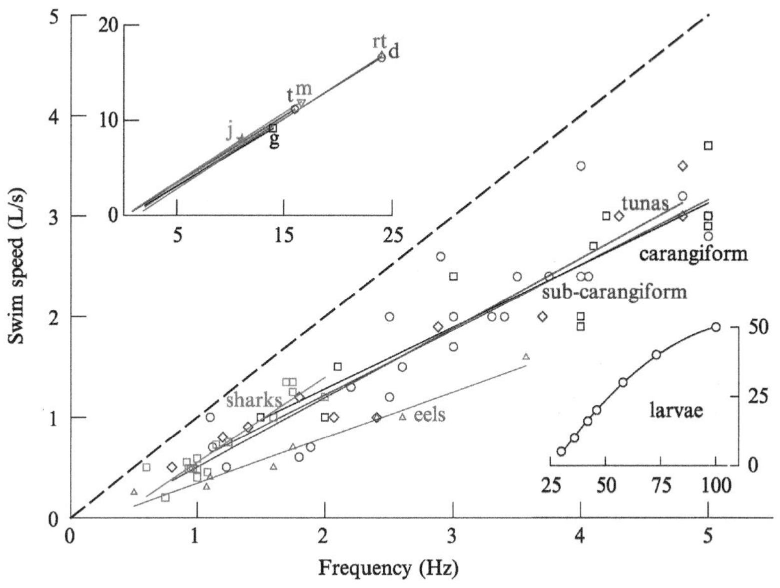

Fig. 7.6. Summary of data for a wide variety of fishes performing undulatory swimming at slow and fast speeds, with speed normalized to body length for comparison. Measurements are from fish swimming against flow in a water tunnel or free swimming in pools. Except for eels, most species follow a similar linear relationship, where speed is a function of tail-beat frequency, and all points fall below the line of equality (dashed), indicating that stride lengths ($\lambda_s = u/f$) are <1.0 and relatively constant across speeds. Eels perform at much lower stride lengths. Eel data: *Anguilla rostrata*, Gillis (1998); *Anguilla anguilla*, D'Août and Aerts (1999); Ellerby *et al.* (2001). Shark data: *Triakis semifasciata* and *Negaprion brevirostris*, Hunter and Zweifel (1971); Graham *et al.* (1990); *Sphyrna lewini*, Lowe (1996); *Charcharhinus*, Webb and Keyes (1982); *Isurus oxyrinchus*, C. Sepulveda, unpublished. Subcarangiform data: *Salvelinus fontinalis*, McGlinchey *et al.* (2001), *Oncorhynchus mykiss*, Webb *et al.* (1984); Coughlin (2000); *Micropterus salmoides*, Jayne and Lauder (1995a); *Dicentrarchus labrax*, Herskin and Steffensen (1998); *Chanos chanos*, Katz *et al.* (1999); *Hypomesus transpacificus*, Swanson *et al.* (1998); *Gadus morhua*, Videler and Wardle (1978); *Mugil cephalus*, Pyatetskiy (1970); *Cyprinus carpio*, Rome *et al.* (1990). Carangiform data: *Stenotomus chrysops*, Rome *et al.* (1992); *Sarda chiliensis*, Magnuson and Prescott (1966); *Trachurus symmetricus*, Hunter and Zweifel (1971); *Trachurus japonicus*, Xu *et al.* (1993); *Scomber japonicus*, Donley and Dickson (2000). Tuna data: *Katsuwonus pelamis* and *Thunnus albacares*, Knower *et al.* (1999); *Thunnus thynnus*, Wardle *et al.* (1989); *Euthynnus affinis*, Dewar (1993). The upper insert shows regression lines for several species that include burst speeds above 10 Ls^{-1} . j = jack, *Trachurus symmetricus*, Hunter and Zweifel (1971); d = dace, *Leuciscus leuciscus*, g = goldfish, *Carassius auratus*, rt = rainbow trout, *Oncorhynchus mykiss* Bainbridge (1958); m = mackerel, *Scomber scombrus*, Wardle (1985); t = tuna, *Thunnus albacares* and *Katsuwonus pelamis*, Yuen (1966), *Euthynnus affinis*, Dewar (1993). The lower insert shows the performance of 4 day, 4 mm zebrafish (*Danio rerio*) larvae in undulatory swimming. (Data from Müller and van Leeuwen, 2004.)

undulation is fundamentally the same among species, and that faster speeds are achieved by faster cyclic contractions of the lateral muscle. At all speeds, from slow aerobic swimming to maximal bursts, the speed in Ls^{-1} is always less than the tail-beat frequency in Hz, thus corresponding to stride lengths ($\lambda_s = u/f$) <1.0. Eels swim with lower λ_s, as reflected by the lower slope of their speed/frequency relation (Figure 7.6). Similarly, larval fish have eel-like kinematics; recent studies of 4 mm zebrafish larvae have documented large-amplitude body waves and relatively low stride lengths (0.3–0.5*L*), but at high swimming velocities with record-breaking tail-beat frequencies of up to 100 Hz (Figure 7.6; Müller and van Leeuwen, 2004).

IV. MUSCLE DYNAMICS ALONG THE BODY IN STEADY SWIMMING

A. Muscle Strains and Body Curvature

Typically, the increase in lateral amplitude of the propulsive wave as it travels along the body is correlated, not surprisingly, with larger body curvature and increasing axial muscle shortening. This reflects the finding that, in most cases, local muscle shortening, or strain amplitude, is spatially and temporally linked to changes in local body bending (Katz and Shadwick, 2000). In early studies (e.g., Videler and Hess, 1984; van Leeuwen *et al.*, 1990; Rome and Sosnicki, 1991; Johnson *et al.*, 1994; Jayne and Lauder, 1995a,b; Long *et al.*, 1996), a swimming fish body was modeled as a simple bending beam; thus amplitude and phase of muscle strain were calculated as the local curvature of the neutral axis (the body midline) multiplied by the lateral distance of the muscle from this axis (Figure 7.7). Curvature calculations can be made by digitizing dorsal or ventral images of the swimming fish, generating coordinate pairs for midline points in an x-y plane, fitting curves to the midlines and then calculating curvature κ (equal to the inverse of radius of curvature) at specific axial locations according to:

$$\kappa(x) = y''(x)/(1 + y'(x)^2)^{3/2} \qquad (1)$$

where y(x) describes the fish midline as a function of x, and y′ and y″ are first and second derivatives with respect to x (e.g., Rome and Sosnicki, 1991). Alternatively, curvature can be calculated geometrically from midline points by considering the angular changes between pairs of midline segments defined by adjacent midline points, according to:

$$\kappa = 2\cos(\varphi/2)/M, \text{or } \kappa = 2\sin(\beta/2)/M, \qquad (2)$$

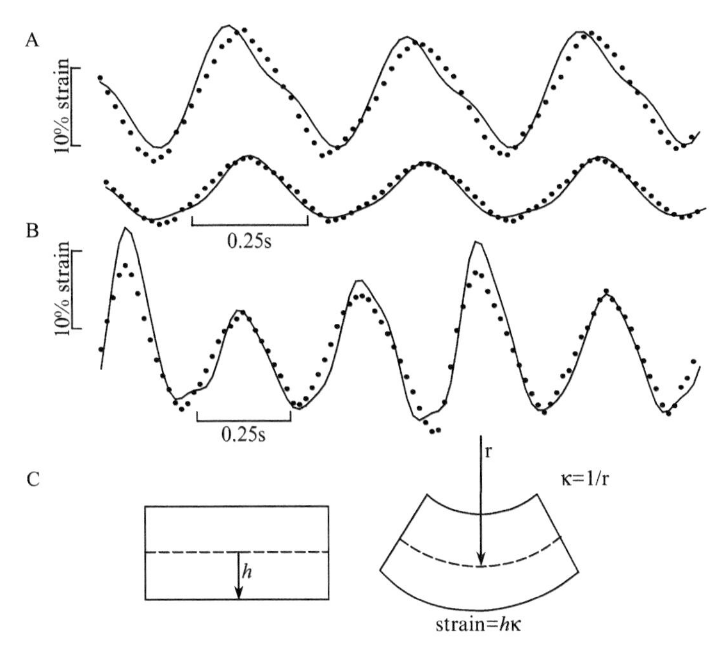

Fig. 7.7. (A) Comparison of muscle strain amplitude in a milkfish (*Chanos chanos*) swimming at 2.3 Ls^{-1} measured by sonomicrometry (solid traces) and calculated from midline curvature (dotted traces). Upper traces are from superficial red muscle, lower traces are from medial white muscle recorded simultaneously at 0.53*L*. (B) Strains in superficial red muscle (measured and calculated, as in (A)) in a milkfish sprinting at 2.8 Ls^{-1}. (C) Diagram to illustrate calculation of strain at the outer surface of a bending beam, where curvature κ is the inverse of the radius of curvature (measured in cm^{-1}), and strain (unitless) is the product of κ and h, the distance from the neutral axis of bending to the surface. In application to strain in a fish body, the neutral axis is the body midline (backbone) and h is the distance from the backbone to the lateral muscle layer. Modified from Katz *et al.*, 1999.

where φ is the angle formed between adjacent segments each of length M (e.g., Long *et al.*, 1996); alternatively β is the "flexion" angle of a segment relative to an adjacent one (i.e., $\pi - \varphi$ radians; Jayne and Lauder, 1993). When segments are relatively short and $\beta < 0.3$ rad, then the approximation $\kappa = \beta / M$ may be used (D'Août and Aerts, 1999).

More recently, direct measurements of muscle strain *in vivo* using X-ray videography (Shadwick *et al.*, 1998) or sonomicrometry, coupled with digital curvature analysis (Coughlin *et al.*, 1996; Katz *et al.*, 1999; Donley and Shadwick, 2003), showed the simple bending model to be generally valid for steady as well as burst swimming. Thus, despite the complex geometry of the myomeres and myoseptal linkages, muscle strain in both superficial

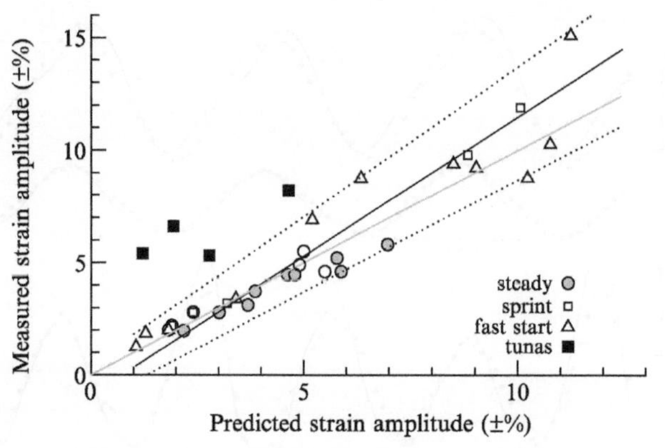

Fig. 7.8. Comparison of muscle strain amplitude measured by sonomicrometry and predicted from midline curvature. Grey is line of equality. Regression line (black) is fitted to data for steady swimming, sprints, and fast starts, but excludes tunas that deviate significantly from the rest and fall outside 95% confidence intervals (dotted lines). (Redrawn from Long *et al.*, 2002; Goldbogen *et al.*, 2005.)

red fibers and medial white fibers can be reasonably well predicted from the amplitude and phase of midline curvature (Figures 7.7 and 7.8; see also Chapter 9). It seems likely that structures such as myoseptal tendons transmit muscle forces through the cross-section in order to keep muscle strain tightly coupled to local curvature in most cases (Long *et al.*, 2002). Exceptions to this generalization have been reported for two cases in which measured strains exceed those predicted: (i) deep red muscle fibers of endothermic tunas, due to their specialized myoseptal anatomy (see Section V and Figure 7.8), and (ii) posterior white muscle fibers of trout during strong escape responses, where curvature is limited by hydrodynamic resistance (Goldbogen *et al.*, 2005).

B. Why Curvature and Lateral Displacement Are Not Synchronous

An important result arising from kinematics studies is that the relationship between curvature of the midline and the shape of the propulsive wave (i.e., lateral displacement) is complex (Katz and Shadwick, 1998). If the waveform of lateral displacement on the fish body was equivalent to a sinusoid of constant amplitude, then the peaks of curvature (and muscle strain) would coincide in time and space with the peaks of lateral displacement [i.e., Eq.(1) would become $\kappa(x) = y''(x)$. This situation is closely approximated in some types of locomotion, such as terrestrial undulation

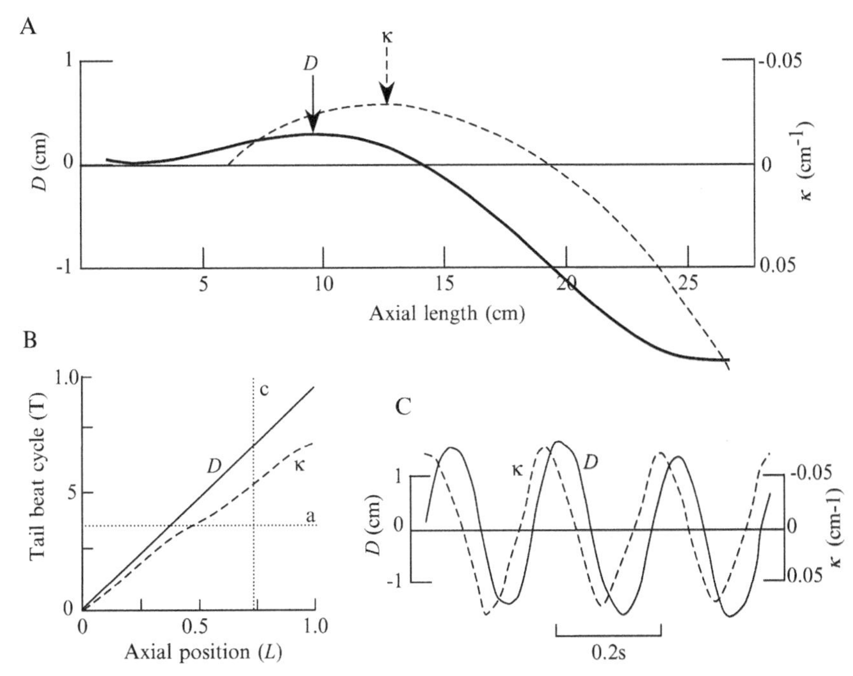

Fig. 7.9. The phase relationship between lateral displacement (*D*: solid lines) and curvature (*κ*: dashed lines) of the midline of a swimming fish. Convex up curvature (displacement to the right side) coincides with negative values of *κ*, and vice versa. (A) Plot of *D* and *κ* as a function of axial position from one video field of a 27 cm mackerel swimming at 3.25 Ls^{-1}. At this time the peak in *D* to the right side occurs at about 10 cm, or 0.37*L*, while the peak in *κ* occurs at about 12.5 cm, or 0.1*L* more posterior. (B) A plot of the progression of the peak of *D* to the right and the corresponding *κ* along the body in time, showing they diverge caudally. Slopes are equal to the inverse of velocity of D and *κ* waves along the body. The horizontal line (a) represents a profile along the body at one time, equivalent to the plot in (A); vertical line (c) represents a time sequence at one axial location, equivalent to the plot in (C). (C) A plot of *κ* and *D* as a function of time at 0.74*L*, showing that *κ* precedes *D* at this location by about 30 ms, or 0.15T (=54°). (Redrawn from Katz and Shadwick, 1998.)

by eels (Gillis, 1998) or nematodes (Gray and Lissman, 1964), but in all aquatic undulators examined, the increase in amplitude of the propulsive wave along the length of the body resulted in phase shifts between curvature and lateral displacement (see Figure 7.9). An analysis of saithe and mackerel kinematics by Videler and Hess (1984) first provided data for lateral deflection and curvature of the midline that revealed a significant phase shift between these two parameters, with curvature leading, of up to 0.18T (or 64°) at the most posterior site. Jayne and Lauder (1995a) compared lateral

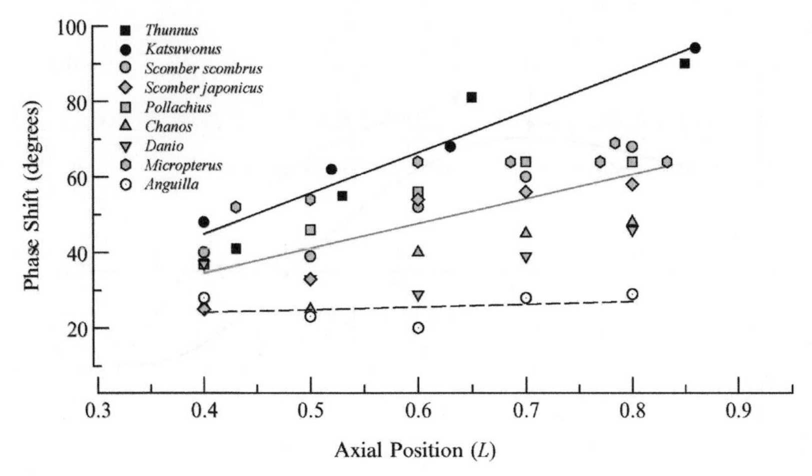

Fig. 7.10. Summary of the phase shift, in degrees of a tail-beat cycle, between midline curvature and lateral displacement as a function of axial position along the body for steady swimming in a variety of species. In all cases κ precedes D in time (as in Figure 7.9C). Separate regressions are fit to data for tunas (black symbols), carangiform and subcarangiform (gray symbols), and eels (open symbols). Data for yellowfin (*T. albacares*) skipjack (*K. pelamis*) tunas from Knower (1998); Atlantic mackerel *S. scombrus* from Videler and Hess (1984); chub mackerel *S. Japonicus* from Katz and Shadwick (1998); saithe *P. virens* from Videler and Hess (1984); milkfish *C. chanos* from Katz *et al.* (1999); zebrafish larvae (*Danio rerio*) from Müller and van Leeuwen (2004); largemouth bass *M. salmoides* from Jayne and Lauder (1995a); eel *A. anguilla* from Videler (1993).

displacement to flexion angles (a proxy for curvature) of the body midline in swimming bass and found phase shifts of 0.13–0.25T, again with curvature leading. Subsequent studies on other species revealed much the same result, with the largest phase shifts occurring in tunas and the smallest in the eel (Figure 7.10), consistent with the lateral amplitude envelop being very nonlinear in the former and much more uniform in the latter (see Altringham and Shadwick, 2001; Chapter 11, Figure 11.1). Indeed, an analytical study (Katz and Shadwick, 1998) showed that the phase relation between midline curvature and the lateral displacement is dependent on the shape of the amplitude envelope, and the preceding observations are to be expected. Specifically, curvature propagates faster along the body than does the lateral displacement (Figure 7.9B), because the phase shift is an increasing function of axial position, for example, phase shift is dependent on the local slope of the midline, which increases posteriorly with the larger lateral displacement amplitude.

C. Muscle Strain, Activation Timing, and Speed

The results of numerous studies on dynamic muscle function in swimming fishes indicate that red muscle strain consistently increases along much of the body, nearly doubling from about ±3–4% at 0.35*L* to ±5–8% at 0.7*L* in many species (Figure 7.11). This leads to an expectation that cyclic work and power, which should increase with muscle strain amplitude, might be greater in posterior positions (e.g., see Figure 6.7 in Chapter 6). However, axial differences in other parameters such as activation timing (see later) and contraction kinetics (see Chapter 6) also have major influences on muscle performance, so regional variations in power output among different species appear complex; their significance is still an intriguing question to be resolved (Altringham and Ellerby, 1999; Coughlin, 2002; Chapter 6). Red muscle strain amplitude in the most-posterior locations may actually

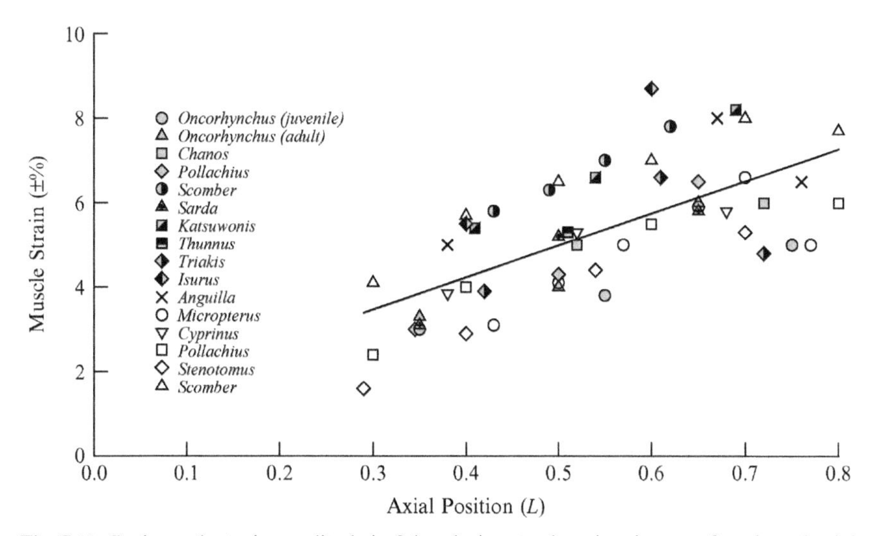

Fig. 7.11. Red muscle strain amplitude in fishes during steady swimming as a function of axial position. Direct measurements (X-ray videography in mackerel *Scomber*, and sonomicrometry in others) are shown by filled symbols. Data for rainbow trout *O. mykiss* juvenile from Coughlin (2000), adult from Hammond *et al.* (1998); milkfish *C. chanos* from Katz *et al.* (1999); saithe *P. virens* from Hammond (1996); mackerel *S. japonicus* from Shadwick *et al.* (1998); bonito *S. chiliensis* from Ellerby *et al.* (2000); skipjack tuna *K. pelamis* from Shadwick *et al.* (1999); yellowfin tuna *T. albacares* from Katz *et al.* (2001); leopard shark *T. semifasciata* from Donley and Shadwick (2003); mako shark *I. oxyrinchus* from Donley *et al.* (2005). Strains calculated from midline curvature are open symbols. Data for eel *A. anguilla* from D'Août and Aerts (1999); bass *M. salmoides* from Jayne and Lauder (1995b); carp *C. carpio* from Rome *et al.* (1990); saithe *P. virens* from Hess and Videler (1984); scup *S. chrysops* from Rome *et al.* (1993); mackerel *S. japonicus* from Shadwick *et al.* (1998). Regression is fit to all data.

decrease because body tapering may be greater than increases in curvature, so $h\kappa$ will fall (Hess and Videler, 1984; Shadwick *et al.*, 1998; D'Aoút and Aerts, 1999; Donley and Shadwick, 2003). Specific examples of this can be seen in Figure 7.11 (*Anguilla, Scomber, Micropterus, Triakis*).

Like the lateral amplitude of the body and caudal fin, red muscle strain amplitude is generally independent of swim speed, although few studies have systematically examined this. Notable exceptions are at speed extremes; for example, in slow swimming bass ($<1\ Ls^{-1}$) and eels ($<0.5\ Ls^{-1}$), curvature and muscle strain decreased slightly (Jayne and Lauder, 1995a; D'Aoút and Aerts, 1999), in milkfish sprinting at $>3\ Ls^{-1}$, curvature and red muscle strain along the body increased significantly (Katz *et al.*, 1999), while muscle strain increased only at $0.65L$ in rainbow trout performing high-frequency sprints. As described previously, muscle fibers participating in undulatory movements are activated once in each tail-beat cycle (e.g., Figure 7.4). Regional variations in muscle activation during swimming at different speeds have been noted in a few cases, for example, a posterior-to-anterior recruitment of red muscle fibers with increasing speed (Jayne and Lauder, 1995b; Gillis, 1998; Coughlin and Rome, 1999; Coughlin *et al.*, 2004), and a superficial-to-medial recruitment of white muscle fibers during sprint swimming (Jayne and Lauder, 1995c; Ellerby and Altringham 2001). With increasing speed, the activation duration (as measured by EMG burst times) decreases in proportion to the decreasing tail-beat period, such that the duty cycle (i.e., the fraction of T when muscle is active) at each axial location is relatively constant across a range of swimming speeds in fishes in which this has been examined (see Figure 7.12), even in sprints powered by white muscle (Ellerby and Altringham, 2001). An exception to this pattern was reported by Rome and Swank (1992), who found that EMG duty cycle in the anterior red fibers of scup decreased significantly with increasing swim speed (from 1.5 to 4 Ls^{-1}). Actually, such a decrease is predicted from *in vivo* contractile work loop experiments of fish muscle, in which stimulus duration was optimized to produce maximal power across a range of cycle frequencies, corresponding to those experienced in steady swimming (e.g., Rome and Swank, 1992; Johnston *et al.*, 1993), suggesting that fishes are not necessarily maximizing muscle power output in all circumstances, including in response to changes in temperature, perhaps as a tradeoff to maximizing contractile efficiency (see Chapter 6 for discussion).

In anguilliform fishes and sharks, EMG burst durations remain relatively constant along the body because activation onset and termination propagate at near-equal rates (see Figure 7.12; Grillner and Kashin, 1976; Williams *et al.*, 1989; Gillis, 1998; Donley and Shadwick, 2003; Donley *et al.*, 2005). In other species, EMG offset travels much faster than onset, and duty cycles may decrease substantially along the body (Figure 7.12). This

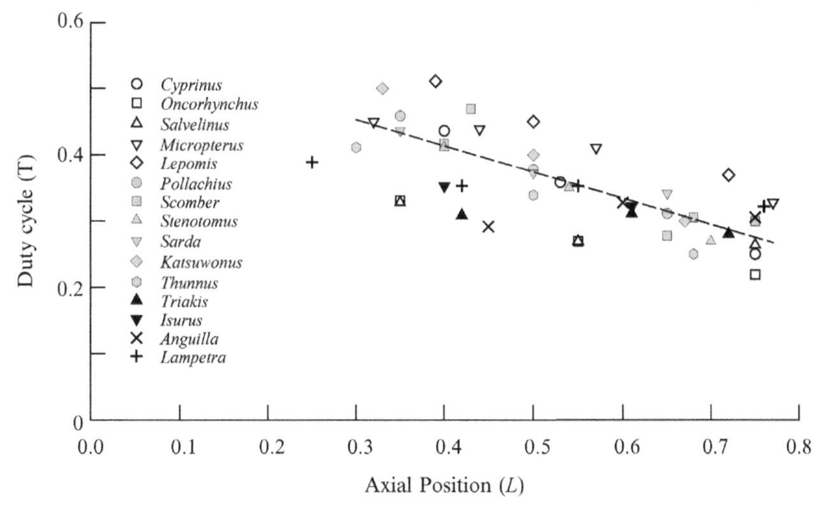

Fig. 7.12. EMG duty cycle (i.e., fraction of tail-beat cycle T that muscle is active) at different axial locations for a number of species. Regression is fit to data for subcarangiform (open symbols), carangiform, and thunniform teleosts (gray symbols), showing a steady decline in duty cycle from anterior to posterior. In contrast, duty cycle in sharks (black symbols) and anguilliform fish (×, +) is not correlated to axial position. Data sources: *C. carpio*, van Leeuwen *et al.* (1990); *O. mykiss*, Coughlin (2000); *S. fontinalis*, McGlinchey (2001); *M. salmoides*, Jayne and Lauder (1995b); *L. macrochirus*, Jayne and Lauder (1993); *P. viriens*, Hammond (1996); *S. japonicus*, Shadwick *et al.* (1998); *S. chrysops*, Rome *et al.* (1993); *S. chilensis*, Ellerby *et al.* (2000); *K. pelamis* and *T. albacares*, Knower *et al.* (1999); *T. semifasciata*, Donley and Shadwick (2003); *I. oxyrinchus*, Donley *et al.* (2005); *A. rostrata*, Gillis (1998); *L. fluviatilis*, Williams *et al.* (1989).

result, that muscle activation is longer in anterior than in posterior sites, may be reconciled with the observation that anterior fibers often have faster twitch kinetics; thus they deactivate and relax more rapidly at the end of shortening, so they can sustain longer duty cycles that generally enhance work output (see Chapter 6). Furthermore, species in which duty cycle is relatively constant along the body (eel and sharks) have no apparent axial variation in muscle contractile properties (D'Août *et al.*, 2001; Donley, 2004).

Another important finding from EMG studies is that the onset of the muscle activation propagates along the body faster than the wave of curvature or muscle shortening (summarized in Altringham and Ellerby, 1999, Coughlin, 2002), demonstrating that the temporal relation between muscle strain and neuronal activation varies with position along the body in the same fiber type (e.g., Figure 7.13). The extent to which this occurs is

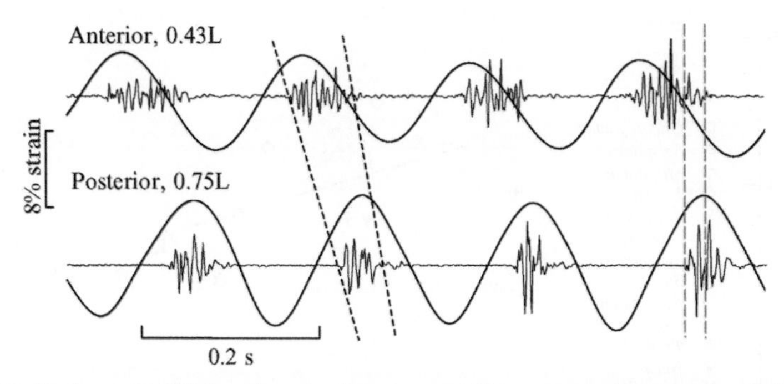

Fig. 7.13. Red muscle EMG and strain (based on midline curvature) at anterior (0.43*L*) and posterior (0.75*L*) locations in a mackerel swimming at 3.0 Ls^{-1} (R. E. Shadwick, unpublished). Diagonal dashed lines indicate the onset and offset of EMG, which occur later in time posteriorly, but earlier relative to the peaks of strain. Vertical dashed lines indicate a period in one tail-beat cycle when anterior muscle is actively shortening while posterior muscle is stretched as it is activated. Data of this form are used to determine the EMG-muscle strain phase relationships, shown in Figure 7.14.

summarized in Figure 7.14 for a variety of species, based on experiments like that shown in Figure 7.13, for steady swimming with muscle strain either measured directly or calculated from body midline curvature. It is important to note that the EMG burst reflects the period when muscle is stimulated by the nervous system, and that this differs from the periods of force development or shortening, which are delayed relative to EMG (see Figure 7.13 and Figures 6.5 and 6.7 in Chapter 6; also Altringham and Shadwick, 2001). Despite specific differences in activation patterns, some general features are evident.

i. All muscle is activated before reaching peak length *in vivo*. This suggests that muscle fibers are initially active while being lengthened (by shortening of antagonistic, contralateral fibers), a condition that should enhance force and work output (see Chapter 6).
ii. All muscle is deactivated while shortening. This is important in ensuring that relaxation is complete before the next cycle begins and in allowing cyclic contractions at higher frequencies than those predicted by isometric twitch kinetics (Altringham and Johnston, 1990).
iii. In all species, except sharks and skipjack tuna, posterior muscle fibers are activated and deactivated earlier in their strain cycle than are anterior fibers, despite being activated later in time (e.g., Figure 7.13). In many species this appears to be correlated with differences in twitch kinetics between anterior and posterior muscle (see Chapter 6;

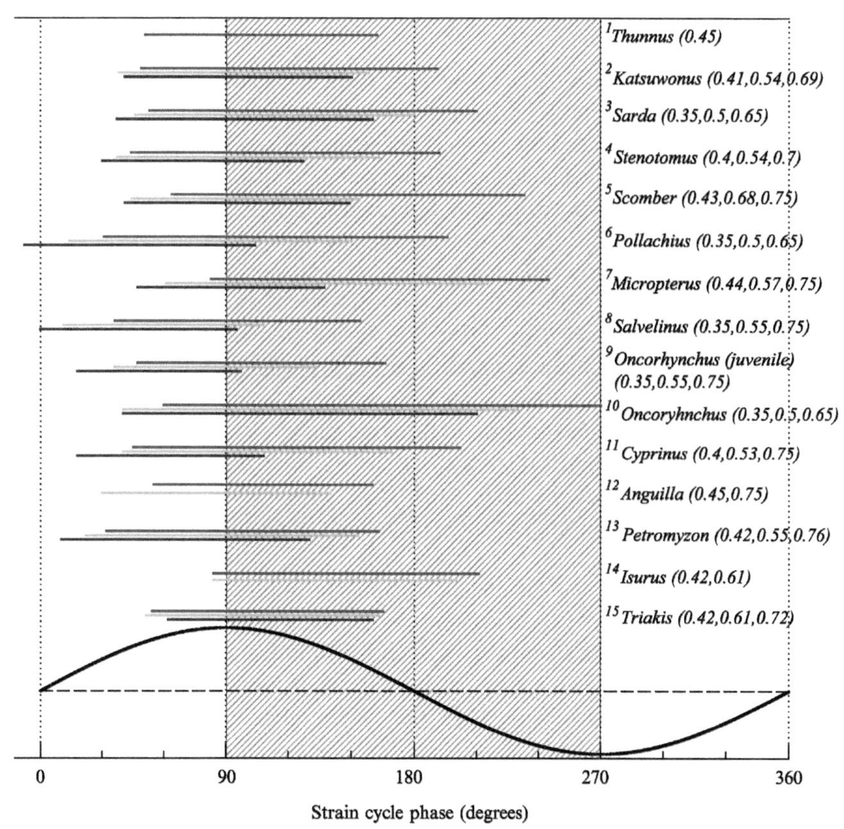

Fig. 7.14. Summary of the activation timing of red muscle in steady swimming. The lower curve represents one cycle of muscle strain, with peak length occurring at 90°, and shortening occurring between 90 and 270° (shaded area). Horizontal bars represent the duration of EMGs and their phase relation with strain. Anterior, middle, and posterior positions are represented by red, green, and blue bars, respectively. The axial locations (L) of EMGs are given after each species. Data sources: 1, yellowfin tuna, R. E. Shadwick, unpublished; 2, skipjack tuna, Shadwick *et al.* (1999); 3, bonito, Ellerby *et al.* (2000); 4, scup, Rome *et al.* (1993); 5, chub mackerel, Shadwick *et al.* (1998); 6, saithe, Hammond (1996); 7, bass, Jayne and Lauder (1995b); 8, brook trout, McGlinchey *et al.* (2001); 9, rainbow trout, juvenile, Coughlin (2000), 10, rainbow trout, adult, Hammond *et al.* (1998); 11, carp, van Leeuwen *et al.* (1990); 12, eel, Gillis (1998); 13, lamprey, Williams *et al.* (1989); 14, mako shark, Donley *et al.* (2005); 15, leopard shark, Donley and Shadwick (2003).

Altringham and Ellerby, 1999; Coughlin, 2002). On the other hand, the similar activation onset phase in anterior and posterior red muscle of sharks correlates with uniform twitch kinetics of these muscles (Donley, 2004).

D. Axial Variations in Muscle Function

The phase and duration of muscle activation are primary determinants of how much work muscle fibers produce in cyclic contractions. Significant variations in activation timing among different species and, more importantly, along the rostral-caudal axis of individual fishes suggest that there may be important differences in the way lateral muscles are used during swimming. This problem has been studied largely by *in vitro* simulations of "swimming" muscle function, in which cyclic contractions or "work loops" are performed by isolated bundles of live muscle fibers, while strain amplitude, frequency, and activation timing are varied to maximize power output or to mimic *in vivo* operating conditions (e.g., Altringham and Johnston, 1990; Rome and Swank, 1992; Rome *et al.*, 1993; Hammond *et al.*, 1998). Various models of muscle function in different groups of fishes have resulted, and are discussed at length in several reviews (van Leeuwen, 1995; Wardle *et al.*, 1995; Shadwick *et al.*, 1998; Altringham and Ellerby, 1999; Coughlin, 2002) as well as in Chapter 6 of this volume. For example, Coughlin (2002) categorized muscle function for steady swimming on the basis of whether the majority of power is produced by muscle anteriorly, posteriorly, or uniformly along the body. Altringham and Ellerby (1999) pointed out that, despite axial differences in power production, notably a larger portion of energy absorbing lengthening or negative work in posterior myomeres, all work loop simulations of slow and fast swimming have shown that net power output is positive all along the body. In addition, in most fishes, except tunas and lamnid sharks, anterior muscles shorten with high power at the same time that posterior muscles are being stretched while active (see Figure 7.13), a feature that could have important implications for force transmission along the body. The possibility that stretching of posterior muscles while they are active may be used to control body stiffness and facilitate increases in the speed of the propulsive wave, and thus the fish, has been investigated by Long and colleagues (McHenry *et al.*, 1995; Long, 1998; Long *et al.*, 2002).

V. SPECIALIZATIONS IN THUNNIFORM SWIMMERS

The tunas (comprising 15 species of Scombrids in the tribe Thunnini) and the lamnid sharks (5 species in the family Lamnidae, including great white and mako) share several morphological and physiological specializations related to a fast and continuous swimming ability that underlies their emergence as apex pelagic predators over the last 60 million years or so (Bernal *et al.*, 2001a). In addition to endothermy that enhances muscle power output

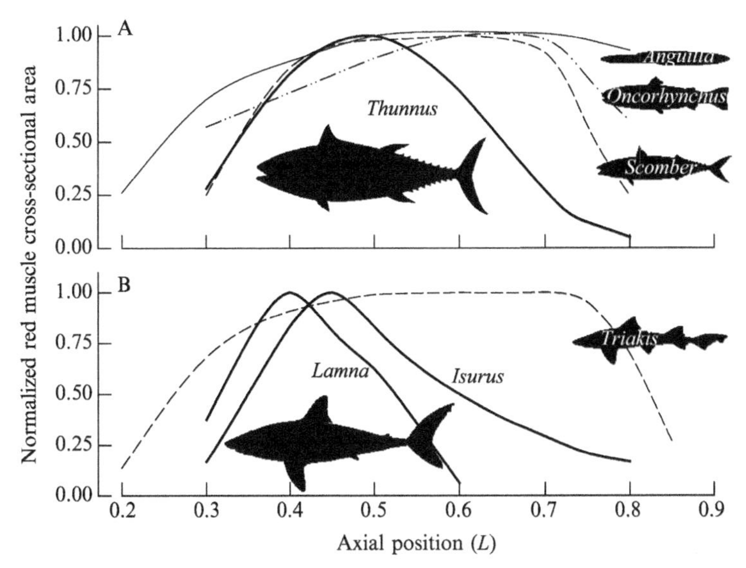

Fig. 7.15. Axial distribution of red muscle in bony fishes (A) and sharks (B). Red muscle cross-sectional area as a function of axial position is normalized so the maximum for each species is 1.0. Thus, these curves provide a comparison of red muscle distribution, but not total muscle mass. Most undulatory swimmers have a relatively uniform quantity of red muscle along much of the body, extending to the caudal peduncle. In contrast, thunniform swimmers (tunas and lamnid sharks) have highly tapered bodies with red muscle internalized and concentrated in the mid-body region. Data for *Thunnus* from Bernal *et al.* (2001a); *Scomber* from Shadwick *et al.* (1998); *Oncorhynchus* and *Anguilla* from Ellerby *et al.* (2000); *Lamna*, *Isurus*, and *Triakis* from Bernal *et al.* (2003).

(Altringham and Block, 1997; Syme and Shadwick, 2002; Donley, 2004; Bernal *et al.*, 2005), these fishes have thick, muscular bodies with a highly tapered posterior region ending in a narrow peduncle and stiff hydrofoil-like caudal fin that produces thrust by a hydrodynamic lift-based mechanism (see Figure 7.15; reviewed in Bernal *et al.*, 2001a; Altringham and Shadwick, 2001). This shape results in the bulk of the locomotor muscle being more centrally located, compared to most other fishes, and reduces the mass of the posterior portion of the body, where the lateral motion is greatest (Figure 7.15). In spite of the anterior shift in muscle mass, lateral motion of the mid-body region is greatly reduced, providing the kinematic definition of the "thunniform" swimming mode (Figure 11.1, Chapter 11). Moreover, the aerobic red fibers are shifted medially to occupy a position deep in the body, rather than superficially, as in other fishes (Figure 7.16). Recent studies on steady swimming and muscle-tendon structure have investigated

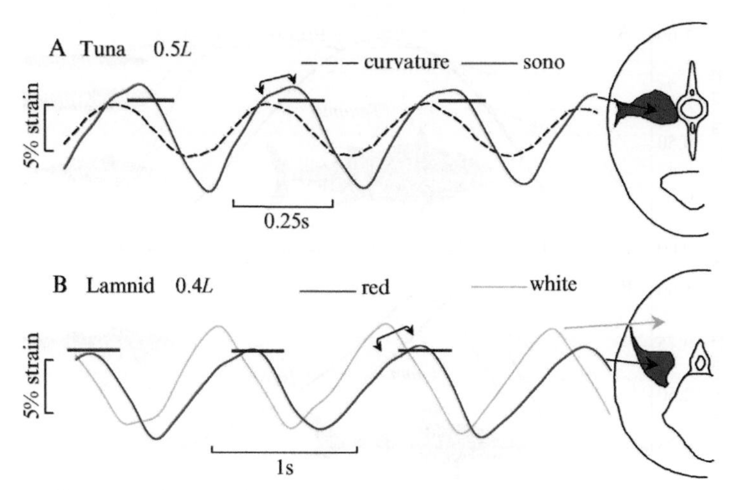

Fig. 7.16. (A) Muscle dynamics in thunniform swimmers, yellowfin tuna (*T. albacares*), and shortfin mako shark (*I. oxyrinchus*). Cross-sections at mid-body of each species indicate the internalized location of the red muscle loin. (A) In the tuna, strain in deep red muscle measured by sonomicrometry (red curve) is larger and occurs later in time than strain predicted from local body curvature (dashed curve). (B) In the lamnid mako shark, strain in active deep red muscle is delayed relative to strain in adjacent inactive white muscle, both measured by sonomicrometry. Blue bars indicate the periods of EMG activity. Bracket arrows indicate a period where red muscle is still lengthening while adjacent white muscle is shortening. (Adapted from Katz *et al.*, 2001; Donley *et al.*, 2005.)

the problem of how the anterior and medial positioning of red muscle in tunas and lamnids can result in lateral motion that is restricted to the caudal region, where little muscle is found (see Katz *et al.*, 2001; Westneat and Wainwright, 2001; Donley *et al.*, 2004). In fact, in both groups, the strain cycle of red muscle in the mid-body coincides with the lateral sweep of the tail tip, for example, shortening on one side begins when the tail tip is at its extreme to the opposite side and finishes when the tail tip is at its extreme on the side of contraction (Shadwick *et al.*, 1999; Donley *et al.*, 2005).

Initial measurements of red muscle EMG and body kinematics of yellowfin and skipjack tuna (Knower, 1998) yielded a paradoxical result when red muscle strain was assumed to be in phase with local midline curvature, as it is in other fishes. Specifically, activation onset at all axial locations appeared only after muscle shortening began (e.g., Figure 7.16A). Subsequently, direct measurements of red muscle strain by sonomicrometry demonstrated that, in fact, shortening of tuna deep red muscle is significantly phase delayed relative to local midline curvature (Figure 7.16A), such that shortening in mid body muscle is actually in phase with curvature 8–10 vertebral segments (or 0.15–0.2*L*) more posterior (Shadwick *et al.*, 1999; Katz *et al.*, 2001; Katz, 2002). Furthermore, strain amplitude of the deep red muscle is considerably larger than predicted from the curvature (Figures 7.8 and 7.16A). Because this muscle is physically uncoupled from local body curvature (i.e., from direct connections with adjacent skin, white muscle, and backbone) its close proximity to the backbone does not bestow a mechanical disadvantage, as might be expected, that would limit is ability to shorten and produce power (Bernal *et al.*, 2001a; Katz *et al.*, 2001). These measurements also showed that red muscle activation does begin before peak length and continues well into shortening, as is typical of other fishes (Figures 7.14 and 7.16A). Experiments on deep red muscle contractile mechanics using *in vitro* work loops (see Chapter 6) showed that the activation patterns observed *in vivo* yield maximal power output at both anterior and posterior locations (Syme and Shadwick, 2002), highlighting the specialization of tunas for continuous and steady locomotion.

Recent studies of kinematics and muscle dynamics in a lamnid shark, the shortfin mako (Donley *et al.*, 2004, 2005), revealed strong similarities to tunas that represent a remarkable evolutionary convergence for biomechanical design between these two groups of fishes. In particular, the deep red muscle fibers of the mako appear to function with the same physical uncoupling from local body bending as is found in tunas (Figure 7.16B), but slightly more pronounced in the more posterior locations. In slow swimming makos, deep red muscle contractions in the mid-body region are delayed by about 0.1–0.15T relative to the surrounding inactive white muscle and local curvature; shortening at 0.4*L* is in phase with midline curvature about 0.2*L* more posterior, while shortening at 0.6*L* is nearly 0.5T out of phase with local curvature (Donley *et al.*, 2005). As in tunas, activation timings recorded *in vivo* match those that produce maximal power in anterior and posterior red muscle, which have uniform contractile properties (Donley, 2004).

An interesting consequence of the physical uncoupling of deep red muscle from local curvature in tunas and lamnid sharks is that there are significant portions of each contraction cycle in which deep red muscle is still lengthening while the adjacent white muscle is being shortened, and vice versa (see Figure 7.16). Thus, a substantial degree of shearing between these muscle layers occurs, probably made possible by the rather loose intervening connective tissue, most pronounced in the lamnids (Bernal *et al.*, 2001a; Donley *et al.*, 2004; S. Gemballa *et al.*, submitted).

A major feature of the convergent design of tunas and lamnids is the unique muscle-tendon architecture that facilitates their thunniform swimming mode, allowing mid-body red muscle to effect large amplitude lateral motion in the caudal region rather than cause local bending. In principle,

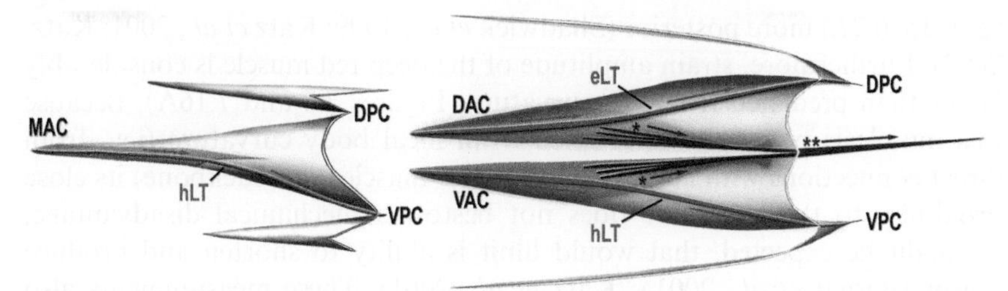

Fig. 7.17. Morphological basis for long-distance force transmission in thunniform swimmers, lateral view, anterior to the left. Two myosepta from the posterior body (0.55–0.80*L*) are shown. Pink color indicates insertion area of red muscle to myoseptum. Red muscle fibers insert either to anterior ends of lateral tendons (red; eLT, epaxial lateral tendon; hLt, hypaxial lateral tendon) or to anterior cone tendon (*, blue). Arrows indicate transmission of red muscle forces as suggested by these muscle-tendon associations. (A) Lamnid shark (*Isurus oxyrinchus*). The hLt is the only tendon associated with red muscle fibers. It spans a distance of 0.19*L* (see also Figure 7.2) between main anterior cone (MAC) and ventral posterior cone (VPC; Donley *et al.*, 2004; S. Gemballa *et al.*, submitted). (B) Tunas (based on *Euthynnus alletteratus*; S. Gemballa, unpublished). Tunas have a bifid anterior cone (dorsal and ventral anterior cone, DAC, VAC). The span of the eLT and hLT is 0.25*L* (see also Figure 7.2). In addition, red muscle forces in tunas might be directed posteriorly along the anterior cone tendons (*, blue) and the posterior oblique tendon (POT, blue, **) that connect to the vertebral column. Red arrows, LT pathway; blue arrows, ACT-POT pathway (S. Gemballa, unpublished).

this long-reaching force transmission seems to be facilitated by extreme myomere elongation (including myoseptal tendons) and, most importantly, by situating red muscle fibers deep in the anterior pointing myoseptal cones, where tendinous attachments, unavailable for superficial fibers, can span large numbers of body segments (Figure 7.17A,B; also Figure 7.2 and Sections II.A and II.B). Additionally, in tunas and lamnids, mediolateral tendinous fibers of the myosepta (e.g., ENT in Figure 7.1B) are poorly developed or lacking when compared to non-thunniform swimmers (S. Gemballa, unpublished).

Despite this convergence in the musculotendinous design of tunas and lamnids, the red muscle-tendon associations that provide force transmission over a large portion of the body are of different anatomical origin. In tunas, deep red muscle inserts onto the anterior portion of two tendons, the anterior cone tendon, and the lateral tendon (Figure 7.17B). These two red muscle-tendon associations are present both in epaxial and hypaxial parts. The hypaxial and epaxial anterior cone tendons join to POTs in the horizontal septum that are directed obliquely to posterior vertebrae (see Figure 7.17B). This is in contrast to the arrangement of superficial red muscle fibers in other teleosts, which insert onto the posterior portion of much shorter lateral tendons (Figures 7.2 and 7.3A) and link into POTs with much steeper angles to the vertebrae. A second pathway of deep red muscle force trajectory in tunas is via the myoseptal lateral tendons, which run from anterior to posterior cone tips (Figure 7.17B; for comparison see Figures 7.1B and 7.3). Again, due to their deep location in anterior cones, red muscle fibers insert onto the anterior portion of these long lateral tendons and may direct force posteriorly via subsequent posterior cone linkages. In the most posterior myomeres of tunas, the anterior cone tendons and lateral tendons coalesce to form the robust tail tendons that insert directly onto the caudal fin rays (Fierstine and Walters, 1968).

A striking similarity between tunas and lamnids is the elongation of the anterior myoseptal cone (including lateral tendons) in the posterior portion of the body (Figures 7.2 and 7.17). However, in lamnid sharks, the anterior cone is not bifid so red muscle inserts only onto hypaxial lateral tendons rather than onto both epaxial and hypaxial lateral tendons (Figure 17A). Internalized red muscle inserts onto the anterior portion of these tendons, which project caudally through white muscle to the posterior cones and into skin in the region of the peduncle (Donley *et al.*, 2004; S. Gemballa *et al.*, submitted). This is unlike the arrangement of superficial red muscle in non-lamnid sharks, and it seems likely to provide the morphological basis for transfer of red muscle forces to the caudal region, as the kinematic data suggest. Furthermore, lamnids differ from tunas in having no equivalent of POTs posteriorly. Though the hypaxial lateral tendons are involved in force transmission in both thunniform swimmers, the remaining tendons in tunas result in a more dorsoventrally symmetric musculotendinous design than in lamnid sharks.

VI. SUMMARY AND FUTURE DIRECTIONS

Our knowledge of fish swimming biomechanics has increased substantially in recent years. Apart from significant advances in quantifying external forces and flow fields involved in thrust production (see Chapter 11), other studies have attempted to link body kinematics with internal dynamics of the muscle, primarily the aerobic red fibers used in steady undulatory swimming. This has led to new ideas about how fishes control swimming, how muscle in different axial positions is activated, and how regional and species differences in muscle function contribute to the net power needed for swimming. New techniques in morphological investigations have yielded a wealth of information on the three-dimensional organization of the muscle-–tendon system in the lateral muscle of various fishes, to the extent that we can now see the anatomical basis for force transfer from muscle to the axial skeleton and caudal fin. A notable example is the recent work on lamnid sharks (Donley *et al.*, 2004), in which pathways of force transmission have been described by a combination of kinematics, red muscle dynamics, and morphology of the musculotendinous system. But experimental evidence for muscle-tendon interactions is almost completely lacking, mechanical properties of myoseptal tendons remain largely unknown (see Chapter 5), and direct force measurements on myoseptal tendons remain intractable. Future efforts to address these deficiencies are needed. Furthermore, new studies that integrate muscle dynamics, patterns of muscle activity in context with their insertion to myoseptal tendons, and kinematics at high spatial resolutions may lead to refined hypotheses or models of muscle-tendon interactions that drive locomotion. The least-studied problem is the use of the white muscle in fast and burst swimming. While considerable effort to analyze white muscle performance in fast starts has been made (see Chapter 9), very little data exist for white muscle regional activation patterns and strain during straight-line swimming (Katz *et al.*, 1999; Ellerby and Altringham, 2001). In particular, the three-dimensional recruitment (e.g., Jayne and Lauder, 1995c) and deformation patterns of white fibers as a function of swim speed, the contribution, if any, by red muscle at fast speeds, and the relation between power production in white fibers and maximal swimming speeds have not been investigated sufficiently. These important aspect of swimming biomechanics present major challenges for future investigators.

ACKNOWLEDGMENTS

R.E.S. thanks Doug Syme and Jeremy Goldbogen for helpful input, and the National Science Foundation for financial support. S.G. thanks John Long and Beth Brainerd for many valuable discussions, and the Deutsche Forschungsgemeinschaft and Wilhelm-Schuler Stiftung for financial support.

REFERENCES

Alexander, R. M. (1969). The orientation of muscle fibres in the myomeres of fishes. *J. Mar. Biol. Assoc. UK* **49,** 263–290.

Altringham, J. D., and Block, B. A. (1997). Why do tuna maintain elevated slow muscle temperatures? Power output of muscle isolated from endothermic and ectothermic fish. *J. Exp. Biol.* **200,** 2617–2627.

Altringham, J. D., and Ellerby, D. J. (1999). Fish swimming: Patterns in muscle function. *J. Exp. Biol.* **202,** 3397–3403.

Altringham, J. D., and Johnston, I. A. (1989). The innervation pattern of fast myotomal muscle in the teleost *Myoxocephalus scorpius*: A reappraisal. *Fish Physiol. Biochem.* **6,** 309–313.

Altringham, J. D., and Johnston, I. A. (1990). Modelling muscle power output in a swimming fish. *J. Exp. Biol.* **148,** 395–402.

Altringham, J. D., and Shadwick, R. E. (2001). Swimming and muscle function. *In* "Tuna. Physiology, Ecology and Evolution, Fish Physiology Vol. 19" (Block, B. A., and Stevens, E. D., Eds.), pp. 313–344. Academic Press, San Diego, CA.

Azizi, E., Gillis, G. B., and Brainerd, E. L. (2002). Morphology and mechanics of myosepta in a swimming salamander (*Sirenia lacertina*). *Comp. Biochem. Physiol.* **133A,** 967–978.

Bainbridge, R. (1958). The speed of swimming of fish as related to size and to the frequency and amplitude of the tail beat. *J. Exp. Biol.* **35,** 109–133.

Bernal, D., Dickson, K. A., Shadwick, R. E., and Graham, J. B. (2001a). Review: Analysis of the evolutionary convergence for high performance swimming in lamnid sharks and tunas. *Comp. Biochem. Physiol.* **129A,** 695–726.

Bernal, D., Sepulveda, C., and Graham, J. B. (2001b). Water-tunnel studies of heat balance in swimming mako sharks. *J. Exp. Biol.* **204,** 4043–4054.

Bernal, D., Sepulveda, C., Mathieu-Costello, O., and Graham, J. B. (2003). Comparative studies of high performance swimming in sharks I. Red muscle morphometrics, vascularization and ultrastructure. *J. Exp. Biol.* **206,** 2831–2843.

Bernal, D., Donley, J. M., Shadwick, R. E., and Syme, D. A. (2005). Mammal-like muscle powers swimming in a cold water shark. *Nature* (in press).

Blight, A. R. (1976). Undulatory swimming with and without waves of contraction. *Nature* **264,** 352–354.

Boddeke, R., Slijper, E. J., and van der Stelt, A. (1959). Histological characteristics of the body musculature of fishes in connection with their mode of life. *Proc. K. Ned. Akad. Wet. Ser. C* 576–588.

Bone, Q. (1966). On the function of the two types of myotomal muscle fibre in elasmobranch fish. *J. Mar. Biol. Assoc. UK* **46,** 321–349.

Bone, Q., Kiceniuk, J., and Jones, D. R. (1978). On the role of the different fibre types in fish myotomes at intermediate swimming speeds. *Fish. Bull. USA* **76,** 691–699.

Brill, R. W., and Dizon, A. E. (1979). Red and white muscle fibre activity in swimming skipjack tuna, *Katsuwonus pelamis* (L.). *J. Fish Biol.* **15,** 679–685.

Coughlin, D. J. (2000). Power production during steady swimming in largemouth bass and rainbow trout. *J. Exp. Biol.* **203,** 617–629.

Coughlin, D. J. (2002). Aerobic muscle function during steady swimming in fish. *Fish Fisheries (Oxford)* **3,** 63–78.

Coughlin, D. J., and Rome, L. C. (1999). Muscle activity in steady swimming scup, *Stenotomus chrysops*, varies with fiber type and body position. *Biol. Bull.* **196,** 145–152.

Coughlin, D. J., Valdes, L., and Rome, L. C. (1996). Muscle length changes during swimming in scup: Sonomicrometry verifies the anatomical high-speed cine technique. *J. Exp. Biol.* **199,** 459–463.

Coughlin, D. J., Spieker, A., and Shiavi, J. (2004). Red muscle recruitment during steady swimming correlates with rostral-caudal patterns of power production in trout. *Comp. Biochem. Physiol.* **137A,** 151–160.

D' Aoû t, K. D., and Aerts, P. (1999). A kinematic comparison of forward and backward swimming in the eel *Anguilla anguilla*. *J. Exp. Biol.* **202,** 1511–1521.

D' Aoû t, K., Curtin, N. A., Williams, T. L., and Aerts, P. (2001). Mechanical properties of red and white swimming muscles as a function of the position along the body of the eel *Anguilla anguilla*. *J. Exp. Biol.* **204,** 2221–2230.

Dewar, H. (1993). Studies of tropical tuna swimming performance: Thermoregulation, swimming mechanics and energetics. Ph.D. Thesis, University of California, San Diego, CA.

Dewar, H., and Graham, J. B. (1994). Studies of tropical tuna swimming performance in a large water tunnel III. Kinematics. *J. Exp. Biol.* **192,** 45–59.

Donley, J. M. (2004). Mechanics of steady swimming and contractile properties of muscle in elasmobranch fishes. Ph.D. thesis, University of California, San Diego, CA.

Donley, J. M., and Dickson, K. A. (2000). Swimming kinematics of juvenile kawakawa tuna (*Euthynnus affinis*) and chub mackerel (*Scomber japonicus*). *J. Exp. Biol.* **203,** 3103–3116.

Donley, J. M., and Shadwick, R. E. (2003). Steady swimming muscle dynamics in the leopard shark *Triakis semifasciata*. *J. Exp. Biol.* **206,** 1117–1126.

Donley, J. M., Sepulveda, C. A., Konstantinidis, P., Gemballa, S., and Shadwick, R. E. (2004). Convergent evolution in mechanical design of lamnid sharks and tunas. *Nature* **429,** 61–65.

Donley, J. M., Shadwick, R. E., Sepulveda, C. A., Konstantinidis, P., and Gemballa, S. (2005). Patterns of red muscle strain/activation and body kinematics during steady swimming in a lamnid shark, the shortfin mako (*Isurus oxyrinchus*). *J. Exp. Biol.* **208,** 2377–2387.

Ellerby, D. J., and Altringham, J. D. (2001). Spatial variation in fast muscle function of the rainbow trout *Oncorhynchus mykiss* during fast-starts and sprinting. *J. Exp. Biol.* **204,** 2239–2250.

Ellerby, D. J., Altringham, J. D., Williams, T., and Block, B. A. (2000). Slow muscle function of Pacific Bonito (*Sarda chilensis*) during steady swimming. *J. Exp. Biol.* **203,** 2001–2013.

Ellerby, D. J., Spierts, I. L. Y., and Altringham, J. D. (2001). Slow muscle power output of yellow- and silver-phase European eels (A*nguilla anguilla* l.): Changes in muscle performance prior to migration. *J. Exp. Biol.* **204,** 1369–1379.

Fierstine, H. L., and Walters, V. (1968). Studies in locomotion and anatomy of scombroid fishes. *Mem. South Cal. Acad. Sci.* **6,** 1–31.

Gemballa, S. (2004). Musculoskeletal system of the caudal fin in basal actinopterygians: Heterocercy, diphycercy, homocercy. *Zoomorphology* **123,** 15–30.

Gemballa, S., and Britz, R. (1998). Homology of intermuscular bones in acanthomorph fishes. *Am. Mus. Novitat.* **3241,** 1–25.

Gemballa, S., and Ebmeyer, L. (2003). Myoseptal architecture of sarcopterygian fishes and salamanders with special reference to *Ambystoma mexicanum*. *Zoology* **106,** 29–41.

Gemballa, S., and Hagen, K. (2004). The myoseptal system of Chimaera monstrosa: Collagenous fiber architecture and its evolution in the gnathostome stem lineage. *Zoology* **107,** 13–27.

Gemballa, S., and Röder, K. (2004). From head to tail: The myoseptal system in basal actinopterygians. *J. Morphol.* **259,** 155–171.

Gemballa, S., and Treiber, K. (2003). Cruising specialists and accelerators—Are different types of fish locomotion driven by differently structured myosepta? *Zoology* **106,** 203–222.

Gemballa, S., and Vogel, F. (2002). Spatial arrangement of white muscle fibers and myoseptal tendons in fishes. *Comp. Biochem. Physiol.* **133A,** 1013–1037.

Gemballa, S., Hagen, K., Röder, K., Rolf, M., and Treiber, K. (2003a). Structure and evolution of the horizontal septum in vertebrates. *J. Evol. Biol.* **16,** 966–975.

Gemballa, S., Ebmeyer, L., Hagen, K., Hoja, K., Treiber, K., Vogel, F., and Weitbrecht, G. W. (2003b). Evolutionary transformations of myoseptal tendons in gnathostomes. *Proc. Roy. Soc. Lond. B* **270,** 1229–1235.

Gemballa, S., Weitbrecht, G. W., and Sánchez-Villagra, M. R. (2003c). The myosepta in *Branchiostoma lanceolatum* (Cephalochordata): 3D reconstruction and microanatomy. *Zoomorphology* **122,** 169–179.

Gillis, G. B. (1998). Environmental effects on undulatory locomotion in the American eel *Anguilla rostrata*: Kinematics in water and on land. *J. Exp. Biol.* **201,** 949–961.

Goldbogen, J. A., Shadwick, R. E., Fudge, D. S., and Gosline, J. M. (2005). Fast-start dynamics in the rainbow trout *Oncorhynchus mykiss*: Phase relationship of white muscle shortening and body curvature. *J. Exp. Biol.* **208,** 929–938.

Graham, J. B., Dewar, H., Lai, N. C., Lowell, W. R., and Arce, S. M. (1990). Aspects of shark swimming performance determined using a large water tunnel. *J. Exp. Biol.* **151,** 175–192.

Gray, J. (1933a). Studies in animal locomotion I. The movement of fish with special reference to the eel. *J. Exp. Biol.* **10,** 88–104.

Gray, J. (1933b). Studies in animal locomotion II. The relationship between waves of muscular contraction and the propulsive mechanism of the eel. *J. Exp. Biol.* **10,** 386–390.

Gray, J. (1933c). Studies in animal locomotion III. The propulsive mechanism of the whiting (*Gadus merlangus*). *J. Exp. Biol.* **10,** 391–400.

Gray, J. (1968). "Animal Locomotion." W. W. Norton and Company Inc., New York.

Gray, J., and Lissman, H. W. (1964). The locomotion of nematodes. *J. Exp. Biol.* **41,** 135–154.

Greene, C. W., and Greene, C. H. (1913). The skeletal musculature of the king salmon. *Bull. U. S. Bur. Fish.* **33,** 21–60.

Grillner, S., and Kashin, S. (1976). On the generation and performance of swimming in fish. *In* "Neural Control of Locomotion" (Herman, R. M., Grillner, S., Stein, P. S. G., and Stuart, D. G., Eds.), pp. 181–201. Plenum Press, New York.

Hammond, L. (1996). Myotomal muscle function in free swimming fish. Ph.D. thesis, University of Leeds, Leeds, UK.

Hammond, L., Altringham, J. D., and Wardle, C. S. (1998). Myotomal slow muscle function of rainbow trout *Oncorhynchus mykiss* during steady swimming. *J. Exp. Biol.* **201,** 1659–1671.

Herskin, J., and Steffensen, J. F. (1998). Energy savings in sea bass swimming in a school: Measurements of tail beat frequency and oxygen consumption at different swimming speeds. *J. Fish Biol.* **53,** 366–376.

Hess, F., and Videler, J. J. (1984). Fast continuous swimming of saithe (*Pollachius virens*): A dynamic analysis of bending movements and muscle power. *J. Exp. Biol.* **109,** 229–251.

Hunter, J. R., and Zweifel, J. R. (1971). Swimming speed, tail beat frequency, tail beat amplitude and size in jack mackerel, *Trachurus symmetricus*, and other fishes. *Fish. Bull.* **69,** 253–267.

Jayne, B. C., and Lauder, G. V. (1993). Red and white muscle activity and kinematics of the escape response of the bluegill sunfish during swimming. *J. Comp. Physiol. A* **173,** 495–508.

Jayne, B. C., and Lauder, G. V. (1994). How swimming fish use slow and fast muscle fibers: Implications for models of vertebrate muscle recruitment. *J. Comp. Physiol. A* **175,** 123–131.

Jayne, B. C., and Lauder, G. V. (1995a). Speed effects on midline kinematics during steady undulatory swimming of largemouth bass, *Micropterus salmoides*. *J. Exp. Biol.* **198,** 585–602.

Jayne, B. C., and Lauder, G. V. (1995b). Red muscle motor patterns during steady swimming in largemouth bass: Effects of speed and correlations with axial kinematics. *J. Exp. Biol.* **198,** 1575–1587.

Jayne, B. C., and Lauder, G. V. (1995c). Are muscle fibers within fish myotomes activated synchronously? Patterns of recruitment within deep myomeric musculature during swimming in largemouth bass. *J. Exp. Biol.* **198,** 805–815.

Johnson, T. P., Syme, D. A., Jayne, B. C., Lauder, G. V., and Bennett, A. F. (1994). Modeling red muscle power output during steady and unsteady swimming in largemouth bass. *Am. J. Physiol.* **267,** R481–R488.

Johnston, I. A., Franklin, C. E., and Johnson, T. P. (1993). Recruitment patterns and contractile properties of fast muscle fibres isolated from rostral and caudal myotomes of the short-horned sculpin. *J. Exp. Biol.* **185,** 251–265.

Kafuku, T. (1950). Red muscles in Fishes I: Comparative Anatomy of the Scombroid Fishes of Japan.. *Japanese J. Ichth.* **1,** 89–100.

Katz, S. L. (2002). Design of heterothermic muscle in fish. *J. Exp. Biol.* **205,** 2251–2266.

Katz, S. L., and Shadwick, R. E. (1998). Curvature of swimming fish midlines as an index of muscle strain suggests swimming muscle produces net positive work. *J. Theor. Biol.* **193,** 243–256.

Katz, S. L.., and Shadwick, R. E. (2000). *In vivo* function and functional design in steady swimming fish muscle. *In* "Skeletal Muscle Mechanics" (Herzog, W., Ed.), pp. 475–502. John Wiley and Sons, Ltd., Chichester, UK.

Katz, S. L., Shadwick, R. E., and Rapoport, H. S. (1999). Muscle strain histories in swimming milkfish in steady as well as sprinting gaits. *J. Exp. Biol.* **202,** 529–541.

Katz, S. L., Syme, D. A., and Shadwick, R. E. (2001). High speed swimming: Enhanced power in yellowfin tuna. *Nature* **410,** 770–771.

Knower, T. (1998). Biomechanics of thunniform swimming. Ph.D. Thesis. University of California, San Diego, CA.

Knower, T., Shadwick, R. E., Katz, S. L., Graham, J. B., and Wardle, C. S. (1999). Red muscle activation patterns in yellowfin (*Thunnus albacares*) and skipjack (*Katsuwonus pelamis*) tunas during steady swimming. *J. Exp. Biol.* **202,** 2127–2138.

Long, J. H. (1998). Muscles, elastic energy, and the dynamics of body stiffness in swimming eels. *Amer. Zool.* **38,** 771–792.

Long, J. H., Adcock, B., and Root, R. G. (2002). Force transmission via axial tendons in undulating fish: A dynamic analysis. *Comp. Biochem. Physiol. A* **133,** 911–929.

Long, J. H., Jr., Hale, M. E., McHenry, M. J., and Westneat, M. W. (1996). Functions of fish skin: Flexural stiffness and steady swimming of longnose gar *Lepisosteus osseus*. *J. Exp. Biol.* **199,** 2139–2151.

Lowe, C. (1996). Kinematics and critical swimming speed of juvenile scalloped hammerhead sharks. *J. Exp. Biol.* **199,** 2605–2610.

Magnuson, J. J., and Prescott, J. H. (1966). Courtship, locomotion, feeding, and miscellaneous behavior of pacific bonito (*Sarda chiliensi*). *Anim. Behav.* **14,** 54–67.

McGlinchey, S. M., Saporotti, K. A., Forry, J. A., Pohronezny, J. A., Coughlin, D. J. (2001). Red muscle function during steady swimming in brook trout *Salvelinus fontinalus*. *Comp. Biochem. Physiol.* **129,** 727–738.

McHenry, M. J., Pell, C. A., and Long, J. H. (1995). Mechanical control of swimming speed: Stiffness and axial wave form in undulating fish models. *J. Exp. Biol.* **198,** 2293–2305.

Müller, U. K., and van Leeuwen, J. L. (2004). Swimming of larval zebrafish: Ontogeny of body waves and implications for locomotory development. *J. Exp. Biol.* **207,** 853–868.

Müller, U. K., van den Huevel, B. L. E., Stamhuis, E. J., and Videler, J. J. (1997). Fish foot prints: Morphology and energetics of the wake behind a continuously swimming mullet (*Chelon labrosus,* Risso). *J. Exp. Biol.* **200,** 2893–2906.

Nursall, J. R. (1956). The lateral musculature and the swimming of fish. *Proc. Zool. Soc. Lond.* **126,** 127–143.

Patterson, C., and Johnson, G. D. (1995). The intermuscular bones and ligaments of teleostean fishes. Smiths. *Contr. Zool.* **559,** 1–85.

Pyatetskiy, V. Y. (1970). Kinematic swimming characteristics of some fast marine fish. *In* "Hydrodynamic Problems in Bionics, Bionika, Vol. 4," Kiev (translated from Russian, Joint Publications Research Service (JPRS) 52605, pp. 12–23. *Natl. Tech. Inf. Serv.*, Springfield, VA, 1971).

Raso, D. S. (1991). A study of the peripheral innervation and the muscle fibre types of *Ictalurus nebulosus* (Lesueur) and *Ictalurus punctatus* (Rafinesque). *J. Fish Biol.* **39,** 409–419.

Rayner, M. D., and Keenan, M. J. (1967). The role of red and white muscle in the swimming of the skipjack tuna. *Nature* **214,** 392–393.

Rome, L. C., and Sosnicki, A. A. (1991). Myofilament overlap in swimming carp. 2. Sarcomere length changes during swimming. *Am. J. Physiol.* **260,** C289–C296.

Rome, L. C., and Swank, D. (1992). The influence of temperature on power output of scup red muscle during cyclic length changes. *J. Exp. Biol.* **171,** 261–282.
Rome, L. C., Loughna, P. T., and Goldspink, G. (1984). Muscle fibre recruitment as a function of swim speed and muscle temperature in carp. *Am. J. Physiol.* **247,** R272–R279.
Rome, L. C., Funke, R. P., Alexander, R. McN., Lutz, G., Aldridge, H, Scott, F., and Freadman, M. (1988). Why animals have different muscle fibre types. *Nature* **335,** 824–827.
Rome, L. C., Funke, R. P., and Alexander, R. McN. (1990). The influence of temperature on muscle velocity and sustained performance in swimming carp. *J. Exp. Biol.* **154,** 163–178.
Rome, L. C., Sosnicki, A., and Choi, I. H. (1992). The influence of temperature on muscle function in the fast swimming scup. 2. The mechanics of red muscle. *J. Exp. Biol.* **163,** 281–295.
Rome, L. C., Swank, D., and Corda, D. (1993). How fish power swimming. *Science* **261,** 340–343.
Shadwick, R. E., Steffensen, J. F., Katz, S. L., and Knower, T. (1998). Muscle dynamics in fish during steady swimming. *Amer. Zool.* **38,** 755–770.
Shadwick, R. E., Kate, S. L., Korsmeyer, K. E., Knower, T., and Covell, J. W. (1999). Muscle dynamics in skipjack tuna: Timing of red muscle shortening in relation to activation and body curvature during steady swimming. *J. Exp. Biol.* **202,** 2139–2150.
Shadwick, R. E., Rapoport, H. S., and Fenger, J. M. (2002). Structure and function of tuna tail tendons. *Comp. Biochem. Physiol. A* **133,** 1109–1125.
Swanson, C., Young, P. S., and Cech, J. J. (1998). Swimming performance of delta smelt: Maximum performance, and behavioral and kinematic limitations on swimming at submaximal velocities. *J. Exp. Biol.* **201,** 333–345.
Syme, D. A., and Shadwick, R. E. (2002). Effects of longitudinal body position and swimming speed on mechanical power of deep red muscle from skipjack tuna (*Katsuwonus pelamis*). *J. Exp. Biol.* **205,** 189–200.
Tsukamoto, K. (1984). The role of red and white muscles during swimming of the yellowtail. *Bull. Jap. Soc. Scientif. Fish.* **50,** 2025–2030.
van Leeuwen, J. L. (1995). The action of muscles in swimming fish. *Exp. Physiol.* **80,** 177–191.
van Leeuwen, J. L., Lankheet, M. J. M., Akster, H. A., and Osse, J. W. M. (1990). Function of red axial muscles of carp (*Cyprinus carpio*)—Recruitment and normalized power output during swimming in different modes. *J. Zool.* **220,** 123–145.
Videler, J. J. (1993). "Fish Swimming." Chapman and Hall, London.
Videler, J. J., and Hess, F. (1984). Fast continuous swimming of two pelagic predators: Saithe (*Pollachius virens*) and mackerel (*Scomber scombrus*): A kinematic analysis. *J. Exp. Biol.* **109,** 209–228.
Videler, J. J., and Wardle, C. S. (1978). New kinematic data from high speed cine film recordings of swimming cod (*Gadus morhua*). *Neth. J. Zool.* **28,** 465–484.
Videler, J. J., and Wardle, C. S. (1991). Fish swimming stride by stride: Speed limits and endurance. *Rev. Fish Biol. Fisheries* **1,** 23–40.
Vogel, F., and Gemballa, S. (2000). Locomotory design of "cyclostome" fishes: Spatial arrangement and architecture of myosepta and lamellae. *Acta Zool.* **81,** 267–283.
Wardle, C. S. (1985). Swimming activity in marine fish. *In* "Physiological Adaptations of Marinc Animals, Symposia of the Society for Experimental Biology, Vol. XXXIX" (Laverack, M. S., Ed.), pp. 521–540. Company of Biologists Ltd., Cambridge.
Wardle, C. S., and He, P. (1988). Burst swimming speeds of mackerel, *Scomber scombrus* L. *J. Fish Biol.* **32,** 471–478.
Wardle, C. S., Videler, J. J., Arimoto, T., Franco, J. M.., and He, P. (1989). The muscle twitch and the maximum swimming speed of giant bluefin tuna, *Thunnus thynnusc. J. Fish Biol.* **35,** 129–137.

Wardle, C. S., Videler, J. J., and Altringham, J. D. (1995). Tuning in to fish swimming waves: Body form, swimming mode and muscle function. *J. Exp. Biol.* **198,** 1629–1636.

Webb, P. W. (1975). Hydrodynamics and energetics of fish propulsion. *Bull. Fish. Res. Bd. Can.* **190,** 1–59.

Webb, P. W., and Blake, R. W. (1985). Swimming. *In* "Functional Vertebrate Morphology" (Hildebrand, M., Bramble, D. M., Liem, K. F., and Wake, D. B., Eds.), pp. 110–128. Belknap Press of Harvard University Press, Cambridge, MA.

Webb, P. W., and Keyes, R. S. (1982). Swimming kinematics of sharks. *Fish. Bull.* **80,** 803–812.

Webb, P. W., Kostecki, P. T., and Stevens, E. D. (1984). The effect of size and swimming speed on locomotor kinematics of rainbow trout. *J. Exp. Biol.* **109,** 77–95.

Westerfield, M., McMurray, J. V., and Elsen, J. S. (1986). Identified motor neurons and their innervation of axial muscles in the zebrafish. *J. Neurosci.* **6,** 2267–2277.

Westneat, M. W., and Wainwright, S. A. (2001). Mechanical design for swimming: Muscle, tendon, and bone. *In* "Fish Physiology 19: Tuna—Physiology, Ecology, and Evolution" (Block, B. A., and Stevens, E. D., Eds.), pp. 272–313. Academic Press, San Diego, CA.

Westneat, M. W., Hoese, W., Pell, C. A., and Wainwright, S. A. (1993). The horizontal septum: Mechanisms of force transfer in locomotion of scombrid fishes (Scombridae, Perciformes). *J. Morphol.* **217,** 183–204.

Westneat, M. W., Hale, M. E., McHenry, M. J., and Long, J. H. (1998). Mechanics of the fast start: Muscle function and the role of intramuscular pressure in the escape behavior of *Amia calva* and *Polypterus senegalus*. *J. Exp. Biol.* **201,** 3041–3055.

Williams, T. L., Grillner, S., Smoljaninov, V. V., Wallen, P., Kashin, S., and Rossignol, S. (1989). Locomotion in lamprey and trout: The relative timing of activation and movement. *J. Exp. Biol.* **143,** 559–566.

Xu, G., Arimoto J., and Inoue, M. (1993), Red and white muscle activity of the jack mackerel. Trachurus japonicus during swimming. *Nippon. Suisan Gakkaishi* **59,** 745–751.

Yuen, H. S. H. (1966). Swimming speeds of yellowfin and skipjack tuna. *Trans. Am. Fish. Soc.* **95,** 203–209.

8

STABILITY AND MANEUVERABILITY

PAUL W. WEBB

I. INTRODUCTION

Studies of the relationships between fish form and swimming performance go back several millennia, an interest stimulated by the high speeds achieved by some fishes (Alexander, 1983). More recently, the agility of fishes and other aquatic animals has caught the attention of both biologists and those who seek inspiration from nature for human-designed vehicles (Bandyopadhyay, 2002; Fish, 2003). Not only are fishes highly maneuverable, but they also hover and move smoothly in spite of numerous perturbations arising from the surrounding water as well as an intrinsic tendency by many species to roll belly-up.

Fish Biomechanics: Volume 23
FISH PHYSIOLOGY

DOI: 10.1016/S1546-5098(05)23008-X

This high maneuverability and impressive stability result from the large number and diversity of control surfaces shared for both functions (Webb, 1997b, 2002a). Multiple control surfaces were established early in the evolution of fishes (Webb, 1997b, 2002a; Lauder and Drucker, 2003), and this in turn reflects the physical properties of the water (Daniel and Webb, 1987; Webb, 1988b). The density and viscosity of water are high compared to air. As a result, the mechanical power required for swimming increases rapidly with swimming speed, u, approximately as u^3. No single motor system can efficiently power organisms over the wide range of speeds and acceleration rates typical of fishes (Alexander, 1989; Webb, 1993, 1994, 1997a), which necessitates performance range fractionation with motor-effector "gears" or gaits. Each gait works over a portion of the total performance range, and successive gaits are recruited as needed to provide for the increasing power requirement as speed increases. Because the performance power range for fishes is large, many gaits driven by many propulsors are necessary (Webb, 1993; Webb and Gerstner, 2000). At the same time, fish carcass density is close to that of water. As a result, the net weight of a fish in water is always small compared to that on land. Thus, the high density of water relaxes the pervading role of gravity experienced in air and the need to provide body support during movement. As a result, the body and all appendages of fishes can be used for movement, including maneuvers, and controlling posture and swimming trajectories (Webb and Blake, 1985; Lauder and Drucker, 2003).

Sharing the large number of control surfaces for stability and maneuvering is one area in common between these functions. In addition, both stability and maneuvering primarily involve changes of state. These overlaps provide the starting point for this chapter. Nevertheless, stability and maneuverability also differ in terms of the special problems posed by each, including the ease of study and hence the state of current knowledge. These aspects are considered, as is, finally, the question of what is entailed in being both stable and maneuverable.

II. GENERAL PRINCIPLES

A. Definitions

Webster's Dictionary defines a stable system as one "designed so as to resist forces tending to cause motion or change of motion" and "designed so as to develop forces that restore the original condition when disturbed from a condition of equilibrium or steady motion" (Anonymous, 1971). There are two major approaches for returning a system to an original condition. The first is self-correction, requiring only "unconscious attention" (Weihs, 1993). Here, a change in posture or trajectory results in modification of forces acting on the system, returning it to the original state. The second approach to corrections involves active creation of correction forces via sensory-motor regulatory pathways (see Table 8.1). Stability of fishes is ultimately dependent on the creation of control forces to make corrections as well as to help resist changes. As a result, stability is essentially dynamic equilibrium.

The distinctions among ways in which stability is achieved are important. For example, it is well known that a dead fish (or an anesthetized fish) tends to float belly-up, reflecting destabilizing hydrostatic forces acting on the body (see Figure 8.2). Therefore, a dead fish in a dorsal-side-up posture is unstable. As a result, it may be thought that fishes, dead or alive with the dorsal surface uppermost, are unstable. However, a *live* fish *creates* forces to resist changes in posture, including those due to destabilizing hydrostatic body forces, and to restore posture to its original condition should there be an unwanted change. As such, the *live* fish meets the criteria for a stable system. It is not self-correcting (*sensu* Weihs, 1993) and hence is clearly a controlled dynamic state.

Most fish behaviors involve maneuvers, which may be defined as "a series of changes in direction and position for a specified purpose (as in changing course ... or in docking)" (Anonymous, 1971). Maneuvers may be executed at a variety of rates, and animals making higher rate maneuvers in smaller volumes are defined as being more agile.

These definitions lead to the expectation that high stability and high maneuverability are mutually exclusive, a situation typical of human-engineered vehicles. Thus, the trajectory of a highly stable vehicle is hard to change, and, conversely, vehicles that are highly maneuverable tend to be less stable (von Mises, 1945; Goldberg, 1988; Marchaj, 1988). Such mutual exclusion is antithetical to the lives of many animals, and like them, fishes appear both stable and highly maneuverable.

B. Frames of Reference

Most disturbances, stabilizing corrections, and maneuvers involve changes of state. State is described by velocity, with magnitude u, and $0 \geq u \geq 0$. Changes occur in one or more of the following: position or speed as a result of linear acceleration, $du/dt \neq 0$; direction of the swimming trajectory due to an angular acceleration, $d\omega/dt \neq 0$, where ω is angular velocity; and change in position by rotation, $d\theta/dt \neq 0$, where θ is rotation angle.

State changes are defined in a frame of reference within the organism, the head and forward motion in the +x direction. Changes in state involve

Table 8.1
A Simple Classifications of Perturbations and Control Forces Affecting Stability and Maneuverability, and the Effectors Associated with Them[a]

Types of perturbations	
Hydrostatic	Hydrodynamic
Self-generated:	Self-generated:
Negative metacentric height	Ventilatory flow
Longitudinal separation of centers of mass and buoyancy	Locomotor movements
Gas inclusions	
External abiotic:	External biotic:
Density gradients (haloclines, thermoclines)	Other animals, e.g., school members
	External abiotic:
	Gravity and wind currents, and turbulence from their interactions with surfaces and projecting structures (including plants)

Types of control forces used in stability and maneuvers			
Hydrostatic forces	Hydrodynamic forces (drag, lift, and acceleration reaction)		
	Trimming: modification of flow arising from the environment or due to translocation of the body to which the control surface is attached		Powered: creation of flow over a control surface independent of environmental flow or that due to whole body translocation
	Self-correction (inherent stability)	Active modification of control surface orientation or shape for correction or steering maneuvers	Active whole-control-surface motion

Effectors: control surfaces	
Hydrostatic	Hydrodynamic
Proportions and distribution of body components, including gas inclusions	Body shape, ornamentation and head angle
	Dorsal and anal fins
	Pectoral and pelvic fins
	Caudal peduncle rotation and caudal fin

[a]Based on Webb, 2000.

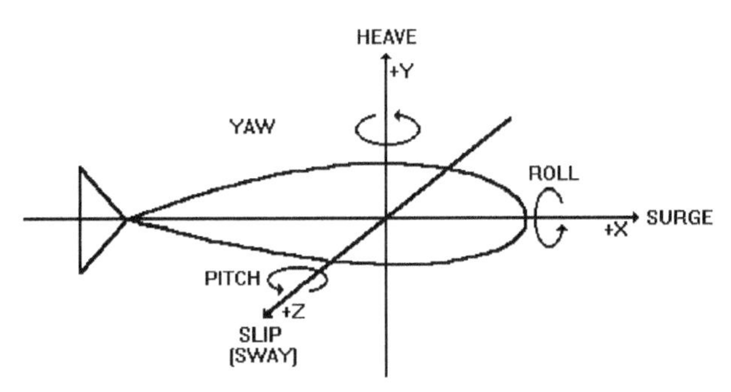

Fig. 8.1. The coordinate system in which changes of state are measured is fixed in the fish, with the head along the +x axis. Six degrees of freedom of motions are comprised of three translational motions, slip, heave, and surge, and three rotational motions, yaw, pitch, and roll.

whole body motions, for which free-body analysis of the forces and torques should be resolved about the center of mass. Changes in state for the center of mass in the organism frame of reference are defined in three translational planes and about three rotational axes (Figure 8.1). Translational changes in state are heave (a vertical displacement), slip (lateral), and surge (anterior-posterior). Rotational changes are pitch (vertical, head up/down rotation), yaw (left/right lateral horizontal rotation), and roll (rotation about the head-to-tail longitudinal axis).

Certain behaviors are considered to be maneuvers but do not involve changes of state. These are backward swimming (du/dt, $d\omega/dt$, and $d\theta/dt$ all may be zero) and hovering ($u = 0$, $du/dt = 0$, $d\omega/dt = 0$, $d\theta/dt = 0$). Both are major components of routine swimming, require coordination of multiple propulsors, are often slow and appear clumsy, and are energetically costly (Blake, 1983; Webb, 1997a). These features are shared with other stabilizing and maneuvering behaviors. Consequently, although it creates some awkwardness in terms of basic physical principles, biologists consider backward swimming and hovering to be maneuvers (Webb, 2002a, 2003; Walker, 2003).

C. Perturbations and Disturbances

Forces and torques tending to cause unwanted changes of state are defined as perturbations. A maneuver may be considered the result of an intentional change causing a destabilization but with desired magnitude and direction. Unwanted changes in state due to perturbations are defined as disturbances.

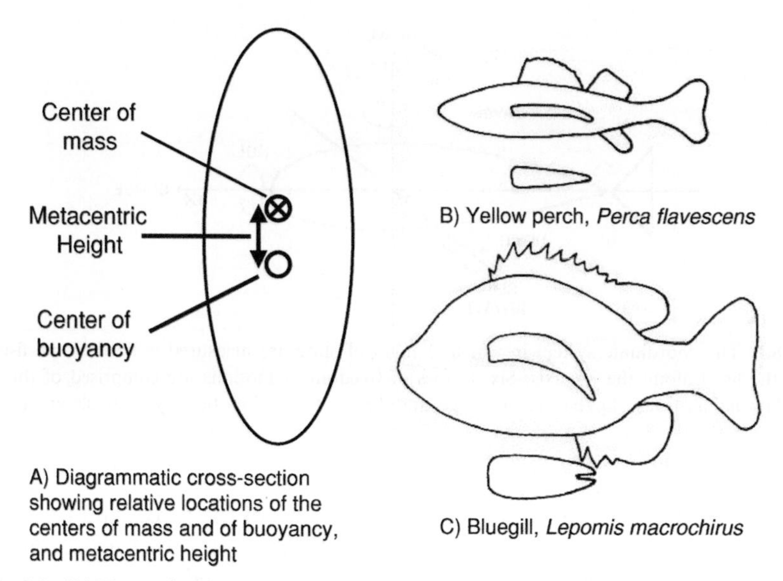

Fig. 8.2. (A) The vertical locations of the center of mass (enclosed cross), dorsal to the center of buoyancy (circle) are shown schematically in a fish cross-section. The vertical distance separating these centers is the metacentric height, which takes negative values when the center of mass is above the center of buoyancy. This arrangement is destabilizing because a rolling disturbance will create a rolling torque that adds to the perturbation. The longitudinal locations and shapes of the swimbladder of (B) yellow perch (*Perca flavescens*) and (C) bluegill (*Lepomis macrochirus*) are expanded anteriorly but taper and bend ventrally posteriorly. The swimbladder is restricted to the abdominal cavity, and its shape minimizes pitching torques. A large head-down hydrostatic torque is created by the dense, bony head that is opposed by the hydrostatic lift of the anterior, expanded portion of the swimbladder. Parts B and C from Webb and Weihs (1994).

Perturbations, disturbances, and the control forces used to achieve equilibrium and to drive maneuvers may be hydrostatic or hydrodynamic in origin (Table 8.1). Hydrostatic perturbations arise from density differences in the environment, for example, at thermoclines and haloclines, and from body composition. The latter has received considerable attention. Hydrostatic perturbations arising from the distribution of tissues with varying densities through the body have affected the evolution of fishes and the behavior and habitat use of modern fishes. Body composition determines the locations of the centers of the centers of mass and of buoyancy (Figure 8.2). These are separated by a vertical distance defined as the metacentric height (Goldberg, 1988; Marchaj, 1988). The center of mass typically lies above the center of buoyancy when the metacentric height is negative (Weihs, 1993; Webb and Weihs, 1994; Ullén *et al.*, 1995; Webb, 2002a; Eidietis *et al.*, 2003). In this configuration hydrostatic forces amplify rolling disturbances. The centers of mass and buoyancy also usually are separated longitudinally (Webb and Weihs, 1994) so that posture is stable only with the body pitched.

Gas inclusions are often used to regulate density, but these too are sources of destabilizing hydrostatic perturbations. Gas volume is inversely proportional to pressure. If a fish sinks, pressure increases, gas volume decreases, and overall density increases. Therefore, in the absence of intervention by a fish, a depth change is amplified.

Hydrodynamic perturbations are associated with flow, which generates forces and torques associated with drag, acceleration reaction, and lift (Daniel and Webb, 1987). These can cause disturbances or can be used for maneuvers.

Two general categories of hydrostatic and hydrodynamic perturbations can be recognized by their origins. Thus, perturbations may be self-generated or external to a fish (Table 8.1). Self-generated perturbations arise from body composition, gill ventilation, and movements of the body and fins for propulsion as well as from the production of control forces. External flows may be biotic or abiotic in origin. Biotic hydrodynamic perturbations occur during social behaviors in which a fish interacts with one or more fishes or other organisms (Webb and Gerstner, 2000; Webb, 2002a). Abiotic hydrodynamic perturbations arise from gravity and wind-driven currents, and particularly the turbulence these generate where flow interacts with fluid interfaces, surfaces, and projecting structures (Denny, 1988; Vogel, 1994; Bellwood and Wainwright, 2001; Fulton and Bellwood, 2002a,b; Webb, 2002a). Many structures built by humans are sources of turbulence, for example, water intakes, propeller wash, and ships in narrow channels.

D. Turbulence

External hydrodynamic perpetuations arise from turbulence. Studies of relationships between turbulence and fishes are relatively few, with the notable exception of many years of recently summarized Russian research (Pavlov *et al.*, 2000; see also Odeh *et al.*, 2002). Turbulence spans a large range of energies, amplitudes, and periodicities, and often is unpredictable. As a result, turbulence probably creates the greatest stability challenges for fishes. However, currents and turbulence may also provide benefits, such as mixing to make available oxygen and nutrients driving primary production, and for transport. Consequently, turbulence is beginning to receive more attention because unsteady flow is increasingly recognized to affect fish behavior, habitat choices, and distributions (Potts, 1970; Hobson, 1974; Fletcher, 1990, 1992; Hinch and Rand, 1998; Cada 2001; Hinch *et al.*, 2002).

1. Describing Turbulence

Turbulence is defined as "an irregular motion which in general makes its appearance in fluids ... when they flow past solid surfaces or even when neighboring streams of the same fluid flows past or over one another" (Hinze, 1975, p. 1, based on Taylor and Von Kármán, 1937). "The irregularities in the velocity field are certain spatial structures known as eddies" (Panton, 1984, p. 706). A special feature of turbulent flows is a continuous distribution of eddy sizes from the largest to the smallest eddy, called the Kolmogorov eddy or Kolmogorov microscale. The resulting eddy spectrum is known as the Kolmogorov spectrum.

In principle, turbulence can be described by the Navier-Stokes equations, but in practice the ability to do so for turbulence regimes experienced by fishes is limited (Sanford, 1997). The amount of turbulence is related to Reynolds number, Re:

$$\mathrm{Re} = ul/\upsilon, \tag{1}$$

where u = flow velocity, l = characteristic length, usually depth in lotic situations, and υ = kinematic viscosity (Sanford, 1997). Because of the chaotic nature of turbulence, it is often quantified in statistical terms (Gordon *et al.*, 1992; Nezu and Nakagawa, 1993; Vogel, 1994; Odeh *et al.*, 2002). At large Re, the statistical properties of turbulent flow are considered to depend on the rate of energy dissipation from large high-energy eddies with maximum size delineated by the physical dimensions of a system (e.g., river banks), to the smallest size, the Kolmogorov microscale at which inertial and viscous effects are equal (Kolmogorov, 1941). At this microscale, the energy of turbulence is finally dissipated as heat in the water (Sanford, 1997; Pavlov *et al.*, 2000).

The statistical properties of turbulence are becoming easier to measure in field situations using Doppler devices (Figure 8.3) and more recently ultrasound (Johari and Durgin, 1998; Johari, and Moreira, 1998; Desabrais and Johari, 2000). For example, observations of flow on an apparently calm day on a beach show unappreciated velocity variations. In the situation illustrated in Figure 8.3, velocity peaks occur along the x, y, and z axes of approximately^{-10} cm s^{-1}, with maximum resultant speeds as high as 40 cm s^{-1}, all around an average of 2.3 cm s^{-1}. The most common statistic description of this variation in flow velocity due to turbulence is the turbulence intensity, *TI*, which may be determined for flow in x, y, and z directions, or more commonly for the resultant flow (Sanford, 1997; Pavlov *et al.*, 2000; Odeh *et al.*, 2002):

$$TI = \sigma/u_{average}, \tag{2}$$

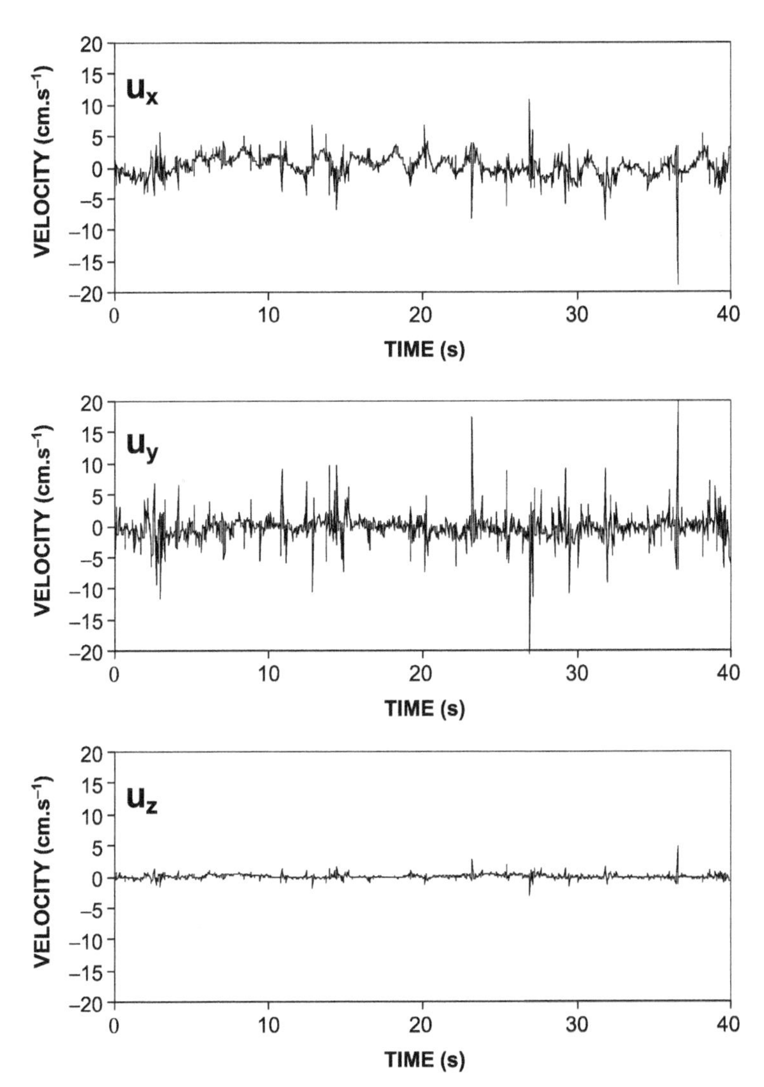

Fig. 8.3. Instantaneous velocities measured at 25 Hz using acoustic Doppler velocimetry (ADV) along x, y, and z axes, 2.2 cm above the bottom on a beach in water 25 cm deep. Measurements were made on a calm day, with wind $\ll 1$ km h^{-1} (P. W. Webb, unpublished observations).

where σ = standard deviation and u_{average} = mean speed. For the data shown in Figure 8.3, $TI = 1.0$. TI varies with the driving force, bathymetry, and distance from the substratum. For example, a windy day on a beach may increase the mean current speed by a relatively small amount, but the periodic flow due to wave action will greatly increase *TI* (Denny, 1988). In a small trout stream, current measured over typical transects varied from 0 to 41 cm s^{-1} and *TI* varied from 0.32 to 1.1. Low speeds and low *TI* occur at the bottom of pools, while the highest values of both are found in riffles (P. W. Webb, unpublished observations).

The effects of turbulence on fishes depend on variation in both the magnitude and the direction of flow. *TI* measures variation in velocity, but there is no standardized method for describing variation in the direction of the flow vector. Variation in azimuth and altitude may be useful (Batschelet, 1965).

Turbulence may result in regions of high shear stress that are sufficient to damage fishes (Pavlov *et al.*, 2000; Odeh *et al.*, 2002). Shear stress, τ, in turbulence may be written (Odeh *et al.*, 2002)

$$\tau = (\mu + \epsilon) du/dy, \tag{3}$$

where μ = viscosity, ϵ = eddy viscosity, and *du/dy* = velocity gradient. Eddy viscosity is not a physical property of a fluid but depends on the intensity of turbulence, and must be found by experiment.

Determining the forces and torques to which a fish is subject in turbulence requires more detailed information than that provided by statistical descriptions. Turbulent scale is related to Re, but understanding the mechanistic basis for fish responses and behavior in turbulence requires detailed analysis and modeling of the underlying vorticity. This necessitates measurement of flow over a finer grid than is possible with Doppler and ultrasound methods. Digital particle image velocimetry (DPIV) gives the necessary resolution. This method has been very successful in revealing details of the flows around swimming fishes (see most recently Müller *et al.*, 2001; Drucker and Lauder, 2002; Zhu *et al.* 2002; Lauder and Drucker, 2003; Tytler and Lauder, 2004; also see Chapters 10 and 11 in this volume). Currently, this method is restricted to the laboratory. Modeling has potential, and recent methods based on Boltzmann kinetic equations hold promise (Chen *et al.*, 2003).

Although turbulence is generally considered to be chaotic, this does not mean that there is no structure to such flows. For example, eddies often recur in fairly predictable ways; notable examples range from gyres in oceanic systems to vortex streets shed behind obstacles. At Re typical of flows experienced by juvenile and adult fishes, regular vortices are shed from elongate objects normal to the flow at a frequency, f_v, given by (Vogel, 1994) the following:

$$f_v = St\, u/d \tag{4}$$

where d = diameter, and St = the Strouhal number, which approximates 0.2 at high Re. At low Re, the Kolmogorov spectrum is smaller than at high Re. These eddies shed alternately from each side of many protuberances in a flow form a regular array stretching downstream, called a Kármán vortex street. These vortices are well defined with sharp transitions in velocity at the edges at lower Re. At large Re, micro-eddies within larger vortices increase the overall turbulent variation, and eddies tend to be less well defined.

2. Amplitude and Frequency

The importance of turbulence for fishes depends on the amplitude and period of perturbations. The importance of amplitude depends on eddy size relative to fish size, while the importance of period depends on eddy frequency relative to the response latency of the sensory-neural-muscular correction system. Amplitude and period are correlated, so that small eddies have small periods (high frequencies) and vice versa.

Small amplitude and period eddies cancel out over the body and are probably ignored (Shtaf *et al.*, 1983; Pavlov *et al.*, 1988, 2000). Indeed, this assumption is made when inducing microturbulence to create a rectilinear flow profile in flumes (Bell and Terhune, 1970).

By definition, intermediate eddies have sizes comparable to the dimensions of the body of a fish (Pavlov *et al.*, 2000; Odeh *et al.*, 2002). They are especially likely to cause disturbances in posture and are known to reduce swimming performance (Pavlov *et al.*, 1982, 1983). The shear stresses of intermediate-sized eddies are likely to cause injuries (Pavlov, 2000; Odeh *et al.*, 2002). At body-size scales, rotational disturbances appear to present larger control problems than translational disturbances (Webb, 2002a, 2004; Eidietis *et al.*, 2003). Motions following a rotational perturbation are more likely to be amplified (Kermack, 1943; Hoerner, 1975; Cruickshank and Skews, 1980; Bunker and Machin, 1991) and can quickly lead to "tumbling." In contrast, translations do not amplify, and correction is rarely pressing.

Turbulence-generated disturbances may occur simultaneously in several translational and rotational directions. As a result, a great deal of information must be integrated and appropriate coordinated responses sent to the many effectors generating control forces. The response latency between the onset of a disturbance and a damping or corrective motion is substantial, typically on the order of 60 ms for rolling and 200 ms for translational disturbances (Webb, 2004). These values are on the same order as latencies for initiating a maneuver to a novel situation, such as changing direction to chase fleeing prey (Webb, 1984). If the response latency approaches or exceeds half the period of a disturbance, an attempted correction will amplify that disturbance. The resulting destabilization quickly grows and is described as "pilot-induced error" (Anderson and Eberhardt, 2001).

Fish experience self-generated locomotor disturbances and external disturbances with periods at which pilot-induced error could result. For example, tail beats with frequency F increase with length-specific swimming speed, u/L, as (Bainbridge, 1958)

$$F = 1.33(u/L + 1) \tag{5}$$

Each beat of the tail generates two vortices, one at the start of each half of the stroke, so that disturbances are expected at $2F$. At a swimming speed of 3 Ls^{-1} (body lengths per second), when $F = 5.3$ Hz, thrust-related vorticity is shed with periodicity of 94 ms, and at 10 Ls^{-1}, the periodicity decreases to 34 ms. Similarly, median diameters of large woody debris in a small stream ranged from 4 to 14 mm in diameter (P. W. Webb, unpublished observations). Typical currents around such material were on the order of 30 cm s^{-1}. From Eq. (4), large woody debris with a diameter of 7.5 mm would shed vortices with a periodicity of 125 Hz, when pilot-induced error might occur, especially for translational disturbances.

The largest-sized eddies have large amplitudes and long periods and carry the most energy, but affect location rather than the stability of a fish. Turbulent eddies, for example, have important impacts on larval feeding, affecting contingency rates for this life history stage and its prey (Dower *et al.*, 1997). Many eddies with diameters much larger than fish body size include various currents. Some of these currents are repeating wave-induced surges. They may cause fishes to move backward and forward relative to the ground, but are essentially ignored by the fishes (P. W. Webb, unpublished observations). Sometimes currents are used to generate negative lift to stabilize posture on the substratum (Arnold *et al.*, 1991; Wilga and Lauder, 2001a). Choosing appropriate currents among those generated in turbulent flow can increase ground speed and/or reduce transport costs. For example, eddies in streams and open-water gyres assist migrations (Brett, 1995; Hinch and Rand, 1998; Hinch *et al.*, 2002). Other flow perturbations with large amplitudes and periods occur as spring freshets (Poff and Ward, 1989). These have high impacts on fitness, displacing individuals, washing out nests, or locally eliminating populations. Such events probably are not very predictable, and are likely to be avoided through seasonal behaviors or accommodated through evolution of adaptive life history traits (Matthews, 1998).

Overall, for a fish of a given size, there will be a range of eddy sizes on the same order as fish body size with the potential to cause large disturbances. Fishes may respond in two ways: they may correct for these disturbances or they may avoid them by choosing among habitats, to minimize energy costs of stability control and/or avoid perturbations causing disturbances exceeding their correction abilities. At the same time, there are ranges of eddy sizes that are beneficial to fishes, especially in providing transport, but also in bringing food and mates and dispersing gametes. Therefore, among intermediate-sized eddies, there is a range of sizes for which corrections can be made, and fishes would be expected to choose habitats within this range. The actual choice is expected to depend on the balance of costs and benefits so that turbulence features of chosen habitats vary among species, sex, life history stage, stress level, and other aspects of functional status (Dower *et al.*, 1997; Pavlov *et al.*, 2000). For example, Russian studies show that when given choices of flows with different *TI*, roaches, *Rutilus rutilus*, choose lower turbulence in the light than in the dark. Starved fishes choose higher *TI* than well-fed fishes, probably because higher flows with larger *TI* are likely to transport more food (Dower *et al.*, 1997; Pavlov *et al.*, 2000). Prolonged and sustained swimming speeds of perch, *Perca fluviatilis*, from more turbulent systems are less affected by turbulence than those of roaches, which in turn are less affected than those of gudgeon, *Gobio gobio*, typically found in least turbulent conditions among these species (Pavlov *et al.*, 2000).

E. Control Forces

Hydrostatic forces are sources of instability that tend to cause the body to roll and pitch. These disturbances are unavoidable and must be corrected. This necessitates the use of hydrodynamic forces to control stability, and the same forces that drive maneuvers.

Hydrodynamic forces arise from flow over the body and appendages and/or dynamically similar movement of an appendage through the water (Table 8.1). Hydrodynamic forces may be classified into two categories based on how a control surface moves relative to the body of which it is a part (Table 8.1). First, flow may arise from environmental currents, coasting, gliding, or other whole-body translocation due to operation of another propulsor. The flow generates forces defined as trimming forces (Webb, 1997b, 2002a), created by the orientation of a control surface redirecting momentum in the fluid (Webb, 1997b, 2002a; Wilga and Lauder, 2000, 2001a). Trimming forces may contribute self-correction. In these situations, a disturbance causes a change in the force generated by a trimmer that opposes the disturbance, thereby negating and correcting that disturbance (see Figure 8.6). The magnitudes and directions of trimming forces also may be changed actively by muscles altering the size, shape, and orientation of effectors either to correct disturbances or to drive maneuvers.

Second, control surfaces may move independently of the overall motion of the body and hence actively change the momentum of the surrounding fluid to generate forces. These actively generated forces are defined as powered forces (Webb, 1997b, 2002a). They may be generated by independent motions of control surfaces separate from a propulsor, or by changes in size, shape, orientation, and motions of the propulsor itself.

Various methods have been used to estimate the magnitude of control force production, primarily for maneuvering, but principles also apply to stabilizing systems. Traditionally, quasistatic principles have been used. This approach is still employed, with increased computing power permitting modeling of increasingly sophisticated motions and inclusion of some dynamic effects (see Blake, 1983; Videler, 1993; Ramamurti *et al.*, 2002; Walker, 2002). Such analyses highlight a challenge faced by researchers studying stability and maneuverability (Lauder and Drucker, 2003): the diversity of ways in which fishes can create control forces. For example, consider a lift-based control force, F_{L}, (Weihs, 2002):

$$F_{\mathrm{L}} \propto u^2(t)S(t)C_{\mathrm{L}\alpha}(t)\alpha(t)\beta(t), \tag{6}$$

where S = reference area, $C_{\mathrm{L}\alpha}$ = slope of the relationship between the lift coefficient and the angle of attack, α = angle of attack, and β = Wagner factor. The parameters of Eq. (6) are affected by many factors: material properties and muscular tuning of control surface size and shape, acceleration, speed, and orientation of the control surface as a whole, all of which can vary within a beat cycle, between beat cycles, among control surfaces, and during corrections and maneuvers (Walker and Westneat, 2000, 2002; Walker, 2002; Weihs, 2002). Many relevant variables have not been measured and/or are very difficult to measure, and/or the relevant equations are difficult to solve. The same factors may also affect the magnitudes of momentum transfers driving rectilinear swimming (Lauder and Drucker, 2003). However, these effects are small compared to driving swimming itself, and hence traditionally have been ignored. In contrast, subtle changes in finer-scale spatial and temporal features in the shapes and motions of propulsors and other control surfaces often are the critical factors leading to stability or the initiation of a maneuver.

For such reasons, new methods will probably prove important for understanding stability and maneuverability. DPIV holds promise for measuring net control forces for inherently unsteady motions of corrections and maneuvers (Stamhuis and Videler, 1995; Ferry and Lauder, 1996; Drucker and Lauder, 1999, 2002; Anderson *et al.*, 2001; Lauder and Drucker, 2003; see also Chapter 10 in this volume), while computational fluid dynamics provides the most useful modeling option (Triantafyllou *et al.*, 1993, 2000; Mittal *et al.*, 2002; Ramamurti *et al.*, 2002; Schultz and Webb, 2002).

F. Resistance

There are some whole-body momentum losses (resistance) associated with an initial state (except in hovering), and control for stability and maneuvers adds to these losses. Because stability and most maneuvers involve a change in state, the additional momentum costs are primarily inertial. As with all force production, there is a cost associated with changing the magnitude or the direction of momentum in the fluid to create the control force. These costs are well studied for lift generation by airfoils and hydrofoils, and are attributed to induced drag. Energy losses due to the creation of any control force by a fish may be large, so the concept may be generalized so that "induced drag" encompasses energy losses associated with generation of all types of momentum transfer, lift, drag, and acceleration reaction.

G. Effectors

Control forces and torques for stability and maneuverability are generated by variable control surfaces, moved in various ways, together constituting effectors (Table 8.1). The properties, numbers, and complexity of effectors have increased over evolutionary time, especially for those actively generating hydrodynamic forces.

In principle, self-generated hydrostatic forces due to body composition could be effectors for maneuvers, determining tilt when swimming at low speeds or the attack posture of an archerfish near the water surface. For example, seahorses can make a moderate change to the location of the swimbladder (Hans, 1951). Similarly, the amplification of depth changes of fishes with gas inclusions might be used to initiate maneuvers. It is not known if this potential of hydrostatic forces is realized. In general, hydrostatic forces are small compared with hydrodynamic forces, so it is unlikely that they make large contributions to maneuvering.

Hydrodynamic control forces are generated by all parts of the body, body shape, ornamentation, and appendages. The deployment of effectors to control motion in various planes and rotational axes is best known (qualitatively) for maneuvers (Table 8.2), but the same principles are expected to apply to stability.

The earliest chordates were characterized by a tail, supplemented by a fin fold in early fishes, but additional control surfaces were lacking. Therefore, the tail presumably generated all necessary control forces (Clarke, 1964). Many extant fishes have reduced appendages, including fishes specialized for burrowing, for example, hagfish, lampreys (Ullén *et al.* 1995; Deliagina,

Table 8.2
Fish Use Various Propulsors to Swim in Various Locomotor Patterns[a]

Locomotor pattern	Dynamic force type	Maneuver	Effectors and kinematics
Hovering and rotation. $u = 0$, $du/dt = 0$, $d\omega/dt = 0$, $d\theta/dt \geq 0$	Powered	Hovering	Small amplitude undulations of multiple pectoral, pelvic, dorsal, anal, and caudal fin-web effectors are necessary to avoid translocation. Hovering may use symmetrical effector use, but effector use is asymmetrical in rotation
Swimming with translocation	Trimming	Yawing turns	Steering torques from the head, tail, dorsal and anal fins, asymmetrical paired fin braking
$u > 0$, $0 \geq du/dt \geq 0$, $d\omega/dt \geq 0$, $d\theta/dt \geq 0$		Pitching	Steering torques from the paired fins, asymmetrical deployment of dorsal and anal fins and caudal fin web
		Rolling	Torques from the asymmetrical deployment of median and paired fins. Negative paired fin dihedral angle
		Heave and slip	Lift forces from the body and paired fins balanced about the center of mass.
		Deceleration surge	Braking by extension and expansion of paired and median fins. Curvature of median fins
	Powered	Yawing turns, pitching and rolling	Asymmetrical tail motions, including fast-start-type kinematics, asymmetrical median fin and pectoral fin beats (the inside fin beats more slowly, stops, or is furled). Orientation of propulsors, and motions. Changes in the properties of tail lobes
		Acceleration surge, heave and slip	Increased beat rates and changes in orientation of propulsors.
Backward swimming. $u < 0$, $0 \geq du/dt \geq 0$, $d\omega/dt \geq 0$, $d\theta/dt \geq 0$	Powered		Reversal of motions used for forward swimming
Fast starts. $u > 0$, $du/dt >> 0$, $d\omega/dt >> 0$, $d\theta/dt >> 0$	Powered	Rapid small radius yawing turns	Large amplitude, asymmetrical, non-repeating tail beats

[a]The table summarizes various types of effectors and their motions for different types of maneuver. It is expected that similar effector use, although less apparent, underlies control for stability. Based on Webb, 2000; Webb and Gerstner, 2000; Gordon *et al.*, 2001; Lauder and Drucker, 2003.

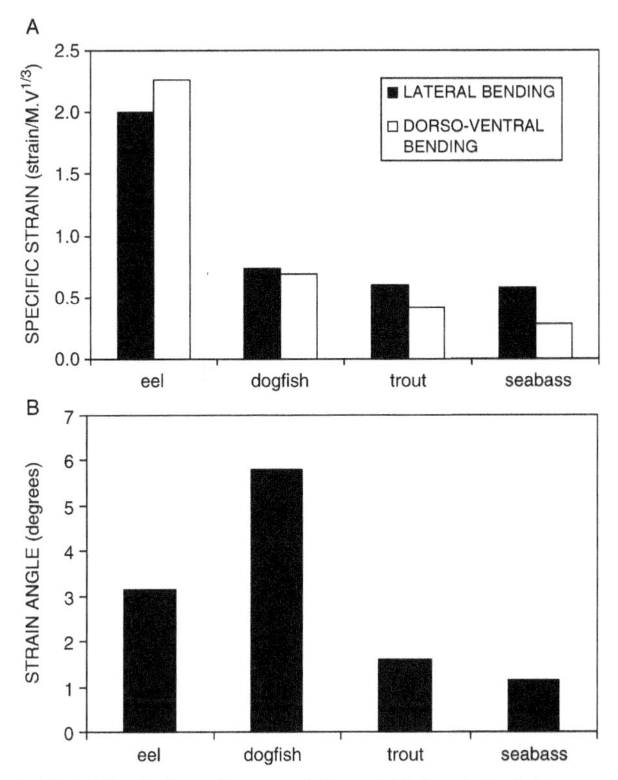

Fig. 8.4. Caudal flexibility is show for several fishes differing in body/fin organization. (A) Lateral and dorso-ventral flexibility are shown as the strain measured at the base of the caudal fin skeleton, normalized by volume$^{1/3}$ for a specific stress of 0.1. Stress was applied to the base of the tail, and the imposed stress was normalized by dividing by mass $\times$ volume$^{1/3}$. (B) Rotation was similarly determined for an imposed torque with stress of 0.01, using weights hung on a lever attached to the base of the caudal fin. Data are shown for eel, *Anguilla anguilla* (LeSeur), representative of fishes with reduced appendage control surfaces, hence reliant on the caudal fin for most control and maneuvers; lesser spotted dogfish, *Scyliorhinus canicula* (Linnaeus), representative of fishes with limited flexibility in appendage control surfaces; trout, *Oncorynchus mykiss*, representative of less-derived, soft-rayed teleosts; and North Sea bass, *Dicentrarchus labrax* (Linnaeus), representative of a more-derived acanthopterygian, spiny-rayed fish (P. W. Webb, unpublished data).

1997a,b), and eels, and larvae (Webb, 1994b). Such fishes probably face control problems similar to those of early chordates and fishes (Graham *et al.* 1987; Ullén *et al.* 1995; Deliagina 1997a,b) and are associated with an especially flexible caudal body/fin in bending and twisting (Figure 8.4) that can direct forces and couples in all directions.

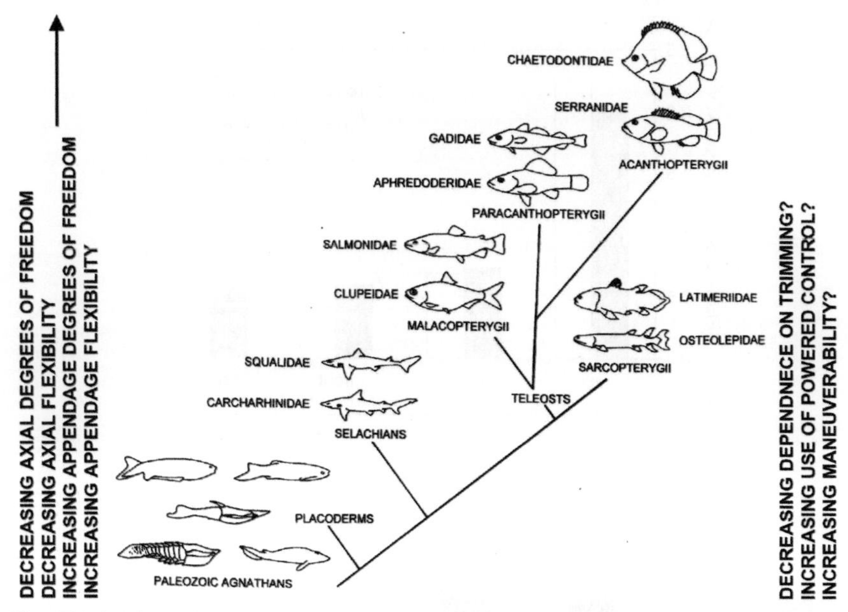

Fig. 8.5. A schematic representation of some major evolutionary patterns among fishes and postulated trends for general properties of control systems. Phylogeny based on Lauder and Liem (1983).

Early in their evolution, fishes evolved armor. Paleozoic agnathans (Figure 8.5) presumably had body densities larger than that of the surrounding water (Moy-Thomas and Miles, 1971). High density facilitates controlled station holding in flow (see later) and hence is most commonly associated with benthic living in current-swept habitats (Aleyev, 1977; Arnold and Weihs, 1978; Matthews, 1998). Among its many potential functions, high density associated with armor probably facilitated expansion of early fishes into highly productive lotic habitats.

Armor also heralded a conspicuous property of fishes: multiple surfaces capable of creating hydrodynamic control forces, often including ornamentation in early fishes. Recent studies on modern trunkfishes have shown that carapace edges and ridges can generate large self-correcting trimming forces during swimming (Bartol *et al.*, 2002, 2003), and the ornamentation on the armor of early fishes may have served a similar function.

Ornamentation gave rise to more discrete appendages (Moy-Thomas and Miles, 1971) that were distributed about the center of mass in orthogonal planes. Some of these surfaces became capable of varying their orientation independent of the body, leading to the evolution of mobile paired and median fins (Nursall, 1962; Jarvik, 1965; Moy-Thomas and Miles, 1971; Hobson, 1974; Aleyev, 1977; Bunker and Machin, 1991; Webb, 1994b).

By the early gnathostomes, the diversity of appendages became simplified into a basic body/fin plan of a fusiform, somewhat compressed or depressed body, with paired antero-ventral pectoral fins and postero-ventral pelvic fins, one or two dorsal median fins, an anal fin, and the caudal fin. The role of the various body and fin control surfaces is strongly affected by their locations relative to the center of mass; they probably functioned similarly to modern elasmobranchs (Harris, 1936, 1938; Alexander, 1965, 1967; Aleyev, 1977; Weihs, 1989, 1993, 2003; Ferry and Lauder, 1996; Wilga and Lauder, 1999, 2000, 2001a; Fish and Shannahan, 2000; Lauder and Drucker, 2003). Thus, the posterior paired fins and the caudal fin were posterior to the center of mass (Figure 8.6). Control surfaces in this location create self-correcting trimming forces in the same manner as the flight of an arrow. Thus, if a perturbation causes a head-up pitching disturbance in a swimming fish (Figure 8.6A), the posterior paired pelvic fins subtend a larger angle to the incident flow. This creates lift, causing the body to pitch head-down, correcting the disturbance. Similarly, the tail and other posterior median fins of early fishes could self-correct for yawing disturbances (Figure 8.5C). These control surfaces, especially the caudal fin, are especially distant from the center of mass in fusiform fishes, ensuring that trimming forces for steering maneuvers are large.

The opposite applies to anterior control surfaces, the pectoral fins, and the head. In these, a perturbation causing a head-up pitching disturbance would increase the angle of the pectoral fins to the flow. This would increase the lift generated by these fins and would add to the pitching disturbance (Figure 8.6B). Thus, the disturbance is amplified and such a system is unstable, similar to canard wings on human-engineered vehicles (Hoerner, 1975). The head similarly destabilizes fishes in yaw (Figure 8.6D). The degree to which the head acts as a control surface is not clear. Turns are often initiated by the head (Eaton *et al.*, 1977; Eaton and Hackett, 1984; Casagrand *et al.*, 1999; Webb and Fairchild, 2001), and head steering is likely to be more important among fishes with more flexible heads, typically more elongate species. Overall, control surfaces anterior to the center of mass in early fishes would cause self-amplifying disturbances promoting maneuverability (Weihs, 1989, 1993, 2002).

The arrangement of pairs of fins also affects their roles as effectors. Paired fins angled upward subtend a positive dihedral angle to the horizontal (von Mises, 1945). When such a fin pair rolls, for example, clockwise as in Figure 8.6E, the projected area in the horizontal plane of the fin on the side of the roll increases, and the lift force rotates toward the vertical plane. This increases the lift force counteracting the roll, promoting self-correction. The

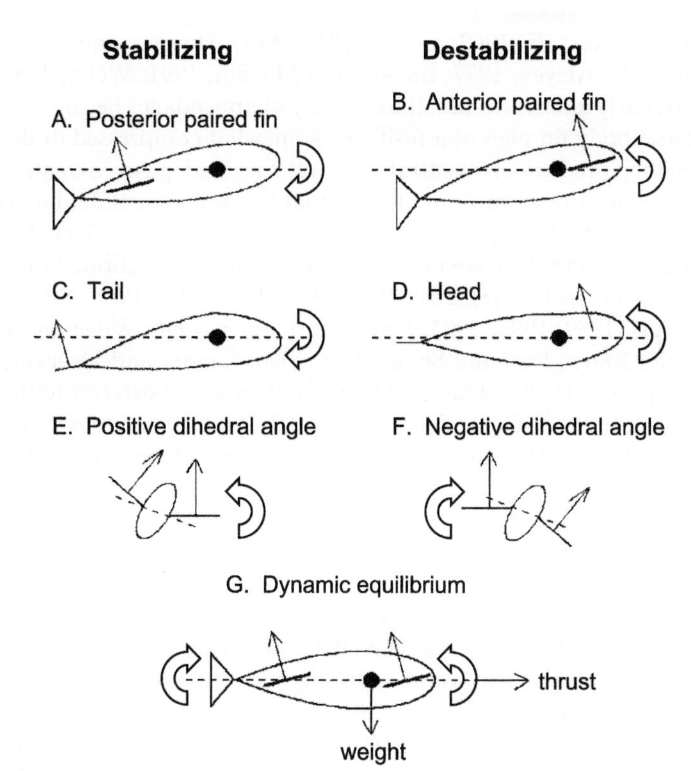

Fig. 8.6. Trimming roles of the body and median and paired fins in stabilizing or destabilizing swimming trajectories. In (A) and (B), a disturbance has caused the body to pitch head-up, increasing the lift produced by pelvic (A) or pectoral (B) fins. The body posterior or anterior to the center of mass functions in the same way as these fins. In both situations, lift is created or increases for each fin, as shown by the arrows. The lift of a posterior fin (A) creates a torque opposite in direction to the disturbance (block arrow), opposing the perturbation force causing that disturbance, thereby correcting the disturbance. This situation, with negative feedback, is stabilizing. The lift of an anterior fin (B) is in the same direction as the disturbance (block arrow), adding its torque to the perturbation, thereby amplifying the disturbance. This situation, with positive feedback, is destabilizing. The same principles apply to median fins and the body posterior (C) and anterior (D) to the center of mass. Here, the caudal fin, caudal peduncle, and often the dorsal and anal fins are stabilizing while the head is destabilizing. In (E) and (F), angling of pairs of fins, usually paired fins, here viewed from the posterior of the fish, affects control. In (E), a disturbance causes the body to roll. Because the fins angle upward (positive dihedral), the roll increases the projected area (the area seen from below) in the horizontal plane of the right fin on the side of the roll, decreasing that on the left fin. In addition, the lift of the right fin becomes more vertical, while that of the left fin subtends a larger angle to the vertical. As a result, the vertical component of lift of the right fin is larger than that of the left fin. This generates a net torque opposing the rolling perturbation (block arrow), correcting the disturbance. This situation is stabilizing. The opposite applies to fins with negative dihedral (F). Then the lift component on the left fin increases while that of the right fin decreases, generating a net

opposite occurs when the fins bend downward in negative dihedral or anhedral, which therefore tends to be destabilizing (Figure 8.6F).

Effectors also work together in groups. For example, the paired fins usually lie in the horizontal plane, orthogonal to median fins in the vertical plane. This organization, together with any body compression or depression, creates a large resistance to translational disturbances in heave and slip. While not self-correcting, this organization would damp the growth of disturbances (Aleyev, 1977). Jayne *et al.* (1996) and Lauder and Drucker (2003) recently showed that median as well as paired fins can generate yawing, rolling, and pitching torques for maneuvering. Anhedral angles of fin pairs also work together to amplify rolling disturbances and hence promote rolling maneuvers (Figure 8.6F). It also is well known that the placement of control surfaces can generate forces that balance the weight of water of a negatively buoyant fish. In this case, head-up pitching couples resulting from anterior control surfaces are cancelled out by head-down pitching couples produced by control surfaces posterior to the center (Figure 8.6G). This situation is expected for early fishes, as well as many modern selachians and acanthopterygian thunnids (Harris, 1936, 1938; Alexander, 1965, 1967; Bunker and Machin, 1991; Fish and Shannahan, 2000; Lauder and Drucker, 2003).

Because the fins and body form of early fishes, as well as many modern representatives, have control surfaces both anterior and posterior to the center of mass, posture and swimming trajectories depend on the balance between destabilizing and stabilizing forces. Therefore, swimming fishes again must be in dynamic equilibrium, and stability depends on (1) damping the growth of disturbances, facilitated by the body/fin organization, plus active tuning of trimming forces and (2) active generation of powered correction forces. Increased flexibility for active trimming and powering control forces is a hallmark of more-derived fishes, especially among the actinopterygians.

The earliest lifting appendages probably were not very flexible, and their range of movement was undoubtedly limited, except for the caudal body and tail (Moy-Thomas and Miles, 1971). A subsequent major evolutionary trend was reduction in the size of the fin base, greatly increasing fin mobility and expanding the degrees of freedom of motion of each control surface. Two

---

torque adding to the disturbance (block arrow). This situation is destabilizing. Fins may work in sets. In (G), the lift from anterior and posterior fins balances the weight of a fish in water, and because they work together, there is no net torque on the body (block arrows). Note, however, that lift of each paired fin is oriented posteriorly, creating a drag force, which is balanced by thrust production.

major approaches are found to increasing appendage mobility (Moy-Thomas and Miles 1971; Romer and Parsons, 1977; Fricke and Hissman, 1992; Clack, 2002). In sarcopterygians (Latimeridae and Osteolepidae in Figure 8.5), leading toward terrestrial vertebrates, the fin is supported by a limited number of basal elements projecting into the fin itself (Romer and Parsons, 1977; Clack, 2002). In contrast, the control surfaces of actinopterygians are supported by bony rays articulating on basal elements within the body (Rosen, 1982; Lauder and Liem, 1983). However, the many elements of the fin base are now known to provide substantial flexibility in the deployment of a fin (Drucker and Lauder, 2003; Lauder and Drucker, 2003; see also Chapter 10 this volume). For example, trout can rotate the fin base through angles of up to 30° (Lauder and Drucker, 2003).

The actinopterygian fin is a thin web supported by numerous fin rays that in turn are activated by one or more pairs of muscles (Arita, 1971; Gosline, 1971; Geerlink and Videler, 1974; Winterbottom, 1974; Harder, 1975; Videler, 1975; Aleyev, 1977; Lauder, 1989). This results in a highly flexible control surface. The shape of the compliant surface can adjust to fluid forces on the surface, providing self-correction described as "self-cambering" (McCutcheon, 1970). The fin muscles can tune the elasticity of the fin web (Geerlink and Videler, 1974; Videler, 1975; Lauder and Drucker, 2003), affecting self-cambering. As such, it is possible that differences in compliance and self-cambering are possible in fins anterior and posterior to the center of mass. This might restore some self-correction to the otherwise dynamic equilibrium implied by fin placement.

An impressive array of highly flexible, versatile effectors for stability and maneuvering is found among actinopterygians, illustrated by the soft-rayed teleosts (illustrated by the Salmonidae and Clupeidae malacopterygians in Figure 8.5). These fishes typically had paired antero-ventral pectoral fins, postero-ventral pelvic fins, dorsal, anal, and caudal fins, each with numerous degrees of freedom provided by the fin base and fin web (Eaton, 1945; Geerlink, 1987; Winterbottom, 1974; Harder, 1975; Lauder, 1989; Lauder and Drucker, 2003). However, starting with the Paracanthopterygii (illustrated by the Aphredoderidae and Gadidae in Figure 8.5), the pectoral fins moved to a lateral position and the pelvic fins moved forward (Rosen, 1982). The deep body of many acanthopterygian teleosts (illustrated by the Serranidae and Chaetodontidae in Figure 8.5) further permitted expansion of the dorsal and anal fins (Breder, 1926; Aleyev, 1977; Rosen, 1982; Gans *et al.*, 1997). This acanthopterygian body/fin organization is associated with larger, more mobile fins in orthogonal planes around and close to the center of mass. These features are associated with greater effectiveness in an expanded range of swimming gaits, especially in hovering and at low speeds (Lauder and Liem, 1983; O'Brien *et al.*, 1986, 1989, 1990; Walker, 2000;

Webb, 2002a). Each gait supports a range of maneuvers, so that the gait expansion also increased the range of maneuvers. In addition, more-derived actinopterygians may place greater reliance on powered dynamic stability control. For example, smallmouth bass, *Micropterus dolomieu*, attempt to stabilize the body while holding station in the wake of cylinders using powered corrective movements of the median and paired fins, while the soft-rayed cyprinid, chub, *Nocomis micropogon*, appear to rely more on trimming forces (Webb, 1998).

The evolutionary and developmental increase in the range and versatility of control surfaces in bony fishes are associated with decreased caudal flexibility (Figure 8.4). Bending and rotation of the caudal peduncle remains important in selachians (Thomson, 1976; Thomson and Simanek, 1977; Ferry and Lauder, 1996) and in cetaceans (Fish, 2002), which also have reduced and stiff control surfaces. In contrast, teleosts have numerous flexible appendicular control surfaces and show an evolutionary loss of tail flexibility through modification of the posterior skeleton (Alexander, 1967; Gosline, 1971; Moy-Thomas and Miles, 1971; Lauder and Liem, 1983; McHenry *et al.*, 1995; Long and Nipper, 1996). This trend is reversible, however, as teleosts with reduced control surfaces show high caudal flexibility, for example, eels (Ullén *et al.*, 1995; Deliagina, 1997a, 1997b) and knifefishes (Kasapi *et al.* 1993),

III. STABILITY

Fishes continuously experience perturbations necessitating control to correct the resulting disturbances. Among these, three categories of self-generated disturbances (Table 8.1) have received much attention because of their impacts on fish evolution and biology. These categories are control of body orientation (posture), depth in the water column, and trajectory.

A. Posture

Controlling posture ensures a stable base for sensory systems and minimizes energy costs during swimming by orienting the body to minimize resistance (Weihs, 1993; Eidietis *et al.*, 2003). Fishes usually hold the body vertically, dorsal side up, although pleuronectiform flatfishes (e.g., plaice in Figure 8.7) are well-known exceptions. In addition, posture varies during routine activity, for example, with feeding, breeding, gill irrigation at the air–water interface, and refuging. (Matthews, 1998).

As described previously, hydrostatic forces and torques arising from body composition make many fishes hydrostatically unstable. Eidietis *et al.* (2003)

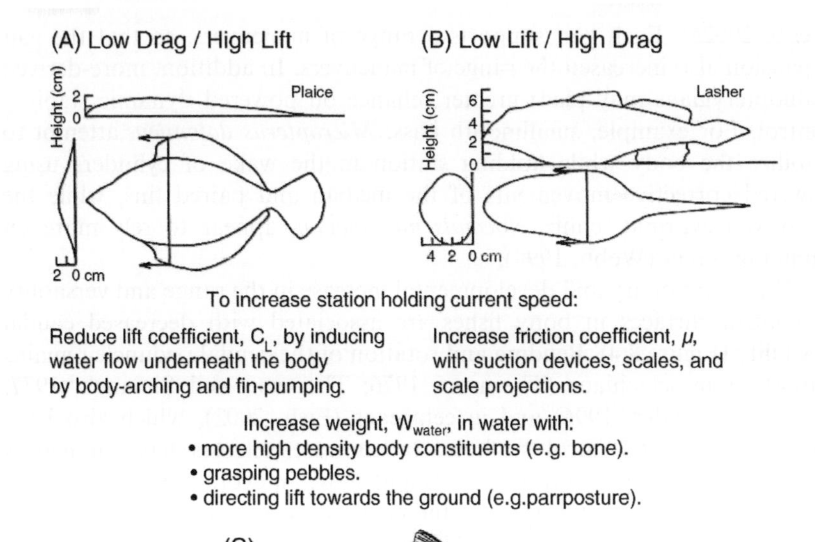

(A) Low Drag / High Lift
(B) Low Lift / High Drag
Plaice
Height (cm)
2
0
2 0 cm
Lasher
Height (cm)
6
4
2
0
4 2 0 cm
To increase station holding current speed:
Reduce lift coefficient, C_L, by inducing water flow underneath the body by body-arching and fin-pumping.
Increase friction coefficient, μ, with suction devices, scales, and scale projections.
Increase weight, W_{water}, in water with:
• more high density body constituents (e.g. bone).
• grasping pebbles.
• directing lift towards the ground (e.g.parrposture).
(C)

Fig. 8.7. Factors affecting station holding by benthic fishes on the substratum (based on Webb, 1989; Arnold *et al.*, 1991). Side, front, and plan views are shown for (A) plaice, a low-drag/high-lift form, and (B) lasher, a low-lift/high-drag form. Each of these modal forms uses different behavioral mechanisms to reduce lift or increase friction, respectively, thereby increasing the speed to which a fish can remain on the bottom without being displaced. All forms can further increase performance by modifying weight in water or creating lift directed toward the ground, increasing the ground reaction and hence fiction force resisting displacement. The parr posture (C) is the best-known behavior using up-curved pectoral fins to create such negative lift. Parts A and B from Webb 1989; part C from Arnold *et al.*, 1991.

added neutrally buoyant weight/float combinations to fishes, increasing rolling torques, to examine the limits to which fishes could correct for rolling perturbations. The threshold at which creek chub, *Semotilus atromaculatus,* were no longer able to correct rolling disturbances occurred with a 78% increase in the natural body rolling torque. This was significantly greater than the 43% and 34% increase in torque at which largemouth bass, *Micropterus salmoides*, and bluegill, *Lepomis macrochirus*, respectively, were no longer able to make corrections. The ability to control posture following increased rolling perturbations ranked in the same way as the ability of chub and bass to control posture and avoid swimming in wakes immediately downstream of vertical and horizontal cylinders (Webb, 1998).

Self-generated hydrostatic postural instability has affected the evolution of gas bladders used to regulate whole-body density (Lagler *et al.*, 1977). Gas inclusions in early osteichthyans were originally ventral to the gut. Modeling metacentric heights and associated rolling torques suggest that those of early fishes would have approached the limits at which corrections could be made (Webb, 1997b, 2002a). The lung migrated early to a more dorsal position to become the swimbladder of actinopterygians (Lauder and Liem, 1983), greatly reducing rolling torques (Webb, 2002a). In addition, the shape of the swimbladder evolved in such a way that pitching torques were reduced, but not eliminated (Figure 8.2). Thus, the gas volume is large anteriorly near the head, which has the highest density, tapering posteriorly (Aleyev, 1977; Webb and Weihs, 1994).

Differences among fishes in the ability to control posture appear to affect habitat choices. Creek chub, as the name indicates, are stream fishes, and like other soft-rayed species, are common in more turbulent riffles and races (Schlosser, 1982; Webb and Fairchild, 2001). Salmonids, also soft-rayed species, are well known for occupying turbulent streams. In lakes, cyprinids are abundant in shallow littoral zones, even during storm conditions (P. W. Webb, unpublished observations). These fishes make greater use of trimming forces for controlling posture and also may have shorter response latencies (Eidietis *et al.*, 2003; Webb, 1998, 2004). Alternatively, more powerful swimmers are likely to be able to marshal larger trimming and powered forces for control. Studies on wrasses (Labridae) have shown that the sustainable power these fishes can produce depends on the aspect ratio of their pectoral fins, the primary propulsors for routine swimming. Species with higher aspect ratio fins are common in more energetic flow situations (Bellwood and Wainwright, 2001; Fulton and Bellwood, 2002a,b). Alternatively, slow swimmers with greater reliance on powered control such as bluegill and ecomorphologically similar species are more common in lakes and ponds (Scott and Crossman, 1973), where flow and turbulence are greatly reduced. In these habitats, percomorphs tend to move offshore and avoid storm-induced turbulent situations (Helfman, 1981). Bass are found in both lotic and lentic habitats, but in slow-moving water in streams (Probst *et al.*, 1984).

B. Density and Depth

Control of depth is important for many activities, such as exploiting productive surface waters, vertical migrations for feeding and predator avoidance, and in benthic living (Allan, 1995; Diana, 1995; Matthews, 1998). The density of fish tissues is greater than that of water, so that a carcass sinks. There is an extensive literature on the use of low-density inclusions (gas, lipids, and in seawater, ion replacement) to reduce overall density. This buoyancy regulation damps, or in the case of neutral buoyancy eliminates, the self-generated sinking (heave) disturbances due to the carcass having a slightly higher density than the surrounding water (Aleyev, 1977; Gee, 1983; Alexander, 1990, 1993). Gas is widely used to control density because it provides the highest net upthrust for a given inclusion volume, but as noted previously, volume of the inclusions follows the gas laws, making fish hydrostatically unstable in depth regulation. Amplification is greatly reduced, essentially avoided, by lipid inclusions, but for a given hydrostatic lift, lipids require more volume, resulting in higher form drag during swimming.

Because inclusions are not self-correcting and often are destabilizing, depth control ultimately depends on hydrodynamic control forces. Body shape and posture (tilt), paired-fin orientation, caudal peduncle shape, and tail orientation all generate trimming forces (Harris, 1937a,b, 1953; Fierstine and Walters, 1968; Lauder, 1989; Bunker and Machin, 1991; Ferry and Lauder, 1996; Fish and Shannahan, 2000). These are supplemented or replaced by powered forces, especially at low swimming speeds (Ferry and Lauder, 1996; Wilga and Lauder, 2001b; Lauder and Drucker, 2003).

Some of the earliest work on depth and posture control was performed on the trimming function of the pectoral fins and hypochordal lobe of the tail of selachians (Harris, 1936; Kermack, 1943; Alexander, 1965). As noted previously (Figure 8.6), numerous studies have demonstrated how the pectoral fins (and to a lesser extent, the head) anterior to the center of mass, plus the hypochordal lobe of the tail posterior to the center of mass, together provide a lift force balancing weight in water while being neutral in pitching (Alexander, 1965, 1990; Fish and Shannahan, 2000). This "classical" view of trimming control of posture and depth (Ferry and Lauder, 1996; Wilga and Lauder, 2001b; Lauder and Drucker, 2003) has been questioned on the basis of recent studies using DPIV for certain fishes. The "classical" view may be adequate to describe depth and posture control for some fishes such as negatively buoyant thunnids (Gibb *et al.*, 1999). Others, for example, many elasmobranches, have flexible caudal peduncles (Figure 8.4), which facilitates orienting forces generated by the tail through the center of mass. This can provide the lift needed for depth control without causing pitching (Thomson, 1976; Thomson and Simanek, 1977; Ferry and Lauder, 1996). Thus, the DPIV studies show that the pectoral fins of selachians do not always generate lift, and the magnitude and direction of the tail force can be controlled to regulate depth and posture without causing pitching (Ferry and Lauder, 1996). However, rather than trimming and powered control being alternatives, fishes undoubtedly use combinations of both to regulate posture and depth, and the relative importance of each is likely to depend on the current conditions faced by a fish, as well as differences among habitats, species, and phylogenetic history.

Fishes use several hydrostatic and hydrodynamic mechanisms for depth control, and their relative importance correlates with behavior and habitat choices (Gee, 1983; Alexander, 1990). Inclusions increase total volume and hence form drag, which increases with swimming speed. Generation of lift, either as a trimming or a powered force, results in energy losses through induced drag that decreases as speed increases. As a result, many selachians and thunnids that swim continuously have reduced or lack low-density inclusions, and instead hydrodynamic forces are most important for the control of depth (Alexander, 1990).

Gas inclusions give high hydrostatic lift and are common in surface waters among slower swimmers, providing large depth changes are not necessary. Gas bladders incur substantial costs maintaining internal pressure equal to ambient pressure for fishes in very deep water (Gee, 1973). As a result, lipids are more common among slower, deep-water swimmers. Dilution of body tissues to reduce overall density in some deep-water marine species is made at a cost to muscle and skeleton. These fishes are sit-and-wait predators, with a high reliance on inertial suction feeding (Pietsch and Grobecker, 1978, 1990).

Many fishes are benthic when depth control involves to maintaining position on the substratum. In the absence of currents, this requires no more than some negative buoyancy. Fishes are stable when their weight is supported by at least three contact points. For example, in the parr posture (Figure 8.7C), named for salmonid parr but used by many benthic species (Arnold *et al.*, 1991; Wilga and Lauder, 2001a; Drucker and Lauder, 2003), the center of mass lies within the triangle of a posterior ground contact at the ventral surface or anal fin and two anterior contacts usually provided by the pectoral fins spread as props (Keenleyside, 1979; Arnold *et al.*, 1991).

Controlling posture to avoid whole body displacement becomes more difficult in a current. Performance depends on the normal reaction with the ground and its associated friction force opposing drag. The normal reaction depends on weight in water, which is reduced by a lift force resulting from water flow over the body (Arnold and Weihs, 1978). The interaction of these forces determines the speed at which displacement, a surge, is initiated. The maximum speed, u_{swim}, at which displacement occurs and a fish must start swimming to remain in place is given by (Arnold and Weihs, 1978) the following:

$$u_{\text{swim}}{}^2 = 2W_{\text{water}}/\rho S(C_{\text{L}} + C_{\text{D}}/\mu), \tag{7}$$

where W_{water} = weight in water, μ = friction coefficient, ρ = density of water, C_{D} = drag coefficient, and C_{L} = lift coefficient. It follows that station holding in a current is maximized by minimizing C_{D} and C_{L}, the area to which they apply, and increasing both W_{water} and μ (Figure 8.7). It is not possible to simultaneously reduce drag and lift coefficients. C_D is minimized by a streamlined, high aspect ratio "blister" form, such as the plaice (Figure 8.7A) (Hoerner, 1965; Arnold and Weihs, 1978). However, the need to maintain sufficient body volume for other functions results in shapes with low C_D becoming flattened and the S becoming large. This increases lift more than drag. Thus, benthic fishes tend toward either a low-drag/high-lift body form, such as in plaice, or a high-drag/low-lift body form, exemplified by benthic cottids such as the lasher, *Myoxocephalus scorpius* (Figure 8.7B) (Webb, 1989, 1994b). However, W_{water} and μ are increased by behavioral and morphological adaptations to ameliorate the inability to minimize both C_L and C_D. Thus, W_{water} can be increased to offset high C_L by reducing buoyant inclusions, increasing denser body constituents, and grasping pebbles. In addition, the overall C_L can be reversed (Figure 8.7C) by orienting the pectoral fins with the leading edge downward and the trailing edge elevated, thereby generating negative lift (Gee, 1973; Arnold and Weihs, 1978; Webb, 1989, 1990; Arnold *et al.*, 1991; Webb *et al.*, 1996b; Drucker and Lauder, 2003). To offset large C_D, μ is increased by suction devices, scales that may have projections, and grasping the substrate (Hora, 1937; Webb, 1989; MacDonnell and Blake, 1990). Finally, fishes may increase the effectiveness of all these mechanisms by choosing microhabitats with reduced flow, usually adjacent to high flows that bring benefits such as food, mates, and dispersal (Gerstner, 1998; Gerstner and Webb, 1998; Liao *et al.*, 2003).

C. Swimming Speed

Swimming fishes control posture and depth, and also the swimming trajectory itself. Disturbances in any of the three translational planes and about the three rotational axes along the swimming path would waste energy. Swimming speed affects how fishes meet control challenges. In general, as speed decreases, trimming forces become poorly matched to those needed to correct disturbances, while powered forces become energetically expensive. Thus, slow swimming presents the greatest stability problems, a situation familiar to any bicycle rider. First, trimming forces are primarily lift (Weihs, 1989, 1993, 2002), proportional to u^2 and effector area. Thus, trimming forces decrease as swimming speed decreases. Second, many disturbances are little affected by speed. Turbulence encountered by fishes varies greatly, and *TI* can be large even when flow is low (Figure 8.3). Self-generated disturbances may depend on factors independent of speed, such as increased ventilatory flows in hypoxic waters. Third, powered correction must make up for declining effectiveness of trimming forces at low speeds, and this requires the expenditure of additional energy. Control costs may be substantially increased by high induced drag, which is maximal at zero speed, in hovering (Blake, 1979; Lighthill, 1979).

In order to compensate for reduced effectiveness of trimming forces at low speeds, some fishes flex their fins to increase effector area (Bone *et al.*, 1995). Others pitch (tilt) at low speeds (He and Wardle, 1986; Webb, 1993; Ferry and Lauder, 1996; Wilga and Lauder, 1999, 2000), which increases the angle of attack of trimmers to the flow and hence increases lift, generating larger control forces to meet stability challenges at low speeds (Webb, 1993). Indeed, such dynamic control of equilibrium apparently explains an unexpected elevation of energy consumption and kinematics at slow swimming speeds. Thus, tilting increases resistance, which requires an increase in the rate of energy expenditure, and metabolic rates appear elevated at low speeds (Webb, 1993, 2002a). Translational mechanical energy losses are zero at zero speed, and tail-beat frequency would be expected to decline to zero at zero speed. It is well known, however, that this does not occur. The relationship between tail-beat frequency and swimming speed does not pass through the origin, and frequencies at low speeds are higher than expected. An outcome of control costs decreasing with increasing swimming speed and translocation costs increasing with increasing swimming speed is that the relationship between swimming speed and energy cost in any gait tends to be U-shaped, as found for tetrapod gaits (Alexander, 1989). Limited data support a U-shaped curve for fishes (Webb, 1997a).

Slow swimming is apparently challenging for control systems, and many fishes avoid low-speed swimming and rest on the bottom. When a fish rests on the bottom, postural and depth control problems reduce to those of holding station, as described previously.

While slow swimming presents challenges, fast swimming presents opportunities. One of these, described previously, is the benefits of declining induced drag with increasing speed, making trimming forces energetically advantageous for depth control. In addition, bony fishes furl their pectoral fins at higher swimming speeds. This eliminates a major destabilizing source (Figure 8.6) anterior to the center of mass, enhancing the self-correction potential of the fins posterior to the center of mass.

D. Swimming Trajectories

Many studies of swimming are made in flow situations with minimal perturbations (Bell and Terhune, 1970). Kinematic data are smoothed over several beats and selected for minimum speed variation (e.g., Webb, 2002b) to minimize possible effects of random and measurement noise (Harper and Blake, 1989). As a result, consideration of stabilizing swimming trajectories pertains to self-generated disturbances.

Self-generated disturbances arise during swimming because fishes, like other animals, use oscillating propulsors. Each propulsor creates thrust and some translation and/or rotational disturbance, defined as recoil (Lighthill, 1977). Propulsors or their elements reverse every half-beat, reducing the overall amplitude of a disturbance and correcting recoil. Nevertheless, damping is still desirable to minimize energy losses.

The intermittent working of each propulsor is damped in several ways. Momentum transfer involves lift, drag, and acceleration reaction (Daniel and Webb, 1987). Lift and drag are speed dependent, while acceleration reaction depends on acceleration rates. Drag and acceleration are generated parallel to the flow and lift is generated nearly normal to the flow. During a propulsor beat, acceleration and speed vary and are out of phase. In principle, therefore, overall force production can be smoothed over a fin beat, thereby reducing recoil disturbances, by varying the temporal contributions of drag, acceleration reaction, and lift through a beat cycle. This potential is realized in the elimination of heaving disturbances during pectoral-fin swimming in centrarchid sunfish (Gibb *et al.*, 1994).

Fishes have the musculature to control the shape and orientation of propulsor surfaces. Therefore, a second way to smooth force production is to control the relationship between heave and pitch of a propulsor (Lighthill, 1977; Lauder and Drucker, 2003), orienting forces to maximize thrust and minimize recoil. The finlets of many scombrids and thunnids are control surfaces that may also smooth force production by orienting flow upstream of the caudal fin over this propulsor (Nauen and Lauder, 2001). The high degree of freedom for propulsor motions and surface control also may result in the inclusion of many half-waves canceling slip forces within the length of undulatory, long-based pectoral, dorsal, and anal fins, and to some extent the whole body of anguilliform swimmers (Webb, 1975).

Third, propulsors may work in sets. Potential slip and yaw from pectoral fin propulsors cancel by the use of these fins working in pairs (Drucker and Lauder, 2000). Similarly, heave and pitch can cancel when the dorsal and anal fins work as a pair. More advanced control of trajectories is achieved with multiple fins working together, as in some more-derived teleosts. For example, in pufferfishes, boxfishes, and trunkfishes, multiple short-based median and paired fin motions are phased to negate recoil in all directions, contributing to remarkably smooth swimming trajectories (Gordon *et al.*, 1989, 1996, 2001; Hove *et al.*, 2001).

These mechanisms smoothing force production are often not sufficient to eliminate recoil. Low-speed swimming powered by oscillations of the pectoral fins causes periodic surge and often heave (Drucker and Jensen, 1996), while body/caudal fin swimming causes periodic surge, slip, and yaw (DuBois and Ogilvy, 1978). Periodic variations in power output mean that swimming speed varies to some extent during a propulsor beat-cycle, resulting in self-generated surge disturbances. As such, swimming is rarely steady in the true physical meaning of the term (Webb, 1988b); rather, "steady" swimming refers to translocation at a constant mean speed when time-averaged over successive fin beats.

The body/caudal fin is used to swim at the highest speeds, and also generates the largest recoil forces, causing slip and yaw disturbances. These could waste considerable energy, making damping their growth desirable (Lighthill, 1977; Webb, 1992). This is achieved by the combined inertia and hydrodynamic resistance to slip of the anterior of the body. Damping effectiveness is evaluated in terms of the ratio, ξ, between the tail side force, $F_{lateral}$, and the lateral resistance of the anterior body (Lighthill, 1977), resulting in:

$$\xi = F_{lateral}/k\omega\rho {B_{max}}^3 L, \tag{8}$$

where k = constant, ω = radian tail-beat frequency, and B_{max} = maximum depth of the body and median fins. Minimizing ξ, and hence improving damping, is strongly affected by B_{max}. This explains the anterior location of the dorsal fin common in fast swimmers. Compression of body shapes common in bony fishes also contributes to damping.

E. Performance

Few measures of stability control performance have been made. Ideally, known perturbations would be imposed, inducing various translational and rotational disturbances. This has yet to be adequately achieved for fishes. Eidietis *et al.* (2003) increased rolling torques with neutrally buoyant combinations of weights and floats from which they determined the limits at which three species could make corrections. Webb (1998) tested the abilities of two species to stabilize posture and hold position in the turbulence downstream of cylinders. Liao *et al.* (2003) did a similar experiment, in which trout held position between the two sets of vortices in a Kármán vortex street. The ability to hold station in ground contact has been studied by determining threshold speeds using the same increasing velocity tests that are a staple of swimming performance studies (Matthews, 1985; Webb, 1989; MacDonell and Blake, 1990; Webb *et al.*, 1996b).

IV. MANEUVERING

A. Maneuvering Patterns

Most fish behaviors involve maneuvers of various types. In simple swimming behaviors, for example, braking and periodic swimming maneuvers typically occur in a single plane or about a single rotational axis (Webb and Gerstner, 2000). Complex behaviors, typically seen as social behavior, involve more complex maneuvers combining translation(s) and/or rotation(s). The best-studied behavior is the fast start, involving a yawing turn simultaneously with translational acceleration in surge and slip (see Chapter 9 this volume). Fish in space-limited situations also often use complex maneuvers. For example, a fish faced with the end of a narrow blind channel may extricate itself by swimming backward, but many species do not reverse easily and then may roll onto the side, make a yawing turn (now in the vertical plane because of the roll), roll upright again, and finally swim upward to its original path (heaving and/or pitch) (Schrank *et al.*, 1999).

Effectors for maneuvers are the same propulsors used for swimming in each gait. Two gaits at opposite end of the performance range are also maneuvers (Table 8.2): hovering and fast starts. Hovering is a common component of routine swimming behaviors such as saltatory search for food (O'Brien *et al.*, 1986, 1989, 1990), vigilance on territories, nest guarding, and parental care (Keenleyside, 1979; Pitcher, 1986). Fast starts are characterized by peak acceleration rates $\leq 15\ g$, where g = acceleration due to gravity (Domenici and Blake, 1997), and are found in fitness-critical situations such as avoiding predators or entrainment; they are also used to strike at potential prey.

Between these extremes, low-speed swimming is achieved using median and paired fin propulsors, transitioning to body and caudal fin swimming at higher speeds. In both these major gait categories, maneuvers are executed by trimming or powering motions of the various median and paired fins and the body (Table 8.2). The median and paired fins tend to work symmetrically in pairs to execute translocation maneuvers and asymmetrically with or without the head and caudal body in maneuvers with rotation.

The most obvious translational maneuvers involve surge, for example, braking and periodic swimming (Weihs, 1973, 1974; Videler and Weihs, 1982). Braking is used to negotiate structurally complex habitats or to change the direction of a chase (Webb, 1984). Measured braking rates vary from -5 to $-155\ Ls^{-2}$ (0.1 to 1 g) (Webb, 1975; Videler, 1981; Geerlink, 1987; Jayne *et al.*, 1996; Webb and Fairchild, 2001). Until recently, it was thought that the lateral insertion of the pectoral fins and antero-ventral insertion of the pelvic fins of more-derived bony fishes provided superior braking (Harris, 1937a,b, 1953; Geerlink, 1987). Recent measurements of braking rates (Webb and Fairchild, 2001) and observations on fin deployment (Drucker and Lauder, 2002, 2003) have shown that less-derived soft-rayed fishes can decelerate as well as, or even better than, spiny-rayed fishes. However, braking by soft-rayed fishes includes heave and/or pitch disturbances, and these are controlled or eliminated by the spiny-rayed fish fin configuration (Harris, 1953; Drucker and Lauder, 2002, 2003). In this sense, the acanthopterygian body/fin organization is indeed superior, but in providing greater stability during braking. Data are lacking on selachians and cetaceans, whose lower fin flexibility may prevent good braking. It should be noted that braking, like many maneuvers, utilizes areal forces to modify body momentum, a volume force. As a result, braking becomes more difficult with increasing animal size. Large animals are likely to choose habitats, food sources, and complementary behaviors to reduce dependence on braking (Webb and de Buffrenil, 1990; Webb and Gerstner, 2000).

Periodic swimming uses alternating surges of acceleration and deceleration during body/caudal fin swimming. This reduces the average energy costs of swimming (Weihs, 1973, 1974; Videler and Weihs, 1982) because the mechanical costs of active propulsion are three to five times larger than those of coasting or gliding. Thus, the average cost of an acceleration/deceleration cycle proves to be lower than that for active swimming at the equivalent average speed.

Translational maneuvers include swimming backward, an important component of many agonistic behaviors, adjusting posture preparatory to taking food, and docking in small spaces (Breder, 1926; Keenleyside, 1979; Gerstner, 1999). Backward swimming involves reversal of the motions of propulsors used for forward swimming. However, morphological features promoting performance in forward swimming, such as stiffness distribution along the body length and the orientation of the fin base of bony fishes, have evolved largely for forward swimming. While backward swimming is not excluded, such propulsor morphology may not be ideal for reversing. As a result, backward swimming often appears awkward and usually is slow compared to forward swimming in the same gait (Figure 8.8). For example, goldfish, *Carrasius auratus*, can swim backward using median and paired fins at 0.6 Ls^{-1}, compared to forward swimming at 1.4 Ls^{-1}. Some fishes are adept at swimming backward. These fishes typically pass propulsive waves from the tail to the head as seen in anguilliform swimming, rostrally along the pectoral fins in rajiform swimming, and rostrally along the dorsal and/or ventral fins in amiiform, gymnotiform, and balistiform swimming. The antero-lateral location of the pectoral fins and more-vertical fin base of

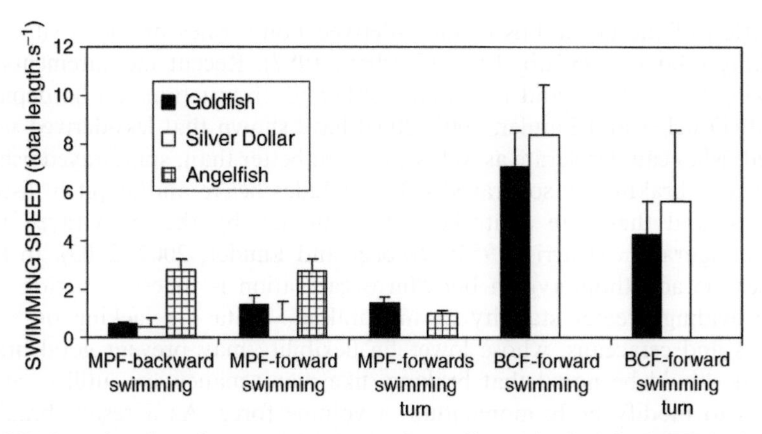

Fig. 8.8. Examples of maximum maneuver rates for median and paired fin (MPF) backward swimming and turns, and maximum forward speed in the absence of maneuvers. A similar comparison is made for body/caudal fin (BCF) swimming and turns. Data for three species, goldfish, *Carassius auratus*, silver dollars, *Metynnis hypsauchen*, and angelfish, *Pterophyllum scalare,* were taken from Webb and Fairchild (2001). Speeds in MPF turns are smaller than those during the BCF turns, as are the rectilinear speeds attained. Maximum swimming speed in a turn is usually reduced compared to that in the absence of a turn.

more-derived teleosts readily permits reversal of power and recovery strokes of paddles, and their backward pectoral fin swimming may be as fast as forward swimming (Webb and Fairchild, 2001). For example, angelfish, *Pterophyllum scalare*, swim backward at 2.8 Ls^{-1}, similar to maximum forward speeds of 2.7 Ls^{-1} (Figure 8.8).

Surging translational maneuvers such as braking and backward swimming are prominent in routine swimming. In addition, fishes may execute heaving maneuvers (Liao and Lauder, 2000; Wilga and Lauder, 2000; Drucker and Lauder, 2003) and are capable of beating their (usually) paired fins to slip. However, large or rapid changes in depth (heave) or slip involve rotational maneuvers.

Less is known about rotational maneuvers, except fast starts. Rotational maneuvers range from hovering with rotation to the turning in fast starts (Domenici and Blake, 1997; Walker, 2000; Domenici, 2001; Webb and Fairchild, 2001). Performance has recently been summarized by Walker (2003), and as with translational components, low rates occur in low-speed gaits. Turning rates increase from low values in median and paired fin, through higher values in body caudal fin gaits (Figure 8.8), to the highest rates of turn in fast starts.

B. Performance and Scaling

The maximum performance possible with a given propulsor system has been important in understanding the functional design of locomotion systems. Measuring performance for most control motions is proving challenging. Maneuvers are intentional instabilities, whose amplitude is large compared to unwanted, damped, and usually rapidly corrected disturbances. Therefore, measuring maneuvering performance has received the most attention.

1. Whole-Body Performance

Performance in changes of state in maneuvers are quantified about the center of mass by one or more of linear acceleration, $du/dt \neq 0$, angular acceleration, $d\omega/dt \neq 0$, and rotation, $d\theta/dt \neq 0$. This list of possible metrics for changes of state shows that maneuvers often cannot be quantified by a single measure, as is usual, for example, for swimming speed (Weihs, 2002). Braking, backward swimming, and rotation can be quantified by the magnitude of $-du/dt$, $-u$, and $d\theta/dt$, respectively. Otherwise, multiple metrics are the norm for describing both the rate of maneuver (often the swimming speed at which a maneuver is executed) and the space (e.g., turning radius) within which it is made (Webb, 1997b).

In addition, performance should be compared for equivalent limiting physiological and mechanical situations (Drucker, 1996; Drucker and Jensen, 1996; Webb and Gerstner, 2000). These occur at gait transitions (Alexander, 1989; Webb, 1994b; Drucker, 1996; Drucker and Jensen, 1996; Webb and Fairchild, 2001), for example, where fishes shift from paired fin swimming to other propulsors (Drucker, 1996; Drucker and Jensen, 1996) or from cruising to sprint speeds using body/caudal fin propulsors (Bainbridge, 1958).

In order to permit comparisons of maneuvering performance among gaits and species, performance should be expressed in nondimensional terms (Bandyopadhyay *et al.*, 1997; Bandyopadhyay, 2002). The appropriate normalization metric or product group will, as usual, depend on the nature of the question (Daniel and Webb, 1987). The simplest and most appropriate normalization metric for many maneuvers is mass, a volume force, which recognizes the role of inertial resistance in changes of state. Volume has some biological attractiveness because space is important for accommodating sensory systems, food acquisition and treatment systems, power supply systems, and space for gametes (Weihs, 1977, 1981; Vogel, 1994). Animal and water densities tend to be similar, so that the numerical value of mass can be substituted as an approximation for volume with little loss in accuracy. Spatial measures such as turning radius in complex maneuvers are often normalized using a linear reference. To be consistent with other maneuvers, volume$^{1/3}$ ($\approx$ mass$^{1/3}$) would be appropriate (Webb *et al.*, 1996a; Schrank *et al.*, 1999; Walker, 2003), but in practice total length is usually used. It should be noted that various normalization options do not affect the ranking of maneuvering performance within gaits among species (Webb *et al.*, 1996a). Finally, some maneuvers are independent of size (Daniel and Webb, 1987), at least for the size range of most juvenile and adult fishes. Absolute performance values have been reported for these, notably acceleration rates in fast starts (Webb, 1994a; Domenici and Blake, 1997; Domenici, 2001).

The paucity of performance measurements, with the exception of fast starts and braking (see Geerlink, 1987; Webb, 1994a; Videler, 1993; Domenici and Blake, 1997; Webb and Fairchild, 2001), reflects difficulties in studying maneuvers. Furthermore, even though fast starts have been extensively studied, there is no uniformity of methods. As a result, comparisons among species must be treated with caution (Webb, 1994a). Studying other maneuvers is complicated because behavior is not easily controlled; maneuvers are difficult to replicate at best, and results are sometimes reported for single events. In keeping with the successful methodology of initiating fast starts with a threat-related stimulus, maneuvers have been initiated with some sort of threat stimulus requiring avoidance (Ferry and Lauder, 1996; Webb *et al.*, 1996a; Schrank and Webb, 1998; Schrank *et al.*, 1999; Webb and Fairchild, 2001). Webb and Fairchild (2001) sought to achieve repeatability using threat stimuli, and although they had some success, the techniques were tedious and required much practice, traits that are not desirable for an experimental design. Alternatively, a range of maneuvers may be obtained during routine activities (Webb and Keyes, 1981, 1982; Bandyopadhyay *et al.*, 1997; Walker, 2000; Domenici, 2001; Bandyopadhyay, 2002).

2. Force Balance

While whole body performance is the easiest way of quantifying maneuvers, understanding the physical basis underlying performance, the basis for differences among species, and accommodating stability all require a more detailed approach. A common starting point is a force balance. For whole body maneuvers measured at the center of mass, the simplest form of the force balance may be approximated as (Webb, 1991) the following:

$$F_{motor} = \sqrt{[(aSu^2 + bMdu/dt)^2 + (cMu^2/r)^2]}, \tag{9}$$

where F_{motor} = motor force, a, b, and c = constants, M = mass, and r = turning radius. The simplest force balance applies to braking, when thrust is zero, and F_{motor} is determined by the momentum loss of the fish (Gray, 1968; Howland, 1974; Webb, 1975; Geerlink, 1986, 1987; Videler, 1993). Analysis of fishes during braking shows that both body and fin drag contribute to deceleration, and the relative importance of these components varies among species and size. Angelfish, a species with a deep, compressed body, may rely on the drag associated with their resultant large wetted body and median fin area, while fusiform goldfish may rely more on extending the pectoral fins to create high drag (Webb and Fairchild, 2001). For larger individuals of other species, mackerel, *Scomber scombrus*, cod, *Gadus morhua*, and saithe, *Pollachius virens*, about 15% of the braking force is provided by extension of the pectoral fins, but the body is bent to increase drag and provide most of the braking (Videler, 1981; Geerlink, 1987).

Models such as Eq. (9) also can lead to important conclusions on tradeoffs in maneuvering performance. For a given force, such as at a gait transition, the highest speed is attained for rectilinear motion. Therefore, adding a maneuver within a gait decreases the maximum speed attainable. (Fishes lack sufficient weight in water to use potential energy to substantially increase speed in a diving maneuver.) Accelerating to surge ($du/dt > 0$) or turn ($r > 0$) can occur only with a reduction in speed for a given force. Thus, the maximum speed during a turn using median and paired fin swimming or in body/caudal fin swimming is lower than the speed in rectilinear swimming (Figure 8.8). The tradeoffs among acceleration rate, speed, and turning radius also are illustrated in fast starts for which muscle input to the body/caudal fin propulsor is considered to be maximized. Fast starts commonly start from rest. Following activation of the myotomal muscle, the acceleration rates is high, at $\leq 15\ g$, and the radius of the turn of the center of mass is small, $\geq 0.01\ L$. As speed increases through the fast start, acceleration rate declines while turning radius increases (see illustrations in Eaton *et al.*, 1977; Webb, 1978; Harper and Blake, 1990, 1991).

Because of tradeoffs among acceleration, speed, and turning, there is no "maximum" maneuver rate. Instead, each gait supports a range of translation rate components inversely related to rotational rate components. Similarly, there is no single "best" design for maneuvering (Walker, 2003), but different maneuvers have different design optima. For example, high rotation while hovering is facilitated by a short, deep body to minimize the moment of inertia (Alexander, 1967), plus suites of highly flexible median and paired fins with multiple degrees of freedom working in concert to provide torque while slip and surge cancel out among the effectors (Walker, 2000). In contrast, fast starts require a large caudal area, distant from the center of mass, and hence an elongate body. The body also should be flexible to achieve large-amplitude motions (Domenici and Blake, 1997).

As noted previously, performance measures should be nondimensional to facilitate comparisons. One approach to quantifying maneuverability in nondimensional form, suggested by Jindrich and Full (1999), is the linear maneuverability number (LMN). The LMN compares the magnitude of maneuvering forces to the momentum of the system to be disturbed as the ratio of the normal (turning) force impulse to that of the forward momentum in the pre-maneuver state. Drucker and Lauder (2001) calculated this ratio for pectoral fin turns by bluegill sunfish. LMN values were very variable, ranging from 0.12 to 0.74, the higher values being comparable with those for cockroaches (Jindrich and Full, 1999). Drucker and Lauder (2001) considered their analysis of LMN to be tentative: turns were submaximal, in fish swimming at low speeds. This reinforces the need to examine performance under limiting conditions.

In another approach, Bandyopadhyay (2002) sought to normalize turning performance for fishes to compare them with small autonomous underwater vehicles. Like Howland (1974) and Webb (1976), he used a coefficient of normal acceleration, C_g, based on centripetal resistance in turns. Thus:

$$C_g = (u^2 r)/g \cdot \tag{10}$$

He first compared C_g with specific turning radius, r/L, showing that C_g was inversely proportional to r/L, consistent with known size relationships for acceleration for animals moving at higher Reynolds numbers (Daniel and Webb, 1987; Webb and de Buffrenil, 1990). Substantial differences in maneuverability remained among vehicles, and some nonlinearity occurred, suggesting C_g also was underestimated at small r/L. Bandyopadhyay (2002) extended his analysis by recognizing that turning involves inertial, viscous, and gravity forces. He computed a reduced coefficient, proving to be the ratio of Froude and Reynolds numbers, and again found inverse relationships between the reduced coefficient and r/L. Separation among vehicles was reduced, but some nonlinearity remained at smaller r/L.

3. Minimum Maneuver Volume

Whole body and force balance approaches do not capture all biologically important aspects of maneuvers, and cannot include hovering at all. Using r to quantify the spatial dimensions of a turn neglects the reality that fishes have bulk, and that effectors protruding from the body need space in which to work (Walker, 2000, 2003). One way to consider the reality of bulk is to consider the minimum volume for a maneuver, V_{maneuver}, which could be normalized as specific minimum maneuver volume, $V_{\text{maneuver}}/V_{\text{fish}}$, where V_{fish} is the volume of the fish. Appropriate data are lacking. Schrank and Webb (1998) and Schrank *et al.* (1999) chased fishes through various tubes that are akin to limited volume situations. They determined the smallest widths of mazes that could be negotiated. Normalizing these with volume$^{1/3}$ showed that goldfish, a soft-rayed teleost, could maneuver in smaller spaces than silver dollar, *Metynnis argenteus*, and angelfish, a spiny-rayed teleost. This was contrary to expectations and was attributed to greater overall body flexibility of goldfish. In general, the axial skeletal modifications of more-derived teleosts reduce the degrees of freedom of the caudal region of the body to at least some extent compared to soft-rayed representatives (Figure 8.4).

4. Energetics

Another biologically important measure of performance is energy expenditure (Daniel and Webb, 1987). Some theoretical estimates have been made of the costs of fast starts (Webb, 1975; Frith and Balke, 1995) as well as measurements of resistance (Webb, 1982). Otherwise, there are no data for the costs of specific maneuvers. Various studies have shown that routine swimming is dominated by maneuvers (Nursall, 1958, 1973; Webb, 1991; Block *et al.,* 1992, 2001; Boisclair, 1992; Boisclair and Tang, 1993; Nilsson *et al.,* 1993; Krohn and Boisclair, 1994; Hughes and Kelly, 1996; Tang *et al.*, 2000). Flow visualization shows that maneuvers often involve substantial energy losses in vortex rotation (McCutcheon, 1971; Lauder and Drucker, 2003; Tytler and Lauder, 2004), and hence high induced drag should be expected. Mechanics and metabolic studies of routine swimming show over 10-fold increases in energy costs compared with those expected for rectilinear swimming at the same mean speed (Weatherley *et al.*, 1982; Puckett and Dill, 1984, 1985; Weatherley and Gill, 1987; Webb, 1991; Boisclair and Tang, 1993; Krohn and Boisclair, 1994). Similarly, energy costs of hovering are large (Blake, 1979; Lighthill, 1979). Models of fish foraging behavior have yet to incorporate the costs of maneuvering into their cost–benefit analyses.

V. FUTURE DIRECTIONS

Studies of stability and maneuverability of fishes and other animals are in their infancy. A uniform conceptual framework is lacking (Fish, 2003). There are no standard techniques or approaches to determine key performance variables in reproducible and controlled situations. Although marine mammals can be trained to execute specific maneuvers (Fish, 2002), it seems unlikely that sufficient control over fish behavior will be possible in experimental situations. It seems more likely that normal variation in fish behavior, perhaps forced by social interactions such as agonistic interactions and feeding, will be required to provide variation in maneuvers and control problems for analysis. If so, the ability to reduce problems in both stability and maneuverability to nondimensional coefficients or product groups will be essential (Bandyopadhyay, 2002).

The need to describe systems used in stability and maneuvers argues for better models. The stability/maneuverability equations are far from unknown, especially for terrestrial and aerial systems, in which gravitational forces dominate the system (Thomas and Taylor, 2001). These equations need modification for submerged aquatic systems at or near neutral buoyancy. This is the purview of naval architects, whose experience can be better drawn upon by biologists. Nevertheless, the relevant equations are complex and are difficult to apply to systems such as fishes that have many ways to perform tasks. Thus, high flexibility and redundancy present special challenges.

Stability and maneuverability issues are considered to be important in understanding where fishes live and why (Goldstein and Pinshow, 2002). Studies of fish biomechanics are rarely explicitly validated in field situations, but the exceptions in which field studies have been performed are notable for the contributions they have made (see Denny, 1988; Vogel, 1988, 1994; Davis *et al.*, 1999; Block *et al.*, 2001, 2002). Very little is known about the perturbations faced by fishes in natural habitat (Figure 8.3). Such knowledge is essential for understanding stability needs, the types of necessary control, and how various effectors meet these requirements. Measurement and modeling techniques are available, and are being developed, to achieve field studies in fish biomechanics that have not previously been possible.

Finally, there is growing interest in capturing the maneuvering capabilities of fishes in human-designed underwater autonomous vehicles (Bandyopadhyay, 2002; Fish, 2003). The success of such biomimetics, however, depends on being able to overcome undesirable correlates of high maneuverability. First, high maneuverability of fishes reflects diverse control surfaces. A comparable level of complexity would not be desirable for many human applications. Second, if control surfaces were combined with propulsors, as in fishes, propulsion would incur high induced energy losses that accompany any reciprocating propulsor system. These costs contribute to costs of transport for fishes, which are much higher than those associated with rotating propulsors. Third, high maneuverability reflects the dynamic equilibrium that for fishes is the solution to hydrostatic and hydrodynamic self-generated and external destabilizing forces. Stabilization for such forces also appears to substantially increase the costs of swimming, especially at low speeds. Domenici and Blake (1997) noted that many fishes have uncoupled slow swimming using multiple propulsors from body/caudal fin swimming. Biomimicry for maneuverability probably will need to use this biological principle too, using efficient, high-endurance rotary systems for routine translocation, with separate rapid-responding, trimming, and powered control surfaces, in a body that is neutrally stable for hydrostatic forces.

ACKNOWLEDGMENTS

Much of my work included in this chapter has been funded by the National Science Foundation, whose support, especially grants IBN 9507197 and IBN 9973942, is gratefully acknowledged. I thank the editors and Frank Fish for their helpful critique of the manuscript.

REFERENCES

Alexander, R. M. (1965). The lift produced by the heterocercal tails of Selachii. *J. Exp. Biol.* **43,** 131–138.

Alexander, R. M. (1967). "Functional Design in Fishes." Hutchinson, London.

Alexander, R. M. (1983). The history of fish biomechanics. *In* "Fish Biomechanics" (Webb, P. W., and Weihs, D., Eds.), pp. 1–35. Praeger, New York.

Alexander, R. M. (1989). Optimization and gaits in the locomotion of vertebrates. *Physiol. Rev.* **69,** 1199–1227.

Alexander, R. M. (1990). Size, speed and buoyancy adaptations in aquatic animals. *Amer. Zool.* **30,** 189–196.

Alexander, R. M. (1993). Buoyancy. *In* "The Physiology of Fishes" (Evans, D. H., Ed.), pp. 75–97. CRC Press, Boca Raton.

Aleyev, Y. G. (1977). "Nekton." Junk, The Hague.

Allan, D. A. (1995). "Stream Ecology." Chapman and Hall, London.

Anderson, D. F., and Eberhardt, S. (2001). "Understanding Flight." McGraw-Hill, New York.

Anderson, E. J., McGillis, W. R., and Grosenbaugh, M. A. (2001). The boundary layer of swimming fish. *J. Exp. Biol.* **204,** 81–102.

Anonymous (1971). "Webster's 3rd New International Dictionary." Merriam Press, Springfield, MA.

Arita, G. S. (1971). A re-examination of the functional morphology of the soft-rayes in teleosts. *Copeia* **1971,** 691–697.

Arnold, G. P., and Weihs, D. (1978). The hydrodynamics of rheotaxis in the plaice (*Pleuronectes platessa*). *J. Exp. Biol.* **75,** 147–169.

Arnold, G. P., Webb, P. W., and Holford, B. H. (1991). The role of the pectoral fins in station-holding of Atlantic salmon (*Salmo salar* L). *J. Exp. Biol.* **156,** 625–629.

Bainbridge, R. (1958). The speed of swimming of fish as related to size and to the frequency and the amplitude of the tail beat. *J. Exp. Biol.* **35,** 109–133.

Bandyopadhyay, P. R. (2002). Maneuvering hydrodynamics of fish and small underwater vehicles. *Integ. Comp. Biol.* **42,** 102–117.

Bandyopadhyay, P. R., Castano, J. M., Rice, J. Q., Philips, R. B., Nedderman, W. H., and Macy, W. K. (1997). Low-speed maneuvering hydrodynamics of fish and underwater vehicles. *Trans. Amer. Soc. Struct. Mech. Engin.* **119,** 136–144.

Bartol, I. K., Gordon, M. S., Gharib, M., Hove, J. R., Webb, P. W., and Weihs, D. (2002). Flow patterns around the carapaces of rigid-bodied, multi-propulsor boxfishes (Teleostei: Ostraciidae). *Integ. Comp. Biol.* **42,** 971–980.

Bartol, I. K., Gharib, M., Weihs, D., Webb, P. W., Hove, J. R., and Gordon, M. S. (2003). Role of the carapace in dynamic stability of the smooth trunkfish *Lactiphyrs triqueter* (Teleostei: Ostraciidae). *J. Exp. Biol.* **206,** 725–744.

Batschelet, E. (1965). "Statistical Methods for the Analysis of Problems in Animal Orientation and Certain Biological Rhythms." American Institute of Biological Sciences, Washington, D.C.

Bell, W. H., and Terhune, L. D. B. (1970). Water tunnel design for fisheries research. *Fish. Res. Bd. Canada Tech. Rept.* **195,** 1–69.

Bellwood, D. R., and Wainwright, P. C. (2001). Locomotion in labrid fishes: Implications for habitat use and cross-shelf biogeography on the Great Barrier Reef. *Coral Reefs* **20,** 139–150.

Blake, R. W. (1979). The energetics of hovering in the mandarin fish (*Synchropus picturatus*). *J. Exp. Biol.* **82,** 25–33.

Blake, R. W. (1983). "Fish Locomotion." Cambridge Univ. Press, Cambridge.

Blake, R. W. (2000). The biomechanics of intermittent swimming behaviours in aquatic vertebrates. *In* "Biomechanics in Animal Behaviour" (Domenici, P., and Blake, R. W., Eds.), pp. 79–103. Bios Scientific Publishers Ltd., Oxford.

Block, B. A., Booth, D., and Carey, F. G. (1992). Direct measurements of swimming speeds and depth of blue marlin. *J. Exp. Biol.* **166,** 267–284.

Block, B. A., Dewar, H., Blackwell, S. B., Williams, T. D., Prince, E. D., Farwell, C. J., Boustany, A., Teo, S. L. H., Seitz, A., Walli, A., and Fudge, D. (2001). Migratory movements, depth preferences, and thermal biology of Atlantic bluefin tuna. *Science* **293,** 1310–1314.

Boisclair, D. (1992). An evaluation of the stereocinematographic method to estimate fish swimming speed. *Can. J. Fish. Aquat. Sci.* **49,** 523–531.

Boisclair, D., and Tang, M. (1993). Empirical analysis of the swimming pattern on the net energetic cost of swimming in fishes. *J. Fish. Biol.* **42,** 169–183.

Breder, C. M. (1926). The locomotion of fishes. *Zoologica* **4,** 159–297.

Brett, J. R. (1995). Energetics. *In* "Physiological-Ecology of Pacific Salmon" (Brett, J. R., and Clark, C., Eds.), pp. 3–68. Gov. Canada, Dept. Fish. Oceans, Ottawa, Ontario.

Bunker, S. J., and Machin, K. E. (1991). The hydrodynamics of cephalaspids. *Soc. Exp. Biol. Seminar Ser.* **36,** 113–129.

Cada, G. F. (2001). The design of advanced hydroelectric turbines to improve fish passage survival. *Fisheries* **26,** 14–23.

Casagrand, J. L., Guzik, A. L., and Eaton, R. C. (1999). Mauthner cell and reticulospinal neuron responses to onset of acoustic pressure and acceleration stimuli. *J. Neurophys.* **82,** 1422–1437.

Chen, H., Kandasamy, S., Orszag, S., Shock, R., Succi, S., and Yakhot, V. (2003). Extended Boltzmann kinetic equation for turbulent flows. *Science* **301,** 633–636.

Clack, J. A. (2002). "Gaining Ground: The Origin and Evolution of Tetrapods." Indiana Univ. Press, Bloomington, IN.

Clark, R. B. (1964). "Dynamics in Metazoan Evolution." Oxford Univ. Press, Oxford.

Cruickshank, A. R. I., and Skews, B. W. (1980). The functional significance of nectridean tabular horns. *Proc. Roy. Soc. Lond.* **209B,** 512–537.

Daniel, T. L., and Webb, P. W. (1987). Physics, design, and locomotor performance. *In* "Comparative Physiology: Life in Water and on Land" (Dejours, P., Bolis, L., Taylor, C. R., and Weibel, E. R., Eds.), pp. 343–369. Liviana Press, Springer-Verlag, NY.

Davis, R. W., Fuiman, L. A., Williams, T. M., Collier, S. O., Hagey, W. P., Kanatous, S. B., kohin, S., and Horning, M. (1999). Hunting behavior of a marine mammal beneath the antartic fast-ice. *Science* **283,** 993–996.

Deliagina, T. G. (1997a). Vestibular compensation in lampreys: Impairment and recovery of equilibrium control during locomotion. *J. Exp. Biol.* **200,** 2957–2967.

Deliagina, T. G. (1997b). Vestibular compensation in lampreys: Role of vision at different stages of recovery of equilibrium control. *J. Exp. Biol.* **200,** 1459–1471.

Denny, M. (1988). "Biology and the Mechanics of the Wave-Swept Environment." Princeton University Press, Princeton, NJ.

Desabrais, K. J., and Johari, H. (2000). Direct circulation measurement of a tip vortex. *AIAA J.* **38,** 2189–2191.

Diana, J. S. (1995). "Biology and Ecology of Fishes." Cooper Publ. Group, Carmel, IN.

Domenici, P. (2001). The scaling of locomotor performance in predator-prey encounters: From fish to killer whales. *Comp. Biochem. Physiol.* **131A,** 169–182.

Domenici, P., and Blake, R. W. (1997). The kinematics and performance of fish fast-start swimming. *J. Exp. Biol.* **200,** 1165–1178.

Dower, J. F., Miller, T. J., and Leggett, W. C. (1997). The role of microscale turbulence in the feeding ecology of larval fish. *Adv. Mar. Biol.* **31,** 169–220.

Drucker, E. G. (1996). The use of gait transition speed in comparative studies of fish locomotion. *Amer. Zool.* **36,** 555–566.

Drucker, E. G., and Jensen, J. S. (1996). Pectoral fin locomotion in the striped surfperch. I. Kinematic effects of swimming speed and body size. *J. Exp. Biol.* **199,** 2235–2242.

Drucker, E. G., and Lauder, G. V. (1999). Locomotor forces on a swimming fish: Three-dimensional vortex wake dynamics quantified using digital particle image velocimetry. *J. Exp. Biol.* **203,** 2393–2412.

Drucker, E. G., and Lauder, G. V. (2000). A hydrodynamic analysis of fish swimming speed: Wake structure and locomotor force in slow and fast labriform swimmers. *J. Exp. Biol.* **203,** 2379–2393.

Drucker, E. G., and Lauder, G. V. (2001). Wake dynamics and fluid forces of turning maneuvers in sunfish. *J. Exp. Biol.* **204,** 431–442.

Drucker, E. G., and Lauder, G. V. (2002). Experimental hydrodynamics of fish locomotion: Functional insights from wake visualization. *Integ. Comp. Biol.* **42,** 243–257.

Drucker, E. G., and Lauder, G. V. (2003). Function of pectoral fins in rainbow trout: Behavioral repertoire and hydrodynamic forces. *J. Exp. Biol.* **206,** 813–826.

DuBois, A. B., and Ogilvy, C. S. (1978). Forces on the tail surface of swimming fish: Thrust, drag and acceleration in bluefish (*Pomatomus saltatrix*). *J. Exp. Biol.* **77,** 225–241.

Eaton, R. C., and Hackett, J. T. (1984). The role of the Mauthner cell in fast-starts involving escape in teleost fishes. *In* "Neural Mechanisms of Startle Behavior" (Eaton, R. C., Ed.), pp. 213–266. Plenum Press, New York.

Eaton, R. C., Bombardieri, R. A., and Meyer, D. L. (1977). The Mauthner-initiated startle response in teleost fish. *J. Exp. Biol.* **66,** 65–81.

Eaton, T. H. (1945). Skeletal supports of the median fins of fishes. *J. Morphol.* **76,** 193–212.

Eidietis, L., Forrester, T. L., and Webb, P. W. (2003). Relative abilities to correct rolling disturbances of three morphologically different fish. *Can. J. Zool.* **80,** 2156–2163.

Ferry, L. A., and Lauder, G. V. (1996). Heterocercal tail function in leopard sharks: A three-dimensional kinematic analysis of two models. *J. Exp. Biol.* **199,** 2253–2268.

Fierstine, H. L., and Walters, V. (1968). Studies in locomotion and anatomy of scombroid fishes. *Mem. S. California Acad. Sci.* **6,** 1–31.

Fish, F. E. (2002). Balancing requirements for stability and maneuverability in cetaceans. *Integ. Comp. Biol.* **42,** 85–93.

Fish, F. E. (Ed.), (2003). Morphology and experimental hydrodynamics of piscine control surfaces. *In* "Biology-Inspired Maneuvering Hydrodynamics for AUV Application." Proceedings of the 13th International Symposium on Unmanned Untethered Submersible Technology, pp. C1–C26. Autonomous Undersea Systems Institute, Durham, NH.

Fish, F. E., and Shannahan, L. D. (2000). The role of the pectoral fins in body trim of sharks. *J. Fish Biol.* **56,** 1062–1073.

Fletcher, R. I. (1990). Flow dynamics and fish recovery experiments: Water intake systems. *Trans. Amer. Fish. Soc.* **119,** 393–415.

Fletcher, R. I. (1992). The failure and rehabilitation of a fish-conserving device. *Trans. Amer. Fish. Soc.* **121,** 678–679.

Fricke, H., and Hissman, K. (1992). Locomotion, fin coordination and body from of the living coelacanth *Latimeria chalumnae*. *Environ. Biol. Fish.* **34,** 329–356.

Frith, H. R., and Balke, R. W. (1995). The mechanical power output and hydromechanical efficiency of northern pike (*Esox lucius*) fast-starts. *J. Exp. Biol.* **198,** 1863–1873.

Fulton, C. J., and Bellwood, D. R. (2002a). Patterns of foraging in labrid fishes. *Mar. Ecol. Prog. Ser.* **226,** 135–142.

Fulton, C. J., and Bellwood, D. R. (2002b). Ontogenetic habitat use in labrid fishes: An ecomorphological perspective. *Mar. Ecol. Prog. Ser.* **236,** 255–262.

Gans, C., Guant, A. S., and Webb, P. W. (1997). Vertebrate locomotion. *In* "Handbook of physiology" (Danzler, W. H., Ed.), pp. 55–213. American Physiological Society, Oxford Univeristy Press, Oxford.

Gee, J. H. (1983). Ecological implications of buoyancy control in fish. *In* "Fish Biomechanics" (Webb, P. W., and Weihs, D., Eds.), pp. 140–176. Praeger, New York.

Geerlink, P. J. (1986). Pectoral fins. Aspects of propulsion and braking in teleost fishes. Ph.D. Thesis, University of Groningen.

Geerlink, P. J. (1987). The role of the pectoral fins in braking of mackerel, cod and saithe. *Neth. J. Zool.* **37,** 81–104.

Geerlink, P. J., and Videler, J. J. (1974). Joints and muscles of the dorsal fin of *Tilapia nilotica* L. (Fam. Cichlidae). *Neth. J. Zool.* **24,** 279–290.

Gerstner, C. L. (1998). Use of substratum ripples for flow refuging by Atlantic cod, *Gadus morhua*. *Environ. Biol. Fish.* **55,** 455–460.

Gerstner, C. L. (1999). Maneuverability of four species of coral-reef fish that differ in body and pectoral-fin morphology. *Can. J. Zool.* **77,** 1–9.

Gerstner, C. L., and Webb, P. W. (1998). The station-holding by plaice, *Pleuronectes platessa* on artificial substratum ripples. *Can. J. Zool.* **76,** 260–268.

Gibb, A. C., Jayne, B. C., and Lauder, G. V. (1994). Kinematics of pectoral fin locomotion in the bluegill sunfish *Lepomis macrochirus*. *J. Exp. Biol.* **189,** 133–161.

Gibb, A. C., Dickson, K. A., and Lauder, G. V. (1999). Tail kinematics of the chub mackerel *Scomber japonicus*: Testing the homocercal tail model of fish propulsion. *J. Exp. Biol.* **202,** 2433–2447.

Goldberg, L. L. (1988). Intact stability. *In* "Principles of Naval Architecture. I: Stability and Strength" (Lewis, E. V., Ed.), pp. 63–142. The Society of Naval Architects and Marine Engineers, Jersey City, NJ.

Goldstein, D. L., and Pinshow, B. (2002). Taking physiology into the field: An introduction to the symposium. *Integ. Comp. Biol.* **42,** 1–2.

Gordon, M. S., Chin, H. G., and Vojkovich, M. (1989). Energetics of swimming in fishes using different modes of locomotion: I. Labriform swimmers. *Fish. Physiol. Biochem.* **6,** 341–352.

Gordon, M. S., Plaut, I., and Kim, D. (1996). How puffers (Teleostei: Tetraodontidae) swim. *J. Fish. Biol.* **49,** 319–328.

Gordon, M. S., Hove, J. R., Webb, P. W., and Weihs, D. (2001). Boxfishes as unusually well controlled autonomous underwater vehicles. *Physiol. Biochem. Zool.* **73,** 663–671.

Gordon, N. D., McMahon, T. A., and Finlayson, B. L. (1992). "Stream Hydrology." John Wiley and Sons, New York, NY.

Gosline, W. A. (1971). "Functional Morphology and Classification of Teleostean Fishes." Univ. Press of Hawaii, Honolulu, HI.

Graham, J. B., Lowell, W. R., Rubinoff, I., and Motta, J. (1987). Surface and subsurface swimming of the sea snake *Pelamis platurus*. *J. Exp. Biol.* **127,** 27–44.

Gray, J. (1968). "Animal Locomotion." Weidenfeld and Nicolson, London.

Hans, P. M. (1951). Beitrage zur okologischen physiologie des seepfredes) *Hippocampus brevirostris*). *Zeit. ver. Physiol.* **33,** 207–265.

Harder, W. (1975). "Anatomy of Fishes." E. Schweizerbart'sche Verlagsbuchhandling, Stuttgart.

Harper, D. G., and Blake, R. W. (1989). On the error involved in high-speed film when used to evaluate maximum acceleration of fish. *Can. J. Zool.* **67,** 1929–1936.

Harper, D. G., and Blake, R. W. (1990). Fast-start performance of rainbow trout *Salmo gairdneri* and northern pike *Esox lucius*. *J. Exp. Biol.* **150,** 321–342.

Harper, D. G., and Blake, R. W. (1991). Prey capture and the fast-start performance of northern pike *Esox lucius*. *J. Exp. Biol.* **155,** 175–192.

Harris, J. E. (1936). The role of fins in the equilibrium of swimming fish. I. Wind-tunnel tests on a model of *Mustelus canis* (Mitchell). *J. Exp. Biol.* **13,** 476–493.

Harris, J. E. (1937a). The mechanical significance of the position and movements of the paired fins in the teleostei. *Papers Tortugas Lab., Carnegie Inst.* **31,** 173–189.

Harris, J. E. (1937b). The role of fin movements in the equilibrium of fish. *Ann. Rep. Tortugas Lab., Carnegie Inst.* **11936–37,** 91–93.

Harris, J. E. (1938). The role of the fins in the equilibrium of swimming fish II. The role of the pelvic fins. *J. Exp. Biol.* **15,** 32–47.

Harris, J. E. (1953). Fin patterns and mode of life in fishes. *In* "Essays in Marine Biology" (Marshall, S. M., and Orr, P., Eds.), pp. 17–28. Oliver and Boyd, Edinburgh.

He, P., and Wardle, C. S. (1986). Tilting behavior of the Atlantic mackerel, *Scomber scombrus*, at low swimming speeds. *J. Fish. Biol.* **29A,** 223–232.

Helfman, G. S. (1981). Twilight activities and temporal structure in a freshwater fish community. *J. Fish. Res. Board Can.* **38,** 1405–1420.

Hinch, S. G., and Rand, P. S. (1998). Swim speeds and energy use of upriver-migrating sockeye salmon (*Oncorhynchus nerka*): Role of local environment and fish characteristics. *Can. J. Fish. Aquat. Sci.* **55,** 1821–1831.

Hinch, S. G., Standen, E. M., Healey, M. C., and Farrel, A. P. (2002). Swimming patterns and behaviour of upriver migrating adult pink salmon (*Oncorhynchus gorbuscha*) and sockeye (*O. nerka*) salmon as assessed by EMG telemetry in the Fraser River, British Columbia. *Hydrobiologia* **483,** 147–160.

Hinze, J. O. (1975). "Turbulence." McGraw-Hill, New York.

Hobson, E. S. (1974). Feeding relationships of teleostean fishes on coral reefs in Kona, Hawaii. *Fish. Bull. (U.S.)* **72,** 915–1031.

Hoerner, S. F. (1965). "Fluid-Dynamic Drag." Hoerner Fluid Dynamics, Brick Town, NJ.

Hoerner, S. F. (1975). "Fluid-Dynamic Lift." Hoerner Fluid Dynamics, Brick Town, NJ.

Hora, S. L. (1937). Comparison of the fish fauna of the northern and southern faces of the Great Himalayan Region. *Rec. Indian Mus.* **39,** 1–250.

Hove, J. R., O'Bryan, L. M., Gordon, M. S., Webb, P. W., and Weihs, D. (2001). Boxfishes (Teleostie: Ostraciidae) as a model system for fishes swimming with many fins: I. Kinematics. *J. Exp. Biol.* **204,** 1459–1471.

Howland, H. (1974). Optimal strategies for predator avoidance: The relative importance of speed and manoeuverability. *J. Theor. Biol.* **47,** 333–350.

Hughes, N. F., and Kelly, L. H. (1996). A hydrodynamic model for estimating the energetic cost of swimming maneuvers from a description of their geometry and dynamics. *Can. J. Fish. Aquat. Sci.* **53,** 2484–2493.

Jarvik, E. (1965). On the origin of girdles and paired fins. *Israel J. Zool.* **14,** 141–172.

Jayne, B. C., Lozada, A., and Lauder, G. V. (1996). Function of the dorsal fin in bluegill sunfish: Motor patterns during fours locomotor behaviors. *J. Morphol.* **228,** 307–326.

Jindrich, D. L., and R.J., Full (1999). Many-legged manoeuvrability: Dynamics of turning in hexapods. *J. Exp. Biol.* **202,** 1603–1623.

Johari, H., and Durgin, W. W. (1998). Direct measurements of circulation using ultrasound. *Exper. Fluids* **25,** 445–454.

Johari, H., and Moreira, J. (1998). Direct measurement of delta-wing circulation. *AIAA J.* **36,** 2195–2203.

Kasapi, M. A., Domenici, P., Blake, R. W., and Harper, D. G. (1993). The kinematics and performance of knifefish *Xenomystus nigri* escape responses. *Can. J. Zool.* **71,** 189–195.

Keenleyside, M. H. A. (1979). "Diversity and Adaptation in Fish Behaviour." Springer-Verlag, Berlin.

Kermack, K. A. (1943). The functional significance of the hypocercal tail in *Pteraspis rostrata. J. Exp. Biol.* **20,** 23–27.

Kolmogorov, A. N. (1941). Local structure of turbulence in an incompressible viscous fluid at very high Reynolds numbers, Dolk. *Akad. Nauk SSSR* **30,** 299; reprinted in *Usp. Fix. Nauk* **93,** 476–481 (1967), transl. in *Sov. Phys. Usp.* **10**(6), 734–736 (1968).

Krohn, M., and Boisclair, D. (1994). The use of a stereo-video system to estimate the energy expenditure of free-swimming fish. *Can. J. Fish. Aquat. Sci.* **51,** 1119–1127.

Lagler, K. F., Bardach, J. E., Miller, R. R., and Pasino, D. R. M. (1977). "Ichthyology." Wiley and Sons, New York.

Lauder, G. V. (1989). Caudal fin locomotion in ray-finned fishes: Historical and functional analysis. *Amer. Zool.* **29,** 85–102.

Lauder, G. V., and Drucker, E. G. (2003). Morphology and experimental hydrodynamics of piscine control surfaces. *In* "Biology-Inspired Maneuvering Hydrodynamics for AUV Application" (Fish, F. E., Ed.), pp. C1–C26. Proceedings of the 13th International Symposium on Unmanned Untethered Submersible Technology. Autonomous Undersea Systems Institute, Durham, NH.

Lauder, G. V., and Liem, K. F. (1983). The evolution and interrelationships of the actinopterygian fishes. *Bull. Mus. Comp. Zool. Harvard Univ.* **150,** 95–197.

Liao, J., and Lauder, G. V. (2000). Function of the heterocercal tail in white sturgeon: Flow visualization during steady swimming and vertical maneuvering. *J. Exp. Biol.* **203,** 3585–3594.

Liao, J., Beal, D. N., Lauder, G. V., and Trianyafyllou, M. S. (2003). The Kármán gait: Novel body kinematics of rainbow trout swimming in a vortex street. *J. Exp. Biol.* **206,** 1059–1073.

Lighthill, J. (1977). Mathematical theories of fish swimming. *In* "Fisheries Mathematics" (Steele, J. H., Ed.), pp. 131–144. Academic Press, New York.

Lighthill, J. (1979). A simple fluid-flow model of ground effect on hovering. *J. Fluid Mech.* **93,** 781–797.

Long, J. H., and Nipper, K. S. (1996). The importance of body stiffness in undulatory propulsion. *Amer. Zool.* **36,** 678–694.

Marchaj, C. A. (1988). "Aero-Hydrodynamics of Sailing." International Marine Publishing, Camden, NJ.

Matthews, W. J. (1985). Critical current speeds and microhabitats of the benthic fishes *Percina roanoka* and *Etheostoma flabellare*. *Environ. Biol. Fish* **12,** 303–308.

Matthews, W. J. (1998). "Patterns in Freshwater Fish Ecology." Chapman and Hall, New York.

McCutcheon, C. W. (1970). The trout tail fin: A self-cambering hydrofoil. *J. Biomech.* **3,** 271–281.

McCutcheon, C. W. (1971). Froude propulsive efficiency of a small fish, measured by wake visualization. *In* "Scale Effects in Animal Locomotion" (Pedley, T. J., Ed.), pp. 339–363. Academic Press, New York.

McHenry, M. J., Pell, C. A., and Long, J. H., Jr. (1995). Mechanical control of swimming speed: Stiffness and axial wave form in undulating fish models. *J. Exp. Biol.* **198,** 2293–2305.

Mittal, R., Utturkar, Y., and Udaykumar, H. S. (2002). Computational modeling and analysis of biomimetic flight mechanisms. *AIAA J.* **2002,** 0865.

Moy-Thomas, J. A., and Miles, R. S. (1971). "Palaeozoic Fishes." Saunders Company, Philadelphia, PA.

Müller, U. K., Smit, J., Stamhuis, E. J., and Videler, J. J. (2001). How the body contributes to the wake in undulatory fish swimming: Flow fields of a swimming eel (*Anguilla anguilla*). *J. Exp. Biol.* **204,** 2751–2762.

Nauen, J. C., and Lauder, G. V. (2001). Locomotion in scombrid fishes: Visualization of flow around the caudal peduncle and finlets of the chub mackerel *Scomber japonicus*. *J. Exp. Biol.* **204,** 2251–2263.

Nezu, I., and Nakagawa, H. (1993). "Turbulence in Open-Channel Flows." Balkema, Totterdam, The Netherlands.

Nilsson, G. E., Rosén, P., and Johansson, D. (1993). Anoxic depression of spontaneous locomotion activity in crucian carp quantified by a computerized imaging technique. *J. Exp. Biol.* **180,** 153–162.

Nursall, J. R. (1958). A method of analysis of the swimming of fish. *Copeia* **1958,** 136–141.

Nursall, J. R. (1973). Some behavioral interactions of spottail shiners (*Notropis hudsonius*), yellow perch (*Perca flavescens*), and northern pike (*Esox lucius*). *J. Fish. Res. Bd. Can.* **30,** 1161–1178.

Nursall, J. R. (1962). Swimming and the origin of paired appendages. *Amer. Zool.* **2,** 127–141.

O'Brien, W. J., Browman, H. I., and Evans, B. I. (1990). Search strategies of foraging animals. *Amer. Scient.* **78,** 152–160.

O'Brien, W. J., Evans, B. I., and Browman, H. I. (1989). Flexible search tactics and efficient foraging in saltatory searching animals. *Oecologia* **80,** 100–110.

O'Brien, W. J., Evans, B. I., and Howick, G. L. (1986). A new view of the predation cycle of a planktivorous fish, white crappie (*Pomoxis annularis*). *Can. J. Fish. Aquat. Sci.* **43,** 1894–1899.

Odeh, M., Noreika, J. F., Haro, A., Maynard, A., Castro-Santos, T., and Cada, G. F. (2002). Evaluation of the effects of turbulence on the behavior of migratory fish. Final Report 2002, Report to Bonneville Power Administration, Contract No. 00000022, Project No. 200005700, pp. 1–55. http://www.epa.bpa.gov/publication/D00000022-1.pdf.

Panton, R. (1984). "Incompressible Flow." Wiley-Interscience, New York.

Pavlov, D. S., and Tyurukov, S. N. (1988). The role of hydrodynamic stimuli in the behavior and orientation of fishes near obstacles. *Voprosy Ikhtiologii* **28,** 303–314.

Pavlov, D. S., Skorobagatov, M. A., and Shtaf, L. G. (1982). The critical current velocity of fish and the degree of flow turbulence. *Rep. USSR Acad. Sci.* **267,** 1019–1021.

Pavlov, D. S., Skorobagatov, M. A., and Shtaf, L. G. (1983) Threshold speeds for rheoreaction of roach in flows with different degrees of turbulence. *Rep. USSR Acad. Sci.* **268,** 510–512.

Pavlov, D. S., Lupandin, A. I., and Skorobogatov, M. A. (2000). The effects of flow turbulence on the behavior and distribution of fish. *J. Ichthyol.* **40**(Suppl. 2), S232–S261.

Pietsch, T. W., and Grobecker, D. B. (1978). The compleat angler: Aggressive mimicry in an antenariid anglerfish. *Science* **201,** 369–370.

Pietsch, T. W., and Grobecker, D. B. (1990). Frogfishes. *Sci. Am.* **262,** 96–103.

Pitcher, T. J. (1986). "The Behavior of Teleost Fishes." John Hopkins Univ. Press, Baltimore, MD.

Poff, N. L., and Ward, J. V. (1989). Implications of streamflow variability and predictability for lotic community structure: A regional analysis of streamflow patterns. *Can. J. Fish. Aquat. Sci.* **46,** 1805–1818.

Potts, J. A. (1970). The schooling ethology of *Lutianus monostigma* (Pisces) in the shallow reef environment of Aldabra. *J. Zool. (Lond.)* **161,** 223–235.

Probst, W. E., Rabeni, C. F., Covington, W. G., and Marteney, R. E. (1984). Resource use by stream-dwelling rock bass and smallmouth bass. *Trans. Amer. Fish. Soc.* **113,** 283–294.

Puckett, K. J., and Dill, L. M. (1984). Cost of sustained and burst swimming to juvenile coho salmon (*Oncorhynchus kisutch*). *Can. J. Fish. Aquat. Sci.* **41,** 1546–1551.

Puckett, K. J., and Dill, L. M. (1985). The energetics of feeding territorially in juvenile coho salmon (*Oncorhynchus kisutch*). *Behaviour* **92,** 97–111.

Ramamurti, R., Sandberg, W. C., Lohner, R., Walker, J. A., and Westneat, M. (2002). Fluid dynamics of aquatic flapping flight in the bird wrasse: Three-dimensional unsteady computations with fin deformation. *J. Exp. Biol.* **205,** 2997–3008.

Romer, A. S., and Parsons, T. S. (1977). "The Vertebrate Body." Saunders Company, Philadelphia, PA.

Rosen, D. E. (1982). Teleostean interrelationships, morphological function and evolutionary inference. *Amer. Zool.* **22,** 261–273.

Sanford, L. P. (1997). Turbulent mixing in experimental ecosystem studies. *Mar. Biol. Prog. Ser.* **161,** 265–293.

Schlosser, I. J. (1982). Fish community structure and function along two habitat gradients in a headwater stream. *Ecol. Monogr.* **52,** 395–414.

Schrank, A. J., and Webb, P. W. (1998). Do body and fin form affect the abilities of fish to stabilize swimming during maneuvers through vertical and horizontal tubes? *Environ. Biol. Fishes.* **53,** 365–371.

Schrank, A. J., Webb, P. W., and Mayberry, S. (1999). How do body and paired-fin positions affect the ability of three teleost fishes to maneuver around bends? *Can. J. Zool.* **77,** 203–210.

Schultz, W. W., and Webb, P. W. (2002). Power requirements of swimming: Do new methods resolve old questions? *Integ. Comp. Biol.* **42,** 1018–1025.

Scott, W. B., and Crossman, E. J. (1973). Freshwater fishes of Canada. *Bull. Fish. Res. Bd. Canada* **184,** 1–966.

Shtaf, L. G., Pavlov, D. S., Skorobogativ, M. A., and Baryekian, A. S. (1983). Fish behavior as affected by the degree of flow turbulence. *Vopr. Ikhtiol.* **3,** 307–317.

Stamhuis, E. J., and Videler, J. J. (1995). Quantitative flow analysis around aquatic animals using laser sheet particle image velocimetry. *J. Exp. Biol.* **198,** 283–294.

Tang, M., Boisclair, D., Ménard, C., and Downing, J. A. (2000). Influence of body weight, swimming characteristics, and water temperature on the cost of swimming in brook trout (*Salvelinus fontinalis*). *Can. J. Fish. Aquat. Sci.* **57,** 1482–1488.

Taylor, G. I., and von Kármán, T. (1937). The statistical theory of isotropic turbulence. *J. Aeronaut. Sci.* **4,** 311.

Thomas, A. L. R., and Taylor, G. K. (2001). Animal flight dynamics I. Stability in gliding flight. *J. Theor. Biol.* **212,** 399–424.

Thomson, K. S. (1976). On the heterocercal tail in sharks. *Palaeobiology* **2,** 19–38.

Thomson, K. S., and Simanek, E. D. (1977). Body form and locomotion of sharks. *Amer. Zool.* **17,** 343–354.

Triantafyllou, G. S., Triantafyllou, M. S., and Gosenbaugh, M. A. (1993). Optimal thrust development in oscillating foils with application to fish propulsion. *J. Fluids Struct.* **7,** 205–224.

Triantafyllou, M. S., Triantafyllou, G. S., and Yue, D. K. P. (2000). Hydrodynamics of fishlike swimming. *Ann. Rev. Fluid Mech.* **32,** 33–53.

Tytler, E. D., and Lauder, G. V. (2004). The hydrodynamics of eel swimming. I. Wake structure. *J. Exp. Biol.* **207,** 1825–1841.

Ullén, F., Deliagina, T. G., Orlovsky, G. N., and Grillner, S. (1995). Spatial orientation in the lamprey. I. Control of pitch and roll. *J. Exp. Biol.* **198,** 665–673.

Videler, J. J. (1975). On the interrelationships between morphology and movement in the tail of the cichlid fish *Tilapia nilotica* L. *Neth. J. Zool.* **25,** 144–194.

Videler, J. J. (1981). Swimming movements, body structure and propulsion in cod *Gadus morhua*. *Symp. Zool. Soc. Lond.* **48,** 1–27.

Videler, J. J. (1993). "Fish Swimming." Chapman and Hall, New York.

Videler, J. J., and Weihs, D. (1982). Energetic advantages of burst-and-coast swimming of fish at high speed. *J. Exp. Biol.* **97,** 169–178.

Vogel, S. (1988). "Life's Devices." Princeton Univ. Press, Princeton, NJ.

Vogel, S. (1994). "Life in Moving Fluids." Princeton Univ. Press, Princeton, NJ.

Von Mises, R. (1945). "Theory of Flight." Dover, New York.

Walker, J. A. (2000). Does a rigid body limit maneuverability? *J. Exp. Biol.* **203,** 3391–3396.

Walker, J. A. (2002). Functional morphology and virtual models: Physical constraints on the design of oscillating wings, fins, legs, and feet at intermediate Reynolds numbers. *Integ. Comp. Biol.* **42,** 232–242.

Walker, J. A. (2003). Kinematics and performance of maneuvering control surfaces in teleosts fishes. *In* "Biology-Inspired Maneuvering Hydrodynamics for AUV Application" (Fish, F. E., Ed.), pp. D1–D14. Proceedings of the 13th International Symposium on Unmanned Untethered Submersible Technology. Autonomous Undersea Systems Institute, Durham, NH.

Walker, J. A., and Westneat, M. W. (2000). Mechanical performance of aquatic rowing and flying. *Proc. Roy. Soc. Lond. Ser. B* **267,** 1875–1881.

Walker, J. A., and Westneat, M. W. (2002). Performance limits of labriform propulsion and correlates with fin shape and motions. *J. Exp. Biol.* **205,** 177–187.

Weatherley, A. H., and Gill, H. S. (1987). "The Biology of Fish Growth." Academic Press, New York.

Weatherley, A. H., Rogers, S. C., Pinock, D. G., and Patch, J. R. (1982). Oxygen consumption of active rainbow trout, *Salmo gairdneri* Richardson, derived from electromyograms obtained by radiotelemetry. *J. Fish Biol.* **20,** 479–489.

Webb, P. W. (1975). Hydrodynamics and energetics of fish propulsion. *Bull. Fish. Res. Bd. Canada* **190,** 1–159.

Webb, P. W. (1976). The effect of size on the fast-start performance of rainbow trout (*Salmo gairdneri* Richardson) and a consideration of piscivorous predator-prey interactions. *J. Exp. Biol.* **65,** 157–177.

Webb, P. W. (1978). Fast-start performance and body form in seven species of teleost fish. *J. Exp. Biol.* **74,** 211–226.

Webb, P. W. (1982). Fast-start resistance of trout. *J. Exp. Biol.* **96,** 93–106.

Webb, P. W. (1984). Chase response latencies of some teleostean piscivores. *Comp. Biochem. Physiol.* **79A,** 45–48.

Webb, P. W. (1989). Station holding by three species of benthic fishes. *J. Exp. Biol.* **145,** 303–320.

Webb, P. W. (1990). How does benthic living affect body volume, tissue composition, and density of fishes? *Can. J. Zool.* **68,** 1250–1255.

Webb, P. W. (1991). Composition and mechanics of routine swimming of rainbow trout, *Oncorhynchus mykiss*. *Can. J. Fish. Aquat. Sci.* **48,** 583–590.

Webb, P. W. (1992). Is the high cost of body/caudal fin undulatory propulsion due to increased friction drag? *J. Exp. Biol.* **162,** 157–166.

Webb, P. W. (1993). Is tilting at low swimming speeds unique to negatively buoyant fish? Observations on steelhead trout, *Oncorhynchus mykiss*, and bluegill, *Lepomis macrochirus*. *J. Fish. Biol.* **43,** 687–694.

Webb, P. W. (1994a). Exercise performance of fish. *In* "Advances in Veterinary Science and Comparative Medicine" (Jones, J. H., Ed.), **38B,** pp. 1–49. Academic Press, Orlando, FL.

Webb, P. W. (1994b). The biology of fish swimming. *In* "Mechanics and Physiology of Animal Swimming" (Maddock, L., Bone, Q., and Rayner, J. M. V., Eds.), pp. 45–62. Cambridge University Press, Cambridge.

Webb, P. W. (1997a). Swimming. *In* "The Physiology of Fishes" (Evans, D. D., Ed.), 2nd edn., pp. 3–24. CRC Press, Marine Science Series, Boca Raton, FL.

Webb, P. W. (1997b). Designs for stability and maneuverability in aquatic vertebrates: What can we learn? *In* "Tenth International Symposium on Unmanned Untethered Submersible Technology," pp. 86–108. Proceedings of the Special Session on Bio-Engineering Research Related to Autonomous Underwater Vehicles. Autonomous Undersea System Institute, Durham, NH.

Webb, P. W. (1998). Entrainment by river chub, *Nocomis micropogon*, and smallmouth bass, *Micropterus dolomieu*, on cylinders. *J. Exp. Biol.* **201,** 2403–2412.

Webb, P. W. (2000). Maneuverability versus stability? Do fish perform well in both? Proceedings of the 1st International Symposium on Aqua Bio-Mechanisms/ International Seminar on Aqua Bio-Mechanisms (ISABMEC 2000), August, Tokai University Pacific Center, Honolulu, HI.

Webb, P. W. (2002a). Control of posture, depth, and swimming trajectories of fishes. *Integ. Comp. Biol.* **42,** 94–101.

Webb, P. W. (2002b). Kinematics of plaice, *Pleuronectes platessa*, and cod, *Gadus morhua*, swimming near the bottom. *J. Exp. Biol.* **205,** 2125–2134.

Webb, P. W. (2003). Maneuverability – definitions and general issues. *In* "Biology-Inspired Maneuvering Hydrodynamics for AUV Application" (Fish, F. E., Ed.), pp. B1–B9. Proceedings of the 13th International Symposium on Unmanned Untethered Submersible Technology. Autonomous Undersea Systems Institute, Durham, NH.

Webb, P. W. (2004). Response latencies to postural disturbances in three species of teleostean fishes. *J. Exp. Biol.* **207,** 955–961.

Webb, P. W., and Blake, R. W. (1985). Swimming. *In* "Functional Vertebrate Morphology" (Hildebrand, M., Bramble, D. M., Liem, K. F., and Wake, D. B., Eds.), pp. 110–128. Harvard University Press, Cambridge, MA.

Webb, P. W., and de Buffrénil, V. V. (1990). Locomotion in the biology of large aquatic vertebrates. *Trans. Amer. Fish. Soc.* **119,** 629–641.

Webb, P. W., and Fairchild, A. (2001). Performance and maneuverability of three species of teleostean fishes. *Can. J. Zool.* **79,** 1866–1877.

Webb, P. W., and Gerstner, C. L. (2000). Swimming behaviour: Predictions from biomechanical principles. *In* "Biomechanics in Animal Behaviour" (Domenici, P., and Blake, R. W., Eds.), pp. 59–77. Bios Scientific Publishers Ltd., Oxford.
Webb, P. W., and Keyes, R. S. (1981). Division of labor between median fins in swimming dolphin fish. *Copeia* **1981,** 901–904.
Webb, P. W., and Keyes, R. S. (1982). Swimming kinematics of sharks. *Fish. Bull. US.* **80,** 803–812.
Webb, P. W., and Weihs, D. (1994). Hydrostatic stability of fish with swimbladders: Not all fish are unstable. *Can. J. Zool.* **72,** 1149–1154.
Webb, P. W., La Liberte, G. D., and Schrank, A. J. (1996a). Does body and fin form affect the maneuverability of fish traversing vertical and horizontal slits? *Environ. Biol. Fish.* **46,** 7–14.
Webb, P. W., Gerstner, C. L., and Minton, S. T. (1996b). Station holding by the mottled sculpin, *Cottus bairdi* (Teleostei: Cottidae), and other fishes. *Copeia* **1996,** 488–493.
Weihs, D. (1973). Mechanically efficient swimming techniques for fish with negative buoyancy. *J. Mar. Res.* **31,** 194–209.
Weihs, D. (1974). The energetic advantages of burst swimming. *J. Theor. Biol.* **49,** 215–229.
Weihs, D. (1977). Effects of size on sustained swimming speeds of aquatic organisms. *In* "Scale Effects in Animal Locomotion" (Pedley, T. J., Ed.), pp. 333–338. Academic Press, New York.
Weihs, D. (1981). Effect of swimming path curvature on the energetics of fish. *Fish. Bull. US* **79,** 171–176.
Weihs, D. (1989). Design features and mechanics of axial locomotion in fish. *Amer. Zool.* **29,** 151–160.
Weihs, D. (1993). Stability of aquatic animal locomotion. *Contemp. Math.* **141,** 443–461.
Weihs, D. (2002). Stability *versus* maneuverability in aquatic animals. *Integ. Comp. Biol.* **42,** 127–134.
Wilga, C. D., and Lauder, G. V. (1999). Locomotion in the sturgeon: Function of the pectoral fins. *J. Exp. Biol.* **202,** 2413–2432.
Wilga, C. D., and Lauder, G. V. (2000). Three-dimensional kinematics and wake structure of the pectoral fins during locomotion in Leopard sharks, *Triakis semifasciata. J. Exp. Biol.* **203,** 2261–2278.
Wilga, C. D., and Lauder, G. V. (2001). Function of the heterocercal tail in sharks: Quantitative wake dynamics during steady horizontal swimming and vertical maneuvering. *J. Exp. Biol.* **205,** 2365–2374.
Wilga, C. D., and Lauder, G. V. (2001). Functional morphology of the pectoral fins in bamboo sharks, *Chiloscyllium plagiosum*: Benthic vs. pelagic station-holding. *J. Morphol.* **249,** 195–209.
Winterbottom, R. (1974). A descriptive synonymy of the striated muscles of the Teleostei. *Proc. Acad. Nat. Sci. Philadelphia* **125,** 225–317.
Zhu, Q., Wolfgang, M. J., Yue, D. K. P., and Triantafyllou, G. S. (2002). Three-dimensional flow structures and vorticity control in fish-like swimming. *J. Fluid Mech.* **468,** 1–28.

FURTHER READING

Drucker, E. G., and Lauder, G. V. (2001a). Locomotor function of the dorsal fin in teleost fishes: Experimental analysis of wake forces in sunfish. *J. Exp. Biol.* **204,** 2943–2958.

Drucker, E. G., and Lauder, G. V. (2001b). Wake dynamics and fluid forces of turning maneuvers in sunfish. *J. Exp. Biol.* **204,** 431–442.

Webb, P. W. (1988b). Simple physical principles and vertebrate aquatic locomotion. *Amer. Zool.* **28,** 709–725.

Foreman, M. B., and Eaton, R. C. (1990). EMG and kinematic analysis of the stages of the Mauthner-initiated escape response. *Soc. Neurosci. Abstr.* **16,** 1328.

Mac Donnell, A. J., and Blake, R. W. (1990). Rheotaxis in *Otocinclus* sp. (Teleostei: Loricariidae). *Can. J. Zool.* **68,** 599–601.

Webb, P. W. (1988a). Steady" swimming kinematics of tiger musky, an esociform accelerator, and rainbow trout, a generalist cruiser. *J. Exp. Biol.* **138,** 51–69.

Weihs, D. (1972). A hydrodynamic analysis of fish turning manoeuvers. *Proc. Roy. Soc. Lond.* **182B,** 59–72.

9

FAST-START MECHANICS

JAMES M. WAKELING

I. INTRODUCTION

Fast starts are high acceleration swimming maneuvers, either starting from rest or imposed upon a period of steady swimming. Fast starts are important for most fishes when escaping predators and for some fishes in

Fish Biomechanics: Volume 23
FISH PHYSIOLOGY

DOI: 10.1016/S1546-5098(05)23009-1

achieving prey capture, and thus they are vital components of a fish's locomotory repertoire. Fast-start performances can be measured in terms of acceleration, velocity achieved, or distance traveled during the initial moments of the start, and such measures of fast-start performance have been reviewed by Domenici and Blake (1997). This chapter considers the various mechanisms that a fish must use in order to achieve a fast-start swimming maneuver.

The mechanics of the fast start are driven by a number of different processes, including force generation by muscles, transmission of those forces to appropriate parts of body, and the transfer of those forces to the water. Studies of fast-start swimming have typically focused on specific aspects of the fast-start maneuver, and so the information presented here is synthesized from a range of different experimental and theoretical observations. Some of the synthesized information points to stereotypical aspects of the swimming behavior. However, other aspects of the fast start show ranges of characteristics observed by the different studies. In such cases we must conclude that there can be considerable variability in the fast-start mechanics between species, and that these particular areas would greatly benefit from further investigation.

Fast starts are typically classified by the shape to which the body bends. The kinematics of S-starts were described for pike *Esox lucius* by Hoogland and co-workers (1956). The body is curved into an S shape prior to a predatory strike; this initial posture is achieved either immediately before the strike, or may persist for a few seconds before the strike (Hoogland *et al.*, 1956; Webb and Skadsen, 1980). Weihs (1973) described L-starts in the trout, which are synonymous with the more commonly referred to C-starts. The start was divided into a number of stages: stage 1, in which the fish initially bends into a tight C shape; stage 2, in which the fish flexes in the opposite direction (this stage may be repeated a number of times); and stage 3, in which the fish either swims off or glides away in a steady fashion. C-starts may involve a change in direction and are commonly used during escape responses, whereas S-starts typically involve little turning and are commonly used for prey capture (Domenici and Blake, 1997). However, S shape body profiles can be observed during escape responses (Webb, 1976; Jayne and Lauder, 1993; Wakeling and Johnston, 1998), and there is some overlap between the muscle dynamics and kinematics (Webb, 1976) of the different types of start. Fast-start escape responses for four different species can be seen in Figure 9.1. This chapter considers the general mechanisms that underlie fast starts of either type. When particular types of fast start are used as examples they will be identified.

In this chapter the mechanics of the fast start are considered in a sequential sequence of events from the neural activation of the muscles,

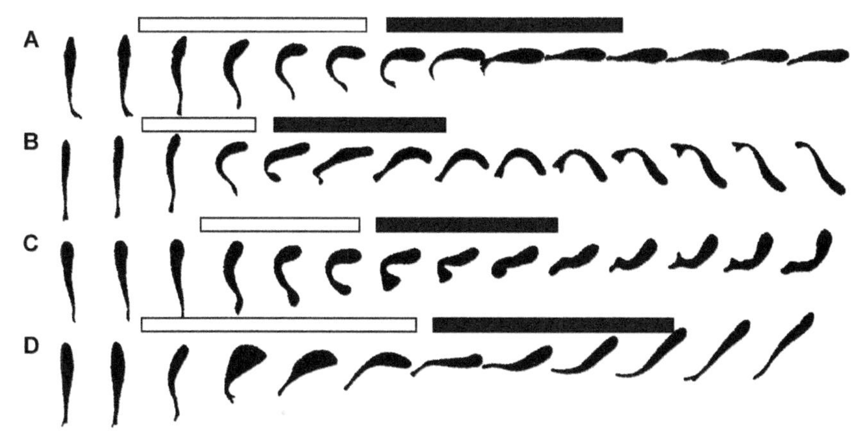

Fig. 9.1. Dorsal view silhouettes of a goldfish (A), zebrafish (B), catfish (C), and marbled hatchetfish (D) during fast-start escape maneuvers. Silhouettes are drawn at 5 ms intervals. Open and closed bars above each sequence indicate stage 1 and stage 2 of the fast start, respectively. Mauthner cells were identified in all these species, and the response latency of the starts was between 5 and 10 ms, corresponding to the Mauthner reflex. (Adapted from Eaton *et al.*, 1977.)

the generation of muscle force that acts to bend the body, the development of the typical kinematic patterns of the fast start, and finally the generation of the hydrodynamic forces that act to accelerate the body.

II. INITIATION OF THE FAST START

Fast starts are typically mediated by the action of the Mauthner cells, or M-cells. These are a pair of reticulospinal neurons that have axons that extend the full length of the spinal cord. M-cell activity is correlated with the escape response in the goldfish *Carassius auratus* (Zottoli, 1977; Eaton *et al.*, 1981), and it is thought to be crucial for predator avoidance. In freely swimming fish, Mauthner-mediated escape responses can be elicited with auditory (Zottoli, 1977; Eaton *et al.*, 1981) and visual (Eaton *et al.*, 1977) stimuli. The M-cell axons are the fastest conducting fibers in the fish central nervous system (Funch *et al.*, 1981) and provide synaptic input to the primary motor neurons (Myers, 1985; Fetcho, 1986). Sensory information is modulated by the M-cells, resulting in large-amplitude body bending and the splaying of some or all of the fins (Eaton *et al.*, 1977). The response latencies recorded in 13 species of teleost fishes were between 5 and 10 ms (Eaton *et al.*, 1977). Activity in the M-cell is not a requirement for the fast start, however, and fast-start escape responses are possible (although less likely) when the M-cells have been ablated (Eaton *et al.*, 1982; Zottoli *et al.*, 1999) and also occur in species where the M-cells do not exist (*Cyclopterus lumpus*; Hale, 2000). Not only are Mauthner cells used for escape responses, but their activity has also been correlated with the final stage of the predatory strike in *C. auratus* (Canfield and Rose, 1993).

Direct stimulation to the M-cells results in a fairly constant duration and final trajectory angles for the stage 1 kinematics (Nissanov and Eaton, 1989); however, freely swimming fishes can show substantial variation in these initial kinematic parameters (Foreman and Eaton, 1993; Tytell and Lauder, 2002). The onset of the stage 1 muscle contraction is mediated by the action of the M-cells and is simultaneous along the length of the body (Jayne and Lauder, 1993; Wakeling and Johnston, 1999a; Ellerby and Altringham, 2001; Tytell and Lauder, 2002); however, the magnitude and duration of this contraction vary between different escape responses. The turning angle achieved during stage 1 depends on the direction at which the startle stimulus is presented in both the goldfish *C. auratus* (Eaton and Emberley, 1991) and the angelfish *Pterophyllum eimekei* (Domenici and Blake, 1993), in which the turning angle is correlated with the duration that the muscle is active during the stage 1 (Eaton *et al.*, 1988). The duration of the stage 1 muscle activity also varies at different positions along the fish, and has been shown to increase in a caudal direction for the bluegill sunfish *Lepomis macrochirus* (Jayne and Lauder, 1993). These observations suggest that other reticulospinal neurons acting in parallel with the active M-cell help determine the swimming trajectory, and, indeed, the existence of parallel pathways has been indicated by the presence of descending interneurons that are also active during the fast start (Fetcho, 1992).

Fast-start kinematics are not stereotypical across species; this is partly due to interspecific variation in the degree of bilateral muscle activation that occurs at the beginning of the maneuver. In the goldfish *C. auratus* and the tench *Tinca tinca*, reciprocal inhibition between the two M-cells results in one M-cell being activated while the other is inhibited (Yasargil and Diamond, 1968). The consequence of this neural inhibition is to elicit muscle contraction down only one side of the body, and indeed little or no contralateral muscle activity is observed during stage 1 of the fast start in *L. macrochirus* (Jayne and Lauder, 1993), the common carp *Cyprinus carpio* (Wakeling and Johnston, 1999a), and the trout *Oncorhynchus mykiss* (Hale *et al.*, 2002). Reciprocal inhibition of the M-cells has been associated with the M-cell axon cap, a structure surrounding the axon hillock (Furukawa and Furshpan, 1963), and more basal species with "primitive" or absent axon caps experience considerable bilateral muscle contraction during stage 1. Indeed, the bowfin *Amia calva* (Westneat *et al.*, 1998) and bichirs

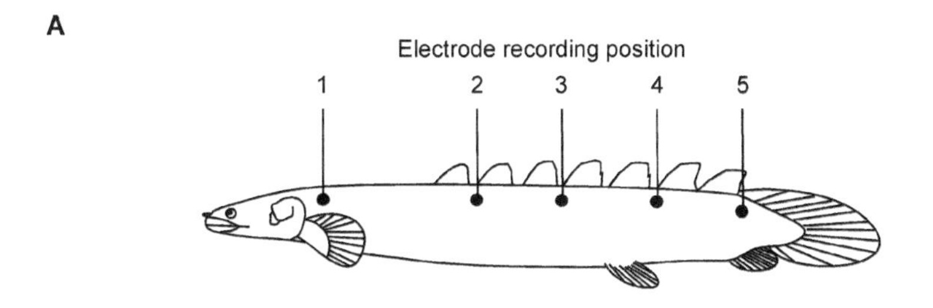

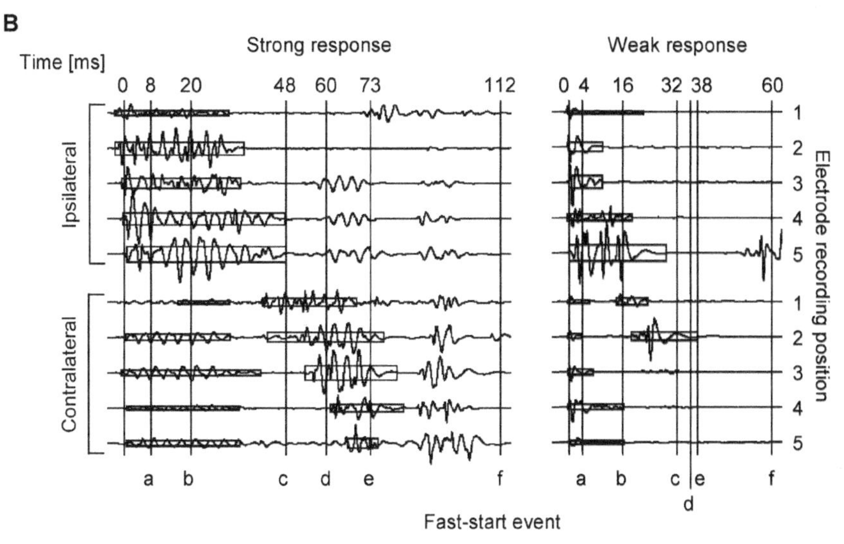

Fig. 9.2. Electromyographic (EMG) recordings from the bichir *Polypterus senegalus* during a strong and a weak fast-start escape response. Recording electrodes were placed at different longitudinal locations on both the left and right sides of the body (A). The duration and magnitude of the EMG is shown by the rectangles (B). Note the simultaneous onset of the EMG, and the bilateral muscle activity in this species. Marked fast-start events indicate (a) first visible motion, usually at the head, (b) maximum acceleration, (c) maximum velocity achieved during stage 1, (d) maximum curvature, (e) end of kinematic stage 1, and (f) maximum overall velocity. The body formed a true "C" shape during stage 1 of the strong response, but showed a double-bend "S" shape during the weak response. (Adapted from Tytell and Lauder, 2002.)

Polypterus palmas (Westneat *et al.*, 1998) and *P. senegalus* (Figure 9.2; Tytell and Lauder, 2002) showed varying degrees of bilateral muscle activity during stage 1 of the fast start that was attributed to the presence of a "primitive" axon cap for *P. senegalus*. At the extreme, the larval lamprey *Petromyzon marinus* lacks an axon cap and shows bilateral muscle activity that results in a withdrawal behavior when startled (Currie and Carlsen, 1987).

The initial body bend of stage 1 is often followed by a contralateral muscle contraction (stage 2) bending the body to the opposite side. The turning angle in stage 2 correlates with the turning angle in stage 1 (Eaton *et al.*, 1988; Domenici and Blake, 1991) and indicates that the underlying neural command is ballistic and can control the swimming trajectory without sensory feedback (Eaton and Emberley, 1991). The stage 2 muscle contraction occurs as a wave of posteriorly propagating muscle activity in *C. carpio* (Wakeling and Johnston, 1999a), *Lepistosteus osseus*, *A. calva*, *O. mykiss* (Jayne and Lauder, 1993; Ellerby and Altringham, 2001; Hale *et al.*, 2002), and *P. senegalus* (Tytell and Lauder, 2002). From stage 2 onward, the muscle dynamics and swimming mechanics of the fast start approach those of steady swimming and are discussed in Chapters 7 and 11 of this volume.

Little is known about the neural activation pattern of S-start maneuvers. In some species, an "S" shape body profile can occur despite a C-start type of muscle activation, and such "S" shapes are the result of passive, inertial bending in the caudal region (Section IV.B; Jayne and Lauder, 1993; Wakeling and Johnston, 1998). However, an S-start muscle activation pattern has been observed in the muskellunge *Esox masquinogy* (Hale, 2003), which is distinct from the C-start pattern described previously. In the muskellunge, the S-start occurs with a simultaneous muscle activation at anterior and central locations of the myotomes on one side with an activation in the contralateral myotome at a posterior site. This pattern suggests that a distinct neural pathway may be responsible for S-start activations in some species.

The initial muscle contractions at the beginning of a C-bend fast start show both similarities and variations across species. The characteristic features of the stage 1 muscle activity are that activity within the M-cells initiates muscle activity simultaneously along the length of the body on at least one side. The duration of this burst of activity is variable, is under the control of parallel neural networks, and dictates the swimming trajectory of the fish. The next section discusses how this muscle activity translates to muscle shortening and bending of the body.

III. MUSCULAR CONTRACTION ACTS TO BEND THE FISH

During steady swimming, the mechanical power output from the myotomal muscle is largely required to overcome the drag on a swimming fish, and the power required is proportional to the third power of the swimming velocity (Bone, 1978; Chapter 11 of this volume). Therefore, in order for fishes to swim at even a moderate range of speeds their muscles must be capable of producing a large range of power outputs. During fast-start swimming, the muscle must additionally be able to generate large forces at a rapid rate. These requirements place conflicting demands on the fish muscle, which must support low-speed cruise economy and high-speed force and power output. Fishes have solved this problem with a common solution, by dividing the locomotor system into specialized regions for these differing functions. This section discusses how the specific adaptations of the muscle are used to power fast-start swimming.

The contractile properties of fish muscle vary from "fast" to "slow." The slow fibers are red in color and contain large numbers of mitochondria, extensive myoglobin, and blood supply. Slow fibers are particularly suited to contracting over a sustained period of time, and indeed are typically used for low-speed swimming (Grillner and Kashin, 1976; Chapters 6 and 7). The fast fibers, on the other hand, are white in color due to poor vascularization and lack of myoglobin. These fibers are fast, but fatigue rapidly. Mechanically, the fast fibers take shorter times to develop and relax twitch forces (Granzier *et al.*, 1983; Akster, 1985; Akster *et al.*, 1985) and can achieve faster maximum strain rates than the slower fibers (Johnston *et al.*, 1985). The fast muscle fibers typically constitute more than 90% of the myotomal muscle, and never less than 74% in 84 species examined (Greer-Walker and Pull, 1975). Fast starts are high acceleration maneuvers and, therefore, require the rapid development of large amounts of muscle force. This task is most suited to the fast fibers due to their faster contractile properties and larger muscle bulk.

A. Muscle Strain During Fast Starts

Muscle fiber strain is a measure of the change in length of the fibers relative to their resting length (Figure 9.3). The muscle fiber strain in the myotomes varies at different positions along the body, and depends on both the position and orientation of the muscle fibers, and the degree of bending experienced by the body. In a simplistic sense, the myotomal muscle can be considered a homogenous beam whereby the muscle fiber strain increases with increasing distance from the spine (see Chapter 7). However, the majority of the fast fibers in teleost myotomal muscle are neither horizontal (Shann, 1914) nor parallel to the median plane (Nursall, 1956), and thus modifications to simple bending beam models should be used when estimating strain in fast fibers. The fiber trajectories in the main, fast, myotomal muscle of teleosts have been considered to approximate segments of helices that form co-axial bundles along the body (Alexander, 1969). One consequence of this helical arrangement is that the muscle strain becomes

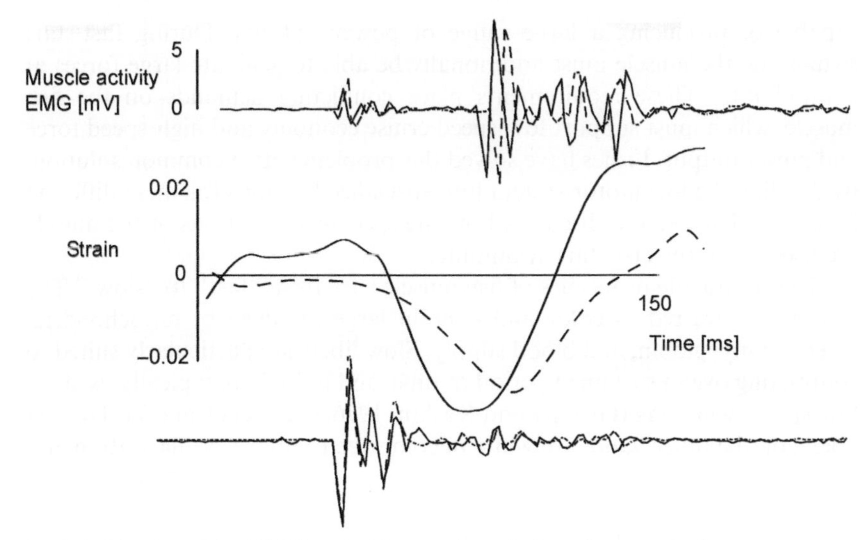

Fig. 9.3. Muscle activity (EMG) and white muscle strain for a fast start in the common carp, *Cyprinus carpio*. The uppermost traces are EMG recordings from the right side of the body, and the lowermost traces are EMGs from the left side of the body. The central traces show the white muscle strain during the fast start, with positive strain occurring during bends to the right. Solid lines denote recordings for longitudinal position 0.43*L*, dashed lines for position 0.56*L*, where *L* is total body length. (From Wakeling and Johnston, 1999a.)

more uniform across the width of the fish than would be predicted from a simple bending beam theory (Alexander, 1969). The slower fibers typically form a superficial layer covering the main bulk of fast myotomal muscle, which gives them a greater lever arm distance from the spine, and so they experience greater strains than the adjacent fast fibers. Indeed, the mean muscle fast fiber strain during fast starts in the carp *C. carpio* was calculated to be 36% of that of the adjacent slow fibers (Wakeling and Johnston, 1999b). Therefore, the rapid, large-amplitude body bending that occurs during stage 1 of the fast start results in higher strains and strain rates in the superficial slow fibers than the fast, deeper myotomal muscle. Muscle activity can be recorded from both fast and slow regions of the muscle during stage 1 of the fast start (Jayne and Lauder, 1993); however, the slow muscle reaches strain rates too great for them to produce force (Rome *et al.*, 1988). It is thus likely that the initial muscle contraction of the fast start is powered mainly by the action of the fast myotomal muscles (Rome *et al.*, 1988) due to their advantageous position and orientation, and the fast fiber specialization for generating stress at a rapid rate during stage 1 contraction.

B. Intrinsic Contractile Properties of the Myotomal Muscle

The myotomal muscle properties vary not only between the lateral locations of the superficial slow muscle and deeper fast muscle, but may also change along the length of the body. The rostral myotomes typically have faster contractile properties than the caudal myotomes. The rostral myotomes have been shown to have faster intrinsic shortening velocities in the short-horned sculpin *Myoxocephalus scorpius* (James *et al.*, 1998), and faster twitch times in *M. scorpius* (Johnston *et al.*, 1993, 1995), the bluefin tuna *Thunnus thynnus* (Wardle *et al.*, 1989), saithe *Pollachius virens* (Altringham *et al.*, 1993), cod *Gadus morhua* (Davies *et al.*, 1995), and mackerel *Scomber scombrus* (Wardle, 1985). However, the evidence suggests that the maximum isometric stress of the fast myotomal muscle varies little with longitudinal position (Altringham *et al.*, 1993; Johnston *et al.*, 1993; James *et al.*, 1998). The initial muscle contraction of the fast start is simultaneous between rostral and caudal locations along the body (Jayne and Lauder, 1993; Wakeling and Johnston, 1999a; Tytell and Lauder, 2002), and so the fast muscle will begin generating stress at the same time along the myotomes. However, the faster contractile properties of the rostral myotomes will result in the rostral muscle quickly generating greater stress than the caudal myotomes.

C. Muscle Bending Moments

The force produced by the muscle is a function of both the muscle fiber stress and the cross-sectional area of active muscle. The fast muscle is more massive in the rostral myotomes than in the caudal region (Johnston *et al.*, 1995; Wakeling and Johnston, 1999a). Therefore, the combined effect of a larger cross-sectional area of muscle and that muscle generating stress at a higher rate results in greater fast muscle forces being initially developed in the rostral than the caudal myotomes of the fish. The muscle generates a bending moment acting on the spine. This moment is a function of the muscle force and the moment arm to the center of bending. For each section of the fish body, the force can be considered to act through the centroid (first moment) of area of the fast-myotomal muscle. With the fish being more massive in its central region, the lateral distance of the centroid of fast muscle area is greater in the rostral than in the caudal myotomes. Indeed, the moment arms are 6-fold greater at a longitudinal position 40% than at 80% from the snout in the carp *C. carpio* (Wakeling and Johnston, 1999a). Due to their generating greater forces and having longer moment arms, the bending moments generated by the rostral myotomes during stage 1 of the fast start can be at least 40-fold greater than those generated by the caudal

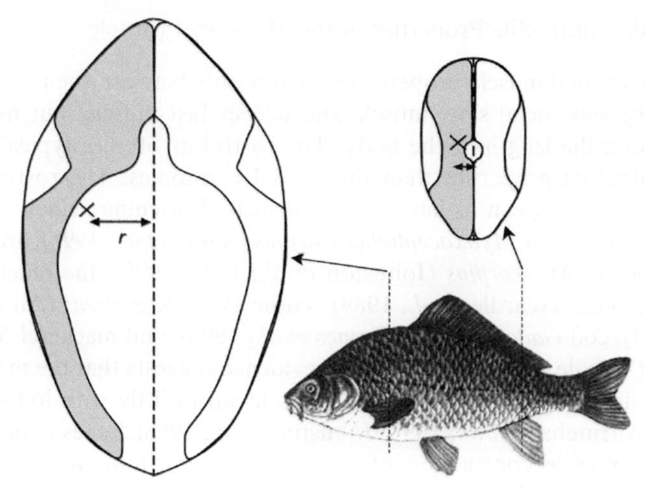

Fig. 9.4. The myotomal muscle generates a bending moment on the spine that is proportional to the product of the muscle stress, cross-sectional area, and moment arm (r). The muscle force can be considered to act through its centroid of cross-sectional area (denoted by the cross). The fast muscle at the rostral site at $0.3L$ (A) has greater cross-sectional area (shaded region) and moment arm than for the more caudal site at $0.7L$ (B). Therefore, the more rostral muscle generates greater bending moments on the spine. Transverse sections are taken from the common carp, *Cyprinus carpio* (C), as indicated by vertical dashed lines and arrows.

myotomes in *C. carpio* (Figure 9.4; Wakeling and Johnston, 1999a). It should be noted that in fishes that exhibit some bilateral muscle activity during stage 1 of the fast start (e.g., *A. calva*, Westneat *et al.*, 1998; *P. palmas*, Westneat *et al.*, 1998; *P. senegalus*, Tytell and Lauder, 2002) the moments generated by the muscle of the two sides of the fish will be opposed, and the net moment will be the result of the contribution from the two sides.

IV. STAGE 1 BODY BENDING OCCURS WITH A TRAVELING WAVE OF CURVATURE

At the beginning of the fast start, the myotomal muscle generates a net bending moment along one side of the fish that is greater at the rostral than at the caudal end of the body. This bending moment will act to cause a flexion in the body whose amplitude is dependent on the bending moments, body stiffness, and external resistance. This section discusses the various factors that influence stage 1 body bending and describes how the initial muscle contraction generates a rearward wave of bending to travel along the body.

A. Internal Stiffness and External Resistance

Body bending during swimming results from the interaction of the bending moments generated by the muscle and the internal stiffness and external resistance to bending of the body. Blight (1977) considered the simple case of flexion in a homogenous rod in which a combination of internal tension and axial incompressibility caused the rod to flex. The shape of the flexion from a straight to a "C" profile depended on the dominance of the internal stiffness to the external resistance. If internal stiffness dominated, then the rod would flex into a "C" shape around the side of internal tension; however, if external resistance dominated, then the rod would show additional flexion to the opposite direction near its ends. Blight argued that fish may act as "hybrid" oscillators, in which stiffness dominated in the rostral region, whereas resistance dominated in the caudal region of the fish. In this way, unilateral tension would result in the rostral end bending into flexure that would then propagate as a wave of bending in a rearward direction along the fish.

The magnitude and propagation of the body bending depend on the balance of internal stiffness and external resistance to bending. The flexural stiffness of the body depends on both passive and active components. The passive stiffness depends on the material and structural properties of the skeletal system. In static bending tests, spine stiffness was constant along the body length for the Norfolk spot *Leiostomus xanthurus* and constant for the first 34 vertebrae for the skipjack tuna *Katsuwonus pelamis* (Hebrank, 1982); this region encompasses the distribution of the myotomal muscle. The angular stiffness of the spine during dynamic lateral bending has been shown to be greater in caudal than in precaudal regions in the blue marlin *Makaira nigricans* (Long, 1992). The titin molecule, which is thought to be responsible for passive tension in muscle, is present as different isoforms in carp that change along the length of the body (Spierts *et al.*, 1997). Passive muscle stiffness and tension increase with increasing sarcomere length, and are greater in rostral muscle than in caudal muscle in carp (Spierts *et al.*, 1997). Furthermore, the intramuscular pressure that is generated during muscle contractions (Wainwright *et al.*, 1978; Westneat *et al.*, 1998) places the skin under tension. This skin stiffness has been shown to contribute to whole body stiffness in the longnose gar *Lepidisteus osseus* (Long *et al.*, 1996), and skin stiffness varies with the skin tension and thus muscle activity in the lemon shark *Negaprion brevirostris* (Wainwright *et al.*, 1978). The passive stiffness of the body is thus affected by a range of tissues that provide different contributions to overall stiffness along the body.

The exact stiffness of the body can be dynamically modulated via muscular contractions. The muscle force may contribute to the flexural stiffness by stressing the skin (as mentioned previously), but also in a more direct manner because a muscle produces greater tension when it is stretched (Aubert, 1956) than when it is shortened (Hill, 1938). A whole body work loop method has been used to quantify the changes in body stiffness with altered muscle activity in a largemouth bass. This method involved the cyclical bending of dead fish while the myotomal muscle was electrically stimulated and the bending moment was measured. It was demonstrated that muscle activity on the inside of the bend during body flexion resulted in a decrease in stiffness, whereas muscle activity on the outside of the bend increased the stiffness (Long and Nipper, 1996). These observations are consistent with the role of stretching and shortening in modulating the muscle force production. However, it should be noted that the bilateral muscle activity in the bichir *Polypterus senegalus* stiffened the body by a constant amount despite variation in the amount of muscle activity (Tytell and Lauder, 2002), and so in this case changes in muscle activity may not change the flexural stiffness.

The resistance to body bending comes in part from the inertial loads of the body mass and added mass of water that must be displaced as the body flexes. The inertial properties of the body can be calculated from the longitudinal mass distribution of the body, and it should be noted that fishes are typically most massive in their rostral region (Hess and Videler, 1984; Frith and Blake, 1991; Wakeling and Johnston, 1998, 1999a). The added mass of water is the mass of water that will be accelerated when a section of body accelerates laterally through the water (Lighthill, 1970). Large values of added mass occur at the tail blade due to the large depth profile of the body (Weihs, 1973; Hess and Videler, 1984; Frith and Blake, 1991; Wakeling and Johnston, 1999a). The large added mass in the caudal region (Figure 9.5) provides a large external resistance that supports Blight's 1977 model of the fish acting as a hybrid oscillator. The nature of the body bending and the propagation of the wave of curvature along the body can be modulated by altering the added mass of the water. This can be achieved by altering the depth profile of the fish by extending the fins and tail. Fin splaying is one of the actions caused by the Mauthner reflex (Eaton *et al.*, 1977), and the degree of muscle activity employed in the erector muscles will alter the speed and swimming trajectory of the fish.

B. Bending against a Hydrodynamic Load

Not only does the muscle provide body stiffness that shapes the body flexion but, as discussed in a previous section (III.C), the muscle generates the force required to bend the body. The body bending depends on the interaction between the bending moments generated by the muscle and the

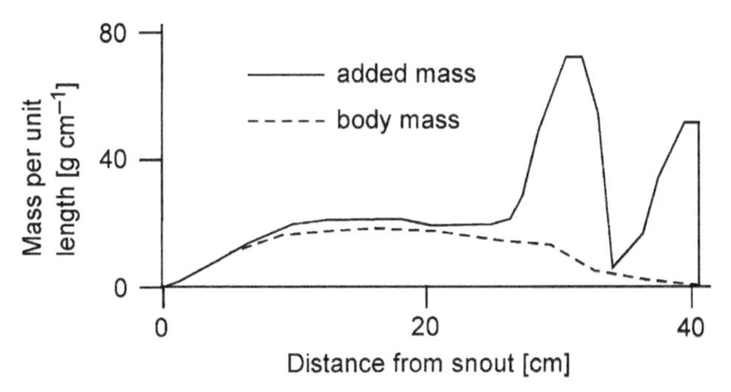

Fig. 9.5. Mass and added mass as a function of longitudinal position along the body of the northern pike *Esox lucius*. Note how the greatest contributions to the mass per unit length are from the added mass in the region of the dorsal and caudal fin. (Adapted from Frith and Blake, 1991.)

internal stiffness and external resistance. It is possible to explain the stage 1 body bending using a simplified model of muscle torque and hydrodynamic resistance, in which the muscle torque drives the body flexion without contributions from the internal stiffness (Wakeling and Johnston, 1999a). Body bending is resisted, in part, by the inertia of the body and the added mass of water. The angular acceleration at each intervertebral junction is a function of the ratio of the bending moment generated by the adjacent myotomal muscle and the inertia of the body mass and added mass of water (Wakeling and Johnston, 1999a). In the rostral region the bending moment is large and rapidly causes spine flexion. However, the myotomal muscle bulk decreases toward the tail and occurs at a closer distance to the spine, resulting in a net decrease in bending moment that is less able to flex the body. These relations were used to predict that during the initial stage 1 muscle contraction in the common carp *C. carpio*, the angular acceleration of bending in the spine could be more than two orders of magnitude greater at rostral than at caudal sites on the myotomal muscle (Wakeling and Johnston, 1999a). This difference was sufficient to explain a greater rate of bending at the rostral sites, with peak body bending occurring sooner at rostral than at caudal sites. Thus, a wave of peak body bending can be explained purely in terms of the muscle torque and the body inertia.

The mechanisms causing stage 1 body bending and the propagation of a wave of body bending involve contributions from both the body stiffness and the inertial resistance to bending. Both lines of argument lead to the same conclusions: that the body initially flexes in its rostral region with a wave of bending propagating along the body to the tail. These conclusions are consistent with the observed kinematics of fast starts (Figure 9.1; Webb,

1975a; Eaton *et al.*, 1977; Jayne and Lauder, 1993; Wakeling and Johnston, 1998; Spierts and van Leeuwen, 1999; Temple *et al.*, 2000; Fernández *et al.*, 2002; Goldbogen *et al.*, 2005). The degree of bending and the propagation of the wave of bending depend on the level of muscle activity and the amount of asymmetry in the muscle activity between the two sides of the body. As discussed in Section II, these factors can vary between types of fast start and between species.

V. MUSCLE POWER PRODUCTION AND FORCE TRANSMISSION TO THE WATER

The bulk of the myotomal muscle resides in the central region of the body, where the majority of the mechanical power is produced for fast-start swimming. However, as is discussed in Section VI, the majority of the hydrodynamic forces during fast starts are imparted to the water from the caudal region (Weihs, 1973; Webb, 1977; Frith and Blake, 1991). Therefore, there must be mechanisms to transmit the muscle power to the caudal region of the body. The transmission can occur through both active and passive structures, and in this section the role of the muscle for active power production and force transmission as well as the role of the passive tendons and skeletal structures in force transmission is discussed.

Mechanical power is the product of force and shortening velocity, and so the production of mechanical power requires a moderate balance between both muscle shortening and force production. On the other hand, force transmission is best achieved through stiff structures that combine high forces with little shortening, such as tendons (see Chapter 5; Shadwick *et al.*, 2002). There has been some debate about the relative role of the muscle in different myotomes in producing power or transmitting force during steady swimming (Wardle *et al.*, 1995), and the variations in longitudinal function are partly determined by the relative timing between muscle activation and changes in length (Chapter 7). However, during stage 1 of the fast start, the muscles are activated simultaneously at the onset of shortening (Figures 9.2 and 9.3) and changes in muscle function along the fish are the result of the intrinsic muscle properties and the dynamics of muscle shortening (e.g., inverse force-velocity relation; see Figure 6.4, Chapter 6).

The myotomal muscle varies with its longitudinal position along the fish both in its intrinsic contractile properties (Section III.B) and with its contraction dynamics; it is the combination of these two factors that determines the balance of force and power produced by the muscle at each region. Several approaches have been described in the literature for determining the muscle fiber strain during swimming. The strain can be calculated from measurements of the spine curvature and the muscle fiber geometry (e.g., Alexander, 1969; Rome *et al.*, 1988; van Leeuwen *et al.*, 1990; van Leeuwen, 1992; Lieber *et al.*, 1992; Johnston *et al.*, 1995; Wakeling and Johnston, 1998, 1999b; Spierts and van Leeuwen, 1999), or measured directly using implanted sonomicrometry crystals (e.g., Covell *et al.*, 1991; Franklin and Johnston, 1997; James and Johnston, 1998a; Wakeling and Johnston, 1998; Ellerby and Altringham, 2001). Most studies that have reported white muscle strains at varying longitudinal positions during fast starts have used the first calculation technique.

During fast starts, the spine flexes into greater curvatures than typically occur during steady swimming (Figure 9.1). The maximum spine curvature achieved during stage 1 of the fast start varies along the length of the body. The maximum curvature has been shown to increase in a caudal direction to a maximum at 0.6 body lengths (L) from the snout for a range of seven species of marine teleost (Wakeling and Johnston, 1998; Fernández *et al.*, 2002), with this curvature remaining steady between 0.6 and 0.8L. For the common carp, *C. carpio*, the maximum spine curvatures have been reported to be greatest between 0.5 and 0.8L for C-starts (Spierts and van Leeuwen, 1999; Wakeling *et al.*, 1999). S-starts in *C. carpio* occur with smaller accelerations than C-starts, and the presumably lower level of muscle activation results in a reduction of the caudal spine curvature, with the maximum curvature occurring at 0.5L (Spierts and van Leeuwen, 1999).

A. Longitudinal Variation in Muscle Strain

Evidence for how the patterns of fast-muscle strain are distributed along the body during stage 1 of the fast start is currently equivocal. The muscle fiber strain depends on the orientation of the fibers (Figure 9.6), their distance from the spine, and the spine curvature. The decreasing width of the fish toward the caudal region is a major factor in contributing to smaller fiber strains in this region (Wakeling and Johnston, 1999b; this is in contrast to the situation in steady swimming; see Chapter 7). Despite the greater spine curvature in the caudal region, the maximum fiber shortening during stage 1 of the fast start has been predicted to decrease between 0.6 and 0.8L for seven species of marine fish (Johnston *et al.*, 1995; Wakeling and Johnston, 1998; Fernández *et al.*, 2002) and the carp *C. carpio* (Wakeling *et al.*, 1999). It is interesting to note that this caudal region that shows decreases in muscle strain corresponds to the region where the helical muscle fiber arrangement (Alexander, 1969) is replaced by more longitudinally oriented fibers (Figure 9.6; Gemballa and Vogel, 2002). In another study on *C. carpio* it was predicted that there was no significant effect of the longitudinal position on the maximum fast fiber strain for C-starts; however, the caudal strain was

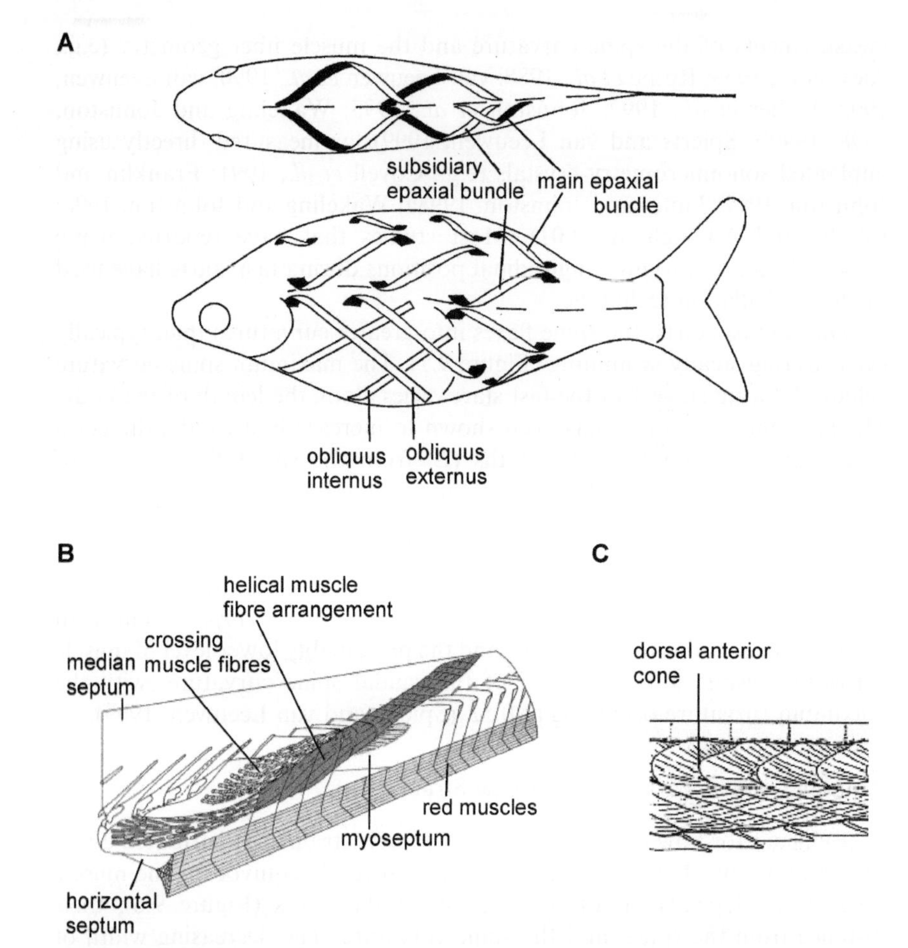

Fig. 9.6. (A) Schematic of the muscle fiber trajectories of the anterior and central myotomes of a typical teleost (Alexander, 1969). (B) Muscle fiber trajectories in the epaxial midbody of the gnathostome fishes (Gemballa and Vogel, 2002). (C) Muscle fiber orientations projected onto a horizontal plane for the posterior myomeres in the gnathostome *Polypterus delhezi* (Gemballa and Vogel, 2002).

significantly smaller during S-starts (Spierts and van Leeuwen, 1999). Strain in fast fibers has been measured by sonomicrometry at different longitudinal sites during fast starts in *O. mykiss*; Ellerby and Altringham (2001) observed that strain was significantly greater at 0.65L than at more rostral sites during stage 1 of the fast start. In contrast, Goldbogen *et al.* (2005) found higher muscle strain at 0.4L than at 0.7L during what they categorized as weak fast-start responses, while posterior stains exceeded anterior strains only in the strongest responses.

B. High Stress in Caudal Region

Decreases in muscle strain toward the caudal region may cause a longitudinal change in muscle function, with the rostral fibers producing greater mass-specific power and the caudal fibers producing greater stress (Wakeling *et al.*, 1999). The general slowing of the intrinsic contractile kinetics toward the caudal region predisposes these fibers to generate high stresses at slow strain rates, achieved by contracting the caudal myotomes in a more isometric fashion (Johnston *et al.*, 1995; Wakeling *et al.*, 1999). Furthermore, it has been observed that the interfacial ratio of the myotendinous junctions increases in a rostral-caudal direction along the fast fibers of *C. carpio*, indicating that the more-caudal fibers generate greater stresses during contractions (Spierts *et al.*, 1996). It has been suggested that the muscle mass-specific power output during the stage 1 contraction was similar between rostral and caudal myotomes in the short-horned sculpin *M. scorpius* (Johnston *et al.*, 1995); however, a simulation approach has suggested that the muscle mass-specific power output should decrease toward the caudal myotomes for *C. carpio* (Wakeling *et al.*, 1999). Regardless of the muscle mass-specific powers generated at the different regions, the majority of the muscle mass occurs in the central region of the fish, ensuring that the absolute muscle power generated is greater there than in the caudal myotomes. In order for the caudal myotomes to effectively transmit force to the tail blade, they must be stiff structures. This stiffness is achieved by the muscles contracting at a higher stress and lower strain rate. It has been predicted that the caudal myotomes in *M. scorpius* produce significant tensile stress at around the same time that the central myotomes produce their maximum power (Johnston *et al.*, 1995), and thus they are able to transmit that power to the tail blade. For fast starts that show a decreasing fast fiber strain in the caudal region (Johnston *et al.*, 1995; Wakeling and Johnston, 1998; Spierts and van Leeuwen, 1999; Fernández *et al.*, 2002), the caudal muscle fibers can act as tensile elements for transmitting force to the tail blade. However, it is unclear how general this situation is during fast starts.

The flexural stiffness of the caudal region of the body, as discussed in Section IV.B, is important in resisting flexion due to the inertial load of the mass and added mass of the caudal fin. However, recent studies have suggested that substantial force transmission and stiffness are additionally provided by the myoseptal tendons. Using a model of swimming in the pumpkinseed fish *Lepomis gibbosus*, which included axially oriented

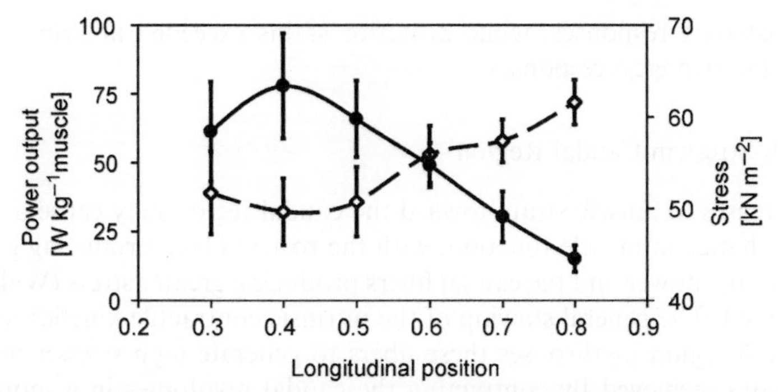

Fig. 9.7. Mean ± S.E.M. ($N = 10$) muscle mass-specific power output (filled circles) and muscle stress (open diamonds) for stage 1 of the fast start for the common carp *Cyprinus carpio* of length 56.6 mm. A longitudinal position of 0 denotes the snout, and 1 the trailing edge of the caudal fin. Note how there is a tradeoff between high power output and high stress between central and caudal regions of the myotome. (Adapted from Wakeling *et al.*, 1999.)

muscle tendons and transverse and axial myoseptal tendons, it was demonstrated that the function of the axial tendons includes stiffening the joints in bending and the intersegmental transmission of axial forces (Long *et al.*, 2002). In a comparative study of the spatial arrangement of the myoseptal tendons in a range of gnathostome fishes, it was shown that the caudal myotomes exhibit a thickening of the longitudinal tendons, horizontal projections of lateral bands that provide additional attachment sites for fast muscle, and that their tendons are more longitudinal than in the more-anterior myotomes (Gemballa and Vogel, 2002). These structural specializations are all consistent with the caudal region providing force transmission between the bulk of the myotomal muscle and the tail blade.

During the stage 1 muscle contraction of the fast start, large muscle powers are produced in the central myotomes due to both the large muscle mass and the muscle fibers contracting with substantial strains and strain rates. These powers are transmitted to the caudal region of the fish and the tail blade due to stiffness of both the muscle and skeletal structures of the caudal region (Figure 9.7). This stiffness has both passive and active components, and will vary between fast starts and species.

VI. HYDRODYNAMIC FORCES ACCELERATE THE BODY

The basic principles of generating hydrodynamic force are common to both steady swimming and fast starts, and many of the hydrodynamic principles discussed in Chapter 11 are also appropriate for considerations

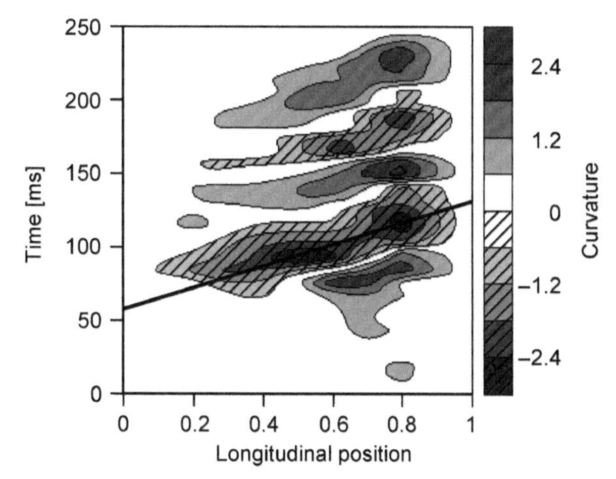

Fig. 9.8. Body curvature for a fast-start escape response for the short-horned sculpin *Myoxocephalus scorpius*. Curvature is the body length divided by the radius of curvature of the spine at each longitudinal position. A longitudinal position (L) of 0 denotes the snout, and 1 the trailing edge of the caudal fin. The body bending begins at 65 ms. Note how the initial bending occurs as a double bend with a negative curvature at $0.4L$ and a positive curvature at $0.7L$. The sloping line indicates the wave of curvature, which travels in a caudal direction along the body during stage 1 of the fast start. (Adapted from Wakeling and Johnston, 1998.)

of fast-start swimming. Fast starts are characterized by their high accelerations that rapidly set the fish into motion. The particular hydrodynamic considerations for generating these accelerations are discussed in the following section. During the final stages of the fast start, the swimming approaches either a steady swimming or a gliding phase; these motions are discussed in Chapter 7 and are not discussed further here.

A. Reactive and Circulatory Mechanisms for Force Production

The kinematics of the fast start typically show large amplitude bending, or body curvature, which travels along the length of the body (Figure 9.8). Early definitions by Weihs (1973) described a preparatory stroke (stage 1) followed by a propulsive stroke (phase 2). However, these categories can give a misleading representation of the movement. Careful observation shows that significant movement of the center of mass occurs during stage 1, due to the high forces (Weihs, 1973) and accelerations present during this stage (Wakeling and Johnston, 1998). Therefore, significant forward propulsion is generated even in this stage, the result of substantial hydrodynamic forces.

The most general description of the relationship between force and flow in a fluid is given by the Navier-Stokes equations, which are a set of partial differential equations that are based on the conservation of mass and momentum. For most biological applications there are no solutions to the full form of these equations. However, for the specific case of the nearly nonviscous regime, a more simplified set of equations, Euler's equations, can be used. Euler's equations state that accelerations in each fluid element yield pressure gradients in the fluid (Daniel *et al.*, 1992), and as elements of the body accelerate laterally at some angle with respect to the fluid then there is a reaction in the fluid to that acceleration. The nondimensional Reynolds number ($Re = VL/v$; where V is the forward velocity, L is the fish length, and v is the kinematic viscosity of the water) indicates the ratio of inertial to viscous forces, and adult fish routinely swim in the inertial hydrodynamic regime with $Re > 10^4$ (Wu, 1977), which can be considered to be nearly nonviscid. Therefore, Euler's equations can be used to approximate the hydrodynamics of fast-start swimming.

Lighthill (1960, 1969, 1970) used Euler's equations to develop a "slender body" theory applicable to increasing small amplitude motions during steady fish swimming and then extended the analysis to an "elongated body" approach that included large amplitude motions that are more appropriate to actual fish motions (Lighthill, 1971). Lighthill's equations describe how lateral accelerations of the body impart momentum into the water; this momentum is transferred down the body due to the action of a rearward propagating wave of body curvature, and the momentum is finally shed from the tail. The elongated body theory describes how forward swimming is achieved when the momentum imparted to the water by each body segment increases in a rostral-caudal direction, and when this momentum is propagated toward the tail by the rearward traveling wave of body curvature. During fast starts, the increasing body undulations (Frith and Blake, 1991) and body curvature (Wakeling and Johnston, 1998; Spierts and van Leeuwen, 1999), combined with extended median and caudal fins (Webb, 1977), result in large momentums being imparted to the water in the caudal region of the fish, and waves of body curvature are established even from the beginning of stage 1. These kinematic features are necessary for generating propulsion from reactive forces (Lighthill, 1971). The elongated body theory assumed independence of fluid flow between each section of the fish; these assumptions can be violated with very large amplitude motions that affect flows in the regions upstream of the trailing edge. The theory also assumed that vorticity shedding effects were all concentrated at the rear end of the body, which was equivalent to assuming that the vortex wakes from the body fins and front part of the fish are reabsorbed into the flow around the tail. This type of analysis is most suitable when the curvature of the body is small, and may not be entirely satisfactory for the large curvatures occurring during fast starts.

Weihs (1973) used an alternative approach to calculate the hydrodynamic forces during fast starts in the trout *Salmo trutta* and pike *Esox*. The hydrodynamic forces propelling the fishes were given by both the reactive forces generated in consequence to each body segment imparting momentum to the water and the sum of the momentum shedding circulatory forces occurring from each sharp-edged surface such as fins and lens-shaped sections of the body (Weihs, 1973). This analysis gave a "good agreement" between expected and measured forces, and it was concluded that a combination of reactive and circulatory forces could account for the thrust production during fast starts. During steady swimming, drag can be reduced without too much penalty in thrust reduction by reducing the caudal fin area in relation to the depth of its trailing edge (Lighthill, 1970). This situation is opposite to optimal body profiles for fast starts, in which all laterally moving portions of the body contribute significantly to thrust generation when the lateral movements are of large amplitude (Weihs, 1973). Lateral profiles that maximize the fin area in the caudal region result in enhanced fast-start performances (Webb, 1977), and in the bony fishes the dorsal, anal, and caudal fins are splayed at the beginning and retracted at the end of fast starts to achieve such large lateral profiles (Webb, 1977).

Reactive and circulatory forces were calculated for both "S" and "C" fast starts in the northern pike *Esox lucius* (Frith and Blake, 1991, 1995). On average, the reactive forces dominated during stage 1, while the circulatory forces dominated in stage 2 of the starts (Frith and Blake, 1991). The circulatory forces were calculated using quasi-steady assumptions and were estimated to start reaching significant levels partway through stage 1. However, the actual lift coefficient (for the circulatory force) varies with the unsteady motion of the fins and takes time to develop for an airfoil moving from rest due to a phenomenon known as the Wagner effect (Wagner, 1925). Therefore, the contribution of the circulatory forces may have been overestimated at the beginning of the fast start. It seems likely that during stage 1 of the fast start, when the fish begins its acceleration, the hydrodynamic forces are largely produced by reactive mechanisms.

B. Wake Momentum

As suggested previously, quantifying the hydrodynamic forces experienced by swimming fishes can present significant challenges because direct measurements of the forces applied to the water are not feasible; these forces must instead be modeled using a number of simplifying assumptions. It is arguably more difficult to model hydrodynamic forces for fast-start swimming than for steady swimming because the specific motions that characterize the fast start, namely, large-amplitude body undulations and unsteady movements of the fins, can test the assumptions invoked in hydrodynamic models (Lighthill, 1971; Weihs, 1973; Frith and Blake, 1991). An alternative approach to calculating the hydrodynamic forces during swimming is to calculate the momentum shed into the wake behind the swimming fish (Lighthill, 1969) because the rate of change of momentum in the wake gives the impulsive force acting on the fish, which, in turn, is equal to the forces required to overcome the drag on the fish and to accelerate the body. The assumptions about propulsor shape and kinematics, inherent to mechanical and computational modeling of locomotion, can be circumvented by direct flow visualization behind unrestrained animals (Drucker and Lauder, 1999). Fluid flow can be measured using digital particle image velocimetry (DPIV); these techniques have been recently applied to studies of swimming fishes (e.g., Müller *et al.*, 1997, 2000, 2001; Drucker and Lauder, 1999, 2001; Wolfgang *et al.*, 1999). The hydrodynamic forces during the accelerations involved with burst swimming were studied in the zebra danios *Brachydanio rerio* using a two-dimensional (2-D) DPIV system (Müller *et al.*, 2000). A horizontal section of the flow in the wake revealed two shed vortices shed during the first tail flick that resembled the cross-section through a vortex ring. Using the assumptions that the vortex ring was circular and that the flow velocities in the two vortices were the same, the impulse was calculated for both adult and larvae *B. rerio*. Estimates of force based on the flow in the wake compared to estimates based on the coasting speed differed by 30–210%, and it was suggested that closer estimates could be obtained if the three-dimensional (3-D) wake structure was known (see Chapters 10 and 11 for discussions of this technique). Three-dimensional DPIV has been used to study the vortex wake during steady swimming in the bluegill sunfish *L. macrochirus*, and resulted in wake-derived thrust measurements that were not significantly different from the expected drag on the body (Drucker and Lauder, 1999). The challenge now would be to apply 3-D DPIV techniques to the study of the wake behind fast-starting fish.

C. Inertial Forces to Accelerate Body

Fast starts are characterized by their high accelerations, and a large proportion of propulsive force is dominated by the inertial force requirements to accelerate the body (Figure 9.9). Therefore, the inertial power requirements can be used to calculate a good estimate of the hydrodynamic power. Inertial forces are the product of the acceleration and the mass. The accelerated mass consists of both the mass of the body and the added mass of water that is also accelerated with the body. The added mass of water that is accelerated with the fish during fast starts was estimated to be 20% of the body mass for the trout (Webb, 1982). The inertial forces should be

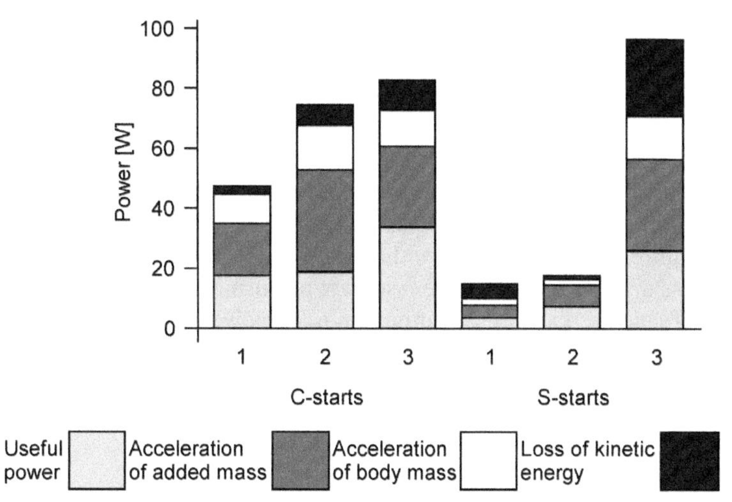

Fig. 9.9. Components of the hydrodynamic power requirements during "C" and "S" fast starts in six northern pike *Esox lucius* (mass 397–430 g). The useful power is the inertial power required to accelerate the fish in its direction of travel. Inertial power terms are also included for the lateral acceleration of the mass and added mass of each body segment. Note how the majority of the hydrodynamic power is comprised by these reactive forces. (Adapted from Frith and Blake, 1995.)

calculated from the accelerations occurring at the center of mass of the body. However, as a fish bends into a C shape, the center of mass moves laterally away from the spine to lie in the space enclosed by the "C" (Wakeling and Johnston, 1998) and so the center of mass cannot be assumed to be fixed to any anatomical position on the body. The first illustration of the motion of the center of mass during a fast start was given by Weihs (1973), and it has since been calculated in a number of more recent studies (Wakeling and Johnston, 1998; Wakeling *et al.*, 1999; Temple *et al.*, 2000; Fernández *et al.*, 2002). The center of mass can be calculated for each fish image from a high-speed film sequence and the accelerations can be calculated from the double differentiation of these position data; however, the choice of experimental and analytical procedures used can greatly influence the calculated values (Harper and Blake, 1989; Wakeling and Johnston, 1998; Walker, 1998). Accelerations can also be measured directly using accelerometers mounted onto the body (Franklin *et al.*, 2003); however, because the center of mass moves with respect to fixed body locations, the accelerometer measurements will not describe the motion of the center of mass. Despite the challenges faced in measuring the inertial power requirements during fast starts, the inertial power requirements for swimming during stage 1 and stage 2 of the fast start have been shown to correlate reasonably well, with no significant differences, to hydrodynamic power requirements calculated from the reactive and circulatory forces integrated along the whole body (Frith and Blake, 1995). Each of the two approaches of calculating the propulsive power from the inertial requirements or from the reactive and circulatory forces involve a number of assumptions and simplifications, but each has its own merits. Using inertial power as an estimate of the total hydrodynamic power requirements, it was shown that the power produced by the myotomal muscle was directly related to the fast-start performance across a range of six species of marine fish: as maximum muscle power output increased, for instance, with increases in environmental temperature, then so too did the hydrodynamic power output and thus the fast-start acceleration (Wakeling and Johnston, 1998).

D. Hydrodynamic Efficiency During Fast Starts

Hydrodynamic forces act not only in the direction of travel but also in a normal direction due to lateral movements of the body (Figure 9.9). The hydrodynamic efficiency during a fast start is the ratio of the hydrodynamic power required to accelerate the fish in its direction of motion to the total hydrodynamic power required for swimming. Hydrodynamic efficiencies during fast starts have been calculated to be between 0.16 and 0.39 for the northern pike *Esox lucius* (Frith and Blake, 1995). These efficiencies are less than those for steady swimming, which can reach 0.7 to 0.9 (Webb, 1975b, 1988; Videler and Hess, 1984). The reduced efficiency of the fast start has been attributed to the costs of accelerating the added mass of water. The energy required to accelerate the added mass is a major component (39% on average) of the total power (Frith and Blake, 1995). Fast starts are characterized by large-amplitude body undulations, and they have relatively high tail-beat frequencies compared to steady swimming. Therefore, fast starts result in greater segmental accelerations and decelerations, and the cost of achieving these results in an inefficiency to the motion. Fast starts are thus not the most efficient form of aquatic locomotion, but it is likely that they are optimized for high hydrodynamic force production and thus high accelerations. It should also be noted that during S-starts in which the fishes swim a number of tail-beat cycles, the hydrodynamic efficiency progressively increases with each subsequent tail-beat as the steady swimming condition is approached (Frith and Blake, 1995). Furthermore, some species adopt an "S" posture before the fast-start prey capture commences, and by deleting the least efficient stage 1 of the start they can achieve higher hydrodynamic efficiencies to maximize their lunging performance (Webb and Skadsen, 1980).

VII. VARIATIONS IN FAST-START PERFORMANCE

Fast-start swimming maneuvers require the integration of many systems within the fish, including neural activation, muscle contractions, force transmission through skeletal elements, and interactions with the water through the body and fins. Many of the traits that result in maximizing fast-start performance are opposite those for efficient steady swimming, and thus a compromise must be reached by each species that meets their ecological demands. It should be noted that two studies that compared different species swimming at a common temperature found no differences in the fast-start accelerations (Webb, 1978b; Law and Blake, 1996); it was suggested that similarly high fast-start performance can be achieved with different suites of morphological characteristics between species. On the other hand, in a comparison of two marine Cottidae, it was found that the maximum swimming velocity achieved within the first two stages of the fast start was greater for *Taurulus bubalis* than for *M. scorpius* (Temple and Johnston, 1998). Therefore, variation in fast-start performance can occur both between starts for an individual, and between different species.

Maximal fast-start performance depends on both muscle activation and body form parameters. Maximizing this performance may be critical to the survival of an individual in a predator–prey encounter. The vast range of structural diversity across species results in a range of fast-start abilities in fishes, and for each given fish the fast-start acceleration and trajectory can be modified by altering the muscle activation patterns and muscle power output. In this section some of the factors that produce this range of swimming behaviors are discussed.

A. Ontogeny

During ontogeny fishes develop from small larvae with a relatively uniform body depth due to their primordial fin into larger adult forms that have irregular profiles due to varying body depth and the development of the medial fins. The changes in size and shape during ontogeny result in changes in swimming hydrodynamics and kinematics, which, in turn, impose different demands on the muscle contractile dynamics (Wakeling *et al.*, 1999). Swimming velocity increases with body size for burst activity and fast starts (Beamish, 1978; Magnuson, 1978; Williams and Brown, 1992; Hale, 1996; Temple and Johnston, 1998; James and Johnston, 1998b), indicating that the larger forms generate greater inertial powers for fast starts (Section VI.C; Frith and Blake, 1995; Wakeling and Johnston, 1998). Hydrodynamic power requirements are limited by the power available from the muscles (Wakeling and Johnston, 1998), which, therefore, must increase for bigger fishes.

Muscle contractile properties scale with increasing body length. The $Q_{10\text{cm}}$ (a Q_{10} value for every 10 cm difference in length; Videler and Wardle, 1991) for twitch contraction frequencies was measured at 0.82–0.94 for cod *Gadhus morhua* (Archer *et al.*, 1990; Videler and Wardle, 1991). A $Q_{10\text{cm}}$ of 0.79 has been measured for the maximum unloaded shortening velocity of fast muscle fibers in the short-horn sculpin *M. scorpius* (James and Johnston, 1998b), but no significant scaling relationship was found for this variable in dogfish *Scyliorhinus canicula* (Curtin and Woledge, 1988). The muscle mass-specific power output during fast starts scales at a much higher rate than for the intrinsic contractile properties of the muscle increases, for instance, muscle mass-specific power output scales with $Q_{10\text{cm}} = 3.94$ in the carp *C. carpio* (Wakeling *et al.*, 1999). It is therefore unlikely that the changes in muscle contractile properties contribute in a major way to the increase in fast-start performance during development.

On the other hand, the principles that determine rates of body bending (Section IV.B) scale to predict that rates of bending should decrease with increasing size during ontogeny (assuming similarity of shape; Wakeling *et al.*, 1999). For the carp *C. carpio* this decrease in the rate of bending occurs as an increase in the time taken for the initial stage 1 bend, and leads to increases in both the muscle force and power production (Wakeling *et al.*, 1999). Furthermore, the development of deep bodies and large areas of the median fins makes the adult form better suited to producing the larger reactive forces needed for the high accelerations of the fast start (Section VI.A; Webb, 1977). Fast-start performance thus increases during ontogeny due to the increases in muscle mass in the larger forms, the increases in the relative proportion of muscle mass, the changes in body profile, the development of medial fins, and the decreases in the rates of body bending that lead to increases in muscle mass-specific power outputs.

B. Neural Drive

Fast starts are mediated by neural control, with the Mauthner neurons providing a fast initial impulse followed by a more extended neural input from parallel pathways (Section II). The initial response time from the onset of a stimulus to the beginning of body flexion decreases with increasing temperature, and for the rainbow trout decreased from 23 ms at 5°C to 6 ms at 25°C (Webb, 1978a). The duration and level of the initial activity encode information about the final trajectory of the fast start in a ballistic manner (Eaton and Emberley, 1991) and must do this by altering the way in which the body flexes and interacts with the water. The degree of bilateral activity varies between species, with more basal species (including sharks) showing considerable bilateral activity (Westneat *et al.*, 1998; Tytell and Lauder, 2002; R. E.

Shadwick, personal communication). Bilateral muscle activity acts to increase the body stiffness, but can reduce the bending moments acting on the spine, and so will alter the body bending and thus the body–water interactions. The magnitude and duration of the muscle activity change the amount of body flexion, with associated swimming performance. Typically, fast starts with lower levels of muscle activity result in the opposite bending occurring at the central and caudal regions of the body because the caudal bending moments are not sufficient to resist the inertial loads at the tail. In such cases, the fishes display an S-shaped flexion, and these starts result in lower accelerations than the faster C-starts (Spierts and van Leeuwen, 1999).

C. Environmental Temperature

The hydrodynamic power requirements for the fast start are closely linked to the mechanical power available from the myotomal muscles (Wakeling and Johnston, 1998). Therefore, factors that increase muscle power output should also increase the fast-start performance. Muscle power output decreases at low temperatures, and the decreased fast-start performance that occurred at lower environmental temperatures could be explained in terms of the lower muscle power outputs for a range of six teleost species (Wakeling and Johnston, 1998). For a given species, the common carp *C. carpio*, acute increases in temperature resulted in decreases in the fast-start body bending and muscle strain rate ϵ. At the same time, acute increases in temperature increase the intrinsic shortening speed of the muscle ϵ_{max}(Johnston *et al.*, 1985) and so the muscle would have operated at lower ϵ/ϵ_{max}ratios during the fast start, which results in greater muscle stress (Wakeling and Johnston, 1999c) and a greater mechanical power output for these contractions (Wakeling *et al.*, 2000). The higher muscle power outputs at the higher acute temperatures result in greater fast-start performances observed for *C. carpio* at higher acute temperatures (Wakeling *et al.*, 2000). More prolonged changes in temperature that last a couple of weeks can induce changes in the isoforms of the muscle proteins, with cyprinid fish showing adjustments in both myofibrillar ATPase (Johnston *et al.*, 1975) and myosin heavy chain composition (Hwang *et al.*, 1990; Watabe *et al.*, 1994). However, the evidence suggests that such alterations extend the range of temperatures over which the fishes can swim rather than changing the fast-start performance at common intermediate temperatures (Wakeling *et al.*, 2000).

D. Muscle Mass

Muscle power output and thus fast-start performance can also be increased by increasing the relative amount of myotomal muscle within the body. This has two effects: first, it increases the power available to accelerate the fish, and second, it reduces the dead weight of nonmuscular structures that must be accelerated (Webb, 1984). Myotomal muscle masses have been recorded at between 29 and 55% of the total body mass for 15 species (Webb, 1978b; Wakeling and Johnston, 1998). The escoid fish (*Esox* sp.) showed the highest proportion of myotomal muscle, and can be considered adept at their fast-start prey capture maneuvers. The sculpins showed the lowest proportion of myotomal muscle, but have bony and spiny skulls that provide passive protection from predation. It is the fast myotomal muscle fibers that are most suited to powering the fast-start acceleration maneuvers, whereas the slow myotomal muscle is better suited for sustained cruising swimming (Rome *et al.*, 1988). A survey of 84 species of fish reported that the fast fibers make up between 74 and 100% of the myotomal muscle (Greer-Walker and Pull, 1975). The species that are more specialized at steady, cruising swimming typically had the lowest proportion of fast myotomal muscle; again, this illustrates the compromises that must be reached for a species to excel at steady swimming or at fast-start acceleration maneuvers. Part of the non-muscle mass comprises the skin, and different species show variability in their proportion of skin mass. A survey of eight centrarchid fishes showed that the four more-piscivorous species had a mean skin mass at 7.6% of their body mass, while the four less-piscivorous species had a mean skin mass of 13.3%. It was suggested that this reduction in skin mass would confer a 5% increase in the fast-start accelerations for the more-piscivorous species (Webb and Skadsen, 1979).

E. Body Flexibility

Fast starts are achieved with large amplitude and rapid body bending. Body flexibility varies between species, with some species showing inflexible bodies for both structural and functional reasons. One example of a relatively inflexible fish is the tuna, in which body bending about the intervertebral joints is limited by the articulation of the zygapophyses (Nursall, 1956) and is limited primarily to the pre- and postpendicular joints at the proximal and distal end of the caudal peduncle. This vertebral structure is a specialization for their thunniform mode of steady swimming (Chapter 7) but restricts bending in the body (Donley and Dickson, 2000). A second example is the box fish (Ostraciontidae), which has rigid bony carapaces that encase the majority of the body. The carapace provides protective armor but limits body movements to locations posterior to the caudal peduncle, and so these fishes propel themselves using motion in their fins (Blake, 1977). In both these examples the lack of body flexibility precludes the fishes from performing fast-start maneuvers. In general, fast-start performance in fishes is limited by lack of body flexibility.

F. Body Profile and Fins

The mechanically optimum lateral body profiles for fast-start and steady swimming performance are mutually exclusive. Large areas of the median fins and a deep body profile are required for high accelerations during fast starts in order to produce large reactive forces in the water (Section VI.A; Webb, 1977). However, this body form is not the most efficient for steady locomotion, for which it has been predicted that the depth of the trailing edge should be maximized while the area of the caudal fin and caudal region of the body should be minimized (Lighthill, 1970). Most fishes are generalists in their behavior and use both unsteady fast starts and steady swimming for their locomotion; therefore, their body profiles must reach a compromise to meet these competing requirements. The mechanical restriction of a fixed body depth has been circumvented in the bony fishes, which have developed flexible and collapsible fins that permit a large variation in lateral body profile. One of the actions of the Mauthner reflex is to splay the fins in the bony fishes (Eaton *et al.*, 1977). In the rainbow trout, fin splaying caused an increase in the depth of the dorsal, anal, and caudal fins within 20 ms of the onset of the stimulus, and the fins showed retraction after the end of stage 2 of the fast start (Webb, 1977). Clearly, variations in the fin depth achieved, and thus the activity in the erector muscles, will alter the reactive forces in the water and thus the fast-start performance.

VIII. CONCLUSIONS

Despite the variability in fast-start mechanics and performance between species, there are a number of general patterns to the fast-start maneuver that occur in most fishes. Fast-start swimming maneuvers in fishes are initiated by the action of the Mauthner neuron and an associated network of parallel neurons. Initially the myotomal muscle is activated simultaneously along one side of the body, and for some species a degree of bilateral muscle activity may also occur. The Mauthner reflex also causes the median fins to be splayed in the bony fish in order to increase the depth of the body.

The myotomal muscle produces bending moments that act on the spine. Body bending depends on the bending moment generated by the muscle, the internal stiffness in the body, and the inertial load of the fish mass and added mass of water. Different body postures, "C" shapes or "S" shapes, occur as a result of different levels of myotomal muscle activity that changes the balance of body bending to inertial resistance for different longitudinal positions. Typically, a large-amplitude wave of bending occurs that travels in a rostral-to-caudal direction along the fish. The bulk of the myotomal muscle occurs in the central region of the fish, where the majority of the mechanical power is produced for the fast start. Changes in muscle function from power production in the central region to force transmission in the caudal region result both from longitudinal variation in the muscle strain and strain rate, and from changes in muscle contractile properties and the geometry of the skeletal structures. The muscle power is transmitted to the caudal region, which is where the majority of the hydrodynamic forces is delivered to the water.

During the initial stage of the fast start, the hydrodynamic forces are predominantly reactive forces from the body pushing on the water. The large and rapid lateral displacements, combined with deep body sections, combine to generate a substantial reactive force with a net rearward component as a result of the wave of body curvature. The reaction to this impulse acts to accelerate the fish in its direction of travel. In general, the fast-start acceleration is limited by the power available from the muscles. It is thought that greater accelerations will confer an advantage to the fish for either prey-capture or escape response behaviors.

IX. FUTURE DIRECTIONS

Advances in technology are expanding the possibilities for studies in fast-start biomechanics. Techniques such as computational fluid dynamics and digital particle image velocimetry are being routinely used to study fish swimming in both 2-D, and, more recently, 3-D cases (e.g., Drucker and Lauder, 1999; Wolfgang *et al.*, 1999; Müller *et al.*, 2001; Ramamurti *et al.*, 2002). It is only a matter of time before 3-D descriptions of the hydrodynamic forces and wake of fast-starting fish are reported. However, there remains much to be determined about the integrative physiology of fast-start swimming, and this is an area that would benefit from the unification of approaches across a range of disciplines. For instance, measurements of muscle activity at multiple points along and across the myotomes show that neural activation may be complex and variable (Tytell and Lauder, 2002), and these should be described further both across and within different species. Transmission of muscle forces through the various skeletal structures of the body (Gemballa and Vogel, 2002) is again little understood and would benefit from further studies of the functional anatomy of the myotomes. These forces create internal bending moments within the fish, but the actual bending of the body also depends on the hydrodynamic interactions between the body and surrounding water (Wakeling and Johnston, 1999a). Each level or organization is only truly relevant when it is functionally integrated into the bigger picture of how fishes fast start, and integrative studies should play a large part of our future research.

REFERENCES

Akster, H. A. (1985). Morphometry of muscle fiber types in the carp (*Cyprinus carpio* L.). Relationships between structural and contractile characteristics. *Cell Tissue Res.* **241,** 193–201.

Akster, H. A., Granzier, H. L. M., and ter Keurs, H. E. D. J. (1985). A comparison of quantitative ultrastructural and contractile properties of muscle fiber types of the perch, *Pera fluviatilis* L. *J. Comp. Physiol. B* **155,** 685–691.

Alexander, R. M. (1969). The orientation of muscle fibers in the myomeres of fishes. *J. Mar. Biol. Ass. UK* **49,** 263–290.

Altringham, J. D., Wardle, C. S., and Smith, C. I. (1993). Myotomal muscle function at different locations in the body of a swimming fish. *J. Exp. Biol.* **182,** 191–206.

Archer, S. D., Altringham, J. D., and Johnston, I. A. (1990). Scaling effects on the neuromuscular system, twitch kinetics and morphometrics of the cod Gadus morhua. *Mar. Behav. Physiol.* **17,** 137–146.

Aubert, X. (1956). Le couplage énergetique de la contraction musculaire. Thesis, Arcsia, Brussels.

Beamish, F. W. H. (1978). Swimming capacity. *In* "Fish Physiology, Vol. 7, Locomotion" (Hoar, W. S., and Randall, D. J., Eds.), pp. 101–187. Academic Press, London.

Blake, R. W. (1977). On ostraciiform locomotion. *J. Mar. Biol. Ass. UK* **57,** 1047–1055.

Blight, A. R. (1977). The muscular control of vertebrate swimming movements. *Biol. Rev.* **52,** 181–218.

Bone, Q. (1978). Locomotor muscle. *In* "Fish Physiology" (Hoar, W. S., and Randall, D. J., Eds.), Vol. VII, pp. 361–424. Academic Press, London.

Canfield, J. G., and Rose, G. J. (1993). Activation of Mauthner neurons during prey capture. *J. Comp. Physiol.* **172,** 611–618.

Covell, J. W., Smith, M., Harper, D. G., and Blake, R. W. (1991). Skeletal muscle deformation in the lateral muscle of the intact rainbow trout Oncorhynchus mykiss during fast start maneuvers. *J. Exp. Biol.* **156,** 453–466.

Currie, S. N., and Carlsen, R. C. (1987). Functional significance and neural basis of larval lamprey startle behavior. *J. Exp. Biol.* **133,** 121–135.

Curtin, N. A., and Woledge, R. C. (1988). Power output and force-velocity relationship of live fibers from white myotomal muscle of the dogfish, *Scyliorhinus canicula. J. Exp. Biol.* **140,** 187–197.

Daniel, T., Jordan, C., and Grunbaum, D. (1992). Hydromechanics of swimming. *In* "Mechanics of Animal Locomotion" (Alexander, R. M., Ed.), pp. 17–49. Springer-Verlag, Berlin.

Davies, M. L. F., Johnston, I. A., and van de Wal, J.-W. (1995). Muscle fibres in rostral and caudal myotomes of the Atlantic cod (*Gadus morhua* L.) have different mechanical properties. *Physiol. Zool.* **68,** 673–697.

Domenici, P., and Blake, R. W. (1991). The kinematics and performance of the escape response in the angelfish (*Pterophyllum eimekei*). *J. Exp. Biol.* **156,** 187–205.

Domenici, P., and Blake, R. W. (1993). Escape trajectories in angelfish (*Pterophyllum eimekei*). *J. Exp. Biol.* **177,** 253–272.

Domenici, P., and Blake, R. W. (1997). Review. The kinematics and performance of fast-start swimming. *J. Exp. Biol.* **200,** 1165–1178.

Donley, J. M., and Dickson, K. A. (2000). Swimming kinematics of juvenile kawakawa tuna (*Euthynnus affinis*) and chub mackerel (*Scomber japonicus*). *J. Exp. Biol.* **203,** 3103–3116.

Drucker, ..E. G., and Lauder, G. V. (1999). Locomotor forces on a swimming fish: Three-dimensional vortex wake dynamics quantified using digital particle image velocimetry. *J. Exp. Biol.* **202,** 2393–2412.

Drucker, E. G., and Lauder, G. V. (2001). Locomotor function of the dorsal fin in teleost fishes: Experimental analysis of wake forces in sunfish. *J. Exp. Biol.* **204,** 2943–2958.

Eaton, R. C., and Emberley, D. S. (1991). How stimulus direction determines the trajectory of the Mauthner-initiated escape response in a teleost fish. *J. Exp. Biol.* **161,** 469–487.

Eaton, R. C., Bombardieri, R. A., and Meyer, D. L. (1977). The Mauthner-initiated startle response in teleost fish. *J. Exp. Biol.* **66,** 65–81.

Eaton, R. C., Di Domenico, R., and Nissanov, J. (1988). Flexible body dynamics of the goldfish C-start: Implications for reticulospinal command mechanicms. *J. Neurosci.* **8,** 2758–2766.

Eaton, R. C., Lavender, W. A., and Wieland, C. M. (1981). Identification of Mauthner-initiated response patterns in goldfish: Evidence from simultaneous cinematography and electrophysiology. *J. Comp. Physiol. A* **144,** 521–531.

Eaton, R. C., Lavender, W. A., and Wieland, C. M. (1982). Alternative neural pathways initiate fast-start responses following lesions of the Mauthner neuron in goldfish. *J. Comp. Physiol. A* **145,** 485–496.

Ellerby, D. J., and Altringham, J. D. (2001). Spatial variation in fast muscle function of the rainbow trout Onchorhynchus mykiss during fast-starts and sprinting. *J. Exp. Biol.* **204,** 2239–2250.

Fernández, D. A., Calvo, J., Wakeling, J. M., Vanella, F. A., and Johnston, I. A. (2002). Escape performance in the sub-Antarctic notothenioid fish *Eleginops maclovinus. Polar Biol.* **25,** 914–920.

Fetcho, J. R. (1986). The organization of the motoneurons innervating the axial musculature of vertebrates. I. Goldfish (*Carassius auratus*) and mudpuppies (*Necturus maculosus*). *J. Comp. Neurol.* **249,** 521–550.

Fetcho, J. R. (1992). Excitation of motoneurons by the Mauthner axon in goldfish: Complexities in a 'simple' retinulospinal pathway. *J. Neurophys.* **67,** 1574–1586.

Foreman, M. B., and Eaton, R. C. (1993). The direction change concept for reticulospinal control of goldfish escape. *J. Neurosci.* **13,** 4101–4133.

Franklin, C. E., and Johnston, I. A. (1997). Muscle power output during escape responses in an Antarctic fish. *J. Exp. Biol.* 703–712.

Franklin, C. E., Wilson, R. S., and Davison, W. (2003). Locomotion at -1.0 °C: Burst swimming performance of five species of Antarctic fish. *J. Thermal Biol.* **28,** 59–65.

Frith, H. R., and Blake, R. W. (1991). Mechanics of the startle response in the northern pike, *Esox lucius. Can. J. Zool.* **69,** 2831–2839.

Frith, H. R., and Blake, R. W. (1995). The mechanical power output and hydromechanical efficiency of northern pike (*Esox lucius*) fast-starts. *J. Exp. Biol.* **198,** 1863–1873.

Funch, P. G., Kinsman, S. L., Faber, D. S., Koenig, E., and Zottoli, S. J. (1981). Mauthner axon diameter and impulse conduction velocity decreases with growth of goldfish. *Neurosci. Lett.* **27,** 159–164.

Furukawa, T., and Furshpan, E. J. (1963). Two inhibitory mechanisms in the Mauthner neurons of goldfish. *J. Neurophysiol.* **26,** 140–176.

Gemballa, S., and Vogel, F. (2002). Spatial arrangement of white muscle fibers and myoseptal tendons in fishes. *Comp. Biochem. Physiol. A* **133,** 1013–1037.

Goldbogen, J. A., Shadwick, R. E., Fudge, D. S., and Gosline, J. M. (2005). Fast-start dynamics in the rainbow trout *Oncorhynchus mykiss*: Phase relationship of white muscle shortening and body curvature. *J. Exp. Biol.* **208,** 929–938.

Granzier, H. L. M., Wiersma, J., Akster, H. A., and Osse, J. W. M. (1983). Contractile properties of a white- and a red-fibre type of the *M. hyohyoideus* of the carp (*Cyprinus carpio* L.). *J. Comp. Physiol.* **149,** 441–449.

Greer-Walker, M., and Pull, G. A. (1975). A survey of red and white muscle in marine fish. *J. Fish Biol.* **7,** 295–300.

Grillner, S., and Kashin, S. (1976). On the generation and performance of swimming in fish. *In* "Neural Control of Locomotion" (Herman, R. M., Grillner, S., Stein, P. S. G., and Stuart, D. G., Eds.), pp. 181–201. Plenum Press, New York.

Hale, M. E. (1996). The development of fast-start performance in fishes: Escape kinematics of the chinook salmon (*Oncorhynchus tshawytscha*). *Am. Zool.* **36,** 695–709.

Hale, M. E. (2000). Startle responses of fish without Mauthner neurons: Escape behavior of the lumpfish (*Cyclopterus lumpus*). *Biol. Bull.* **199,** 180–182.

Hale, M. E. (2003). S- and C-start escape responses of the muskellunge (*Esox masquinongy*) require alternative neuromotor mechanisms. *J. Exp. Biol.* **205,** 2005–2016.

Hale, M. E., Long, J. H., Jr., McHenry, M. J., and Westneat, M. W. (2002). Evolution of behavior and neural control of the fast-start escape response. *Evolution* **56,** 993–1007.

Harper, D. G., and Blake, R. W. (1989). A critical analysis of the use of high-speed film to determine maximum accelerations of fish. *J. Exp. Biol.* **142,** 465–471.

Hebrank, M. R. (1982). Mechanical properties of fish backbones in lateral bending and in tension. *J. Biomech.* **15,** 85–89.

Hess, F., and Videler, J. J. (1984). Fast continuous swimming of saithe (*Pollachius virens*): A dynamic analysis of bending moments and muscle power. *J. Exp. Biol.* **109,** 229–251.

Hill, A. V. (1938). The heat of shortening and the dynamic constraints of muscle. *Proc. R. Soc. B* **136,** 399–420.

Hoogland, R., Morris, D., and Tinbergen, N. (1956). The spines of sticklebacks (*Gasterosteus* and *Pygosteus*) as a means of defence against predators (*Perca* and *Esox*). *Behaviour* **10,** 205–236.

Hwang, G. C., Watabe, S., and Hashimoto, K. (1990). Changes in carp myosin ATPase induce by temperature acclimation. *J. Comp. Physiol. B* **160,** 233–239.

James, R. S., and Johnston, I. A. (1998a). Scaling of muscle performance during escape responses in the fish *Myoxocephalus scorpius* L. *J. Exp. Biol.* **201,** 913–923.

James, R. S., and Johnston, I. A. (1998b). Scaling of intrinsic contractile properties and myofibrillar protein composition of fast-muscle in the fish *Myxocephalus scorpius* L. *J. Exp. Biol.* **201,** 901–912.

James, R. S., Cole, N. J., Davies, M. L. F., and Johnston, I. A. (1998). Scaling of intrinsic contractile properties and myofibrillar protein composition of fast muscle in the fish *Myoxocephalus scorpius* L. *J. Exp. Biol.* **201,** 901–912.

Jayne, B. C., and Lauder, G. V. (1993). Red and white muscle activity and kinematics of the escape response of the bluegill sunfish during swimming. *J. Comp. Physiol. A* **173,** 495–508.

Johnston, I. A., Davison, W., and Goldspink, G. (1975). Adaptations in Mg^{2+}-activated myofibrillar ATPase activity induced by temperature acclimation. *FEBS Lett.* **50,** 293–295.

Johnston, I. A., Sidell, B. D., and Driedzic, W. R. (1985). Force-velocity characteristics and metabolism of carp muscle fibres following temperature acclimation. *J. Exp. Biol.* **119,** 239–249.

Johnston, I. A., Franklin, C. E., and Johnson, T. P. (1993). Recruitment patterns and contractile properties of fast muscle fibres isolated from rostral and caudal myotomes of the short-horned sculpin. *J. Exp. Biol.* **185,** 251–265.

Johnston, I. A., van Leeuwen, J. L., Davies, M. L. F., and Beddow, T. (1995). How fish power predation fast-starts. *J. Exp. Biol.* **198,** 1851–1861.

Law, T. C., and Blake, R. W. (1996). Comparison of the fast-start performances of closely related, morphologically distinct threespine sticklebacks (*Gasterosteus* spp.). *J. Exp. Biol.* **199,** 2595–2604.

Lieber, R. L., Raab, R., Kashin, S., and Edgerton, V. R. (1992). Sarcomere length changes during fish swimming. *J. Exp. Biol.* **169,** 251–254.

Lighthill, M. J. (1960). Note on the swimming of slender fish. *J. Fluid Mech.* **9,** 305–317.

Lighthill, M. J. (1969). Hydromechanics of aquatic animal propulsion: A survey. *Ann. Rev. Fluid Mech.* **1,** 413–446.
Lighthill, M. J. (1970). Aquatic animal propulsion of high hydromechanical efficiency. *J. Fluid Mech.* **44,** 265–301.
Lighthill, M. J. (1971). Large amplitude elongated-body theory of fish locomotion. *Proc. R. Soc. Lond.* **179,** 125–138.
Long, J. H., Jr. (1992). Stiffness and damping forces in the intervertebral joints of blue marlin (*Makaira nigricans*). *J. Exp. Biol.* **162,** 131–155.
Long, J. H., Jr., and Nipper, K. S. (1996). The importance of body stiffness in undulatory propulsion. *Amer. Zool.* **36,** 678–694.
Long, J. H., Jr., Hale, H. A., McHenry, M. J., and Westneat, M. W. (1996). Functions of fish skin: Flexural stiffness and steady swimming of longnose gar Lepisosteus osseus. *J. Exp. Biol.* **199,** 2139–2151.
Long, J. H., Jr., Adcock, B., and Root, R. G. (2002). Force transmission via axial tendons in undulating fish: A dynamic analysis. *Comp. Biochem. Physiol. A* **133,** 911–929.
Magnuson, J. J. (1978). Locomotion by scombriod fishes: Hydrodynamics, morphology and behavior. *In* "Fish Physiology" (Hoar, W. S., and Randall, D. J., Eds.), Vol. 7, pp. 239–313. Academic Press, London.
Müller, U. K., Smit, J., Stamhius, E. J., and Videler, J. J. (2001). How the body contributes to the wake in undulatory fish swimming: Flow fields of a swimming eel (*Anguilla anguilla*). *J. Exp. Biol.* **204,** 2751–2762.
Müller, U. K., Stamhius, E. J., and Videler, J. J. (2000). Hydrodynamics of unsteady fish swimming and the effects of body size: Comparing the flow fields of fish larvae and adults. *J. Exp. Biol.* **203,** 193–206.
Müller, U. K., van den Heuvel, B. L. E., Stamhius, E. J., and Videler, J. J. (1997). Fish foot prints: Morphology and energetics of the wake behind a continuously swimming mullet (*Chelon labrosus*). *J. Exp. Biol.* **200,** 2893–2906.
Myers, P. Z. (1985). Spinal motoneurons of the larval zebrafish. *J. Comp. Neurol.* **236,** 555–561.
Nissanov, J., and Eaton, R. C. (1989). Reticulospinal control of rapid escape turning maneuvers in fishes. *Amer. Zool.* **29,** 103–121.
Nursall, J. R. (1956). The lateral musculature and the swimming of fish. *Proc. Zool. Soc. Lond.* **126,** 127–143.
Ramamurti, R., Sandberg, W. C., Löhner, R., Walker, J. A., and Westneat, M. W. (2002). Fluid dynamics of flapping aquatic flight in the bird wrasse: Three-dimensional unsteady computations with fin deformation. *J. Exp. Biol.* **205,** 2997–3008.
Rome, L. C., Funke, R. P., Alexander, R. M., Lutz, G., Aldridge, H., Scott, F., and Freadman, M. (1988). Why animals have different muscle fibre types. *Nature* **335,** 824–827.
Shadwick, R. E., Rapoport, H. S., and Fenger, J. M. (2002). Structure and function of tuna tail tendons. *Comp. Biochem. Physiol. A* **133,** 1109–1125.
Shann, E. W. (1914). On the nature of the lateral muscle in Teleostei. *Proc. Zool. Soc. Lond.* **1914,** 319–337.
Spierts, I. L. Y., and van Leeuwen, J. H. (1999). Kinematics and muscle dynamics of C- and S-starts of carp (*Cyprinus carpio* L.). *J. Exp. Biol.* **202,** 393–406.
Spierts, I. L. Y., Akster, H. A., Vos, I. H. C., and Osse, J. W. M. (1996). Local differences in myotendinous junctions in axial muscle fibres of carp (*Cyprinus carpio* L.). *J. Exp. Biol.* **199,** 825–833.
Spierts, I. L. Y., Akster, H. A., and Granzier, H. L. (1997). Expression of titin isoforms in red and white muscle fibres of carp (*Cyprinus carpio* L.) exposed to different sarcomere strains during swimming. *J. Comp. Physiol. B* **167,** 543–551.

Temple, G. K., and Johnston, I. A. (1998). Testing hypotheses concerning the phenotypic plasticity of escape performance in fish of the family Cottidae. *J. Exp. Biol.* **201,** 317–331.

Temple, G. K., Wakeling, J. M., and Johnston, I. A. (2000). Seasonal changes in fast-starts in the short-horn sculpin: Integration of swimming behavior and muscle performance. *J. Fish Biol.* **56,** 1435–1449.

Tytell, E. D., and Lauder, G. V. (2002). The C-start escape response of Polypterus senegalus: Bilateral muscle activity and variation in stage 1 and 2. *J. Exp. Biol.* **205,** 2591–2603.

van Leeuwen, J. L. (1992). Muscle function in locomotion. *In* "Mechanics of Animal Locomotion" (Alexander, R. M., Ed.), pp. 191–250. Springer-Verlag, Berlin.

van Leeuwen, J. L., Lankheet, M. J. M., Akster, H. A., and Osse, J. W. M. (1990). Function of red axial muscles of carp (*Cyprinus carpio* L.): Recruitment and normalized power output during swimming in different modes. *J. Zool. (Lond.)* **220,** 123–145.

Videler, J. J., and Hess, F. (1984). Fast continuous swimming of two pelagic predators saithe (*Pollachius virens*) and mackerel (*Scomber scombrus*): A kinematic analysis. *J. Exp. Biol.* **109,** 209–228.

Videler, J. J., and Wardle, C. S. (1991). Fish swimming stride by stride: Speed limits and endurance. *Rev. Fish Biol. Fish.* **1,** 23–40.

Wagner, H. (1925). Über die Entstehung des dynamischen Auftriebes von Tragflügeln. *Z. angew. Math. Mech.* **5,** 17–35.

Wainwright, S. A., Vosburgh, F., and Hebrank, J. H. (1978). Shark skin: Function in locomotion. *Science* **202,** 747–749.

Wakeling, J. M., and Johnston, I. A. (1998). Muscle power output limits fast-start performance in fish. *J. Exp. Biol.* **201,** 1505–1526.

Wakeling, J. M., and Johnston, I. A. (1999a). Body bending during fast-starts in fish can be explained in terms of muscle torque and hydrodynamic resistance. *J. Exp. Biol.* **202,** 675–682.

Wakeling, J. M., and Johnston, I. A. (1999b). White muscle strain in the common carp and red to white muscle gearing ratios in fish. *J. Exp. Biol.* **202,** 521–528.

Wakeling, J. M., and Johnston, I. A. (1999c). Predicting muscle force generation during fast-starts for the common carp *Cyprinus carpio*. *J. Comp. Physiol. B* **169,** 391–401.

Wakeling, J. M., Kemp, K. M., and Johnston, I. A. (1999). The biomechanics of fast-starts during ontogeny in the common carp *Cyprinus carpio*. *J. Exp. Biol.* **202,** 3057–3067.

Wakeling, J. M., Cole, N. J., Kemp, K. M., and Johnston, I. A. (2000). The biomechanics and evolutionary significance of thermal acclimation in the common carp *Cyprinus carpio*. *Am. J. Physiol.* **279,** R657–R665.

Walker, J. A. (1998). Estimating velocities and accelerations of animal locomotion: A simulation experiment comparing numerical differential algorithms. *J. Exp. Biol.* **201,** 981–995.

Wardle, C. S. (1985). Swimming activity in marine fish. *In* "Physiological Adaptations of Marine Animals" (Laverack, M. S., Ed.), pp. 521–540. Company of Biologists, Cambridge.

Wardle, C. S., Videler, J. J., Arimoto, T., Francos, J. M., and He, P. (1989). The muscle twitch and the maximum swimming speed of giant bluefin tuna, *Thunnus Thynnus* L. *J. Fish Biol.* **35,** 129–137.

Wardle, C. S., Videler, J. J., and Altringham, J. D. (1995). Tuning in to fish swimming waves: Body form, swimming mode and muscle function. *J. Exp. Biol.* **198,** 1629–1636.

Watabe, S., Guo, X. F., and Hwang, G. C. (1994). Carp expressed specific isoforms of the myosin cross-bridge head, subfragment-1, in association with cold and warm temperature acclimation. *J. Therm. Biol.* **19,** 261–268.

Webb, P. W. (1975a). Acceleration performance of rainbow trout *Salmo gairdneri* and green sunfish *Lepomis cyanellus*. *J. Exp. Biol.* **63,** 451–465.

Webb, P. W. (1975b). Hydrodynamics and energetics of fish propulsion. *Bull. Fish. Res. Bd Can.* **190,** 1–159.

Webb, P. W. (1976). The effect of size on the fast-start performance of rainbow trout *Salmo cairdneri,* and a consideration of piscivorous predator-prey interactions. *J. Exp. Biol.* **65,** 157–177.

Webb, P. W. (1977). Effects of median-fin amputation on fast-start performance of rainbow trout (*Salmo gairdneri*). *J. Exp. Biol.* **68,** 123–135.

Webb, P. W. (1978a). Temperature effects on acceleration of rainbow trout *Salmo gairdneri. J. Fish. Res. Bd. Canada* **35,** 1417–1422.

Webb, P. W. (1978b). Fast-start performance and body form in seven species of teleost fish. *J. Exp. Biol.* **74,** 211–226.

Webb, P. W. (1982). Fast-start resistance of trout. *J. Exp. Biol.* **96,** 93–106.

Webb, P. W. (1984). Body form, locomotion and foraging in aquatic vertebrates. *Amer. Zool.* **24,** 107–120.

Webb, P. W. (1988). 'Steady' swimming kinematics of tiger musky, an esociform accelerator and rainbow trout, a generalist cruiser. *J. Exp. Biol.* **138,** 51–69.

Webb, P. W., and Skadsen, J. M. (1979). Reduced skin mass: An adaptation for acceleration in some teleost fishes. *Can. J. Zool.* **57,** 1570–1575.

Webb, P. W., and Skadsen, J. M. (1980). Strike tactics of *Esox. Can. J. Zool.* **58,** 1462–1469.

Weihs, D. (1973). The mechanism of rapid starting of a slender fish. *Biorheology* **10,** 343–350.

Westneat, M. W., Hale, M. E., McHenry, M. J., and Long, J. H., Jr. (1998). Mechanics of the fast-start: Muscle function and the role of intramuscular pressure in the escape behavior of *Amia calva* and *Polypterus palma. J. Exp. Biol.* **201,** 3041–3055.

Williams, P. J., and Brown, J. A. (1992). Developmental changes in the escape response of larval winter flounder *Pleuronectes americanus* from hatch through metamorphosis. *Mar. Ecol. Prog. Ser.* **88,** 185–193.

Wolfgang, M. J., Anderson, J. M., Grosenbaugh, M. A., Yue, D. K. P., and Triantafyllou, M. S. (1999). Near-body flow dynamics in swimming fish. *J. Exp. Biol.* **202,** 2303–2327.

Wu, T. Y. (1977). Introduction to the scaling of aquatic animal locomotion. *In* "Scale Effects in Animal Locomotion" (Pedley, J. J., Ed.), pp. 203–232. Academic Press, New York.

Yasargil, G. M., and Diamond, J. (1968). Startle-response in teleost fish: An elementary circuit for neural discrimination. *Nature* **220,** 241–243.

Zottoli, S. J. (1977). Correlation of the startle reflex and Mauthner cell auditory responses in unrestrained goldfish. *J. Exp. Biol.* **66,** 243–254.

Zottoli, S. J., Newman, B. C., Rieff, H. I., and Winters, D. C. (1999). Decrease in occurence of fast startle responses after selective Mauthner cell ablation in goldfish (*Carassius auratus*). *J. Comp. Physiol. A* **184,** 207–218.

10

MECHANICS OF PECTORAL FIN SWIMMING IN FISHES

ELIOT G. DRUCKER
JEFFREY A. WALKER
MARK W. WESTNEAT

I. INTRODUCTION

Fishes use a diverse array of kinematic mechanisms to swim. This diversity arises from different combinations of body and fin use and from different ways of deforming the body and fins during propulsion (Breder, 1926). These swimming mechanisms (or locomotor modes) have been named according to the family of fishes that commonly expresses the mode. To date, nearly 20 distinct locomotor modes have been identified for steadily

Fish Biomechanics: Volume 23
FISH PHYSIOLOGY

DOI: 10.1016/S1546-5098(05)23010-8

swimming fishes (Breder, 1926; Lindsey, 1978; Braun and Reif, 1985; Webb, 1994a). The diversity of swimming modes raises the question: why have so many different propulsive mechanisms arisen? To answer this, we need to know how deforming bodies and fins interact with the surrounding fluid to generate propulsive forces, how variation in body and fin movements and variation in body and fin shape affect the generation of different force components (including lift, thrust, and side forces), which locomotor movements and propulsor shapes generate swimming forces with minimal wasted energy, and which measures of swimming performance (e.g., maximum force versus maximum efficiency) are most important in influencing how fishes exploit resources in their environment. Most of the research addressing these questions to date has concentrated on axial (body and caudal fin) propulsive modes, especially carangiform and thunniform swimming. However, propulsion by the paired pectoral fins is widespread among fishes and has received increasing research attention over the past decade. This chapter reviews recent progress in the study of morphology, kinematics, and force production of the pectoral fins of fishes.

For non-axial modes of propulsion, a large volume of preliminary work was executed by Robert Blake, who described and modeled a broad range of median and paired fin swimming modes (Blake, 1983), including both undulatory and oscillatory pectoral fin mechanisms. In the 20 years following this work, there have been only a few papers expanding this research on pectoral fin undulation (Daniel, 1988; Arreola and Westneat, 1996; Gordon *et al.*, 1996; Rosenberger and Westneat, 1999; Combes and Daniel, 2001; Hove *et al.*, 2001; Rosenberger, 2001). In contrast, there has been a great deal of work on oscillatory pectoral fin propulsion since Blake's treatment. Thus, we focus in this chapter primarily on the mechanics of paired pectoral fins undergoing oscillatory locomotor motions, and supplement this discussion with an overview of current research into undulatory pectoral fin swimming mechanisms.

II. PECTORAL FIN MORPHOLOGY

The morphology of the pectoral fins in fishes, from position and orientation on the body to details of musculoskeletal design, provides insight into the locomotor function and evolution of swimming in fishes. Actinopterygian (ray-finned) fishes have undergone major evolutionary transformations in the anatomical position of the pectoral fin and the orientation of the pectoral fin base (Breder, 1926; Harris, 1937, 1953; Alexander, 1967; Gosline, 1971; Webb, 1982). In basal actinopterygians (Figure 10.1, *Lepisosteus*), as well as in basal teleost fishes (Figure 10.1, *Oncorhynchus*), the pectoral fin base

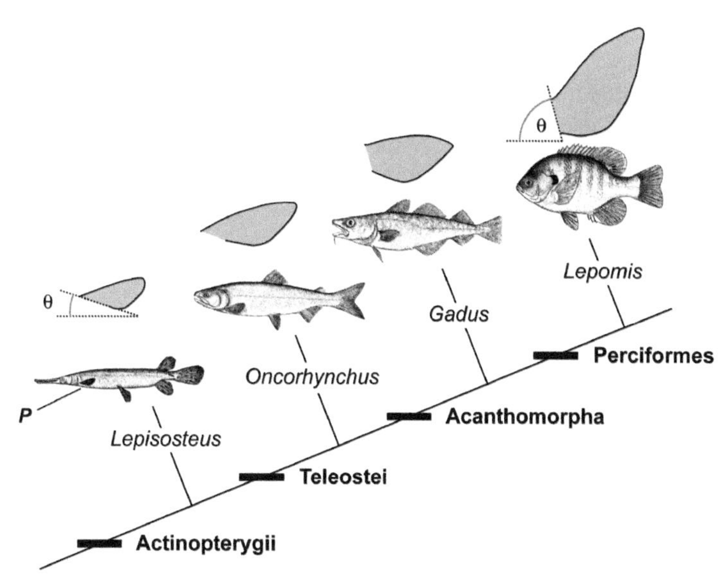

Fig. 10.1. Phylogenetic relationships of selected genera of ray-finned fishes to illustrate two major evolutionary trends in the design of pectoral fins. First, the pectoral fin (labeled *P*) migrates from a ventral body position (shown by basal taxa such as *Lepisosteus* as well as lower teleosts including *Oncorhynchus*) to a derived dorsolateral position (as in *Gadus* and *Lepomis*). Second, the base of the pectoral fin is reoriented from a primarily horizontal to a more vertical inclination. Enlarged images of the pectoral fin shown above each clade have been scaled to the same proportion of body length. θ, angle of inclination of the pectoral fin base measured relative to the longitudinal body axis. (From Drucker and Lauder, 2002a.)

typically has a nearly horizontal orientation (i.e., the fin base lies at a shallow angle relative to the longitudinal body axis). In contrast, more-derived taxa (including acanthomorphs such as *Gadus* and *Lepomis*; Figure 10.1) commonly show pectoral fins with more vertically inclined bases. In addition to this reorientation of the pectoral fin, there is also a trend of change in pectoral fin position. Primitively, the pectoral fin is located low on the body, positioned near the ventral body margin (e.g., Figure 10.1, *Lepisosteus* and *Oncorhynchus*). In acanthomorph fishes (Figure 10.1), the pectoral fin is located higher on the body, at an approximately mid-dorsal position and closer to the center of mass of the fish.

The main features of pectoral fin musculoskeletal structure include the pectoral or shoulder girdle skeleton, a row of four radials that form a basal support for the fin, a fibrocartilage pad upon which the fin rays rotate, a series of fin rays with rotational bases, and a set of muscles that powers the motion of the fin rays (Figure 10.2). Considerable taxonomic variation

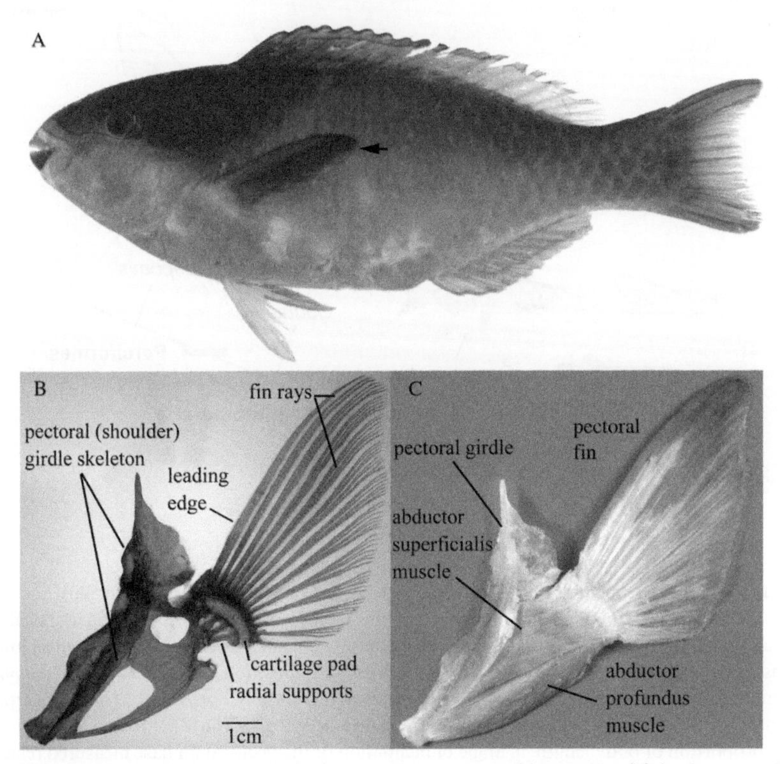

Fig. 10.2. Anatomy of the pectoral fin in a highly maneuverable coral reef fish, the parrotfish. (A) Position of the pectoral fin (arrow) in the parrotfish *Scarus forsteni.* (B) Skeletal structure including shoulder girdle support, fin base, and leading-edge anatomy, and the structure of fin rays of *Scarus frenatus.* (C) Lateral muscles that power the downstroke of the fin, including the broad abductors superficialis and profundus that pull on the bases of many fin rays at once to bring the fin downward and forward. (From Westneat *et al.*, 2004.)

exists in the architecture of the pectoral girdle, the shape of the fin, and the anatomy of the muscles (Figures 10.3 and 10.4). The pectoral girdle (Figures 10.2B and 10.3) is the anchor upon which the pectoral muscles originate. The anteroventral surfaces of the cleithrum, both laterally and medially, as well as scapula and coracoid, are the sites of attachment for abductor and adductor musculature (Figures 10.2C and 10.4). The first pectoral fin ray is typically a short, thick ray that articulates with the scapula in a synovial saddle joint. In the majority of teleost fishes, the first and second pectoral rays are united by connective tissues as a single rotational element that forms the leading edge of the pectoral fin (Figure 10.2B). Most

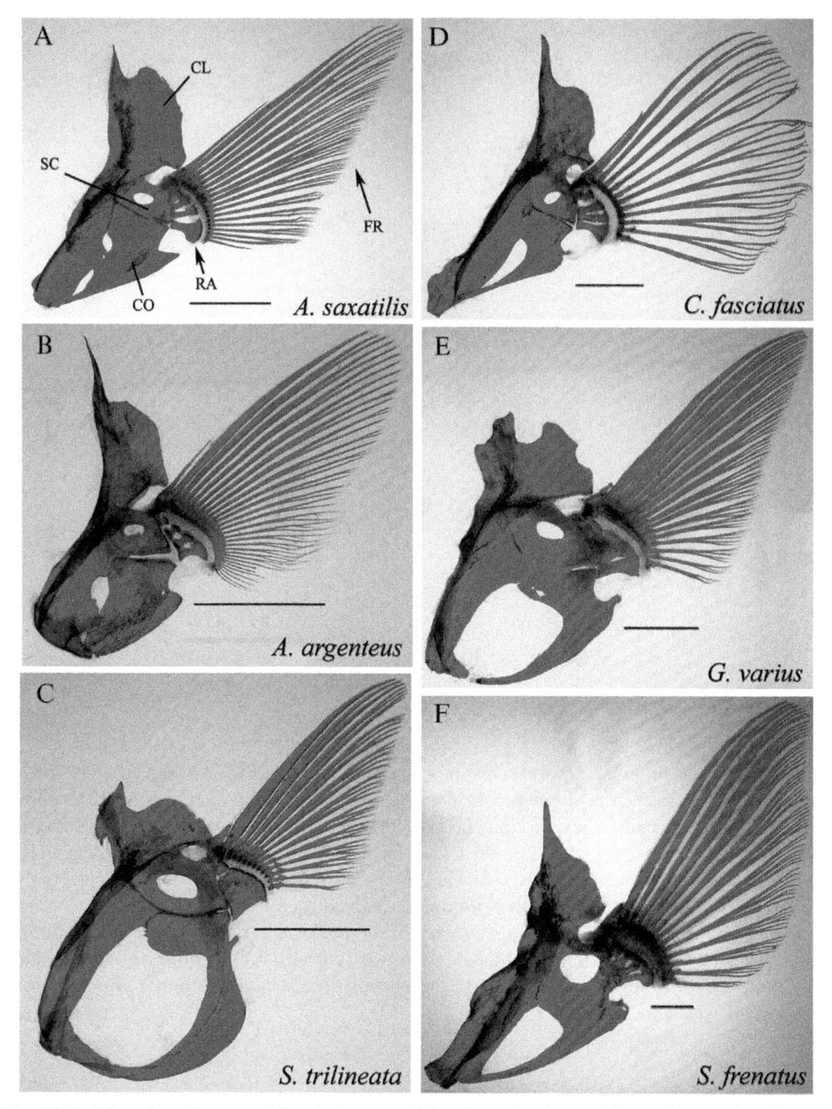

Fig. 10.3. Diversity of pectoral fin skeleton and fin shape in six perciform fish species. Images are cleared and stained pectoral girdles and fins in left lateral view (red: bone; blue: cartilage). (A) *Abudefduf saxatilis*, (B) *Amphistichus argenteus*, (C) *Stethojulis trilineata,* (D) *Cheilinus fasciatus,* (E) *Gomphosus varius*, (F) *Scarus frenatus*. CL, cleithrum; CO, coracoid; FR, fin rays; RA, radials; SC, scapula. Scale bars = 1 cm. (From Thorsen and Westneat, 2005.)

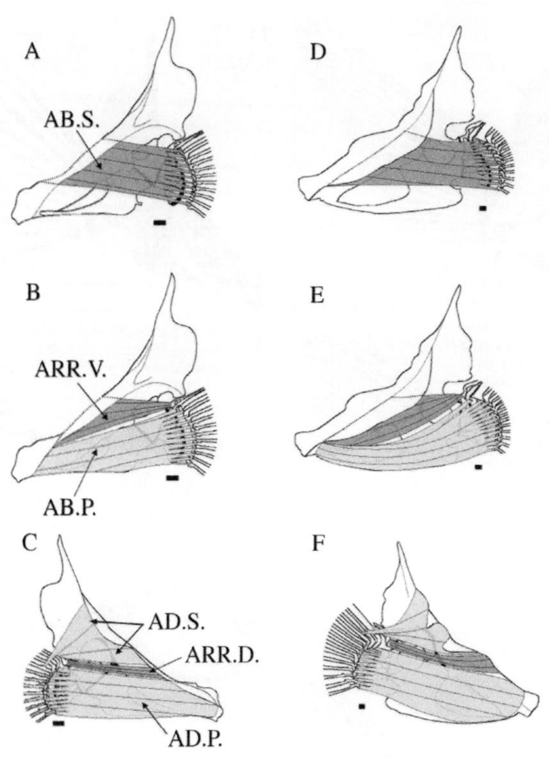

Fig. 10.4. Pectoral fin muscle morphology in a wrasse, *Cheilinus fasciatus* (A–C), which uses a fore–aft rowing stroke, and a parrotfish, *Scarus frenatus* (D–F), which uses a dorsoventral flapping stroke for pectoral locomotion. Left fin illustrated for both species. (A, D) Structure of the abductor superficialis muscle on the lateral face of the pectoral girdle. (B, E) Deep lateral view (abductor superficialis removed) showing the arrector ventralis and abductor profundus muscles. (C, F) Adductor superficialis (with dorsal and ventral divisions), arrector dorsalis, and adductor profundus muscles on the medial face of the pectoral girdle. AB.S., abductor superficialis; AB.P., abductor profundus; ARR.V., arrector ventralis; AD.S., adductor superficialis; AD.P., adductor profundus; ARR.D., arrector dorsalis. Scale bars = 3 mm. (Modified from Thorsen and Westneat, 2005.)

pectoral rays have their bases imbedded in the fibrocartilage pad that separates them from the underlying radials. Pectoral fin shape (Figure 10.3) is determined largely by relative fin ray length—in most fishes the anterodorsal rays are the longest, with other rays tapering in length from dorsal to ventral to form a wing-shaped fin.

Fin shape has important consequences for swimming hydrodynamics and performance, and can be related to habitat use (Section VII). Variables that describe fin shape include aspect ratio, a simple measure of the narrowness of the fin relative to its length, and moments of fin area, which are measures of the base-to-tip distribution of fin area. Long, narrow fins have high aspect ratios, while fins that expand distally have high moments of area. Pectoral fins come in a diverse array of planforms. Within labrids, for example, pectoral fin shape varies from a paddle with a low aspect ratio ($\leq$2.0) and a high standardized second moment of area ($\geq$0.7) to a wing with a high aspect ratio ($\geq$4.5) and a low standardized second moment of area ($\leq$0.5) (Wainwright *et al.*, 2002; Walker and Westneat, 2002b).

Muscle morphology is a key aspect of pectoral fin swimming biomechanics that may differ widely among species. Geerlink (1989) compared the pectoral fin shapes and kinematics of two labrids and a cichlid fish, and concluded that morphological differences did not explain interspecific variation in pectoral fin swimming motion. Although no internal measurements of muscle morphology or electromyographic activity were made, Geerlink (1989) attributed kinematic differences to the neuromuscular control of behavior. A recent study of 12 species of coral reef fishes (Thorsen and Westneat, 2005) showed that muscle mass is proportional to force generation ability, and that pectoral muscle masses can vary widely across species. This research found that the pectoral fin musculature comprises 0.68–8.24% of the total body mass (M_b). Two species examined, *Stethojulis trilineata* and *Zanclas cornutus*, have the largest pectoral musculature so far measured among reef fishes, comprising 6.84 and 8.24% M_b, respectively. Other reef fishes examined ranged from 0.68 to 2.44% M_b. The ratio of abductor (downstroke) to adductor (upstroke) muscle mass ranged from 0.72 to 1.46. The relative mass of abductors and adductors provides an estimate of the total amount of muscle tissue dedicated to downstroke and upstroke power production. Most labrid fishes with a dorsoventral flapping stroke have an abductor to adductor ratio of less than 1, although some flappers in other families have a proportionally larger abductor musculature (Thorsen and Westneat, 2005). These differences highlight the relative importance for thrust production of the upstroke in fishes with more massive adductors, and of the downstroke for some flappers with relatively high abductor mass.

III. MOTOR PATTERNS OF PECTORAL FIN LOCOMOTION

The mechanisms of neural control of pectoral fin locomotion in fishes are relatively unexplored, yet knowledge of muscle contraction patterns, neuroanatomy, and neural circuits is critical to developing models of pectoral fin function. Drucker and Jensen (1997), Westneat (1996), and Westneat and Walker (1997) analyzed the motor patterns of the major pectoral muscles in

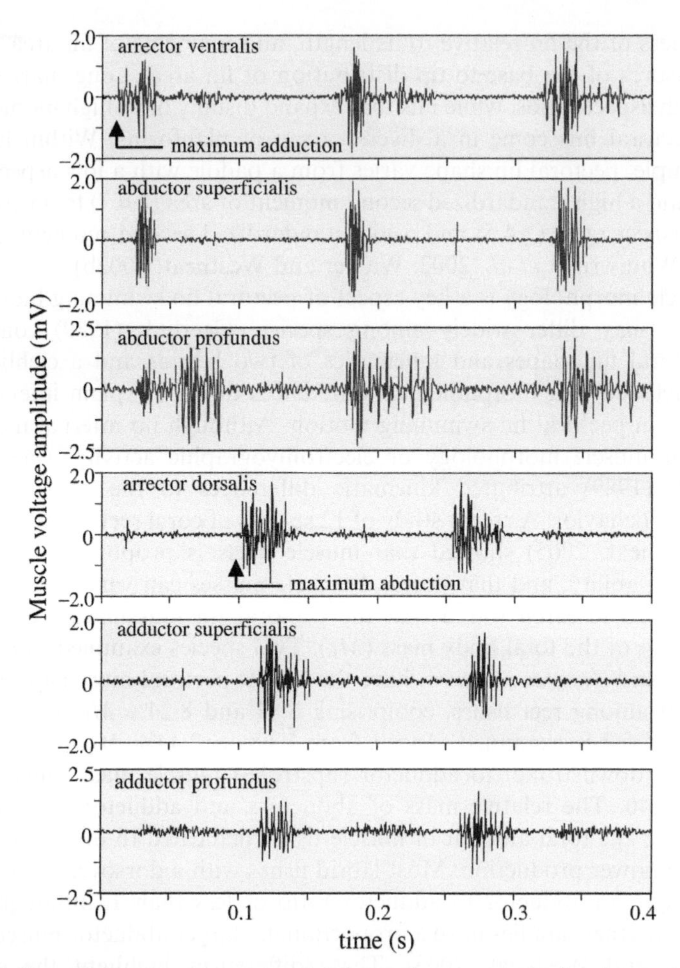

Fig. 10.5. Pectoral electromyograms recorded during three consecutive flapping strokes in the bird wrasse (*Gomphosus varius*) swimming steadily at relatively high speed (3.8 total body lengths per second). Alternating activity of downstroke muscles (top three traces) and upstroke muscles (bottom three traces) powers the pectoral fin locomotor cycle. Average times at which the pectoral fin attains a position of maximal adduction (at the end of the upstroke) and maximal abduction (at the end of the downstroke) are indicated by arrows. (From Westneat and Walker, 1997.)

two labroid groups—the surfperches and wrasses. The basic pattern of muscular contraction during pectoral fin locomotion is one of alternating activity of the antagonistic abductor and adductor groups (Figure 10.5). With the pectoral fins maximally adducted against the body, the abduction (or downstroke) phase of the stroke cycle begins with activity of the arrector ventralis muscle (Figure 10.4B) prior to activity of the other abductors. The arrector ventralis rotates the fin forward to initiate peeling of the leading edge away from the body. The abductors superficialis (Figure 10.4A) and profundus (Figure 10.4B) then become active together with the arrector ventralis to produce the downstroke of the fin. Immediately following maximum fin abduction, adduction (upstroke) begins, with the stroke reversal initiated by activity of the arrector dorsalis muscle (Figure 10.4C). The adductors superficialis and profundus (Figure 10.4C) are then activated synchronously to effect fin adduction (Figure 10.5).

A comparison of pectoral motor patterns among groups of fishes reveals a temporally conservative sequence of muscle activation. The relative timings of onsets and offsets of pectoral muscle activity measured in wrasses (Figure 10.5) with few exceptions also characterize the stroke cycle of surfperches (Figure 10.6). In addition, the locomotor functions served by various divisions of the pectoral musculature are similar in all labroids studied to date, reflecting similar kinematic patterns (Drucker and Jensen, 1997). Divisions of the pectoral musculature inserting on the central and trailing-edge fin rays serve the simple functions of abduction and adduction, whereas muscles controlling the fin's leading edge play more complex roles during the fin stroke, including deceleration of the fin at the downstroke–upstroke transition and rotation of the adducted fin (Figure 10.6). These mechanistic details of pectoral fin swimming are probably widely distributed within the teleosts, although a recent description of muscular variation and novel muscle subdivisions (Thorsen and Westneat, 2005) may indicate a broader range of muscle function and locomotor behavior that warrants investigation.

The frequency and amplitude of pectoral fin electromyographic activity increase with swimming speed (Westneat and Walker, 1997). In wrasses, the duration of electrical activity in abductors is significantly greater than that in adductors, with the abductor profundus having the longest period of activation (Figure 10.5). These electromyogram (EMG) durations do not change significantly in labrids as a function of forward swimming velocity. Rather, the lag time between EMG cycles decreases with increasing speed. The onset times of the abductors profundus and superficialis relative to the onset of arrector ventralis activity do not change significantly with speed, but the onset of the adductors relative to that of the arrector ventralis occurs earlier in the stroke cycle at higher velocities.

The integration of EMG data with swimming kinematics in labrids reveals that activity of pectoral abductor and adductor muscles is largely coincident with the duration of fin abduction and adduction, respectively. In the bird wrasse, the onset of arrector ventralis activity is

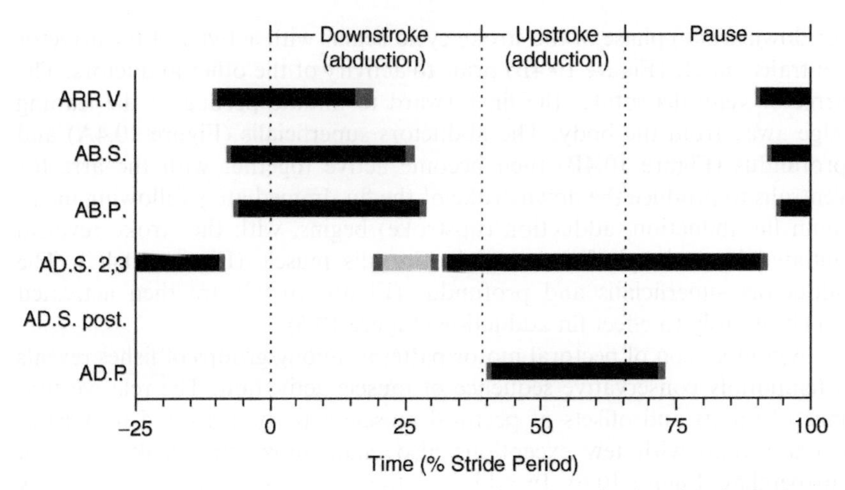

Fig. 10.6. Summary diagram of pectoral muscle activity in striped surfperch (*Embiotoca lateralis*) swimming at 80% of the pectoral-to-caudal gait transition speed. Kinematic phases comprising the pectoral fin stroke cycle are delimited by dotted lines. Mean muscle activity timings, represented by the ends of black bars, are shown with S.E.M. (dark gray bars). The peak of abductor muscle activity (top three bars) precedes the onset of fin abduction itself. The adductor superficialis 2,3 showed activity (light gray bar) that broadly overlaps that of the abductor muscles. Activity in the adductor superficialis posterior was not detected. The kinematic "pause" phase is a nonpropulsive period during which the fully adducted pectoral fin rotates forward against the body in preparation for the following downstroke. AD.S. 2,3, ventral portion of adductor superficialis serving fin rays 2 and 3; AD.S. post., dorsal portion of adductor superficialis serving posterior fin rays. Other muscle abbreviations as in Figure 10.4. (Modified from Drucker and Jensen, 1997.)

nearly simultaneous with the initiation of fin abduction, arrector dorsalis activity begins together with the stroke reversal preceding fin adduction, and the large antagonistic abductor and adductor muscles are active only during the downstroke and upstroke of the fin (Figure 10.5). In striped surfperch, by contrast, the abductor muscles consistently fire before the onset of fin abduction, and early adductor activity typically overlaps the end of abductor activity to decelerate the fin at the transition from downstroke to upstroke (Figure 10.6). This latter pattern of muscle activity preceding the corresponding propulsor motion matches EMG data collected from the axial musculature of undulatory swimmers (Wardle *et al.*, 1995), the pectoral muscles of birds in flight (Dial *et al.*, 1991), and the flight muscles of insects (Tu and Dickinson, 1994). Thus, despite their fixed sequence of activation within the fin stroke cycle, the pectoral fin muscles of fishes vary in the amount of negative work performed during steady swimming. The functional significance of this variation is not well understood, but muscle activity beginning well in advance of fin motion and significant overlap of activity in antagonistic locomotor muscles may provide the fine neuromuscular control of propulsor kinematics required for spontaneous modification of fin movement (e.g., during maneuvering).

IV. PECTORAL FIN KINEMATICS

A. Variation in Movement Patterns

Fishes use the pectoral fins both as nonoscillating, largely passive control surfaces (Wilga and Lauder, 1999, 2000) and as oscillating, active control surfaces. Much of the broad kinematic variation in actively oscillated fins can be summarized by two parameters: the stroke plane angle and the ratio of wavelength to chord length. Oscillating pectoral fin rays sweep back and forth along the surface of a plane called the stroke plane. The stroke plane angle is the angle between the stroke plane and a vertical vector (Figure 10.7D). Oscillations within the sagittal, frontal, and transverse planes, each defined by a distinct stroke plane angle, are called paddling, rowing, and flapping, respectively. The oscillating fins of many fishes pass traveling waves of bending from leading to trailing edge. If the ratio of the length of these waves to the mean fin chord length (distance from leading to trailing edge) is small, the fin is undulating and gives the appearance of having multiple waves traveling along the chord. If this wavelength-to-chord-length ratio is large, there is only a small phase lag in the movement of the leading and trailing edges, and the fin is said to be oscillating or flapping and gives the appearance of a stiff, reciprocating plate. Note, however, that many derived teleosts control the fin rays independently, allowing complex, undulatory motions that are not well described by simple wave parameters (Fish and Lauder, 2006). Rowing, flapping, paddling, oscillation, and undulation are ideal locomotor modes at the extremes of different kinematic axes (Lauder and Jayne, 1996; Walker and Westneat, 2002a,b) that can be useful for classification (Tables 10.1 and 10.2) or modeling (Daniel, 1988; Walker and Westneat, 2000; Combes and Daniel, 2001).

The different styles of pectoral fin motion are broadly distributed among fishes. Nonoscillating pectoral fins are found in sturgeons and sharks (Wilga and Lauder, 1999, 2000). Paddling pectoral fins are found in outgroups to teleosts (except sturgeons) and in many primitive teleosts. Ratfishes, the outgroup to Chondrichthyes, possess beautiful, flapping pectoral fins (Combes and Daniel, 2001). Batoids (skates and rays) present a range of pectoral fin motions between flapping and undulation

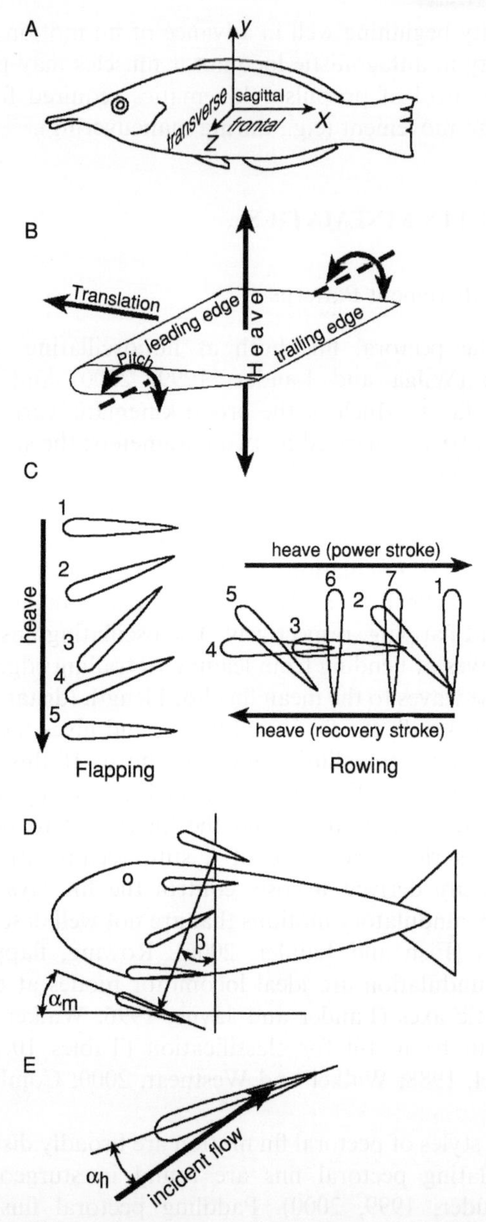

Fig. 10.7. Definition of kinematic parameters. (A) The anatomical planes of a fish. In rowing, flapping, and paddling, the fins oscillate within the frontal (*xz*), transverse (*yz*), and sagittal (*xy*) planes, respectively. (B) Heaving, pitching, and translation of a hydrofoil. (C) Two-dimensional heaving and pitching of a flapping (left) and rowing (right) fin in a horizontal free stream (not

(Rosenberger and Westneat, 1999; Rosenberger, 2001). Within teleosts, flapping appears to be derived from a more primitive rowing or paddling stroke.

Among teleost fishes capable of sustained pectoral fin propulsion at relatively high speeds, the rowing–flapping axis of variation is quite conspicuous. Importantly, rowing and flapping differ in more than the orientation of the stroke plane. Rowing, as well as paddling, is characterized by a distinct recovery (minimal loading) stroke, in which the fin is pulled forward with its surface oriented parallel to fin motion (a feathered orientation), and a distinct power stroke, in which the fin is pulled backward with its surface oriented normal to its motion (a broadside orientation) (Walker and Westneat, 2002a) (Figures 10.7C and 10.8). By contrast, distinct recovery and power strokes do not occur in the flapping fins of fishes. Instead, the fin is pulled either down or up with its surface oriented at a moderate angle (20–40°) to the direction of its motion (Walker and Westneat, 2002a) (Figures 10.7C and 10.8).

B. Speed and Scaling Effects During Steady Swimming

Both swimming speed and body size are important determinants of locomotor behavior. The influence of these variables on patterns of propulsor motion during rectilinear swimming at a relatively constant velocity (steady swimming) has been extensively documented, mainly in fishes relying on axial undulatory propulsion (e.g., Bainbridge, 1958; Hunter and Zweifel, 1971; Webb *et al.*, 1984). Modulation of pectoral fin kinematics with increasing swimming speed has been described for a small number of cartilaginous and bony fishes. Among fishes employing oscillatory pectoral fin locomotion during steady swimming, the speed dependence of pectoral fin motion has been studied both in species that use the paired fins over a narrow range of low speeds prior to initiating axial undulation (Archer and Johnston, 1989; Parsons and Sylvester, 1992; Gibb *et al.*, 1994) and in species that rely on these fins as the primary means of propulsion over a wide range of speeds (Webb, 1973; Drucker and Jensen, 1996a; Walker and Westneat, 1997).

illustrated). In the flapping fin, the downstroke is illustrated in steps 1–5. In the rowing fin, the recovery (1–4) and power (4–7) strokes are illustrated. In steps 3–4 of the rowing fin, the fin is perfectly feathered (oriented parallel to the incident flow). (D) The stroke plane angle, β, is illustrated by the stroke plane tilted β degrees from a transverse plane. The morphological angle of attack, α_m, is the angle between the fin chord and the horizon. (E) The hydrodynamic angle of attack, α_h, is the angle between the local incident flow (thick line) and the fin chord (thin line). (From Walker and Westneat, 2002a.)

Table 10.1
Pectoral Fin Locomotor Modes Used for Steady Swimming

Locomotor mode	Primary pectoral fin motion	Left-right phase difference[a]	Pectoral fin-base width	Taxonomic example	Reference[b]
Labriform	Dorsoventral oscillation (flapping)	0°	Variable	*Hydrolagus*	Combes and Daniel, 2001
				Gomphosus	Walker and Westneat, 1997, 2002b
				Cirrhilabrus	Walker and Westneat, 2002b
				Tautoga, *Scarus*	Breder, 1926
				Lactoria, *Tetrasomus*	Blake, 1977
Gasterosteiform	Anteroposterior oscillation (rowing)	0°	Variable	*Gasterosteus*	Walker, 2004a
				Pterophyllum	Blake, 1979b
				Abudefduf	Breder, 1926
				Pseudocheilinus, *Halichoeres*	Walker and Westneat, 2002b
				Notothenia	Archer and Johnston, 1989
Embiotociform[c]	Dorsoventral oscillation	0°	Narrow	*Cymatogaster*	Webb, 1973
				Amphistichus	Drucker and Jensen, 1997
				Embiotoca	Drucker and Jensen, 1996a
Mobuliform	Dorsoventral oscillation	0°	Broad	Mobulidae, Myliobatidae	Webb, 1994a
				Mobula	Klausewitz, 1964
				Rhinoptera	Rosenberger, 2001

Coelacanthiform	Dorsoventral oscillation	180°	Broad	*Latimeria*	Fricke *et al.*, 1987; Fricke and Hissmann, 1992
Diodontiform	Dorsoventral undulation	–	Broad	*Gasterosteus*	Walker, unpublished
				Hippocampus	Breder and Edgerton, 1942
				Spheroides	Breder, 1926
Rajiform	Anteroposterior undulation	–	Broad	*Raja*	Marey, 1890; Breder, 1926; Daniel, 1988
				Taeniura	Rosenberger and Westneat, 1999
				Dasyatis, *Gymnura*	Rosenberger, 2001
Labriform + diodontiform	Anteroposterior oscillation + dorsoventral undulation	180°	Broad	*Chilomycterus*	Arreola and Westneat, 1996
Tetraodontiform	Dorsoventral undulation (supplementing median fin oscillation)	–	Broad	*Lagocephalus*, *Spheroides*	Breder, 1926
				Arothron	Gordon *et al.*, 1996
Ostraciiform	Anteroposterior rowing (supplementing median fin oscillation)	0° at low speeds; increasing at higher speeds	Broad	Ostraciidae	Breder, 1926
				Ostracion	Hove *et al.*, 2001

[a]For oscillatory swimming modes: 0°, pectoral fins move in phase with each other; 180°, pectoral fins move out of phase with each other by one-half stroke cycle.

[b]For earlier references, consult Table III in Lindsey (1978).

[c]Specialized form of labriform swimming termed by Webb (1994a) to describe "lift-based" pectoral fin propulsion that is used as the primary means of locomotion over a wide range of speeds. Since Webb's review, it has been demonstrated that other groups of fishes (e.g., wrasses) also generate lift-based thrust; embiotocids may equally well be classified as labriform swimmers.

Table 10.2
Pectoral Fin Maneuvering Behaviors

Behavior	Primary pectoral fin motion	Taxonomic example	Reference
Hovering (maintenance of vertical position in water column)	(i) Alternating left- and right-side oscillation	*Oncorhynchus*	Drucker and Lauder, 2003
		Micropterus	Magnan, 1930
		Latimeria	Fricke and Hissmann, 1992
	(ii) Dorsoventral undulation	*Gasterosteus*	Walker, unpublished
		Synchiropus	Blake, 1979a,c
		Rhinecanthus, Odonus	Blake, 1978
Turning	(i) Unilateral (strong-side) pectoral fin propulsion	*Rhinecanthus, Odonus*	Blake, 1978
	(ii) Unilateral (weak-side) pectoral fin pivot	*Lagocephalus*	Breder, 1926
	(iii) Asynchronous, bilateral pectoral fin excursion	*Oncorhynchus*	Drucker and Lauder, 2003
		Lepomis	Drucker and Lauder, 2001
		Tetrasomus	Blake, 1977
		Ostracion	Walker, 2000, 2004b
		Latimeria	Fricke *et al.*, 1987; Fricke and Hissmann, 1992

Braking	Bilateral extension	*Carassius*	Bainbridge, 1963
		Esox	Breder, 1926
		Oncorhynchus	Drucker and Lauder, 2003
		Pollachius	Geerlink, 1987
		Gadus	Videler, 1981; Geerlink, 1987
		Lepomis	Harris, 1938; Drucker and Lauder, 2002b
		Scomber	Geerlink, 1987
Rising/sinking (vertical translation in water column)	Bilateral trailing-edge depression/elevation	*Chiloscyllium*	Wilga and Lauder, 2001
		Triakis	Wilga and Lauder, 2000
		Acipenser	Wilga and Lauder, 1999
Benthic station holding	Bilateral trailing-edge elevation (for negative lift)	*Salmo*	Kalleberg, 1958; Arnold *et al.*, 1991
Backward swimming	Reverse rowing (kinematically similar to forward rowing, but with power and recovery strokes reversed)	*Carassius*, *Metynnis*, *Pterophyllum*	Webb and Fairchild, 2001
		Latimeria	Fricke and Hissmann, 1992

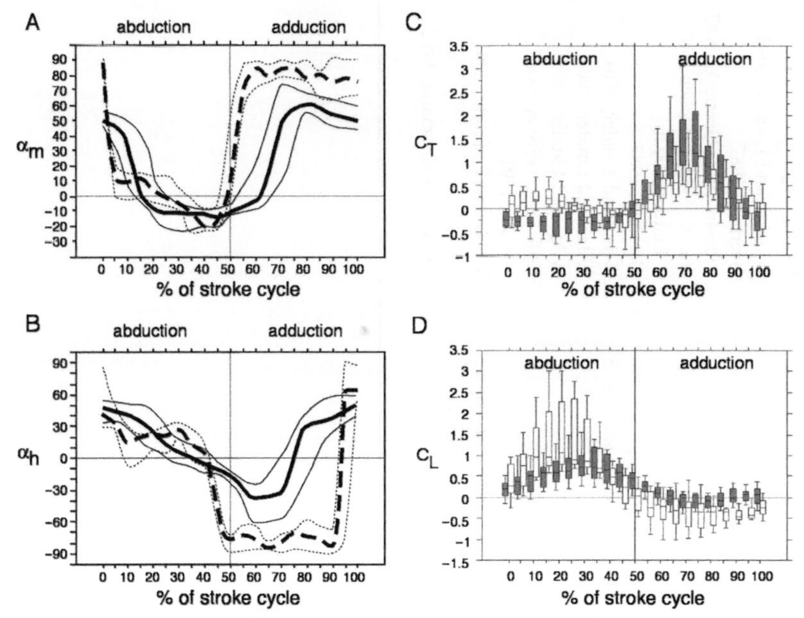

Fig. 10.8. Morphological (α_m) (A) and hydrodynamic (α_h) (B) angles of attack during pectoral fin locomotion by the threespine stickleback, *Gasterosteus aculeatus* (dashed lines), and the bird wrasse, *Gomphosus varius* (solid lines), with associated force coefficients (dimensionless measures of force) (C, D). (A, B) Angles α_m and α_h are defined in Figure 10.7; thick lines represent the mean attack angles while thin lines delimit the 90% confidence intervals. (C) Measured thrust coefficient, C_T, for *G. aculeatus* (shaded boxes) and *G. varius* (open boxes). (D) Measured lift coefficient, C_L, for *G. aculeatus* (shaded) and *G. varius* (open). The line within each box is the median, the box top and bottom delimit the middle 50% of the distribution, and the whiskers represent the middle 90% of the distribution. Kinematic differences between the rowing stroke of *G. aculeatus* and the flapping stroke of *G. varius* are associated with differences in thrust and lift generation, as predicted by a simple rowing–flapping model (see text for details). (From Walker and Westneat, 2002a.)

Oscillatory pectoral fin swimmers typically exhibit several distinct locomotor patterns during steady swimming trials with increasing velocity. At low to intermediate swimming speeds, the pectoral fins move synchronously though repeated cycles of abduction and adduction to generate forward thrust. At higher speeds, the pectoral-to-caudal gait transition occurs: undulation of the trunk and tail begins to supplement oscillation of the pectoral fins. At speeds just above this pectoral-to-caudal gait transition speed ($U_{p\text{-}c}$), the paired and median fins may be used simultaneously (e.g., in sunfish, Gibb *et al.*, 1994) or may be recruited alternately in a "burst-and-flap" gait (e.g., surfperches, wrasses, and parrotfishes), or paired fin oscillation may be abandoned altogether (e.g., in the surgeonfish *Naso literatus*). When fishes employ the burst-and-flap gait in a swim tunnel, the swimming behavior cycles between rapid acceleration upstream powered by the oscillating tail and slow drifting downstream with attempts to maintain position with only pectoral fin oscillation. In stickleback (Walker, unpublished data) and wrasses (Korsmeyer *et al.*, 2002; Walker and Westneat, 2002b), a burst-and-flap gait is sustainable for only a few (<10) minutes.

The stroke plane swept out by the leading edge of an oscillating pectoral fin is not strictly a plane but rather a curved surface that is inclined more vertically during fin abduction and more horizontally during fin adduction (Webb, 1973; Geerlink, 1983; Gibb *et al.*, 1994; Lauder and Jayne, 1996; Walker and Westneat, 1997; Hove *et al.*, 2001; Walker and Westneat, 2002b). The steeper downstroke facilitates thrust production during this phase by creating a large vertical component to the incident flow, which shifts the angle of attack from negative (producing a downward and backward force) to positive (producing an upward and forward force). Fishes with either an extreme rowing stroke, such as the threespine stickleback (*Gasterosteus aculeatus*), or an extreme flapping stroke, such as the bird wrasse (*Gomphosus varius*), present little speed-dependent variation in the stroke plane angle (Walker and Westneat, 2002a). By contrast, fishes whose fin kinematics are more intermediate between pure rowing and flapping may either decrease (Hove *et al.*, 2001) or increase (Walker and Westneat, 2002b) the steepness of the stroke plane as swimming speed increases.

Pectoral fin beat amplitude (dorsoventral fin excursion expressed as a fraction of body or pectoral fin length) has been shown to increase over low to intermediate speeds and plateau at higher speeds in surfperch (Webb, 1973), sunfish (Gibb *et al.*, 1994), and boxfish (Hove *et al.*, 2001). In the striped surfperch, fin amplitude remains a roughly constant proportion of body length at all but the lowest speeds (Drucker and Jensen, 1996a). Walker and Westneat (1997) demonstrated that amplitude measured as the angular excursion increases monotonically rather than asymptotically in the bird wrasse, but noted that this species was studied at speeds below its maximum sustainable pectoral fin swimming speed.

Pectoral fin beat frequency (f) typically increases slowly at low speeds and more rapidly at intermediate speeds, attaining a maximum at the pectoral-to-caudal gait transition. A plateau or even decrease of pectoral fin beat frequency at the onset of tail beating at $U_{p\text{-}c}$ has been documented for a variety of teleosts (e.g., Archer and Johnston, 1989; Parsons and Sylvester, 1992; Gibb *et al.*, 1994; Drucker and Jensen, 1996a; Mussi *et al.*, 2002). The rate of increase in f with speed up to $U_{p\text{-}c}$ varies according to the kinematics of the fin stroke. The pectoral fin beat period of some fishes (e.g., surfperches) contains a nonpropulsive "refractory" or pause period

(Figure 10.6). Accordingly, f reflects the time-averaged frequency of fin oscillation, but not the rate at which forward thrust is generated. In these swimmers, f increases curvilinearly with speed (Webb, 1973; Drucker and Jensen, 1996a). When f is corrected by excluding the pause period, fin beat frequency increases linearly with speed, a pattern observed in other pectoral fin swimmers that lack a kinematic pause period (Westneat, 1996; Hove *et al.*, 2001) as well as in caudal fin swimmers.

Fishes exhibiting undulatory pectoral fin swimming modes employ a variety of kinematic mechanisms for increasing swimming speed. Rajiform swimmers (Table 10.1), for example, modulate three primary gait parameters during steady forward progression: undulatory wave frequency, wavespeed, and length-standardized wavelength (wavelength/chord length, or the reciprocal of the number of waves along the chord). Among batoid fishes, which are typical rajiform swimmers, stingrays in the genera *Dasyatis* and *Taeniura* increase both the frequency and wavespeed of pectoral fin undulation in order to increase forward swimming speed (Rosenberger and Westneat, 1999; Rosenberger, 2001). The skate *Raja*, by contrast, achieves speed increases without modulating undulatory frequency, instead increasing the speed and number of undulatory waves traveling along the pectoral fin. For all batoids studied thus far, pectoral fin amplitude is largely independent of swimming speed (Rosenberger and Westneat, 1999; Rosenberger, 2001).

The influence of body size on pectoral fin kinematics and performance has been the subject of limited experimental study. Drucker and Jensen (1996a,b) examined the relationship between body size and pectoral fin kinematics and swimming speed in an ontogenetic series of striped surfperch ranging 5-fold in body length and over 100-fold in body mass (M_b). Kinematic comparisons were made at the pectoral-to-caudal gait transition speed since $U_{p\text{-}c}$, unlike length-specific speed, represents an equivalent level of exercise for fishes of different sizes (Drucker, 1996). Pectoral fin beat frequency measured at $U_{p\text{-}c}$ shows negative allometry in striped surfperch: small fish switch gaits at higher fin beat frequencies than large fish (Drucker and Jensen, 1996b). The same pattern was observed for shiner surfperch spanning an approximately 2-fold range in body length (Mussi *et al.*, 2002) and for two size classes (juvenile and adult) of Antarctic cod (Archer and Johnston, 1989).

Negative scaling of limb oscillation frequency at equivalent speeds of locomotion is predicted by biomechanical theory (Hill, 1950; McMahon, 1975; Goldspink, 1977) and has been demonstrated empirically for a diversity of vertebrates (e.g., Bainbridge, 1958; Greenewalt, 1962; Heglund and Taylor, 1988; Marsh, 1988; Clark and Fish, 1994). The specific allometric relationship observed for striped surfperch (f at $U_{p\text{-}c} \propto M_b^{-0.12}$) appears also to describe the scaling of pectoral fin beat frequency for other fishes, including centrarchids and notothenioids. Further, this size dependence of fin beat frequency is consistent with that observed for fishes swimming by axial undulation. Caudal fin beat frequency at estimated maximum aerobic and anaerobic swimming speeds for many species scales approximately with $M_b^{-0.1}$ to $M_b^{-0.2}$ (Drucker and Jensen, 1996b).

C. Fin Motions During Maneuvering

Maneuvering behaviors powered entirely or in part by pectoral fins include yawing turns for both fine correction and radical change of heading, body pitching for vertical translation through the water column, rapid acceleration (e.g., during the pursuit of prey), braking, backward swimming, and hovering (i.e., station holding in still water) (Table 10.2). The following section provides a taxonomic review of pectoral fin motions exhibited during maneuvering.

Three-dimensional kinematic analysis has been employed in the study of pectoral fin motion during maneuvering by both cartilaginous and bony fishes. Recent kinematic work on paired fin function in a basal ray-finned fish (white sturgeon, Wilga and Lauder, 1999) and chondrichthyan outgroups (leopard and bamboo sharks; Wilga and Lauder, 2000, 2001) demonstrates that the "primitive" horizontally oriented pectoral fin is a relatively flexible propulsor with a limited but significant range of motion. The posterior portion of this pectoral fin is capable of acting as a simple flap to redirect incident water flow and exert a pitching moment around the body's center of mass. Such fin excursions reorient the body to initiate rising and sinking maneuvers in the water column (Table 10.2).

Teleost fishes possess pectoral fins with considerably greater functional versatility for maneuvering. Drucker and Lauder (2003) examined the behavioral repertoire of the pectoral fins of rainbow trout (*Oncorhynchus mykiss*) (Figure 10.1) performing both steady rectilinear swimming and unsteady maneuvering locomotion. During constant-speed swimming (0.5 and 1.0 body lengths per second), the pectoral fins remain adducted against the body (note that brook trout, by contrast, have been reported to oscillate the pectoral fins during steady swimming; see McLaughlin and Noakes, 1998). The paired fins of rainbow trout are actively recruited, however, for a variety of maneuvering behaviors, including hovering, low-speed turning, and rapid deceleration of the body during braking (Figure 10.9) (see also Figures 2–4 in Drucker and Lauder, 2003). Hovering is commonly employed by salmonids to maintain vertical position in the water column, and involves low-speed sculling of the pectoral fins beneath the body with the left and right fins moving out of phase with each other (Table 10.2). Slow turning in trout is characterized by rapid abduction and elevation of the "strong-side"

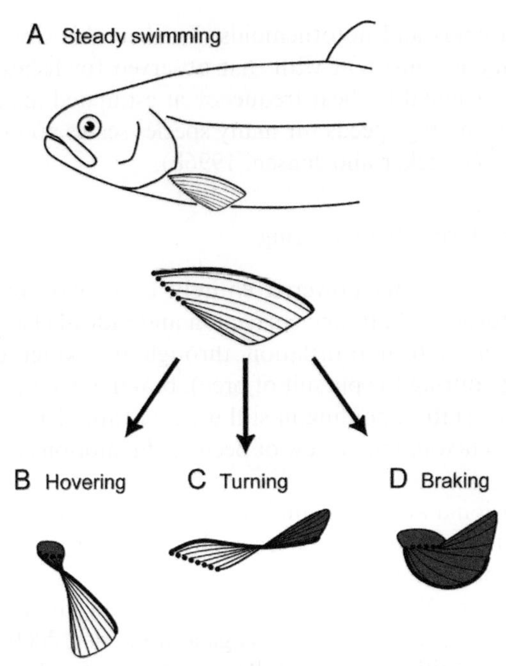

Fig. 10.9. Kinematic repertoire of the pectoral fin of rainbow trout (*Oncorhynchus mykiss*). (A) During steady swimming, the fin remains adducted against the body. In the enlarged image of the fin below the body, the first pectoral fin ray is indicated by a thick line and the fin base by a dotted line. For the maneuvering behaviors illustrated in (B–D), white and gray areas indicate fin surfaces that face laterally and medially, respectively, when the fin is at rest in an adducted position (as in A). (B) While hovering, trout twist the fin along its spanwise axis to enable fore and aft sculling beneath the body. (C) Turning is characterized by rotation of the fin in the opposite direction above the ventral body margin. (D) Braking involves fin rotation in the same direction as during turning, but to a greater degree such that the fin surface that faces medially at rest becomes dorsolaterally oriented. The kinematic versatility of the trout pectoral fin permits a range of locomotor functions comparable to that of more-derived teleost fishes. (Modified from Drucker and Lauder, 2003.)

pectoral fin (closer to the source of the turning stimulus) and slower, delayed abduction of the contralateral "weak-side" fin. During braking, trout simultaneously bend the left- and right-side pectoral fins along their longitudinal axes so that the trailing edges are elevated and protracted. A similar pectoral fin motion has been observed in juvenile salmonid fish during benthic station holding (Kalleberg, 1958; Arnold *et al.*, 1991) (Table 10.2). Despite its traditional categorization as a propulsor of limited functional importance, the salmoniform pectoral fin exhibits a diverse locomotor repertoire comparable to that of higher teleostean fishes.

Paired fin motions during maneuvering locomotion have been most extensively examined in acanthomorph fishes (Figure 10.1). Station holding (hovering) in still water requires fin motions to generate forces that counter both the weight of a fish in water and small forces associated with ventilating the gills. Magnan (1930) published the first kinematic study of paired fin propulsion with an analysis of hovering behavior in bass (*Micropterus*), which uses alternating oscillatory strokes of the left and right pectoral fins to maintain a fixed position in the water column. Breder and Edgerton (1942) made the first high-speed motion picture of an undulating pectoral fin in their study of vertical maneuvering by seahorses. Blake (1978, 1979a,c) documented dorsoventrally propagated waves of bending along the pectoral fins of balistiform fishes and the mandarinfish (Callionymidae) during hovering. During hovering in the threespine stickleback, *Gasterosteus aculeatus*, the pectoral fin rays oscillate within a horizontal plane with a large phase lag between leading and trailing edges (Walker, unpublished observations). An undulatory wave is passed from the leading (dorsal) to trailing (ventral) edge, creating a large positive angle of attack (approximately 45°) relative to the horizontal in both abduction (forward) and adduction (backward) strokes. The large angle of attack and the ventrally propagating wave generate an upward force that balances the weight of the fish, while the symmetry in angle of attack of the abduction and adduction strokes balances fore–aft forces.

Turning and braking have been the focus of most recent kinematic studies of paired fin maneuvering in acanthomorph fishes. Braking is characterized by the simultaneous extension of the left and right paired fins. Utilization of both the pectoral and pelvic fins for this task has been likened to a "four-wheel braking system" (Harris, 1953). Detailed information about fin kinematics during braking, however, has been limited primarily to pectoral fin motions. Many ray-finned fishes have been observed to protract the left and right pectoral fins together in order to exert a drag force for decelerating their bodies (Breder, 1926; Harris, 1938; Bainbridge, 1963; Videler, 1981; Webb and Fairchild, 2001). In short-bodied species, this fin excursion is usually bilaterally symmetrical (e.g., Drucker and Lauder, 2002b); longer-bodied fishes often exhibit simultaneous but asymmetrical left- and right-side pectoral fin motions during braking (e.g., Geerlink, 1987) (Table 10.2).

Kinematic strategies for turning the body with the paired fins include (i) suspending weak-side fin motion while continuing strong-side fin motion as during steady swimming and (ii) suspending strong-side fin motion while holding the weak-side fin in an abducted position to function as a pivot (Table 10.2). The former mechanism has been documented for balistiform swimmers (Blake, 1978) and the latter for fishes that swim by axial undulation (Harris, 1936) or tetraodontiform locomotion (Breder, 1926). In labriform, coelacanthiform, and ostraciiform swimmers (cf. Table 10.1), a third strategy may be used, in which neither fin suspends its motion during the turn, and instead the left and right propulsors undergo simultaneous but distinct motions. During yawing rotation of the body, the labriform swimmer *Lepomis macrochirus* (bluegill sunfish) rapidly abducts the strong-side pectoral fin to initiate the turn, and then abducts the weak-side fin during strong-side adduction in order to pivot the body further and translate it away from the turning stimulus (Drucker and Lauder, 2001). A similar pattern has been observed for the coelacanth *Latimeria chalumnae,* which uses pectoral fin locomotion to swim in open water (Fricke *et al.*, 1987; Fricke and Hissmann, 1992). The coelacanth initiates swimming along a curved trajectory by strokes of the strong-side pectoral fin, while simultaneously abducting the weak-side fin to achieve a "braking" effect. Subsequent adduction of both fins at once results in forward translation of the body.

Boxfishes, representative ostraciiform swimmers, also use both pectoral fins independently during yawing rotation of the body. Blake (1977) described yawing turns of the boxfish *Tetrasomus gibbosus* as involving adduction of the outside (strong-side) pectoral fin while the inside (weak-side) fin is either held against the body or oscillates with the reverse motions of the strong-side fin. Walker (2004b) observed a variation of this pattern in the boxfish *Ostracion meleagris*. During a 195° yaw turn by this species (Figure 10.10), the strong-side fin completed five full fin beat cycles, each a steady forward rowing stroke, while the weak-side fin completed two distinct kinematic cycles. During the first 90° of the turn, the weak-side pectoral fin remained extended in a feathered orientation. Over the last 105° of the turn, the weak-side fin performed two strokes resembling "reverse" or backward rowing. Interspecific variability of this sort in maneuvering kinematics underscores the impressive functional versatility of the teleost pectoral fin.

V. FLUID DYNAMICS

A. Background Theory

The teleost pectoral fin is a thin hydrofoil with active pitch and camber control. To facilitate the description of the fluid dynamics of pectoral fin propulsion in teleosts, we start with an overview of the fluid dynamics of thin hydrofoils in steady motion (i.e., at constant velocity) and unsteady motion (with varying, time-dependent velocities). A thin hydrofoil moving at a constant angle of attack, α, at a constant velocity generates a pressure force

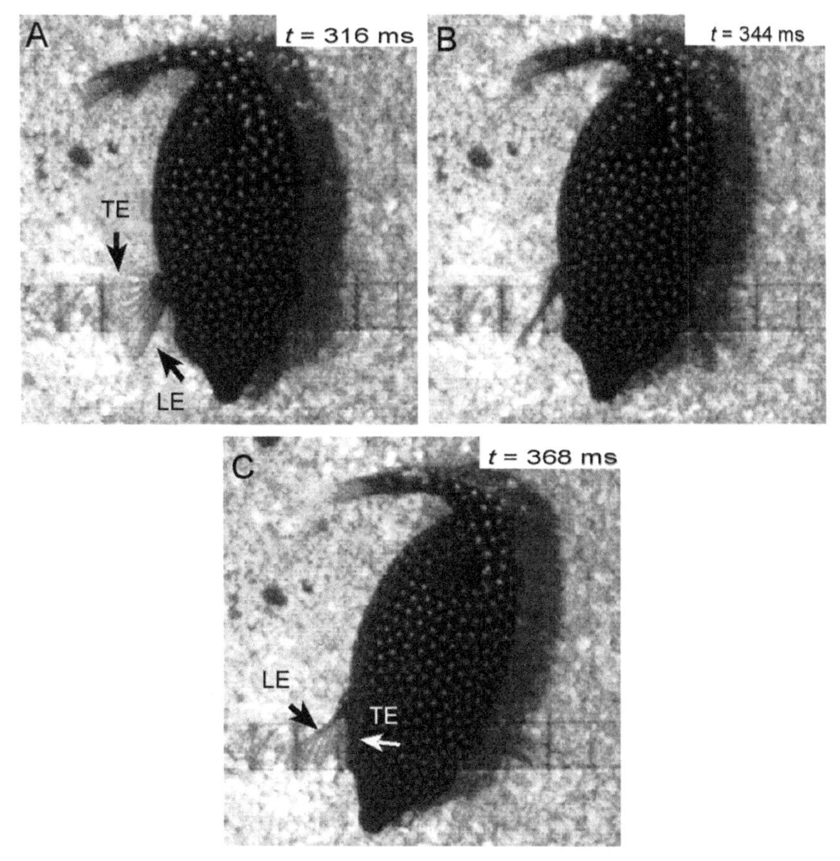

Fig. 10.10. The power stroke of the inside (weak-side) pectoral fin (at left) during a yawing turn in the boxfish *Ostracion meleagris*. Timings indicated from onset of the turn. (A) The fin is in a feathered orientation with the leading-edge ray (LE) maximally abducted. (B) The fin has rotated (supinated) about the leading edge causing the trailing edge to abduct and the fin to enter a broadside orientation. (C) The trailing-edge ray (TE) has continued to abduct to a position anterior to the leading-edge ray. Turning in the boxfish is accomplished by reverse rowing of the weak-side fin and forward rowing of the strong-side fin. (From Walker, 2004b.)

that is normal to its surface and a viscous (skin friction) force that is parallel to its surface. Because the pressure force is much larger than the viscous force above all but very small α, the net force is nearly normal to the hydrofoil's surface. The net force, F (Figure 10.11A), is typically partitioned into a lift component (L') normal to the hydrofoil's motion and a drag component (D') parallel to its motion. For a hydrofoil that is traveling along a horizontal axis, L' and D' will always be directed upward and backward,

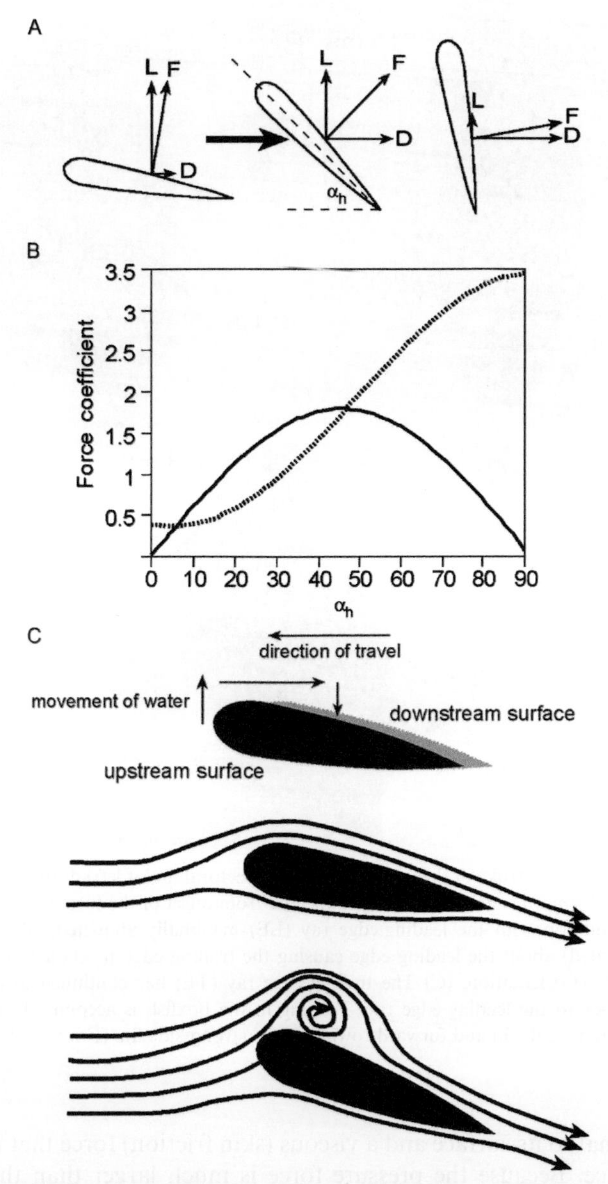

Fig. 10.11. Principles of pectoral fin hydrodynamics. (A) The net (or resultant) force (*F*) on a hydrofoil can be partitioned into lift (*L*), the component normal to the incident flow (represented by the thick horizontal vector), and drag (*D*), the component parallel to the incident flow. As the angle α_h between the fin and the incident flow increases, the proportion of the circulatory force that contributes to lift decreases while that which contributes to drag increases. Nevertheless, actual lift increases until $\alpha_h \sim 45°$ because of the increase in the magnitude of the circulatory

respectively, and are, therefore, equivalent to L (the upward component) and D (the backward component) (Figure 10.11A).

The pressure force arises because the fluid pressure on the hydrofoil's downstream surface is lower than that on its upstream surface (in the traditional representation of a hydrofoil—slightly inclined above a horizontal incident flow—the upstream surface is the "top" surface, but in a heaving and pitching hydrofoil, the upstream surface may be either the dorsal or ventral surface, depending on the pitch and the incident flow velocity). To understand why this pressure differential exists, consider that the volume of space immediately downstream of the translating hydrofoil was previously occupied by the hydrofoil itself (Figure 10.11C, upper panel). As the hydrofoil moves out of this space, a potential vacuum is created, and the resulting drop in pressure is the source of the difference in pressure between downstream and upstream hydrofoil surfaces. The low pressure on the downstream surface sucks fluid over the leading edge, increasing the velocity of the fluid moving past the leading edge and over the downstream surface (Figure 10.11C, middle panel). This phenomenon is described by the classic, inverse relationship between pressure and fluid velocity (the Bernoulli principle). By sucking fluid into the region of low pressure, the pitched foil also necessarily accelerates fluid in a direction normal to the incident flow. For a hydrofoil inclined slightly above the horizontal, this induced flow is oriented down and is called the downwash. Lift is experienced by the hydrofoil as the reaction to the acceleration of this mass of fluid in the downwash.

As α increases from 0 to 90°, the volume of fluid disturbed by the moving hydrofoil increases. Consequently, the pressure difference across the fin increases and a larger resultant force is generated. In addition, the lift component of a unit normal vector decreases with increasing α (Figure 10.11A). The combination of the larger resultant force and smaller lift component with increasing α explains a well-known empirical result: lift increases with α up to about 45° and then decreases, while drag increases monotonically (Figure 10.11B).

---

force with increasing angle of attack. This is shown in (B), which illustrates the coefficients of lift (solid line) and drag (dotted line) for $0° \leq \alpha_h \leq 90°$. (Modified from Dickinson *et al.*, 1999.) (C) Low pressure arises over the downstream surface of the fin as the volume of fluid previously occupied by the translating fin (gray shading, upper panel) is filled with a potential vacuum. This low pressure sucks fluid in from above and in front of the fin (arrows), increasing the speed of the flow over the downstream surface. At low angles of attack ($\alpha_h < 12°$), the fluid smoothly follows the contour of the fin (middle panel) but at angles above approximately 12°, the fluid separates from the fin's surface as it flows by. This separation creates a large vortex bubble attached to the leading edge (lower panel) that augments the circulatory force.

At scales relevant to fish locomotion, fluid has relatively high inertia and moving fluid has high momentum. At low α, the momentum of the fluid sucked over the leading edge is small enough to allow the incident flow to follow the contour of the hydrofoil's surface (Figure 10.11C, middle panel). Above some α (about 12°, depending on factors such as cross-sectional shape and camber), the required change in direction of the incident flow to follow this contour is too large, and, consequently, fluid separates from the hydrofoil's surface as it flows over the leading edge (Figure 10.11C, lower panel). The separation creates a region of low pressure that sucks surrounding fluid in, creating a vortex that is attached to the leading edge of the downstream surface. The presence of the attached, leading-edge vortex contributes an additional region of low pressure on the downstream surface, and the resulting pressure force is larger than what would occur in a fin at the same α that did not support a leading-edge vortex.

The circulating motion of the leading-edge vortex entrains surrounding fluid, causing the vortex to grow. After a small number of chord lengths (five to ten) traveled by the fin following an impulsive start from rest, the vortex grows to an unstable size and separates, or sheds, from the fin into the wake, causing a sharp drop in lift (and drag) called stall. In contrast to traditional fixed airplane wings that are designed to avoid stall, flapping fish fins can exploit the augmented force from an attached vortex as long as the growth of the vortex is slow enough that vortex shedding does not occur until stroke transition, when the rotational sense of the fluid circulating around the fin reverses. Base-to-tip spanwise flow removes accumulating fluid from the leading-edge vortex, preventing the vortex from growing to an unstable size and delaying vortex shedding until stroke reversal (Birch *et al.*, 2004). Spanwise flow occurs because the gradient in flow velocities along the root-flapping fin creates increasingly lower pressures on the downstream surface from base to tip. Spanwise flow may also be a function of wing shape. For example, the triangular wing planform in delta-wing aircraft (similar in shape to the pectoral fins of mobulid rays) creates a spanwise flow that stabilizes leading-edge vortices (Kuethe and Chow, 1986).

At the high angles of attack characteristic of the power stroke in a rowing fin, flow separation over the trailing edge causes the growth of an attached trailing-edge vortex. Growth and shedding of leading- and trailing-edge vortices are out of phase, such that one is shed while the other is growing. While a trailing-edge vortex is attached, the fin will experience both large drag and negative lift (because the vortex sucks fluid from below). The dynamic consequences of leading- and trailing-edge vortex growth and shedding within the power stroke of rowing fins have not been investigated.

Inertial forces resulting from the reaction to the acceleration of the fin and a volume of entrained water also contribute to the force balance. The strength of these inertial (or acceleration reaction) forces is a function of both the magnitude of the acceleration and the volume of water accelerated. A fin accelerating with its surface orientated normal to its motion (a broadside orientation) generates a larger inertial force than a fin accelerating with its surface parallel to its motion (a feathered orientation). Were the fin simply to row back and forth or flap up and down, maximum fluid accelerations, and thus peak inertial forces, would occur at the stroke transitions. When a rowing pectoral fin adducts against the body or is maximally abducted, the fluid acceleration at the following stroke reversal is directed mostly laterally. By contrast, if a rowing pectoral fin reverses stroke when it is abducted 90° from the body axis, the fluid acceleration is directed anteriorly or posteriorly. Consequently, a rowing fin could exploit the acceleration reaction to generate a net forward (thrust) or backward force simply by root-pivoting a broadside fin asymmetrically about a mediolateral axis (Daniel, 1984).

To what extent do fishes actually exploit the acceleration reaction to generate thrust during pectoral fin locomotion? In sticklebacks, the pectoral fins are feathered at the point in the stroke cycle when the acceleration reaction could maximally contribute to thrust and, consequently, these fishes develop negligible thrust at the transition from recovery to power stroke (Walker, 2004a). In general, the importance of the acceleration reaction to thrust generation by oscillating fins is characterized by the dimensionless reduced frequency parameter (Daniel and Webb, 1987). Fishes that oscillate large pectoral fins with high reduced frequencies, such as the burrfish and pufferfish (Arreola and Westneat, 1996; Gordon *et al.*, 1996), are predicted to transition from recovery to power stroke with a mechanism that effectively exploits the acceleration reaction.

Pectoral fin rowing and flapping have traditionally been classified as "drag-based" and "lift-based" propulsive mechanisms, respectively (reviewed by Webb and Blake, 1985; Vogel, 1994). These terms have been used to indicate that thrust on the body results from either fluid dynamic drag or lift on the fins. This classification is satisfactory for simple models of heaving plates with instantaneous rotations at stroke transitions but does not take into account the fact that real oscillating fins are able to generate thrust from both lift and drag, and that the relative contribution of these forces to overall thrust can change with swimming speed. In addition, both rowing and flapping pectoral fins have been shown to generate thrust from numerous fluid dynamic mechanisms, involving viscous stresses, bound circulation, flow separation, fin-vortex interactions, acceleration reaction forces, and jet-like (squeeze) forces. While we do not advocate the unqualified use of the terms "drag-based" and "lift-based," we note that the pattern of measured thrust and lift on fishes at the extremes of the rowing–flapping continuum is similar to the ideal pattern expected of "drag-based" and "lift-based" propulsion (Figure 10.8) (Walker and Westneat, 2002a).

B. Empirical Wake Dynamics

The complexity of unsteady fluid flow around swimming fishes has required that hydrodynamic analyses of locomotion be based to a large extent upon mathematical and mechanical modeling. There is, however, a complementary, experimental approach to studying locomotor hydrodynamics based on direct measurement of water movement in the wake of live swimming animals. Empirical studies of biological wake flow have involved the application of Schlieren imaging (Strickler, 1975) and shadowgraphic projection (McCutchen, 1977), as well as the tracking of streams of dye and buoyant particles introduced into the fluid through which an animal moves (reviewed in Stamhuis and Videler, 1995; Müller *et al.*, 1997). Such flow visualizations provide qualitative information about patterns of fluid movement but are difficult to quantify, especially when flow is three-dimensional.

In the past 10 years, biologists interested in the mechanics of aquatic propulsion have increasingly adopted more quantitative methods for measuring wake flow. Digital particle image velocimetry (DPIV) is one such technique, originally developed for examining human-engineered flows (Willert and Gharib, 1991). DPIV involves laser illumination of densely seeded, reflective particles in the water surrounding a freely swimming animal, high-speed imaging of the particles in one or more video views, and computational resolution of particle displacement in consecutive video frames (for further details of the technique as applied to the study of aquatic animal motion, see Drucker and Lauder, 1999; Lauder, 2000; Stamhuis *et al.*, 2002). The result of each DPIV analysis is a matrix of uniformly distributed velocity vectors describing the average magnitude and orientation of wake flow over the course of the video framing period. Traditional DPIV systems employ a laser light sheet to provide a description of fluid flow in two-dimensional transections of the wake; reorientation of the light sheet in separate experiments enables reconstruction of three-dimensional wake morphology. Newer methods (holographic, stereo, and defocusing DPIV) yield information about instantaneous wake velocity in three dimensions (e.g., Herrmann *et al.*, 2000; Gharib *et al.*, 2002; Nauen and Lauder, 2002).

The advantage of this quantitative flow visualization approach over the manual tracking of suspended particles or dye fronts is that it can be used to make reliable estimates of momentum injected into the wake and, from the rate of change in momentum, propulsive fluid force. In contrast to computational fluid dynamic models, which yield predictions about

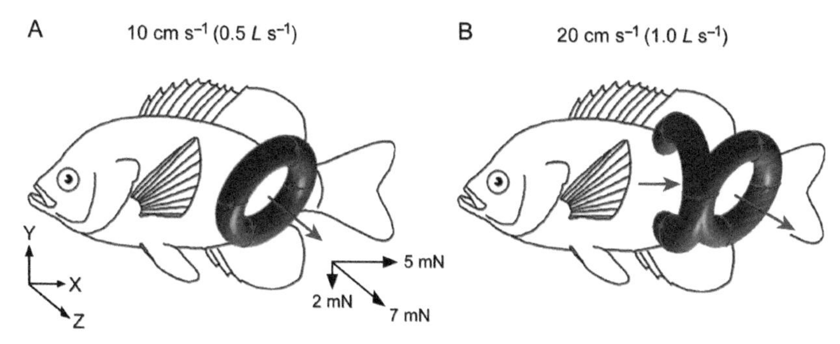

Fig. 10.12. Pectoral fin wake of bluegill sunfish (*Lepomis macrochirus*) reconstructed using quantitative flow visualization (two-dimensional digital particle image velocimetry). (A) At low swimming speeds (e.g., 0.5 body length (*L*) per second), the pectoral fin wake consists of a single vortex ring produced per fin beat cycle. (B) At higher speeds, the wake is composed of one complete vortex ring generated on the downstroke of the pectoral fin and a linked vortex filament produced by the upstroke, which terminates on the flank of the fish to form a second, incomplete ring attached to the body. In these schematic reconstructions of wake morphology, curved red arrows represent centers of vorticity observed in perpendicular flow fields, and straight red arrows indicate the direction of jet flow through the center of each vortex ring. Locomotor forces (shown per fin) derived from wake momentum flux are resolved into thrust, lift, and laterally directed components. (Modified from Drucker and Lauder, 1999.)

instantaneous forces acting on the propulsor, DPIV provides empirical data on stroke-averaged forces.

1. Steady Swimming

An empirical wake dynamic approach was first adopted for the study of paired fin locomotion in fishes by Drucker and Lauder (1999), who analyzed the fluid forces generated by bluegill sunfish (*Lepomis macrochirus*) during steady labriform locomotion. Two-dimensional DPIV was used to quantify flow patterns in each of three perpendicular sections of the wake shed by the pectoral fin. At low swimming speed, DPIV revealed the presence of closely apposed pairs of counterrotating vortices within each flow plane. This pattern is consistent with the production of a single, enclosed vortex ring with central jet flow over the course of each complete fin stroke cycle (Figure 10.12A) (for further detail, see Drucker and Lauder, 2002a). During slow swimming, wake vorticity (i.e., flow rotation) is generated predomintly by the downstroke of the pectoral fin. In contrast, at high swimming speed both the downstroke and upstroke are hydrodynamically active, together resulting in a pair of linked vortex rings, each with a central fluid jet (Figure 10.12B). In sunfish, increasing labriform swimming speed involves increasing the strength of the vorticity and size of the wake, a pattern reflecting the injection of additional momentum into the fluid behind the animal. Such a momentum addition is required to counter increased drag experienced at higher speeds.

The magnitude of fin forces can be evaluated in terms of the rate of change in wake momentum. For pectoral fin swimmers, the total stroke-averaged fluid force generated by each fin can be taken as the momentum of vortex ring structures shed into the wake divided by the time period over which the wake is produced (Drucker and Lauder, 1999, 2000, 2001). According to the average orientation of velocity vectors comprising the central momentum jet, the direction of the total wake force can be resolved into three perpendicular components: thrust, lift, and laterally directed (parallel to the x, y, and z reference axes, respectively; Figure 10.12).

The accuracy of DPIV in estimating wake forces is supported by an experimental analysis of the hydrodynamic force balance on freely swimming fish. Drucker and Lauder (1999) compared stroke-averaged forces calculated from DPIV wake measurements (Figure 10.12A) to empirically determined counter-forces during slow labriform swimming by bluegill sunfish. Lift and thrust derived from wake velocity fields were not significantly different from the resistive forces of body weight and total body drag, respectively. On opposite sides of the animal, medially directed forces on each pectoral fin arising in reaction to lateral wake force are presumed to balance, since sunfish show negligible sideslip during the course of the stroke cycle. Such a force equilibrium matches the theoretical expectation for animals moving straight ahead at constant speed, yet previous attempts to determine such a force balance experimentally have rarely been successful (for an exception, see Spedding, 1987). Empirically determining the force balance on a freely swimming fish indicates that DPIV can successfully detect the major vortical structures shed by swimming animals, and validates the technique for measuring locomotor forces from the wake.

The application of quantitative flow visualization has also shed light on the physical basis of interspecific variation in top pectoral fin swimming speed. Drucker and Lauder (2000) studied the wake of sunfish and surfperch, two fishes of similar adult size and body shape that nonetheless exhibit pronounced differences in steady swimming ability. Comparably sized animals (approximately 20 cm total length) show a 2-fold difference in maximal labriform swimming speed (cf. *Lepomis macrochirus* and *Embiotoca jacksoni* in Table 10.3). DPIV analysis revealed that, when swimming at equal fractions of their respective pectoral-to-caudal gait transition speed ($U_{\text{p-c}}$), the two fishes generate markedly different vortex wake structures. At 50% $U_{\text{p-c}}$, for instance, sunfish shed a single vortex ring with each pectoral fin, while surfperch shed two linked rings, each with a central fluid momentum jet. In addition, as swimming speed increases, surfperch redirect their vortex jets increasingly downstream, whereas sunfish reorient the wake jet laterally (Drucker and Lauder, 2000, 2002a). The ability to increase wake momentum progressively and to orient this momentum in a direction favorable for thrust production are key hydrodynamic factors influencing a species' steady swimming performance, as measured by the maximal speed attained within a gait.

2. Maneuvering

Quantitative flow visualization techniques have also provided insight into the mechanics of unsteady paired fin maneuvering. During rising and sinking of the body in the water column, sharks and sturgeon shed discrete vortical wake structures from the pectoral fin's trailing edge. DPIV analysis has revealed that the forces arising in reaction to these momentum-bearing structures exert a pitching moment around the body's center of mass to assist in vertical maneuvering (Wilga and Lauder, 1999, 2000, 2001). In rainbow trout, a lower teleost, wake dynamics have been quantified for a range of maneuvering behaviors using DPIV (Drucker and Lauder, 2003). During station holding by trout in still water, for instance, alternating anteriorly and posteriorly directed wake flows are generated by each pectoral fin; the forces associated with these low-velocity flows cancel each other on opposite sides of the body over the course of each complete fin stroke cycle. During slow turning and braking, the pectoral fins of trout generate relatively large forces oriented laterally and anteriorly, respectively. In general, the pectoral fins of sharks, sturgeon, and trout do not function as thrust-generating surfaces, instead serving to control body maneuvers through the production of off-axis fluid forces.

The pectoral fins of acanthomorph fishes (Figure 10.1), by contrast, are generally associated with a wider range of motion, and a correspondingly greater hydrodynamic versatility, relative to more primitive fins. Drucker and Lauder (2001) used DPIV to investigate the wake dynamics of pectoral fin turning maneuvers in bluegill sunfish, and found that the left and right fins play functionally distinct roles during yawing of the body. The fin nearer the stimulus inducing the turn (i.e., strong side) generates a laterally oriented vortex ring with a strong central jet whose associated lateral force is four times greater than that produced during steady swimming (Figure 10.13A and B), and thus this fin acts to rotate the body away from the source of the stimulus. The contralateral (weak-side) fin generates a posteriorly oriented vortex ring with a thrust force nine times that produced by the fin during steady swimming (Figure 10.13C and D); this fin acts primarily to translate the body away from the stimulus. The pectoral fins of sunfish, therefore, unlike those of trout (cf. Drucker and Lauder, 2003), are capable of generating both laterally and posteriorly directed (thrust) force during turning.

Table 10.3
Pectoral Fin Swimming Performance Limits

Species	Common name	Water temperature (°C)	Total body length[a], L (cm)	$U_{p\text{-}c}$[b]		$U_{p\text{-}crit}$[c]		Reference
				cms^{-1}	Ls^{-1}	cms^{-1}	Ls^{-1}	
Gasterosteus aculeatus	Threespine stickleback	20	4.8	24.0–33.6	5–7			Whoriskey and Wootton, 1987
Micropterus salmoides	Largemouth bass	Not reported	20	16.0	0.8			Lauder and Jayne, 1996
Pomoxis annularis	White crappie	25	17.0*	10.0	0.6			Parsons and Sylvester, 1992, Figure 2
Lepomis macrochirus	Bluegill sunfish	15	14.0	23.8	1.7			Webb, 1973
		20	17.9	19.7	1.1			Gibb *et al.*, 1994
Lepomis gibbosus	Pumpkinseed sunfish	20	12.7	12.7	1			Brett and Sutherland, 1965
Cymatogaster aggregata	Shiner surfperch	15	14.3	48.6	3.4			Webb, 1973
		12	3.5*	14.5	3.9			Mussi *et al.*, 2002, Figure 10.2
			6.3*	22.5	3.7			
			13.8*	44.1	3.1			
Amphistichus rhodoterus	Redtail surfperch	12	7.8	27.6	3.5			Drucker, 1996, Figure 10.3
			15.7	46.5	3.0			
			25.1	59.1	2.4			
			29.0	55.2	1.9			
			31.6	54.7	1.7			
Embiotoca lateralis	Striped surfperch	12	7.1	19.6	2.8			Drucker and Jensen, 1996b, Tables 1, 2
			15.1	35.0	2.3			
			21.5	42.1	2.0			
			26.2	44.4	1.7			
			29.5	44.6	1.5			
			31.8	44.0	1.4			
Embiotoca jacksoni	Black surfperch	20	20.8	41.6	2.0			Drucker and Lauder, 2000

Dischistodus spp.	Damselfish spp.	26–30	1.1			37.0	33.6	Stobutzki and Bellwood, 1994[d]
			4.4			47.2	10.7	
Pomacentrus amboinensis	Pallid damselfish	26–30	1.5			45.5	30.3	Stobutzki and Bellwood, 1994[d]
			4.6			50.0	10.8	
Neopomacentrus bankieri	Orangetailed damselfish	26–30	1.9			43.7	23.0	Stobutzki and Bellwood, 1994[d]
			3.9			62.7	15.9	
Gomphosus varius	Bird wrasse	25	9			47.4	5.3	Walker and Westneat, 2002b[e]
			12	60.1	5.0			
			16			84.0	5.3	
			17	70.6	4.2			
Halichoeres bivittatus	Slippery dick	25	9			38.2	4.2	Walker and Westneat, 2002b[e]
			12	35.5	3.0			
			16			52.0	3.2	
			17	37.5	2.2			
Cirrhilabrus rubripinnis	Redfin wrasse	25	8.1	61.7	7.6	49.0	6.0	Walker and Westneat, 2002b[e]
Pseudocheilinus octotaenia	Eight-lined wrasse	25	9.2	27.9	3.0	37.3	4.1	Walker and Westneat, 2002b[e]
Scarus schlegeli	Schlegel's parrotfish	26–27	22.7				3.2[f]	Korsmeyer *et al.*, 2002
Notothenia neglecta	Antarctic cod	2	7.5	10.2	1.4			Archer and Johnston, 1989
			29.3	22.9	0.8			
Pagothenia borchgrevinki	Antarctic toothfish	−2	22.5	40.5	1.8			Montgomery and Macdonald, 1984

*Standard body length.

[a]Average total length (unless noted) for size class studied in laboratory swimming trial.

[b]Pectoral-to-caudal gait transition speed, the highest speed powered by pectoral fin propulsion alone before recruitment of the caudal fin.

[c]Pectoral fin-powered critical swimming speed, the maximum sustainable pectoral fin-powered speed (i.e., threshold of fatigue).

[d]Five-minute critical swimming speed achieved by simultaneous pectoral and caudal fin propulsion. Data from Tables 1, 2 and allometric equations on p. 281.

[e]Authors report "$U_{p\text{-}max}$," which is measured by the same protocol as $U_{p\text{-}c}$, and 15-minute critical swimming speed achieved by simultaneous pectoral- and caudal-fin propulsion.

[f]Thirty-minute critical swimming speed achieved by exclusively pectoral fin propulsion.

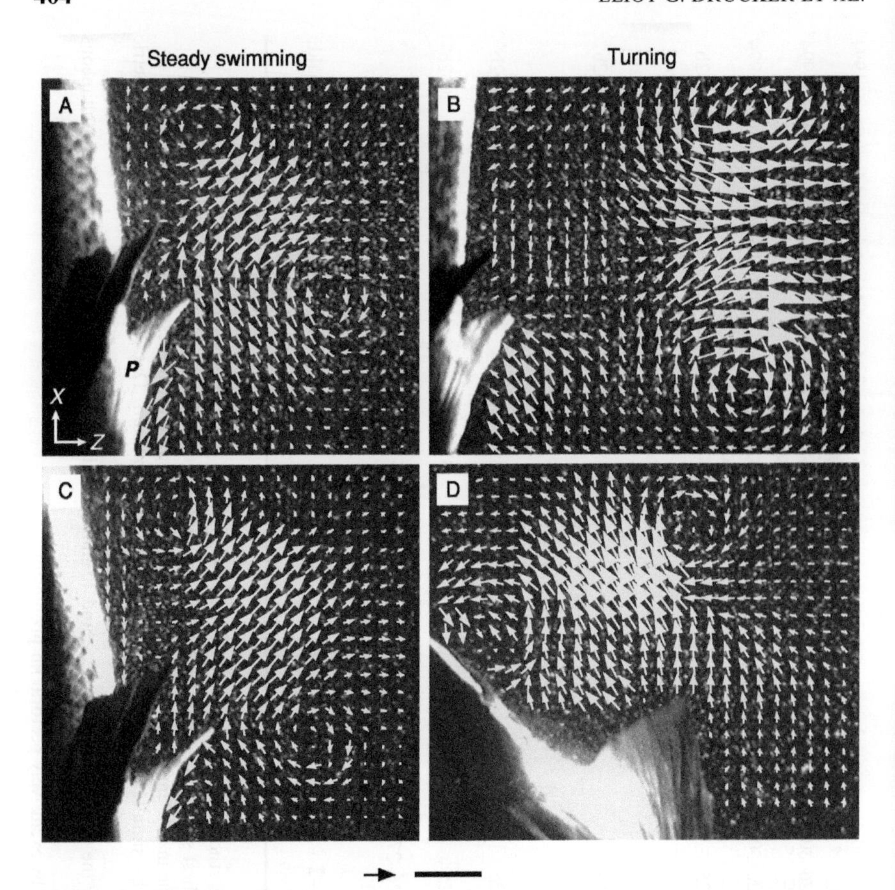

Fig. 10.13. Wake flow patterns during steady pectoral fin locomotion and pectoral fin turning maneuvers in bluegill sunfish (*Lepomis macrochirus*) recorded using digital particle image velocimetry. Water velocity fields in the frontal (*XZ*) plane are shown at the final stage of the upstroke of the pectoral fin (labeled *P*). Free-stream velocity of 0.5 body length (*L*) s^{-1} (11.2 cms^{-1}) has been subtracted from the *X*-component of each velocity vector to reveal vortical wake structures. (A) Pattern of fluid flow in the wake of the pectoral fin during steady swimming at 0.5 Ls^{-1}. (B) Wake flow produced by the pectoral fin during turning elicited by a stimulus issued from the right. This pectoral fin beat cycle immediately follows the steady fin beat illustrated in (A). As this strong-side wake develops, with a laterally directed fluid jet, the body rotates in the opposite direction to the left. (C) Pattern of fluid flow during a separate, steady pectoral fin beat. (D) Turning wake generated subsequent to the fin beat in (C) in response to a stimulus issued from the left. This weak-side wake arises on the upstroke and contains jet flow oriented nearly parallel to the body axis. The reaction to this momentum flow causes the fish to translate upstream (toward the bottom of the page), away from the source of the stimulus. Although wake flow during steady swimming shows little variation in gross morphology among fin beats (A, C), conspicuous differences exist in the orientation of central jet flow in the strong-side (B) and weak-side (D) wakes. Scales: arrow, 10 cm s^{-1}; bar, 1 cm. (From Drucker and Lauder, 2001.)

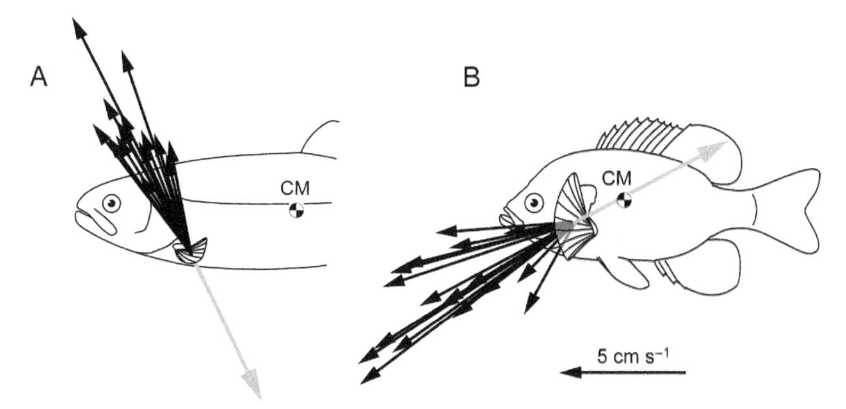

Fig. 10.14. Wake jet velocity and fluid force orientation during braking in (A) rainbow trout and (B) bluegill sunfish. Black arrows originating from the centroid of the pectoral fin signify the mean magnitude and orientation of multiple water velocity vectors comprising the central wake jet for braking maneuvers performed by a representative individual of each species. Gray vectors illustrate the orientation of mean reaction forces. Trout, possessing horizontally inclined, ventrally positioned pectoral fins, orient the braking-force line of action far from the center of mass of the body (CM). Sunfish possess vertically inclined, mid-dorsally positioned pectoral fins with a greater range of motion, and are capable of orienting the braking-force line of action directly through the center of mass of the body. (Modified from Drucker and Lauder, 2002b.)

In general, turning with the paired fins is not simply steady swimming performed unilaterally. Instead, this maneuver commonly involves bilaterally asymmetrical fin movements and fluid forces that are distinct in both direction and magnitude from those used to swim forward at constant speed.

Drucker and Lauder (2002b) used wake visualization to explore the relationship between pectoral fin design and fluid force production during maneuvering in rainbow trout and bluegill sunfish. During slow turning, the vertically oriented pectoral fins of sunfish generate significantly more lateral yawing force (per unit fin area) than the horizontally oriented fins of trout. During abrupt deceleration of the body, both trout and sunfish generate an anteriorly directed momentum flow with the pectoral fins, which in reaction experience a retarding posteriorly directed force. In trout, the line of action of the braking force falls far from the center of mass of the body as a result of the restricted range of motion of the pectoral fins, together with their ventral position on the body (Figure 10.14A). Sunfish, by contrast, possessing more flexible, mid dorsally positioned pectoral fins, are capable of directing the braking-force line of action directly through the center of mass of the body (Figure 10.14B). Accordingly, the two species differ in the degree to which destabilizing pitching moments of the body are generated during braking. Although both fishes recruit fins posterior to the center of mass during braking, presumably to counter the "somersaulting" moment induced by pectoral fin extension, trout undergo marked downward pitching of the body during deceleration while sunfish do not (Drucker and Lauder, 2002b).

VI. PECTORAL FIN SWIMMING PERFORMANCE

A. Steady Swimming

Experimental study of steady, pectoral fin-powered swimming performance has included (1) investigation of the scaling of pectoral fin swimming performance both within and among species, (2) analysis of the trigger for the pectoral-to-caudal gait transition, (3) comparison of endurance performance maxima in pectoral fin and body and caudal fin swimmers, and (4) comparison of endurance performance among pectoral fin rowers and flappers. We discuss points 1–3 in this section and reserve discussion of point 4 for Section VII.

A commonly used endurance performance measure, the critical swimming speed (U_{crit}), is the highest speed that a fish can maintain in an increasing velocity swimming trial without fatiguing, and should equal the upper bound for the maximum sustainable swimming speed. In practice, measures of U_{crit} are dependent on the time interval and velocity increment used in the performance trial, but at time intervals exceeding 15–20 minutes and velocity increments less than $^1/_4 U_{crit}$, there is little variation in measured performance (Hammer, 1995). Although traditionally measured for fishes swimming by axial undulation, U_{crit} is an appropriate measure of swimming endurance performance in fishes that propel themselves with only the pectoral fins up to fatigue velocities. For these fishes, we use $U_{p\text{-}crit}$ to refer to this pectoral fin-powered critical swimming speed. A related performance measure, the pectoral-to-caudal gait transition speed, $U_{p\text{-}c}$, is the highest speed achievable by pectoral fin propulsion alone prior to recruitment of the axial propeller (Drucker, 1996). In fishes that fatigue at velocities immediately above this gait transition speed, $U_{p\text{-}crit}$ and $U_{p\text{-}c}$ will be equivalent. Both measures of pectoral fin swimming performance are presented for a variety of bony fishes in Table 10.3.

1. Scaling of Pectoral Fin-Powered Swimming Performance

To account for body size-related variation in swimming performance, fish biologists have traditionally reported length-standardized maximum speeds expressed in body lengths per second. Within all species of labriform swimmers studied to date, length-standardized $U_{p\text{-}c}$ and $U_{p\text{-}crit}$ decline with body size (Table 10.3). This decrease in length-specific maximum swimming speed has also been observed for axial undulators, both within species (e.g., Webb *et al.*, 1984) and among species (Hammer, 1995). If we pool the $U_{p\text{-}c}$ and $U_{p\text{-}crit}$ data for pectoral fin swimmers that fatigue at or immediately above the pectoral-to-caudal gait transition speed (embiotocids, labrids, and sticklebacks), the among-species maximum sustainable swimming speed scales with body length as $L^{0.56}$. This scaling coefficient closely matches the value of 0.5 estimated for axial powered U_{crit} (Hammer, 1995).

Absolute top speed generally increases with body size for pectoral fin swimmers (e.g., Walker and Westneat, 2002b) (Table 10.3). The fastest speeds, however, are not always exhibited by the largest animals. Drucker and Jensen (1996b) found that the largest size classes of surfperch have lower average absolute $U_{p\text{-}c}$ values than fish of intermediate size (Table 10.3). Such declining performance at the largest body sizes has been observed in both aquatic and terrestrial vertebrates at maximum anaerobic speed and may stem from energetic, structural, or hydrodynamic limitations imposed on large animals (reviewed in Goolish, 1991). In addition, the relationship between swimming speed and body size may be influenced by the allometry of propulsor dimensions. In surfperch, for instance, pectoral fin area increases through ontogeny at a significantly slower rate than expected by models of geometrically similar growth (Drucker and Jensen, 1996b). Disproportionately smaller fins in larger fish may limit peak thrust and, accordingly, maximum labriform swimming speed.

2. The Pectoral-to-Caudal Gait Transition Trigger

Increasing speed in vertebrates typically requires discrete modifications of locomotor patterns, or gait transitions (Alexander, 1989; Webb, 1994b), which may occur for several physiological reasons. In terrestrial mammals, gait transitions may occur because the gait used at one speed is more energetically costly than the gait used at a higher speed (Alexander, 1989). By contrast, fishes swimming by axial undulation switch gaits when the major aerobic muscle fiber type being used (red myotomal) can no longer generate power sufficient to sustain locomotion (Rome *et al.*, 1988). Is the gait transition from pectoral fin to axial swimming triggered by constraints of locomotor cost or power? Drucker and Jensen (1996a) proposed for surfperches that aerobic pectoral muscle fibers of increasing contraction speed may be recruited as labriform swimming speed increases, and that at the pectoral-to caudal gait transition the fastest pectoral fibers recruited approach the upper limit of shortening velocities for peak mechanical power output. To swim faster, these fishes must recruit more powerful aerobic and anaerobic myotomal muscle and initiate tail beating.

The energetics of pectoral-to-caudal gait transition were investigated in the parrotfish *Scarus schlegeli*, which swims with a largely flapping motion of its pectoral fins up to fatigue speeds, at which point it switches to a burst-and-flap gait (Korsmeyer *et al.*, 2002). The net cost of swimming (the rate of oxygen consumption measured at a given speed minus that measured at rest) increased with speed in *S. schlegeli*, but achieving the speed of gait transition to a burst-and-flap mode was characterized by an additional energetic cost above that expected if the fish could maintain the same speed with the pectoral fins alone. While this observation does not indicate that flapping propulsion is less energetically expensive than axial propulsion within the range of sustained speeds, it does suggest that labriform cruisers switch gaits not to reduce energy costs, as proposed for terrestrial mammals, but instead to employ a muscle system that has sufficient power to maintain high swimming speeds.

3. Comparison of Pectoral Fin and Axial Swimming Performance

Most fishes recruit a longitudinal band of slow-twitch, oxidative (red) myotomal muscle fibers to power steady axial propulsion at the highest sustained swimming speeds (Jayne and Lauder, 1994). Such red axial fibers are absent in sticklebacks and wrasses, although the pectoral muscles of these fishes are invested with high concentrations of oxidative fibers (te Kronnie *et al.*, 1983; Davison, 1988, 1994). Does the evolutionary loss of red muscle fibers from the body wall limit the ability of pectoral fin cruisers to achieve and maintain high steady swimming speeds?

This question was recently addressed by comparing maximum sustained swimming speeds in fishes using pectoral fin oscillation versus axial undulation (Walker and Westneat, 2002b). The comparative data (Table 10.3, Figure 10.15A) show that pectoral fin cruisers have maximum sustained speeds comparable to those exhibited by axial undulators of the same size. While certainly not as high as the top axial powered speeds sustained by some pelagic and anadromous fishes, the highest pectoral fin-powered speeds measured suggest that fishes lacking red myotomal muscle fibers are not necessarily limited to life in slow motion. Indeed, the concentration of slow-twitch oxidative fibers in the pectoral muscles enables such fishes to achieve relatively high sustained swimming speeds by pectoral fin propulsion.

B. Maneuvering

For fishes, there is no single most appropriate measure of maneuvering performance. High linear and rotational body velocities and accelerations are critical for maneuvers during predator evasion. The ability to rapidly and precisely change the body's orientation over a broad range of angular

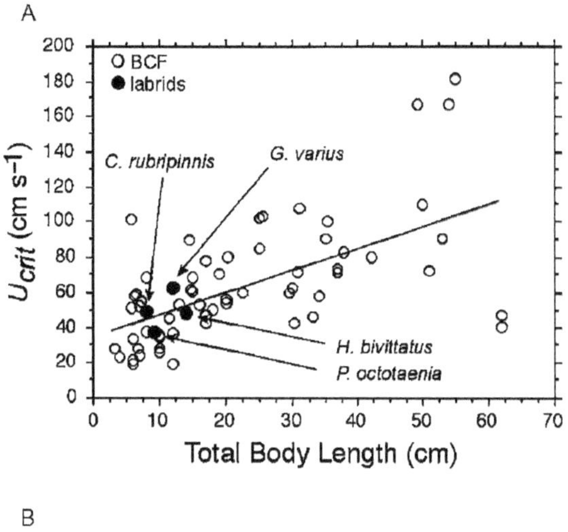

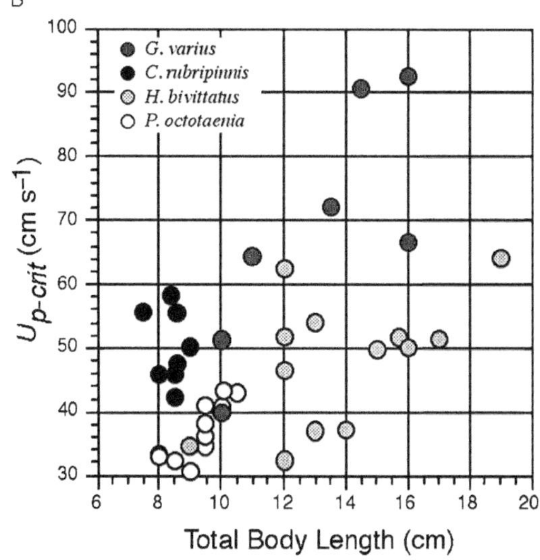

Fig. 10.15. (A) Comparison of maximum sustained swimming speed among fishes that swim in the labriform mode and fishes that swim with body and caudal fin (BCF) undulation. The regression line is through the BCF data only. Pectoral fin and BCF swimmers of the same size show comparable swimming performance. (B) Comparison of the pectoral fin-powered critical swimming speed ($U_{\text{p-crit}}$) among labrid fishes with a steep stroke plane (*G. varius* and *C. rubripinnis*) and a shallow stroke plane (*H. bivittatus* and *P. octotaenia*). The fishes with the steep stroke are closer to the flapping extreme on the rowing–flapping axis, while the fishes with the shallow stroke plane are intermediate between rowing and flapping. Data in (A) from Walker and Westneat (2002b), Videler (1993). Data in (B) from Walker and Westneat (2002a).

displacements should facilitate inspection behavior (scanning a territory, searching for prey) and enable correction for small perturbations to the heading. Minimizing the radius of the turning path should increase the ability of fishes to exploit resources in structurally complex environments. Turning with high efficiency saves energy that could be allocated to other functions or behaviors. To date, few of these maneuvering performance indices have been measured experimentally. Some, such as the ability to modulate body orientation, are poorly defined and difficult to quantify, and accordingly have not been reported in the literature. Others, such as rotational acceleration, are easily defined and measured, but have not yet been reported. Still others, such as turning efficiency, have been measured in swimming behaviors for which other performance variables are more likely optimized (e.g., in fast-start escape responses; Frith and Blake, 1995).

Two types of turning performance measures have been explored: kinematic-based and force-based. The former includes both translational variables, which are a function of the linear displacement of the center of mass of the body, and rotational variables, which are a function of the angular displacement of the body. Rotational variables have been estimated from center of mass displacement, but the duration of body rotation can be much longer than the duration of body translation around a turning arc (Walker, 2000). Translational variables include maximum linear velocity and acceleration, minimum coefficient of normal acceleration, and minimum turning radius. The coefficient of normal acceleration (Bandyopadhay *et al.*, 1997) is the gravity-standardized acceleration normal to the turning path. While turning radius is traditionally reported as a single number, the radius of curvature along a turning path is often not constant in fishes powering turns with pectoral fins. Walker (2000) used numerical differentiation to estimate instantaneous radius of curvature along the turning path of boxfish and reported the minimum radius as the measure of maximum performance. Because the minimum radius does not capture the shape of the turning curve, it may be more desirable to use the major and minor axes of the turning path as performance variables (Walker, 2004b).

Peak rotational acceleration, α_{max}, during a turn is most directly related to pectoral fin performance (net maximum torque generated by the control surfaces), but no measures of α_{max} are available for interspecific comparison. Peak rotational velocity, ω_{max}, has been reported and is closely related to pectoral fin performance if the duration of the turn is greater than that necessary to achieve ω_{max}. The most common measure of rotational performance is the average rotational velocity over a turn, ω_{avg}, which is a function of the duration of the turn, the torque generated by the pectoral fins, and the inertial and drag resistance of the body (Wakeling *et al.*, 1999; Domenici, 2001).

The influence of pectoral fin motions on turning performance was synthesized in a recent review (Walker, 2004b). The turning radius, R, a common measure of maneuverability, can be modeled by:

$$R \propto (m/S)(V^2/U^2),$$

where m is the virtual mass (the sum of the fish and added mass), S is the area of the pectoral fins, V is the tangential velocity of the center of mass of the fish, and U is the average speed of a pectoral fin at its center of force. Two useful sets of predictions can be drawn from this model. First, if the body or nonoscillating pectoral fins are generating the turning moment, then $U \approx V$ and R will be independent of speed, as shown by Webb (1983). But if an oscillating pectoral fin is contributing to the turning moment, then R will decrease to zero as V^2/U^2 decreases to zero (or, effectively, as V decreases to zero). R will also go to zero if the motions of each pectoral fin form a couple that augments rotational moments but cancel translational forces. Second, R will scale with body length, L, and the relative performance, $^R/_L$, will be proportional to V^2/U^2. It has been argued that $^R/_L$ should increase with the flexural stiffness of the body (Domenici and Blake, 1997; Fish, 1999) but, as noted previously, the model suggests that this constraint should only be relevant to turns powered by the kinetic energy of the moving body; fishes that power turns with pectoral fins should be able to turn with effectively zero turning radius. This prediction of the model is supported by turns with extremely low $^R/_L$ documented in coral reef fishes (Gerstner, 1999; Walker, 2000).

Another aspect of maneuvering performance is the ability to move easily within confined spaces, which was compared among goldfish (*Carassius auratus*), silver dollar (*Metynnis hypsauchen*), and angelfish (*Pterophyllum scalare*) negotiating slits and tubes of various geometries (Webb *et al.*, 1996; Schrank and Webb, 1998; Schrank *et al.*, 1999). Relative to goldfish, silver dollar and angelfish, possessing deeper, narrower bodies and more laterally positioned pectoral fins, were able to move though 63 and 45% narrower tubes per cube-root unit body mass, respectively. In contrast, goldfish turned in narrower bent tubes, and the difference in performance between goldfish and the other two species increased with the magnitude of the bend (Schrank *et al.*, 1999). Similarly, the spotted boxfish has a very small turning radius but cannot turn in confined spaces because of its rigid body (Walker, 2000). These experiments demonstrate that optimal body and paired fin designs for open water maneuvering are relatively poor for maneuvering in confined spaces.

A force-based index for quantifying turning performance is the linear maneuverability number (LMN) developed by Jindrich and Full (1999). The

LMN measures the force impulse exerted by a propulsor in a direction perpendicular to the animal's heading as a proportion of the forward momentum of the animal's center of mass. The force impulse is defined as the component of force perpendicular to the heading integrated over the stride period. Drucker and Lauder (2001) approximated this index for pectoral fin turns performed by sunfish according to:

$$\mathrm{LMN} \approx \frac{M_{\perp}}{mU},$$

where $M_{\perp}$ is the stroke-averaged pectoral fin wake momentum oriented perpendicular to the fish's initial heading (i.e., before body rotation), m is the body mass in air, and U is the forward velocity of the body before the turn. In effect, the LMN reflects the ability of an animal to generate laterally oriented momentum in order to deflect its own forward progress. For slow turns in sunfish, LMN ranged from 0.12 to 0.74, indicating that these fishes are able to use the pectoral fins to generate approximately one-tenth to three-quarters of their own forward momentum as laterally oriented wake momentum for changes in heading. The primary benefit of the LMN as a measure of maneuvering performance is that its calculation does not depend directly upon the specific kinematics of any given locomotor system and can therefore serve as an index for cross-taxonomic comparisons.

VII. ECOMORPHOLOGY OF PECTORAL FIN PROPULSION

Ecomorphology addresses the mechanistic relationships among morphology, performance, and habitat. While most ecomorphological studies concentrate on the evolutionary consequences of habitat on morphology (adaptation), Wainwright has emphasized how morphology constrains resource acquisition and microhabitat use (Wainwright, 1988, 1991; Wainwright *et al.*, 2002). The idea that body and fin shape, body and fin motion, locomotor performance, and habitat are causally associated has a long history (Breder, 1926; Gregory, 1928; Greenway, 1965; Keast and Webb, 1966; Aleev, 1969; Davis and Birdsong, 1973). An influential, synthetic model of fish body and fin ecomorphology was developed by Webb (1982, 1984), but few quantitative tests of *a priori* hypotheses resulting from Webb's model have been conducted.

A. Quantitative Model of Pectoral Fin Ecomorphology

The causal relationships among pectoral fin shape, pectoral fin motion, swimming performance, and habitat are the most thoroughly studied within fish locomotor ecomorphology. As emphasized in Sections II and

IV, pectoral fin shape and motion are highly variable, and we have no comprehensive model to explain this variation. The one functional pattern for which there has emerged a fairly specific model is the continuum between distally expanding (fan-shaped), rowing fins and distally tapering (wing-shaped), flapping fins. Breder (1926) was the first to formally recognize this rowing–flapping continuum. Blake (1983) developed fluid dynamic models of rowing and flapping propulsion but never systematically compared the dynamic or energetic consequences of their differences. Broad comparisons across diverse vertebrate taxa suggest that flapping is more effective in high-speed and high-endurance behaviors, while rowing is more effective in low-speed and high-maneuverability behaviors.

A major advance in understanding rowing–flapping performance was the effectively simple computational model of Vogel (1994) of a stiff, heaving, flat plate. The flapping plate heaved up and down with a constant downstroke and upstroke pitch that changed instantaneously at the stroke transitions. The rowing plate heaved back and forth with a fixed, feathered orientation during the forestroke and a fixed, broadside orientation during the backstroke. Neither plate moved with root oscillation (pivoting about the base) but instead reciprocated along an axis. Vogel kept the cycle frequency constant but allowed the speed of travel (of the virtual body) to vary and used a quasi-steady coefficient method to model the resulting thrust. Vogel (1994) found that the rowing plate generated more thrust at low speeds while the flapping model generated more thrust at moderate to high speeds. High thrust at low speeds from one pectoral fin could be used to generate turning forces, suggesting that a rowing motion may be more common in slow maneuvering specialists. By contrast, the greater thrust from flapping at high speeds suggests that flapping motions may be more common in fast, open-water cruisers.

To explore the dynamic and energetic differences between rowing and flapping in more detail, Walker and Westneat (2000) expanded and refined Vogel's analysis by comparing virtual rowing and flapping fins that (1) root oscillate and (2) dynamically twist from base to tip (such that any spanwise section of the fin rotates about a spanwise axis). This fluid dynamic modeling of virtual fins accounted for several important unsteady hydrodynamic effects, including the delay in the development of the circulatory force (the Wagner effect) and the attached vortex force (cf. Section V). Two performance criteria were computed: mean thrust over either a half-stroke (power stroke) or full stroke cycle and mechanical efficiency (the ratio of work performed to overcome drag on the body to the total work done by the fin on the water). Patterns of modeled thrust and efficiency as a function of kinematics were consistent with those generated by both computational fluid dynamic and physical models of rowing and flapping plates (Walker, 2002).

Two important results emerged from the Walker and Westneat (2000) simulation. First, rowing fins generate more thrust per stroke than do flapping fins at low swimming speeds, but this trend is reversed at higher swimming speeds, as predicted by the original model of Vogel (1994). Second, flapping fins are more mechanically efficient than rowing fins across the entire range of biologically relevant swimming speeds. The difference in thrust generated by rowing and flapping was especially large when only a half-stroke was considered. This result suggests that a rowing stroke is the more advantageous motion for maneuvering behaviors such as accelerating and yaw turning that require large fore–aft forces from a single power stroke.

B. Maneuvering Ecomorphology

Experimental data on the relationship between pectoral fin swimming kinematics and maneuvering performance are limited. As expected by the models described previously, boxfishes use bilaterally asymmetrical rowing strokes of the pectoral fins to execute slow "hovering" turns of 360° or more with near-zero turning radii (Walker, 2000, 2004b). By contrast, flapping pectoral fin strokes produce predominantly dorsoventral forces, which may limit the ability to turn precisely in a small space. Indeed, two similarly sized fishes that use a flapping stroke, the bluehead wrasse (*Thalassoma bifasciatum*) and the ocean surgeonfish (*Acanthurus bahianus*), do not perform turns in which all of the power for body rotation is generated by oscillating pectoral fins (Gerstner, 1999). Instead, these fishes use the kinetic energy of forward swimming in combination with pectoral fin oscillation to power the turn. Relative to the "hovering" turns of the boxfish, these "cruising" turns are much faster but with larger radii (>0.05 body length) (Gerstner, 1999).

Rowing fins may also allow fishes to hover in still water more effectively than flapping fins. Boxfishes and threespine stickleback hover by rowing their pectoral fins with large angles of attack on both recovery and power strokes. The bird wrasse, which flaps its fins, hovers by pitching its body nose-up while oscillating its pectoral fins back and forth. Even in this position, a bird wrasse can hover for only a few strokes. Korsmeyer *et al.* (2002) found that the parrotfish *Scarus schlegeli*, which also uses a flapping stroke, does not exhibit hovering at all.

C. Steady Swimming Ecomorphology

A major finding of the Vogel (1994) model—that rowing generates more thrust than flapping at low speeds—could imply that rowing is also more mechanically efficient than flapping at low steady swimming speeds.

The second major simulation result of Walker and Westneat (2000), however, does not support this interpretation. Instead, the simulated flapping fin had higher mechanical efficiency than the simulated rowing fin at all speeds. The twisted rowing fin (i.e., that rotates along its span due to the fixed orientation of the shoulder joint) was especially inefficient (mechanical efficiency <0.05) because the fin necessarily generated large drag on the recovery stroke. Sticklebacks, which row their pectoral fins, avoid this large drag cost by "peeling" the fin rays off the body at the beginning of fin abduction (Walker, 2004a). The untwisted rowing fin (in which the shoulder joint is allowed to rotate) had an efficiency greater than that of the twisted rowing fin but still less than that of the flapping fin. At very low swimming speeds ($<1\ Ls^{-1}$), the mechanical efficiencies of the untwisted rowing fin and the flapping fin are sufficiently close to suggest that an optimized rowing fin (i.e., with a force-optimizing shape or motion) could be more efficient than a typical flapping fin. These computational modeling results have recently been supported by data from a physical model of a stiff pectoral fin driven by three motors (Kato and Liu, 2003).

Simulations of Walker and Westneat (2000) predict that fishes that flap pectoral fins should be able to achieve higher sustained pectoral fin-powered swimming speeds than fishes that row. One way to test this hypothesis is to compare the maximum steady speed of pectoral fin locomotion between rowers and flappers. Consistent with the model, the highest speed powered by pectoral fin propulsion alone ($U_{p\text{-}c}$) is roughly 11.0 root lengths s^{-1} in the rower *Gasterosteus aculeatus* (mean 4.8 cm body length) but 17.3 root lengths s^{-1} in the flapper *Gomphosus varius* (12 cm total length) (Table 10.3).

In a formal test of the hypothesis that fishes that flap can achieve and maintain higher pectoral fin-powered swimming speeds than fishes that row, Walker and Westneat (2002b) compared $U_{p\text{-}crit}$ in two pairs of closely related fishes from the family Labridae (Table 10.3). One species within each pair (*Gomphosus varius* and *Cirrhilabrus rubripinnis*) flapped each pectoral fin along a steep stroke plane at all speeds. The other species within each pair (*Halichoeres bivittatus* and *Pseudocheilinus octotaenia*) rowed each fin along a shallow stroke plane at low speeds and steepened the stroke plane with increasing speed. At fatigue velocities, *G. varius* and *C. rubripinnis* had significantly steeper stroke planes than *H. bivittatus* and *P. octotaenia*, respectively. As predicted by the rowing–flapping model, within each pair the species with the steeper stroke plane had a significantly higher $U_{p\text{-}crit}$ than the species with the shallower stroke plane (Figure 10.15B).

While the rowing–flapping model correctly predicts performance differences between *Gasterosteus aculeatus* and *Gomphosus varius*, and between closely related wrasses, can it also predict differences in behavior and ecology? The rowing–flapping model makes two general predictions. First, the routine speeds of fishes with flapping, wing-shaped fins should be higher than those of fishes with rowing, paddle-shaped fins. Second, if flow velocities typical of high-energy environments exclude species by swimming ability, then fishes that row paddle-shaped fins should be excluded. Recent field studies have shown that labrids with lower aspect ratio, paddle-shaped fins tend to swim more slowly (Wainwright *et al.*, 2002) and occupy less energetic zones on the reef (Bellwood and Wainwright, 2001; Fulton *et al.*, 2001) relative to labrids with higher aspect ratio, wing-shaped fins. Additionally, as summarized previously, labrids with flapping, wing-shaped fins have higher $U_{\text{p-crit}}$ than labrids with rowing, paddle-shaped fins. The combination of these field and laboratory studies gives broad support for the rowing–flapping model as a descriptive ecomorphological tool.

VIII. SUMMARY AND AREAS FOR FUTURE RESEARCH

The existing body of research on paired fin swimming mechanics effectively counters early impressions (reviewed in Lindsey, 1978; Drucker and Summers, 2006) that axial undulation is the only swimming mode of importance to fishes. Up to 20% of living fishes rely on the pectoral fins as the primary means of propulsion (Westneat, 1996), and most of the remainder undoubtedly employ both sets of paired fins on an intermittent basis to control maneuvers and maintain body stability. In spite of the substantial gains made to date in our understanding of how fishes use the paired fins during swimming, a number of important questions about their function remain. Two promising areas for future research include refinement of hydromechanical analysis and synthesis of comparative biomechanical data.

Extending our understanding of paired fin mechanics will be facilitated by interdisciplinary investigation of locomotor force production. We need to integrate computational fluid dynamic (CFD) analysis with quantitative wake visualization to test assumptions inherent in both approaches. Data from CFD and DPIV studies on the same propulsor system will validate the strengths of each technique and identify areas where theory and reality do not match well. New flow visualization methods that provide information about wake velocity in three dimensions are being developed (Herrmann *et al.*, 2000; Kähler and Kompenhans, 2000; Pereira and Gharib, 2002; Malkiel *et al.*, 2003) and should allow improved estimates of wake momentum. In addition, microelectricalmechanical systems (e.g., Abeysinghe *et al.*, 2002) that will measure the pressure distribution over flapping fins *in vivo* may soon be available. Such pressure data are critical for direct determinations of fluid force. It is also vital that we bridge the gap between hydrodynamic and musculoskeletal performance studies. At present, the mechanisms by which internal muscular forces are transmitted, via the fins, to the external fluid environment remain obscure. Combining the techniques of *in vitro* muscle physiology with methods for calculating wake momentum flux will enable estimation of the overall mechanical efficiency of paired fin propulsion.

Synthetic analysis of comparative biomechanical data will also become a fruitful tool for understanding fin function. Fishes are characterized by pronounced morphological diversification of the paired fins. Historical transformations in the shape, orientation, and position of pectoral and pelvic fins are well documented (Drucker and Lauder, 2002b; Lauder and Drucker, 2004), yet the hydrodynamic consequences of this evolutionary variation are poorly understood (theoretical and empirical work on this topic includes Combes and Daniel, 2001; Webb and Fairchild, 2001). Research on taxonomically diverse clades will allow a more broadly comparative analysis of the relationship between propulsor anatomy and force production.

In addition, new insights into paired fin swimming performance may be gained by combining laboratory and field studies of locomotor behavior. The touchstone of a well-engineered flume is the minimization of flow heterogeneity. Yet fishes in the real world seldom if ever experience such constrained flows. Recent studies indicate that trout exhibit different patterns of pectoral fin recruitment when swimming in a flow tank as opposed to their natal streams (McLaughlin and Noakes, 1998; Drucker and Lauder, 2003). Data from the field on paired fin use (Fricke and Hissmann, 1992) and swimming speed (Wainwright *et al.*, 2002) complement observations made in a controlled laboratory setting by expanding both behavioral repertoire and performance range. Future technological and analytic advances will undoubtedly allow further strides in understanding the functional role of the appendicular propulsors in fish locomotion.

REFERENCES

Abeysinghe, D. C., Dasgupta, S., Jackson, H. E., and Boyd, J. T. (2002). Novel MEMS pressure and temperature sensors fabricated on optical fibers. *J. Micromech. Microeng.* **12,** 229–235.

Aleev, Y. G. (1969). "Function and Gross Morphology in Fish." Israel Program for Scientific Translations, Jerusalem.

Alexander, R. McN. (1967). "Functional Design in Fishes." Hutchinson Press, London.

Alexander, R. McN. (1989). Optimization and gaits in the locomotion of vertebrates. *Physiol. Rev.* **69,** 1199–1227.

Archer, S. D., and Johnston, I. A. (1989). Kinematics of labriform and subcarangiform swimming in the Antarctic fish *Notothenia neglecta*. *J. Exp. Biol.* **143,** 195–210.

Arnold, G. P., Webb, P. W., and Holford, B. H. (1991). The role of the pectoral fins in station-holding of Atlantic salmon parr (*Salmo salar* L.). *J. Exp. Biol.* **156,** 625–629.

Arreola, V. I., and Westneat, M. W. (1996). Mechanics of propulsion by multiple fins: Kinematics of aquatic locomotion in the burrfish (*Chilomycterus schoepfi*). *Proc. R. Soc. Lond. B.* **263,** 1689–1696.

Bainbridge, R. (1958). The speed of swimming of fish as related to size and to the frequency and amplitude of the tail beat. *J. Exp. Biol.* **35,** 109–133.

Bainbridge, R. (1963). Caudal fin and body movement in the propulsion of some fish. *J. Exp. Biol.* **40,** 23–56.

Bandyopadhay, P. R., Castano, J. M., Rice, J. Q., Philips, B., Nedderman, W. H., and Macy, W. K. (1997). Low-speed maneuvering hydrodynamics of fish and small underwater vehicles. *J. Fluids Eng.* **119,** 136–144.

Bellwood, D. R., and Wainwright, P. C. (2001). Locomotion in labrid fishes: Implications for habitat use and cross-shelf biogeography on the Great Barrier Reef. *Coral Reefs* **20,** 139–150.

Birch, J. M., Dickson, W. B., and Dickinson, M. H. (2004). Force production and flow structure of the leading edge vortex on flapping wings at high and low Reynolds numbers. *J. Exp. Biol.* **207,** 1063–1072.

Blake, R. W. (1977). On ostraciiform locomotion. *J. Mar. Biol. Ass. U.K.* **57,** 1047–1055.

Blake, R. W. (1978). On balistiform locomotion. *J. Mar. Biol. Ass. U.K.* **58,** 73–80.

Blake, R. W. (1979a). The energetics of hovering in the mandarin fish (*Synchropus picturatus*). *J. Exp. Biol.* **82,** 25–33.

Blake, R. W. (1979b). The mechanics of labriform locomotion. I. Labriform locomotion in the angelfish (*Pterophyllum eimekei*): An analysis of the power stroke. *J. Exp. Biol.* **82,** 255–271.

Blake, R. W. (1979c). The swimming of the mandarin fish *Synchropus picturatus* (Callionyiidae: Teleostei). *J. Mar. Biol. Ass. U.K.* **59,** 421–428.

Blake, R. W. (1983). Median and paired fin propulsion. *In* "Fish Biomechanics" (Webb, P. W., and Weihs, D., Eds.), pp. 214–247. Praeger, New York.

Braun, J., and Reif, W.-E. (1985). A survey of aquatic locomotion of fishes and tetrapods. *Neues Jahrb. Geol. P.-A.* **169,** 307–332.

Breder, C. M., Jr., and Edgerton, H. E. (1942). An analysis of the locomotion of the seahorse, *Hippocampus*, by means of high speed cinematography. *Ann. N.Y. Acad. Sci.* **43,** 145–172.

Breder, C. M., Jr. (1926). The locomotion of fishes. *Zoologica* **4,** 159–297.

Brett, J. R., and Sutherland, D. B. (1965). Respiratory metabolism of pumpkinseed (*Lepomis gibbosus*) in relation to swimming speed. *J. Fish. Res. Bd. Canada* **22,** 405–409.

Clark, B. D., and Fish, F. E. (1994). Scaling of the locomotory apparatus and paddling rhythm in swimming mallard ducklings (*Anas platyrhynchos*): Test of a resonance model. *J. Exp. Zool.* **270,** 245–254.

Combes, S. A., and Daniel, T. L. (2001). Shape, flapping and flexion: Wing and fin design for forward flight. *J. Exp. Biol.* **204,** 2073–2085.

Daniel, T. L. (1984). Unsteady aspects of aquatic locomotion. *Am. Zool.* **24,** 121–134.

Daniel, T. L. (1988). Forward flapping flight from flexible fins. *Can. J. Zool.* **66,** 630–638.

Daniel, T. L., and Webb, P. W. (1987). Physical determinants of locomotion. *In* "Comparative Physiology: Life in Water and on Land" (Dejours, P., Bolis, L., Taylor, C. R., and Weibel, E. R., Eds.), pp. 343–369. Liviana Press, New York.

Davis, W. P., and Birdsong, R. S. (1973). Coral reef fishes which forage in the water column. A review of their morphology, behavior, ecology and evolutionary implications. *Helgolander Wiss. Meer.* **24,** 292–306.

Davison, W. (1988). The myotomal muscle of labriform swimming fish is not designed for high speed sustained swimming. *New Zealand Nat. Sci.* **15,** 37–42.

Davison, W. (1994). Exercise training in the banded wrasse *Notolabrus fucicola* affects muscle fibre diameter, but not muscle mitochondrial morphology. *New Zealand Nat. Sci.* **21,** 11–16.

Dial, K. P., Goslow, G. E., Jr., and Jenkins, F. A., Jr. (1991). The functional anatomy of the shoulder of the European starling (*Sturnus vulgaris*). *J. Morphol.* **207,** 327–344.

Dickinson, M. H., Lehmann, F.-O., and Sane, S. P. (1999). Wing rotation and the aerodynamic basis of insect flight. *Science* **284,** 1954–1960.

Domenici, P. (2001). The scaling of locomotor performance in predator–prey encounters: From fish to killer whales. *Comp. Biochem. Physiol.* **131A,** 169–182.

Domenici, P., and Blake, R. W. (1997). The kinematics and performance of fish fast-start swimming. *J. Exp. Biol.* **200,** 1165–1178.

Drucker, E. G. (1996). The use of gait transition speed in comparative studies of fish locomotion. *Am. Zool.* **36,** 555–566.

Drucker, E. G., and Jensen, J. S. (1996a). Pectoral fin locomotion in the striped surfperch. I. Kinematic effects of swimming speed and body size. *J. Exp. Biol.* **199,** 2235–2242.

Drucker, E. G., and Jensen, J. S. (1996b). Pectoral fin locomotion in the striped surfperch. II. Scaling swimming kinematics and performance at a gait transition. *J. Exp. Biol.* **199,** 2243–2252.

Drucker, E. G., and Jensen, J. S. (1997). Kinematic and electromyographic analysis of steady pectoral fin swimming in the surfperches. *J. Exp. Biol.* **200,** 1709–1723.

Drucker, E. G., and Lauder, G. V. (1999). Locomotor forces on a swimming fish: Three-dimensional vortex wake dynamics quantified using digital particle image velocimetry. *J. Exp. Biol.* **202,** 2393–2412.

Drucker, E. G., and Lauder, G. V. (2000). A hydrodynamic analysis of fish swimming speed: Wake structure and locomotor force in slow and fast labriform swimmers. *J. Exp. Biol.* **203,** 2379–2393.

Drucker, E. G., and Lauder, G. V. (2001). Wake dynamics and fluid forces of turning maneuvers in sunfish. *J. Exp. Biol.* **204,** 431–442.

Drucker, E. G., and Lauder, G. V. (2002a). Experimental hydrodynamics of fish locomotion: Functional insights from wake visualization. *Integr. Comp. Biol.* **42,** 243–257.

Drucker, E. G., and Lauder, G. V. (2002b). Wake dynamics and locomotor function in fishes: Interpreting evolutionary patterns in pectoral fin design. *Integr. Comp. Biol.* **42,** 997–1008.

Drucker, E. G., and Lauder, G. V. (2003). Function of pectoral fins in rainbow trout: Behavioral repertoire and hydrodynamic forces. *J. Exp. Biol.* **206,** 813–826.

Drucker, E. G., and Summers, A. P. (2006). Moving with fins and limbs: An historical perspective on the study of animal locomotion with paired appendages. *In* "Fins and Limbs: Evolution, Development, and Transformation." (Hall, B. K., Ed.). University of Chicago Press, Chicago. In press.

Fish, F. E. (1999). Performance constraints on the maneuverability of flexible and rigid biological systems. *In* "Eleventh International Symposium on Unmanned Untethered Submersible Technology," pp. 394–406. Autonomous Undersea Systems Institute, Durham, NH.

Fish, F., and Lauder, G. V. (2006). Passive and active flow control by swimming fishes and mammals. *Ann. Rev. Fluid Mech.* **38,** 193–224.

Fricke, H., and Hissmann, K. (1992). Locomotion, fin coordination and body form of the living coelacanth *Latimeria chalumnae*. *Env. Biol. Fish.* **34,** 329–356.

Fricke, H., Reinicke, O., Hofer, H., and Nachtigall, W. (1987). Locomotion of the coelacanth *Latimeria chalumnae* in its natural environment. *Nature* **329,** 331–333.

Frith, H. R., and Blake, R. W. (1995). The mechanical power output and hydromechanical efficiency of northern pike (*Esox lucius*) fast-starts. *J. Exp. Biol.* **198,** 1863–1873.

Fulton, C. J., Bellwood, D. R., and Wainwright, P. C. (2001). The relationship between swimming ability and habitat use in wrasses (Labridae). *Mar. Biol.* **139,** 25–33.

Geerlink, P. J. (1983). Pectoral fin kinematics of *Coris formosa* (Teleostei, Labridae). *Netherl. J. Zool.* **33,** 515–531.

Geerlink, P. J. (1987). The role of the pectoral fins in braking of mackerel, cod and saithe. *Netherl. J. Zool.* **37,** 81–104.

Geerlink, P. J. (1989). Pectoral fin morphology: A simple relation with movement pattern? *Netherl. J. Zool.* **39,** 166–193.

Gerstner, C. L. (1999). Maneuverability of four species of coral-reef fish that differ in body and pectoral-fin morphology. *Can. J. Zool.* **77,** 1102–1110.
Gharib, M., Pereira, F., Dabiri, D., Hove, J. R., and Modarress, D. (2002). Quantitative flow visualization: Toward a comprehensive flow diagnostic tool. *Integr. Comp. Biol.* **42,** 964–970.
Gibb, A. C., Jayne, B. C., and Lauder, G. V. (1994). Kinematics of pectoral fin locomotion in the bluegill sunfish *Lepomis macrochirus*. *J. Exp. Biol.* **189,** 133–161.
Goldspink, G. (1977). Mechanics and energetics of muscle in animals of different sizes, with particular reference to the muscle fiber composition of vertebrate muscle. *In* "Scale Effects in Animal Locomotion" (Pedley, T. J., Ed.), pp. 37–55. Academic Press, London.
Goolish, E. M. (1991). Aerobic and anaerobic scaling in fish. *Biol. Rev.* **66,** 33–56.
Gordon, M. S., Plaut, I., and Kim, D. (1996). How puffers (Teleostei: Tetraodontidae) swim. *J. Fish Biol.* **49,** 319–328.
Gosline, W. A. (1971). "Functional Morphology and Classification of Teleostean Fishes." University of Hawaii Press, Honolulu, HI.
Greenewalt, C. H. (1962). Dimensional relationships for flying animals. *Smithson. Misc. Coll.* **144,** 1–46.
Greenway, P. (1965). Body form and behavioral types in fish. *Experient.* **21,** 489–497.
Gregory, W. K. (1928). Studies of the body forms of fishes. *Zoologica* **8,** 325–421.
Hammer, C. (1995). Fatigue and exercise tests with fish. *Comp. Biochem. Physiol.* **112A,** 1–20.
Harris, J. E. (1936). The role of the fins in the equilibrium of the swimming fish. I. Wind tunnel tests on a model of *Mustelus canis* (Mitchell). *J. Exp. Biol.* **13,** 476–493.
Harris, J. E. (1937). The mechanical significance of the position and movements of the paired fins in the Teleostei. *Papers of the Tortugas Laboratories* **31,** 173–189.
Harris, J. E. (1938). The role of the fins in the equilibrium of the swimming fish. II. The role of the pelvic fins. *J. Exp. Biol.* **16,** 32–47.
Harris, J. E. (1953). Fin patterns and mode of life in fishes. *In* "Essays in Marine Biology" (Orr, A. P., Ed.), pp. 17–28. Oliver and Boyd, Edinburgh.
Heglund, N. C., and Taylor, C. R. (1988). Speed, stride frequency and energy cost per stride: How do they change with body size and gait? *J. Exp. Biol.* **138,** 301–318.
Herrmann, S., Hinrichs, H., Hinsch, K. D., and Surmann, C. (2000). Coherence concepts in holographic particle image velocimetry. *Exp. Fluids* **29,** S108–S116.
Hill, A. V. (1950). The dimensions of animals and their muscular dynamics. *Sci. Prog. Lond.* **38,** 209–230.
Hove, J. R., O'Bryan, L. M., Gordon, M. S., Webb, P. W., and Weihs, D. (2001). Boxfishes (Teleostei: Ostraciidae) as a model system for fishes swimming with many fins: Kinematics. *J. Exp. Biol.* **204,** 1459–1471.
Hunter, J. R., and Zweifel, J. R. (1971). Swimming speed, tail beat frequency, tail beat amplitude, and size in jack mackerel, *Trachurus symmetricus*, and other fishes. *Fish. Bull.* **69,** 253–266.
Jayne, B. C., and Lauder, G. V. (1994). How swimming fish use slow and fast muscle fibers: Implications for models of vertebrate muscle recruitment. *J. Comp. Physiol. A* **175,** 123–131.
Jindrich, D. L., and Full, R. J. (1999). Many-legged maneuverability: Dynamics of turning in hexapods. *J. Exp. Biol.* **202,** 1603–1623.
Kähler, C. J., and Kompenhans, J. (2000). Fundamentals of multiple plane stereo particle image velocimetry. *Exp. Fluids* **29,** S70–S77.
Kalleberg, H. (1958). Observations in a stream tank of territoriality and competition in juvenile salmon and trout (*Salmo salar* L. and *S. trutta* L.). *Rep Inst. Freshwater Res. Drottningholm* **39,** 55–98.
Kato, N., and Liu, H. (2003). Optimization of motion of a mechanical pectoral fin. *JSME Int. J. Ser. C.* **46,** 1356–1362.

Keast, A., and Webb, D. (1966). Mouth and body form relative to feeding ecology in the fish fauna of a small lake, Lake Opinicon, Ontario. *J. Fish. Res. Bd. Canada* **23,** 1845–1874.

Klausewitz, W. (1964). Der Lokomotionsmodus der Flügelrochen (Myliobatoidei). *Zool. Anz.* **173,** 110–120.

Korsmeyer, K. E., Steffensen, J. F., and Herskin, J. (2002). Energetics of median and paired fin swimming, body and caudal fin swimming, and gait transition in parrotfish (*Scarus schlegeli*) and triggerfish (*Rhinecanthus aculeatus*). *J. Exp. Biol.* **205,** 1253–1263.

Kuethe, A. M., and Chow, C.-Y. (1986). "Foundations of Aerodynamics." John Wiley & Sons, New York.

Lauder, G. V. (2000). Function of the caudal fin during locomotion in fishes: Kinematics, flow visualization, and evolutionary patterns. *Am. Zool.* **40,** 101–122.

Lauder, G. V., and Drucker, E. G. (2004). Morphology and experimental hydrodynamics of fish fin control surfaces. *IEEE J. Ocean. Eng.* **29,** 556–571.

Lauder, G. V., and Jayne, B. C. (1996). Pectoral fin locomotion in fishes: Testing drag-based models using three-dimensional kinematics. *Am. Zool.* **36,** 567–581.

Lindsey, C. C. (1978). Form, function, and locomotory habits in fish. *In* "Fish Physiology" (Hoar, W. S., and Randall, D. J., Eds.), Vol. VII, pp. 1–100. Academic Press, New York.

Magnan, A. (1930). Les caractéristiques géométriques et physiques des Poissons, avec contribution à l'étude de leur équilibre statique et dynamique. *Ann. Sci. Nat. Zool.* **13,** 355–489.

Malkiel, E., Sheng, J., Katz, J., and Strickler, J. R. (2003). The three-dimensional flow field generated by a feeding calanoid copepod measured using digital holography. *J. Exp. Biol.* **206,** 3657–3666.

Marey, E. J. (1890). La locomotion aquatique étudiée par la chronophotographie. *C. R. Acad. Sci.* **111,** 213–216.

Marsh, R. L. (1988). Ontogenesis of contractile properties of skeletal muscle and sprint performance in the lizard *Dipsosaurus dorsalis*. *J. Exp. Biol.* **137,** 119–139.

McCutchen, C. W. (1977). Froude propulsive efficiency of a small fish, measured by wake visualization. *In* "Scale Effects in Animal Locomotion" (Pedley, T. J., Ed.), pp. 339–363. Academic Press, London.

McLaughlin, R. L., and Noakes, D. L. G. (1998). Going against the flow: An examination of the propulsive movements made by young brook trout in streams. *Can. J. Fish. Aquat. Sci.* **55,** 853–860.

McMahon, T. A. (1975). Using body size to understand the structural design of animals: Quadrupedal locomotion. *J. Appl. Physiol.* **39,** 619–627.

Montgomery, J. C., and Macdonald, J. A. (1984). Performance of motor systems in Antarctic fishes. *J. Comp. Physiol. A* **154,** 241–248.

Müller, U. K., van den Heuvel, B. L. E., Stamhuis, E. J., and Videler, J. J. (1997). Fish foot prints: Morphology and energetics of the wake behind a continuously swimming mullet (*Chelon labrosus* Risso). *J. Exp. Biol.* **200,** 2893–2906.

Mussi, M., Summers, A. P., and Domenici, P. (2002). Gait transition speed, pectoral fin-beat frequency and amplitude in *Cymatogaster aggregata*, *Embiotoca lateralis* and *Damalichthys vacca*. *J. Fish Biol.* **61,** 1282–1293.

Nauen, J. C., and Lauder, G. V. (2002). Quantification of the wake of rainbow trout (*Oncorhynchus mykiss*) using three-dimensional stereoscopic digital particle image velocimetry. *J. Exp. Biol.* **205,** 3271–3279.

Parsons, G. R., and Sylvester, J. L., Jr. (1992). Swimming efficiency of the white crappie, *Pomoxis annularis*. *Copeia* **1992,** 1033–1038.

Pereira, F., and Gharib, M. (2002). Defocusing digital particle image velocimetry and the three-dimensional characterization of two-phase flows. *Meas. Sci. Tech.* **13,** 683–694.

Rome, L. C., Funke, R. P., Alexander, R. McN., Lutz, G., Aldridge, H., Scott, F., and Freadman, M. (1988). Why animals have different muscle fibre types. *Nature* **355,** 824–827.

Rosenberger, L. J. (2001). Pectoral fin locomotion in batoid fishes: Undulation *versus* oscillation. *J. Exp. Biol.* **204,** 379–394.

Rosenberger, L. J., and Westneat, M. W. (1999). Functional morphology of undulatory pectoral fin locomotion in the stingray *Taeniura lymma* (Chondrichthyes: Dasyatidae). *J. Exp. Biol.* **202,** 3523–3539.

Schrank, A. J., and Webb, P. W. (1998). Do body and fin form affect the abilities of fish to stabilize swimming during maneuvers through vertical and horizontal tubes? *Env. Biol. Fish.* **53,** 365–371.

Schrank, A. J., Webb, P. W., and Mayberry, S. (1999). How do body and paired-fin positions affect the ability of three teleost fishes to maneuver around bends? *Can. J. Zool.* **77,** 203–210.

Spedding, G. R. (1987). The wake of a kestrel (*Falco tinnunculus*) in flapping flight. *J. Exp. Biol.* **127,** 59–78.

Stamhuis, E. J., Videler, J. J., van Duren, L. A., and Müller, U. K. (2002). Applying digital particle image velocimetry to animal-generated flows: Traps, hurdles and cures in mapping steady and unsteady flows in Re regimes between 10^2 and 10^5. *Exp. Fluids* **33,** 801–813.

Stamhuis, E. J., and Videler, J. J. (1995). Quantitative flow analysis around aquatic animals using laser sheet particle image velocimetry. *J. Exp. Biol.* **198,** 283–294.

Stobutzki, I. C., and Bellwood, D. R. (1994). An analysis of the sustained swimming abilities of pre- and post-settlement coral reef fishes. *J. Exp. Mar. Biol. Ecol.* **175,** 275–286.

Strickler, J. R. (1975). Swimming of planktonic *Cyclops* species (Copepoda, Crustacea): Pattern, movements and their control. *In* "Swimming and Flying in Nature" (Wu, T. Y., Brokaw, C. J., and Brennen, C., Eds.), vol. 2, , pp. 599–613. Plenum Press, New York.

te Kronnie, G., Tatarczuch, L., van Raamsdonk, W., and Kilarski, W. (1983). Muscle fibre types in the myotome of stickleback, *Gasterosteus aculeatus* L.; A histochemical, immunohistochemical and ultrastructural study. *J. Fish Biol.* **22,** 303–316.

Thorsen, D. H., and Westneat, M. W. (2005). Diversity of pectoral fin structure and function in fishes with labriform propulsion. *J. Morphol.* **263,** 133–150.

Tu, M., and Dickinson, M. H. (1994). Modulation of negative work output from a steering muscle of the blowfly *Calliphora vicina*. *J. Exp. Biol.* **192,** 207–224.

Videler, J. J. (1981). Swimming movements, body structure and propulsion in cod, *Gadus morhua*. *Symp. Zool. Soc. Lond.* **48,** 1–27.

Videler, J. J. (1993). "Fish Swimming." Chapman & Hall, London.

Vogel, S. (1994). "Life in Moving Fluids," 2nd edn., Princeton University Press, Princeton, NJ.

Wainwright, P. C. (1988). Morphology and ecology: Functional basis of feeding constraints in Caribbean labrid fishes. *Ecology* **69,** 635–645.

Wainwright, P. C. (1991). Ecomorphology: Experimental functional anatomy for ecological problems. *Am. Zool.* **31,** 680–693.

Wainwright, P. C., Bellwood, D. R., and Westneat, M. W. (2002). Ecomorphology of locomotion in labrid fishes. *Env. Biol. Fish.* **65,** 47–62.

Wakeling, J. M., Kemp, K. M., and Johnston, I. A. (1999). The biomechanics of fast-starts during ontogeny in the common carp *Cyprinus carpio*. *J. Exp. Biol.* **202,** 3057–3067.

Walker, J. A. (2000). Does a rigid body limit maneuverability? *J. Exp. Biol.* **203,** 3391–3396.

Walker, J. A. (2002). Rotational lift: Something different or more of the same? *J. Exp. Biol.* **205,** 3783–3792.

Walker, J. A. (2004a). Dynamics of pectoral fin rowing in a fish with an extreme rowing stroke: The threespine stickleback (*Gasterosteus aculeatus*). *J. Exp. Biol.* **207,** 1925–1939.

Walker, J. A. (2004b). Kinematics and performance of median and paired fins as control surfaces. *IEEE J. Ocean. Eng.* **29,** 572–584.

Walker, J. A., and Westneat, M. W. (1997). Labriform propulsion in fishes: Kinematics of flapping aquatic flight in the bird wrasse *Gomphosus varius* (Labridae). *J. Exp. Biol.* **200,** 1549–1569.
Walker, J. A., and Westneat, M. W. (2000). Mechanical performance of aquatic rowing and flying. *Proc. R. Soc. Lond. B* **267,** 1875–1881.
Walker, J. A., and Westneat, M. W. (2002a). Kinematics, dynamics, and energetics of rowing and flapping propulsion in fishes. *Integr. Comp. Biol.* **42,** 1032–1043.
Walker, J. A., and Westneat, M. W. (2002b). Performance limits of labriform propulsion and correlates with fin shape and motion. *J. Exp. Biol.* **205,** 177–187.
Wardle, C. S., Videler, J. J., and Altringham, J. D. (1995). Tuning in to fish swimming waves: Body form, swimming mode and muscle function. *J. Exp. Biol.* **198,** 1629–1636.
Webb, P. W. (1973). Kinematics of pectoral fin propulsion in *Cymatogaster aggregata. J. Exp. Biol.* **59,** 697–710.
Webb, P. W. (1982). Locomotor patterns in the evolution of actinopterygian fishes. *Am. Zool.* **22,** 329–342.
Webb, P. W. (1983). Speed, acceleration and manoeuvrability of two teleost fishes. *J. Exp. Biol.* **102,** 115–122.
Webb, P. W. (1984). Body form, locomotion and foraging in aquatic vertebrates. *Am. Zool.* **24,** 107–120.
Webb, P. W. (1994a). The biology of fish swimming. *In* "Mechanics and Physiology of Animal Swimming" (Maddock, L., Bone, Q., and Rayner, J. M. V., Eds.), pp. 45–62. Cambridge University Press, Cambridge.
Webb, P. W. (1994b). Exercise performance of fish. *In* "Advances in Veterinary Science and Comparative Medicine," Vol. 38B, "Comparative Vertebrate Exercise Physiology: Phyletic Adaptations" (Jones, J. H., Ed.), pp. 1–49. Academic Press, San Diego.
Webb, P. W., and Blake, R. W. (1985). Swimming. *In* "Functional Vertebrate Morphology" (Hildebrand, M., Bramble, D. M., Liem, K. F., and Wake, D. B., Eds.), pp. 110–128. Harvard University Press, Cambridge.
Webb, P. W., and Fairchild, A. G. (2001). Performance and maneuverability of three species of teleostean fishes. *Can. J. Zool.* **79,** 1866–1877.
Webb, P. W., Kostecki, P. T., and Stevens, E. D. (1984). The effect of size and swimming speed on locomotor kinematics of rainbow trout. *J. Exp. Biol.* **109,** 77–95.
Webb, P. W., LaLiberte, G. D., and Schrank, A. J. (1996). Do body and fin form affect the maneuverability of fish traversing vertical and horizontal slits? *Env. Biol. Fish.* **46,** 7–14.
Westneat, M. W. (1996). Functional morphology of aquatic flight in fishes: Kinematics, electromyography, and mechanical modeling of labriform locomotion. *Am. Zool.* **36,** 582–598.
Westneat, M. W., Thorsen, D. H., Walker, J. A., and Hale, M. E. (2004). Structure, function and neural control of pectoral fins in fishes. *IEEE J. Ocean. Eng.* **29,** 674–683.
Westneat, M. W., and Walker, J. A. (1997). Motor patterns of labriform locomotion: Kinematic and electromyographic analysis of pectoral fin swimming in the labrid fish *Gomphosus varius. J. Exp. Biol.* **200,** 1881–1893.
Whoriskey, F. G., and Wootton, R. J. (1987). Swimming endurance of threespine sticklebacks, *Gasterosteus aculeatus* L., from the Afon Rheidol, Wales. *J. Fish Biol.* **30,** 335–340.
Wilga, C. D., and Lauder, G. V. (1999). Locomotion in sturgeon: Function of the pectoral fins. *J. Exp. Biol.* **202,** 2413–2432.
Wilga, C. D., and Lauder, G. V. (2000). Three-dimensional kinematics and wake structure of the pectoral fins during locomotion in leopard sharks *Triakis semifasciata. J. Exp. Biol.* **203,** 2261–2278.
Wilga, C. D., and Lauder, G. V. (2001). Functional morphology of the pectoral fins in bamboo sharks, *Chiloscyllium plagiosum*: Benthic vs. pelagic station-holding. *J. Morphol.* **249,** 195–209.
Willert, C. E., and Gharib, M. (1991). Digital particle image velocimetry. *Exp. Fluids* **10,** 181–193.

11

HYDRODYNAMICS OF UNDULATORY PROPULSION

GEORGE V. LAUDER
ERIC D. TYTELL

I. INTRODUCTION

In the years since publication of the major previous reviews addressing various aspects of fish undulatory propulsion (Hertel, 1966; Webb, 1975, 1978, 1993b; Aleev, 1977; Bone, 1978; Lindsey, 1978; Magnuson, 1978; Blake, 1983; Webb and Weihs, 1983; Videler, 1993), one significant new development stands out: the ability to quantify water flow patterns around swimming fishes directly. Until recently, investigations of fish propulsion have had to infer hydrodynamic function from kinematics and theoretical models. Biologists and engineers interested in how fishes interact with their fluid environment have had no quantitative way to visualize this interaction, despite the critical importance of understanding fluid flow patterns produced by swimming fishes for testing theoretical models and for understanding the hydrodynamic effects of different body and fin designs. This predicament

Fish Biomechanics: Volume 23
FISH PHYSIOLOGY

DOI: 10.1016/S1546-5098(05)23011-X

was well expressed by McCutchen (1977) in his chapter in the classic book *Scale Effects in Animal Locomotion* (Pedley, 1977). McCutchen (1977, p. 339) described the current state of research on fish locomotion by stating that "considering the man-hours spent studying fish propulsion, we know precious little about what the fish does to the fluid."

One early attempt to visualize the flow around swimming fishes directly was the influential master's thesis by Rosen (1959). Rosen's ingenious experiments involved swimming fish in a shallow pan of water above a thin layer of milk carefully layered on the bottom. The swimming movements of the fish disturbed the milk layer and revealed general flow patterns which Rosen then photographed. Subsequent attempts that included the use of dye or Schleiren methods (McCutchen, 1977; Arnold *et al.*, 1991; Ferry and Lauder, 1996) gave useful results but provided little information of a quantitative nature.

The combination of high-resolution high-speed video systems, high-powered continuous wave lasers, and an image analysis technique called digital particle image velocimetry (DPIV), developed over the past decade, has permitted the direct visualization of water flow over the surface and in the wake of swimming fishes (e.g., Müller *et al.*, 1997; Wolfgang *et al.*, 1999; Anderson *et al.*, 2000; Lauder, 2000; Müller *et al.*, 2001; Lauder and Drucker, 2002; Lauder *et al.*, 2003; Tytell, 2004a; Tytell and Lauder, 2004). These data have provided a wealth of new information on the fluid flows generated by the body, tail, and fins of freely swimming fishes, and represent a significant new arena of investigation not described in previous reviews.

In this chapter, we focus on recent experimental hydrodynamic data on undulatory locomotion in fishes, and provide as background a general description of the major theoretical model of undulatory propulsion. Hydrodynamic flows are caused by undulatory movements of the body and fins, and so we preface our discussion of models and experiments with an analysis of the classical kinematic modes of undulatory locomotion in fishes.

II. CLASSICAL MODES OF UNDULATORY PROPULSION

One of the most enduring features of the literature on fish locomotion is the classification of fish swimming into general "modes" based on exemplar species (Breder, 1926; Lindsey, 1978). For example, eel locomotion is referred to as anguilliform after the eel genus *Anguilla*, and is used as shorthand for fishes that undulate large portions of their body during propulsion, generally with nearly a full wavelength present on the body at any given time (Gray, 1933; Gillis, 1996). Thunniform locomotion is based on tunas (*Thunnus*), and is verbal shorthand for fish locomotion involving a high aspect ratio tail with relatively little lateral oscillation of the body (Donley

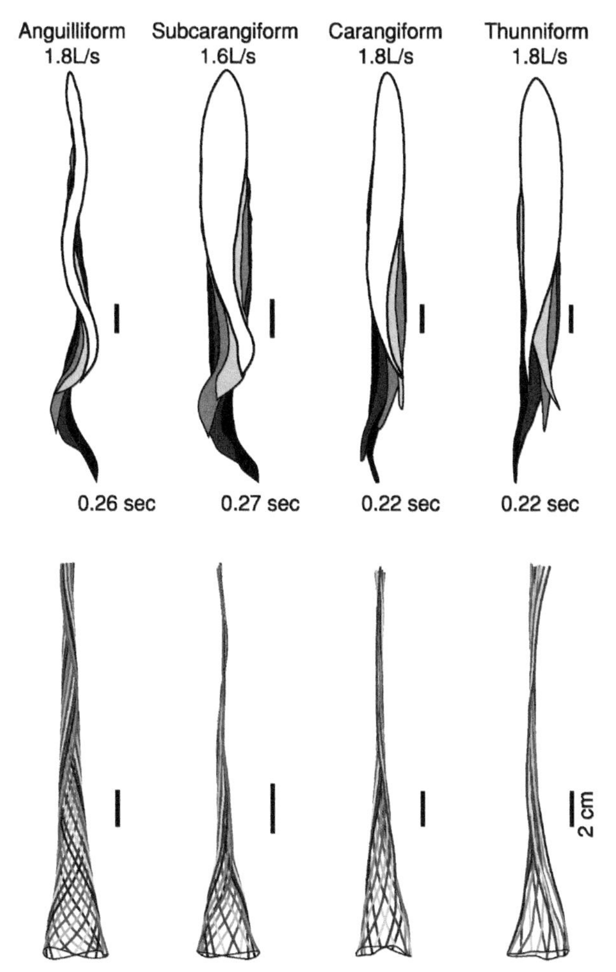

Fig. 11.1. Four classical categories of fish undulatory propulsion illustrated with fish outlines and midlines derived from recent experimental data. This figure updates the classical depiction presented in Lindsey (1978). Outlines of swimming fishes are shown above with displacements that to illustrate forward progression, while midlines at equally spaced time intervals throughout a tail beat are superimposed below, aligned at the tip of the snout; each time is shown in a distinct color. Anguilliform mode based on *Anguilla*, subcarangiform mode based on *Lepomis*, carangiform mode based on *Scomber*, and thunniform mode based on *Euthynnus*. All fishes were between 20 and 25 cm total length (L), and swam at a similar speed of 1.6 to 1.8 Ls^{-1}. Times shown indicate duration of the tail beat. Scale bars = 2 cm. [Fish images are based on data from Tytell and Lauder (2004), Tytell (unpublished), and Donley and Dickson (2000).]

and Dickson, 2000; Graham and Dickson, 2004). Other major categories of undulatory propulsion include carangiform locomotion (up to one half wave on the body) and subcarangiform swimming (more than one half wave but less than one full wave), although additional names have also been used for this mode (Webb, 1975; Webb and Blake, 1985; also see Chapter 7 in this volume).

While these categories endure because of their utility in describing basic patterns of locomotion and because they are easy to visualize with a simple two-dimensional analysis, they are based on older data, sometimes containing inaccuracies or obtained under unsteady swimming conditions. Perhaps the most common misconception in the current literature concerns anguilliform locomotion, as a result of Sir James Gray's classic paper (Gray, 1933), which suggested that large-amplitude undulations occur along the entire eel body at all speeds. Recent data (see data and discussion in Gillis, 1996; Lauder and Tytell, 2004; Tytell and Lauder, 2004), however, indicate that the anterior body only begins undulating at high swimming speeds and during acceleration; at lower speeds, undulation is confined to the posterior region. The eel photographed by Gray (1933) was most likely accelerating (Lauder and Tytell, 2004; Tytell, 2004b), although these images of eel locomotion are often reproduced as representing steady locomotion. Additionally, careful measurements by Donley and Dickson (2000) suggest that thunniform locomotion is not a dramatically different swimming mode, but rather a relatively modest change in the carangiform mode.

In Figure 11.1, we use recent quantitative kinematic data from the literature to produce an update to the classical illustration in Lindsey (1978) of representative modes of undulatory locomotion in fishes. The fishes illustrated in this figure all swam at similar relative speeds (1.6 to 1.8 Ls^{-1}), were all between 20 and 25 cm in total length, and swam in a flow tank under carefully controlled conditions in which data were taken only when fishes were swimming steadily. At this moderately high sustained swimming speed of 1.8 Ls^{-1}, there is remarkable similarity in the undulatory profiles of all four classes of locomotion. Eels show greater lateral oscillations in the front half of the body at this speed, but the amplitude envelopes from subcarangiform to thunniform are extremely similar, with nonlinear amplitude profiles that rapidly increase toward the tail and similar overall patterns of body oscillation.

The similarity becomes more pronounced at slower swimming speeds, particularly at less than 1.0 Ls^{-1} (Figure 11.2). A common characteristic of undulatory propulsion at low speed is that anterior body oscillations are minimal (Lauder, 2005); only when speed exceeds 1.0 Ls^{-1} does the front half of the body begin to oscillate laterally. For example, Figure 11.2 shows data from largemouth bass and American eels, illustrating their similarity at low speeds and the increase in lateral oscillation as speed increases (Jayne

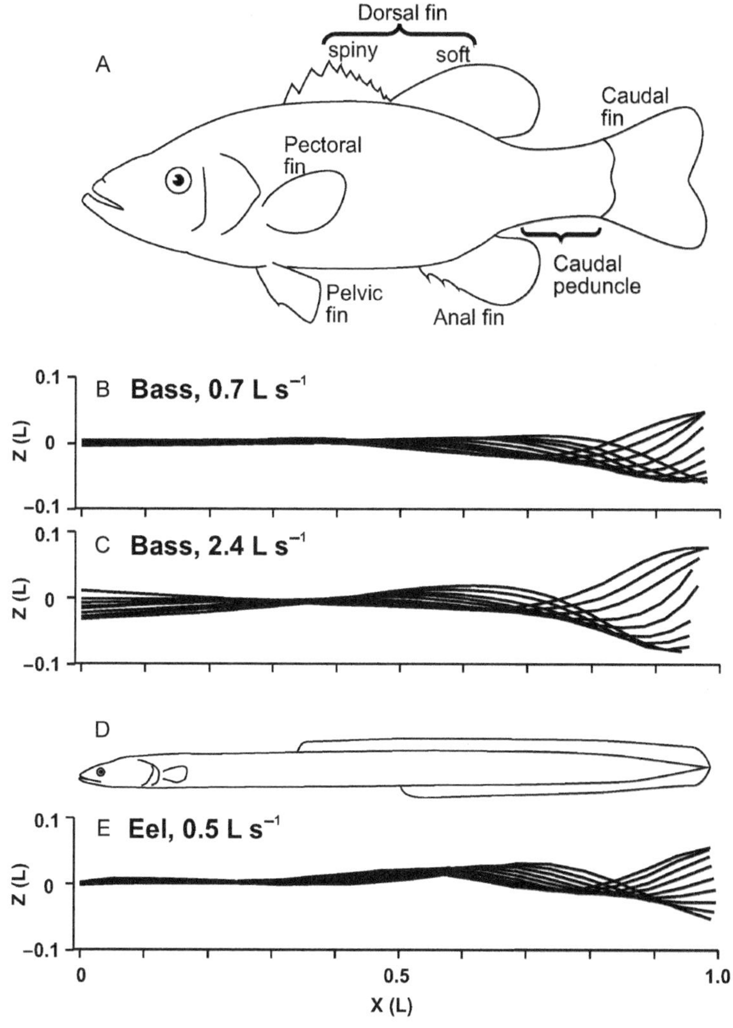

Fig. 11.2. (A) Outline of a largemouth bass (*Micropterus salmoides*) to illustrate the major fins and fin positions along the body and the longitudinal body profile in lateral view. (B–C) Midlines reconstructed from patterns of body bending during steady swimming in largemouth bass at speeds of 0.7 and 2.4 Ls^{-1}. At the slow speed, body bending is confined to the posterior half of the body, but at higher speeds the head begins to oscillate laterally. (D) Outline of an American eel (*Anguilla rostrata*) to show the contrast in body shape with the bass as seen in lateral view. (E) Reconstructed midlines of an eel swimming steadily at 0.5 Ls^{-1}. Note the similarity in amplitude and proportion of the body undulating for low-speed swimming in both the eel and the bass. Successive midlines in all plots are spaced equally in time during a half tail beat cycle. Z, lateral excursion in % body length (L); X, distance along the body, L. [Panels (B) and (C) modified from Jayne and Lauder (1995); panel (E) based on data from Tytell (2004a).]

and Lauder, 1995). Recruitment of increasingly anterior musculature with increased speed results in greater oscillation of the anterior body. At slow speeds, there is no anterior myotomal muscle strain, no body bending, and hence no work done in the front half of the body to power locomotion (e.g., Figure 11.2B and E).

It is our view that recent experimental data increasingly demonstrate that these taxonomically based categories do not represent important hydrodynamic differences, even though they do provide useful verbal shorthand for describing broad types of fish movement. In particular, we are concerned about two aspects of this classification scheme. First, the body amplitude profiles used to define locomotor categories, especially at slow speeds, are remarkably similar across these categories. Even fishes as different as eels and bass show very similar amplitude profiles at slow swimming speeds (Figure 11.2). Plotting midline (two-dimensional) amplitude profiles thus does little to reveal biologically significant hydrodynamic differences among species. Second, and perhaps more importantly, these categories ignore the crucial three-dimensional geometry of fishes. As illustrated in Figure 11.2A and D, most fishes possess numerous fins and regions such as the caudal peduncle that, as we discuss later, shed vortices and accelerate flow. These separation points are not visible in the two-dimensional slice through the fish midline so often used to depict locomotor modes. Many fish species have distinct tails, as well, that function like a propeller, accelerating flow and leaving a thrust signature in the wake (Lauder *et al.*, 2002; Fish and Lauder, 2005; Lauder, 2005). Another key point, which is not evident from examining standard two-dimensional depictions of fish swimming (like Figure 11.1), is that the dorsal and anal fin accelerate and redirect freestream flow; thus, the tail encounters incident flow greatly altered from the freestream (Drucker and Lauder, 2001a; Lauder and Drucker, 2004).

It is likely that the major differences in locomotor hydrodynamics among fishes are primarily a consequence of differences in their three-dimensional geometry and longitudinal area profiles, and not so much from differences in the amplitude profiles of a two-dimensional slice through the middle of the fish. In the following we present data showing a substantial hydrodynamic difference in wake structure between eels and other fish species studied to date. These differences are not attributable to midline amplitude profiles, but rather to the substantial differences in three-dimensional body shapes (Figure 11.2).

III. THEORY OF UNDULATORY PROPULSION

Recent experimental hydrodynamic studies have all been performed in the context of several physical theories of undulatory locomotion, based on midline kinematics such as those we described previously. Although such theories largely pass over some points that we argue are quite important, such as the three-dimensional shape of fish bodies, they provide some useful insights and an important context for the experimental studies that follow.

To begin, we approximate a fish in one of the simplest ways possible: as a flapping hydrofoil. A fish's tail, particularly that of fishes like tunas, is like a section of a hydrofoil moving from side to side. Two main parameters characterize the fluid dynamics of a flapping foil: one, Reynolds number, *Re*, describes the steady motion of the foil through the fluid; and the other, Strouhal number, *St*, describes the flapping motion.

Reynolds number is a dimensionless number that approximates the relative importance of forces due to the fluid's viscosity and forces due to the fluid's inertia. It is often written in two equivalent ways, as

$$Re = \frac{\rho UL}{\eta} \quad or \quad Re = \frac{UL}{\nu}, \tag{1}$$

where U is average forward speed, L is the length of the fish or the chord length of the foil, ρ is the fluid density, and η and ν are the fluid's dynamic and kinematic viscosity, respectively. The definitions of the length and speed are intentionally vague, because small differences in Reynolds number have little physical significance. Typically, only order of magnitude differences in *Re* are informative, although at certain critical points, like the transition from laminar to turbulent flow, small *Re* changes can lead to dramatic changes in the flow patterns. Nonetheless, flows are divided into three broad regimes: the viscous regime, $Re < 1$, in which viscous forces dominate; the inertial regime, $Re > 1000$, in which inertial forces dominate; and an intermediate regime between the two, in which both types of forces are important. For more information, see textbooks such as Denny (1993) or Faber (1995).

Reynolds number involves only the steady motion of the fish (or foil). Strouhal number, in turn, describes how fast the tail is flapping relative to its forward speed. It is defined as

$$St = \frac{fA}{U}, \tag{2}$$

where f is the flapping frequency, A is the flapping amplitude (the total distance from maximum excursion on one side to maximum excursion on the other), and U is again the forward speed. Strictly speaking, *St* describes the wake behind the flapping foil, not the foil itself, but, for biological purposes, the two definitions are often used interchangeably (Triantafyllou *et al.*, 1993). Fishes, marine mammals, and even flying animals seem to flap with a Strouhal number near 0.3 (Triantafyllou *et al.*, 1993; Taylor *et al.*, 2003; Rohr and Fish, 2004), although values from close to 0 to near 1 have been reported (Horton, *et al.*, 2003; Rohr and Fish, 2004).

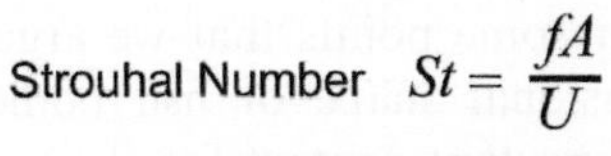
Strouhal Number $St = \frac{fA}{U}$

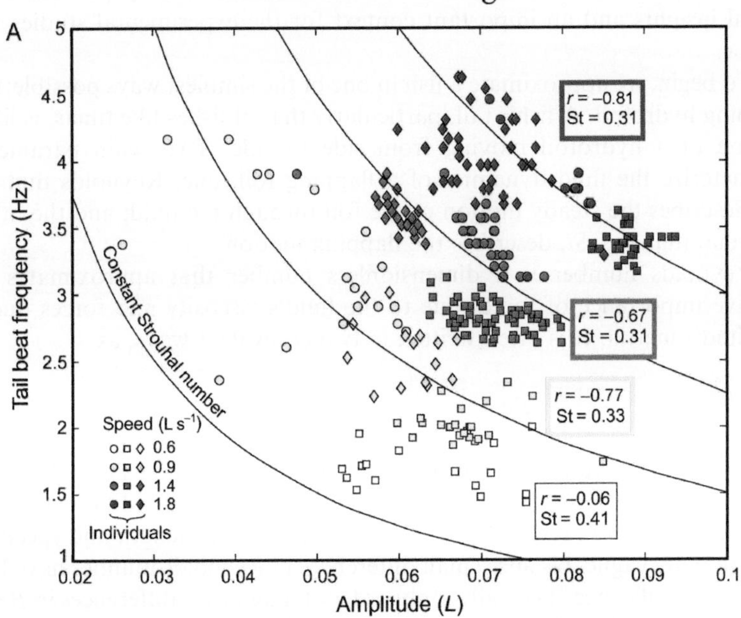
A
Tail beat frequency (Hz)
Amplitude (L)
Constant Strouhal number
r = –0.81
St = 0.31
r = –0.67
St = 0.31
r = –0.77
St = 0.33
r = –0.06
St = 0.41
Speed (L s⁻¹)
0.6
0.9
1.4
1.8
Individuals
5
4.5
4
3.5
3
2.5
2
1.5
1
0.02
0.03
0.04
0.05
0.06
0.07
0.08
0.09
0.1

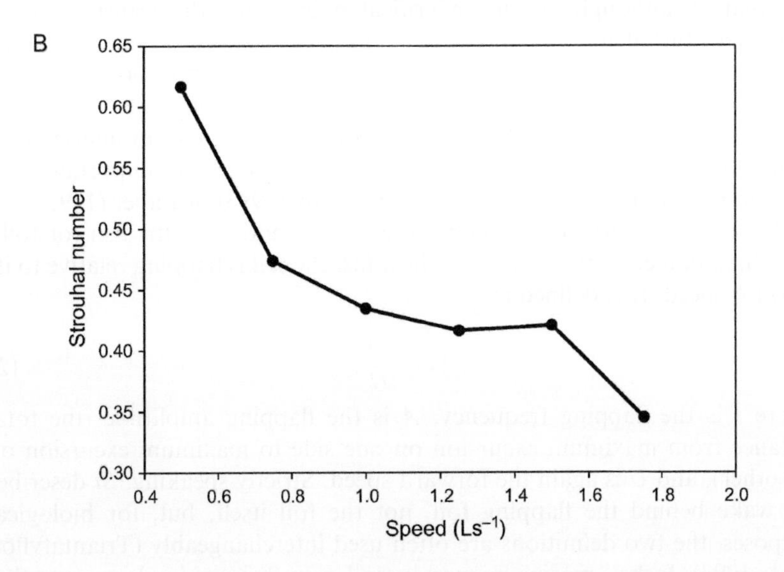
B
Strouhal number
Speed (Ls⁻¹)
0.65
0.60
0.55
0.50
0.45
0.40
0.35
0.30
0.4
0.6
0.8
1.0
1.2
1.4
1.6
1.8
2.0

Figure 11.3A shows data from eels (*Anguilla*) swimming over a speed range of 0.6 to 1.8 Ls^{-1}; St for these eels varied between 0.31 and 0.41, close to the optimal flapping value predicted for swimming fishes (Triantafyllou and Triantafyllou, 1995). For a given tail (or foil) shape, experimental and theoretical studies have shown that its thrust and efficiency are strongly related to Strouhal number (Anderson *et al.*, 1998; Murray and Howle, 2003; Read *et al.*, 2003). Typically, thrust increases with increasing St, but efficiency peaks near 0.3 (Read *et al.*, 2003), which is the value for most swimming fishes.

However, other data show that fishes may swim with apparently non-optimal Strouhal numbers during swimming, especially at lower speeds. Figure 11.3B shows St for Pacific salmon (*Oncorhynchus*) swimming; at slower swimming speeds of 0.5 Ls^{-1}, St is greater than 0.6, much higher than expected for optimal thrust production. These fish may genuinely be inefficient at slow speeds, or Strouhal number may not capture the complexities of fish swimming efficiency at slow speeds.

To proceed beyond this simplistic idea of fish swimming, one must start to examine the swimming motion in more detail. The theories described in the following section treat midline kinematics as a given. In principle, these kinematics result from a complex interaction between the internal forces from muscles and the external forces from the fluid. As a fish swims, muscles deform the body, applying forces to the fluid around the body, but the fluid applies forces back on the body, changing the body shape, which changes the muscle forces, and so on. To model the body deformation accurately, both sets of forces would have to be accounted for simultaneously. Few models attempt to solve this combined problem because of its complexity (but see Ekeberg, 1993; Cortez *et al.*, 2004). Instead, most theories treat the body as a given two-dimensional shape and examine the fluid forces separately.

A. Resistive Models

In one simple model, first proposed by Taylor (1952) and termed a "resistive" model, thrust is estimated from the sum of the drag forces on a

---

Fig. 11.3. Strouhal number and fish locomotion. (A) Tail-beat frequency versus amplitude at a variety of different swimming speeds for eel (*Anguilla*) locomotion; curves indicate a constant Strouhal number of 0.3, and colored symbols from white to red indicate increasing locomotor speed. Note the relative constancy of Strouhal number for eels swimming both over this speed range and within speeds. (B) Data for swimming Pacific salmon (*Oncorhynchus*) over a speed range of 0.5 to nearly 1.8 Ls^{-1}. Note the substantial increase in Strouhal number at slow swimming speeds. [Panel (A) modified from Tytell (2004a); data plotted in panel (B) courtesy of Jacquan Horton, Adam Summers, and Eliot Drucker.]

body with a wave traveling backward at a speed V greater than its forward swimming motion U. The body is divided into many segments down its length, and the drag forces from each segment are summed based on the equation

$$D = \frac{1}{2} C_D \rho S u^2, \tag{3}$$

where C_D is the drag coefficient, a measure of how "draggy" the segment is, ρ is the fluid density, S is an area, typically the presented area or the surface area of the segment, and u is the segment's velocity, including contributions from both the fish's forward motion and its lateral undulations. Typically, D must be broken into components perpendicular and parallel to the body, because C_D differs in different directions. Finding the total force is a somewhat lengthy but mostly geometric exercise of identifying all of the segment angles. The model suffers from other substantial problems, though, most notably because Eq. (3) is valid for steady motions, but may not hold for undulatory motion, because the model assumes that anterior segments do not affect flow as fluid moves past the body, and because estimating C_D can be difficult. Webb (1975) discusses the problems with resistive models in more detail.

Although this model is limited in its ability to estimate resistive forces, this is not to say that these types of forces are unimportant. They may play a role in generating thrust (Tytell, 2004a) and they may contribute to reducing propulsive efficiency (Webb, 1975).

B. Reactive Models

The preferred analytical model of fish swimming is Sir James Lighthill's elongated body theory (1960, 1971), which is based on adding up the forces due to the lateral acceleration and deceleration of the body as it undulates from side to side. This model is often termed "reactive," because it estimates forces based on the acceleration reaction, an effect in which fluids resist changes in velocity of a moving body, as if the body had additional, "virtual," mass (Batchelor, 1973; Childress, 1981; Daniel, 1984). Other forces that could contribute to thrust, including those resulting from the fluid's viscosity, are assumed to be negligible. The following explanation closely follows that of Lighthill (1969) and Webb (1975).

Lighthill (1960) estimated force and power by adding up changes in energy. Without viscosity, there are only three places energy changes while a fish swims (Figure 11.4). First, the total energy added to the system comes from the fish's movements. This energy is consumed in two ways: (1) pushing the fish forward (thrust power) and (2) creating the wake (wake power).

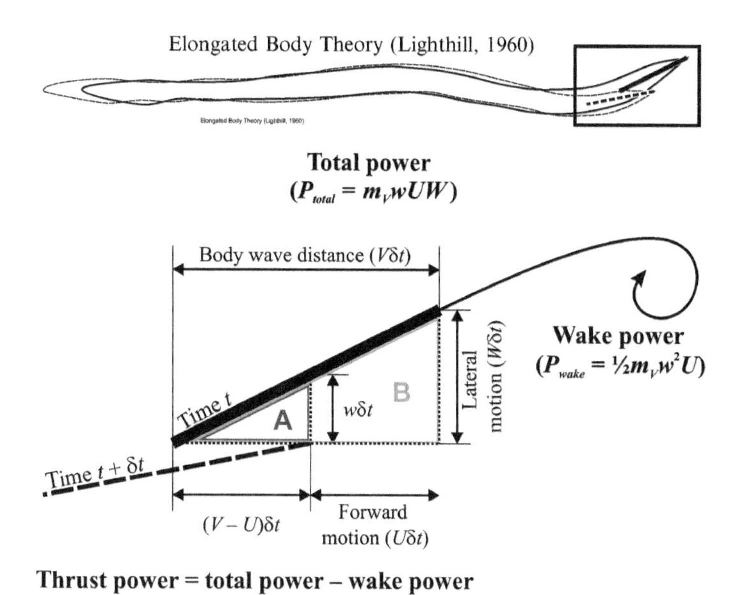

Fig. 11.4. Schematic view of Lighthill's (1960) reactive model showing the geometrical argument used to derive average thrust power. Top: Outlines of an eel at two different points in time (indicated by solid and dashed lines). The tail region is outlined with a box and enlarged below. Bottom: Enlarged view of tail tip segments at two points in time, showing geometrical parameters and the three components of energy in the system (bold equations). The two similar triangles used in the analysis (A and B) are outlined in red and blue, respectively. Total power added to the system, shown at the top, is consumed generating the wake, at right, and propelling the fish forward, at bottom. This figure is discussed in detail in the text. (After Webb, 1975.)

Energy has to be conserved, so these changes in energy must add up to zero. Thus, recalling that power is the rate of change of energy,

$$\text{(total undulatory power)} - \text{(thrust power)} - \text{(wake power)} = 0 \tag{4}$$

If we average these changes in energy over a tail-beat cycle, we can express their mean values based on a geometrical argument examining only the tail's motion. Because undulation is symmetric, the side-to-side motion of the body produces forces that average out over a tail beat. Only the tip of the tail is different: when it changes the fluid's momentum, that momentum is shed into the wake. Thus, to determine average thrust and power, we can restrict our analysis to the tail tip.

First, we examine the tail's motion. The segment at the tip of the tail moves forward and laterally during a time period δt (Figure 11.4). We define the length of the segment (projected on the swimming direction) to be $V\delta t$, where V is the speed of the body wave. During the time interval, the segment moves forward a distance $U\delta t$ and laterally a distance $W\delta t$, where U is the forward swimming speed and W is the lateral velocity of the tail tip. The water, however, is pushed laterally over a smaller distance $w\delta t$, because of how the forward and lateral motions combine (Figure 11.4). This velocity w approximates the motion of the tail segment perpendicular to itself, assuming that the tail does not make a large angle to the swimming direction. The difference between w, the velocity that actually affects the fluid, and the lateral tail velocity W is related to the difference in the body wave speed V and the swimming speed U. If the fish were pinned in place, such that U were zero, all of the lateral motion would affect the fluid and w would equal W. But if U and V were the same, then the tail tip would slip exactly into the spot just vacated by an upstream segment and the fluid would see no lateral motion at all.

To derive w, we examine the two similar triangles **A** and **B** in Figure 11.4:

$$\frac{w\delta t}{W\delta t} = \frac{(V - U)\delta t}{V\delta t} \tag{5}$$

or

$$w = W\frac{V - U}{V}, \tag{6}$$

which is, as we expected, less than W. Because w is the lateral velocity the fluid encounters, the tail increases the lateral momentum of the fluid by $m_V w$, where m_V is the virtual mass per unit length at the tail tip. This momentum is shed off the tail tip into the wake at the same rate the fish swims forward, U, and the tail itself does work on it as it moves laterally at a rate W. Thus, the total power is

$$P_{\text{total}} = m_V w U W. \tag{7}$$

As discussed previously, this input power is divided up into two outputs: the thrust power and the power to create the wake. The wake power is the kinetic energy of the fluid shed into the wake at speed U,

$$P_{\text{wake}} = \frac{1}{2} m_V w^2 U. \tag{8}$$

Subtracting the wake power (8) from the total power (7), according to the energy balance (4), we arrive at the average thrust power:

$$P_{\text{thrust}} = m_V\left(wWU - \frac{1}{2}w^2 U\right). \tag{9}$$

Because power is force multiplied by velocity, we can also write the average thrust force F_{thrust} as P_{thrust} / U.

The virtual mass per unit length m_V is fairly well approximated by the density of water ρ multiplied by the area of the circle that circumscribes the tail tip, including any fins sticking off (Lighthill, 1970). Thus,

$$m_V = \frac{1}{4}\pi\rho d^2, \tag{10}$$

where d is the dorso-ventral height of the caudal fin at the trailing edge.

We can gain some insight into Eq. (9) if we assume that the body has a simple, constant amplitude traveling wave. Averaged over the tail beat period, we find that mean thrust F_{thrust} is proportional to V^2–U^2. Thus we return to Gray's (1933) observation: for positive thrust, the body wave speed should be faster and in the opposite direction as the swimming speed. In fact, to maximize thrust, body wave speed should be as large as possible.

The reason why fish do not use the largest body wave speed possible becomes clear if we calculate efficiency. The Froude propulsive efficiency η is the ratio of useful (thrust) power to total power, or

$$\eta = \frac{P_{\text{thrust}}}{P_{\text{total}}} = 1 - \frac{1}{2}\frac{V - U}{V}, \tag{11}$$

based on Eqs. (6)–(9). Maximum efficiency is achieved when body wave speed is equal to the swimming speed. Clearly, fishes are faced with a tradeoff. At maximum efficiency, thrust is zero, and at high thrust, efficiency is low. Note that the efficiency cannot possibly be lower than 0.5, however, according to Eq. (11).

The reactive model described previously makes the assumption that undulatory amplitudes are small, so that the angles the body makes with the swimming direction are close to zero. When angles become larger, more energy is lost into the wake than Eq. (8) predicts. Lighthill (1971) developed a large-amplitude version of the theory, which corrects for this effect and also predicts lateral forces, which are easier to compare to experimental measurements, as we will discuss later.

Both large- and small-amplitude elongated body theories suffer from an important problem. They make the assumption that the Reynolds number is effectively infinite, meaning that inertial forces completely dominate viscous forces. The Reynolds number, however, is a notoriously fuzzy concept. While the Reynolds number for the whole body may be very large and effectively infinite, the Reynolds numbers for the tip of the tail and the edges of the fins are probably much smaller. In fact, you will almost always be able to find a structure small enough that its Reynolds number is too low to neglect viscosity. These viscous effects on small structures actually can be very important for forces on the overall body. First, they indicate that viscosity can never truly be completely neglected. Thus, resistive forces never become completely negligible. Second, and more importantly, viscous effects around sharp edges, like the trailing edge of an airfoil, can cause large changes in the flow around the body. For example, the lift force develops around an airfoil because of viscous effects around the sharp trailing edge (Faber, 1995). Wu (1971) developed a theory of fish swimming that incorporates an approximation of this effect.

Additionally, as we discussed previously, midline kinematics do not account for important differences in body shape. The virtual mass m_V at the tail accounts for the caudal fin size, but trailing edges of dorsal, anal, pectoral, and pelvic fins, all of which are independently activated, are not represented in the theory. Weihs (1972) accounted for other fins in his model of rapid turns, but three-dimensional computational approaches (e.g., Zhu *et al.*, 2002) show more promise for modeling the diversity of swimming behaviors. Nonetheless, elongated body theory has been used regularly in fish swimming studies, both directly to make predictions of force, and indirectly to inform analyses of parameters such as efficiency (e.g., Weihs, 1972; Webb, 1975; Hess and Videler, 1984; Videler and Hess, 1984; Daniel and Webb, 1987; Müller *et al.*, 2002; Tytell, 2004a; Tytell and Lauder, 2004).

IV. EXPERIMENTAL HYDRODYNAMICS OF UNDULATORY PROPULSION

Within the past 10 years, the ready availability of two new technologies has revolutionized our ability to study fluid flows generated by swimming fishes. Rather than use the theories described earlier to make hydrodynamic predictions based on kinematics, the combination of high-speed video and DPIV has allowed biologists to measure fluid flow directly. First, the new generation of high-speed video systems with resolution of at least 1024 by 1024 pixels, which can operate (at this resolution) at rates above 500 frames per second (fps), permits a whole new level of detail and resolution for imaging swimming fishes. Data obtained from such systems, particularly when cameras are synchronized for two or more simultaneous views of the swimming fish, permit highly accurate quantification of body and fin movements.

The true utility of such systems becomes evident when they are used with high-power continuous-wave lasers for DPIV, an engineering technique to quantify fluid flow by imaging small reflective particles in the flow (Willert and Gharib, 1991; Raffel *et al.*, 1998). Many engineering DPIV studies use pulsed lasers, which typically operate with a maximal temporal resolution of 15 Hz, and often much less, which is sufficient when steady flows are being studied but insufficient to understand the dynamic unsteady character of flows generated by swimming fishes. In contrast, recent biological DPIV work with high-speed video in combination with continuous lasers allows images of fluid flow patterns to be obtained at the full framing rate of the video (500 fps or greater), and can thus provide a detailed picture of flow separation and vortex formation by fins.

The experimental approach used to understand water flow during undulatory locomotion by freely swimming fishes is illustrated in Figure 11.5 (Anderson *et al.*, 2000; Drucker and Lauder, 2001a; Lauder and Drucker, 2002; Lauder and Drucker, 2004; Tytell and Lauder, 2004; Wilga and Lauder, 2004). Fishes swim in a recirculating flow tank and are induced to maintain station with their body or fins projecting into the laser light sheet, generated by a continuous argon ion laser and a series of focusing lenses. Small (12 μm mean diameter) near-neutrally buoyant reflective particles are mixed into the flow to provide reflections from the laser light sheet, which are recorded by a high-speed video camera. Images of fish position in the flow and relative to the light sheet are recorded by one or more additional synchronized video cameras. By reorienting laser optical components, it is possible to produce light sheets in different orthogonal orientations (Figure 11.5B and C) and hence obtain, in separate experiments, a three-dimensional impression of flow patterns. Images obtained from these high-speed videos (Figure 11.5D) are analyzed using standard DPIV cross-correlation techniques (Willert and Gharib, 1991; Raffel *et al.*, 1998) to yield a matrix of velocity vectors that estimate flow in the light sheet (Figure 11.5E).

Standard DPIV using a single video camera orthogonal to the light sheet produces only the two velocity components in the plane illuminated by the laser. It is also possible to use two cameras, offset by a known angle, to record stereo images and hence reconstruct the three components (x, y, and z) of flow velocity in a plane (e.g., Willert, 1997; Veerman and Den Boer, 2000; Westerweel and Oord, 2000). Stereo-DPIV data obtained for steady undulatory locomotion in trout (*Oncorhynchus*) were presented by Nauen and Lauder (2002b), who provided a simple three-dimensional analysis of the wake and calculated locomotor efficiency (see also Sakakibara *et al.*, 2004).

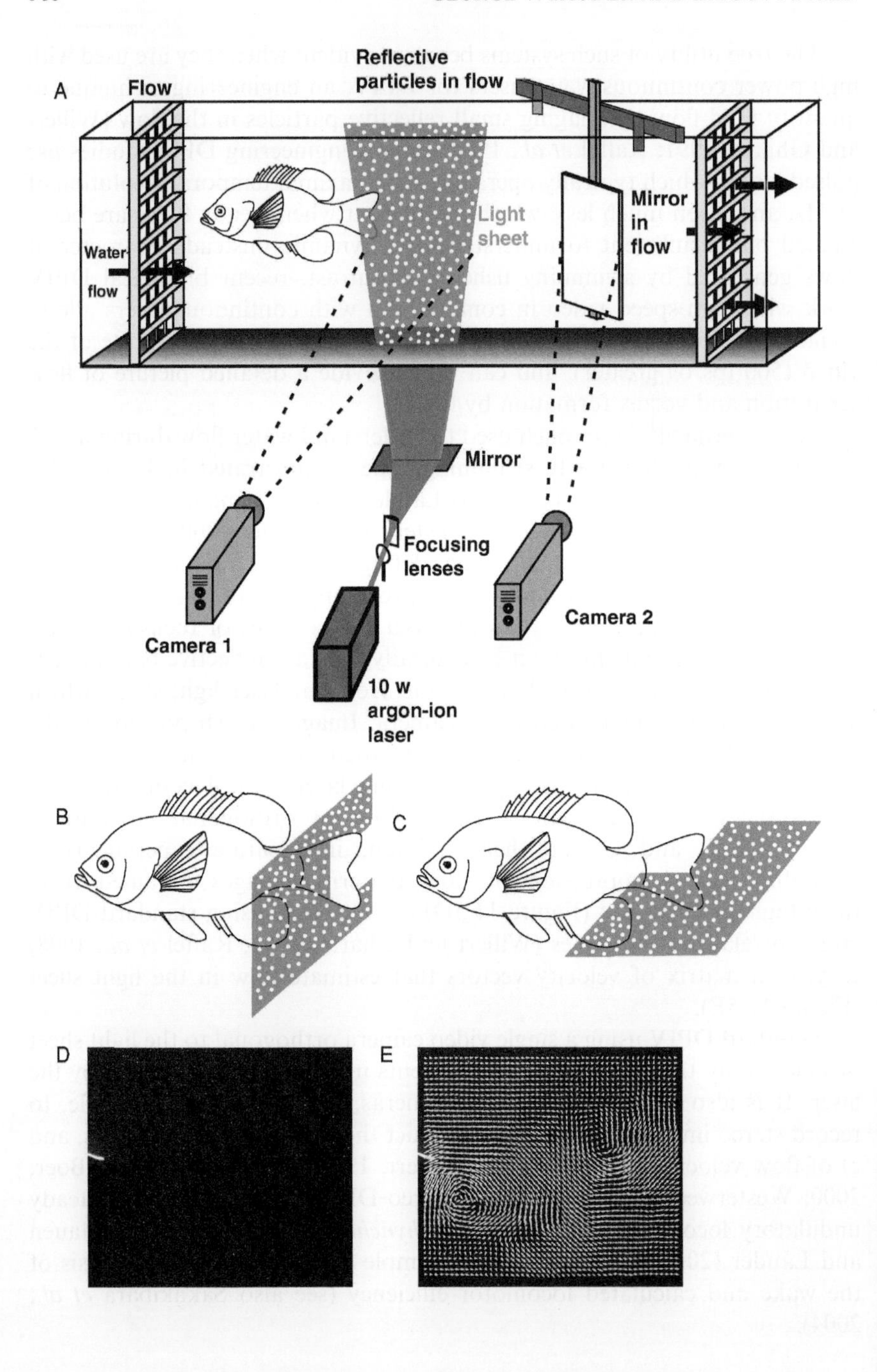
A
Flow
Reflective
particles in flow
Water
flow
Light
sheet
Mirror
in
flow
Mirror
Focusing
lenses
Camera 1
Camera 2
10 w
argon-ion
laser
B
C
D
E

These velocity vectors can then be used to estimate forces by analyzing the vorticity, a measure of the local angular velocity of a fluid element (Faber, 1995). Specifically, the vortex structure in the wake of a fish indicates the magnitude, angle, and, to some extent, timing of the forces the fish applied to the fluid. One can estimate force on the basis of this structure in two ways, both of which actually produce estimates of impulse, I, the average force multiplied by the time over which it was applied. By dividing the impulse by an appropriate time interval (typically half the tail-beat period), one can estimate average force. In the first, most commonly used method, researchers assume that counter-rotating vortices in the wake are connected above and below the plane to form a small-core vortex ring. By using multiple planes to measure the geometry of the ring, impulse can be estimated as

$$I = \rho A \Gamma, \tag{12}$$

where ρ is water density, A is the area circumscribed by the vortex ring, and Γ is the circulation of the vortex cores (a measure of total vorticity in a region) (Batchelor, 1973). The second, and more rigorous, method applies the equation used to derive (12):

$$\mathbf{I} = \frac{1}{2}\rho \int \mathbf{x} \times \boldsymbol{\omega}\, dV, \tag{13}$$

where $\mathbf{I}$ is now a vector impulse, $\mathbf{x}$ is the position vector connecting a fluid element to the place the force was applied, and $\boldsymbol{\omega}$ is the vorticity vector (Batchelor, 1973). This method requires accurate estimation of vorticity, which is often difficult using DPIV, but potentially produces more accurate impulse estimates than Eq. (12) in wakes with complex vortex structure (Rosen *et al.*, 2004; Tytell, 2004a). Finally, Noca *et al.* (1999) derived a method to estimate true instantaneous force production, rather than average

Fig. 11.5. Experimental arrangement used for hydrodynamic analysis of water flow patterns over the body and in the wake of freely swimming fishes. Laser light sheet is shown in blue. (A) Fishes swim in a recirculating flow tank, maintaining station against oncoming flow. Optical components focus the laser beam into a thin (1–2 mm thick) light sheet, and fishes are positioned so that the body and/or tail intersect this light sheet. Small (12 μm mean diameter) reflective, nearly neutrally buoyant particles circulate with the flow. One high-speed video camera films reflections from these particles while a second camera monitors the position of the fish, either through a mirror or from above or below. The laser light sheet may be reoriented into two other orthogonal orientations (B, transverse plane; C, horizontal plane) in separate experiments to quantify flow in these planes. (D) Video images are analyzed using cross-correlation (Willert and Gharib, 1991) to yield matrices of velocity vectors (E, yellow arrows) estimating flow velocities through time.

forces from impulse estimates, but, to our knowledge, this approach has not been applied successfully in any studies of aquatic animal locomotion.

The experimental approach using flow tanks, high-speed video, and DPIV described previously has two significant advantages. First, the swimming speed and position of the fish can be accurately controlled by adjusting speed in the flow tank, and thus true steady swimming can be studied if care is taken during data acquisition. Controlled maneuvers can also be elicited from a known steady swimming posture (e.g., Wilga and Lauder, 1999; Drucker and Lauder, 2001b; Tytell, 2004b). Flow patterns can change considerably with speed or if fishes are accelerating (Drucker and Lauder, 2000; Tytell, 2004b). Second, use of multiple simultaneous cameras allows one or more cameras to record fish position in the laser light sheet (Figure 11.5A). It is difficult to overemphasize the importance of *both* controlling fish swimming speed *and* knowing where the fish is relative to the light sheet. Because of the complex three-dimensional body shape of most fishes, data obtained without knowing the position of the fins or body relative to the light sheet are extremely difficult to interpret. Other studies (Müller *et al.*, 1997, 2001) have applied DPIV to fish swimming in still water, which has the advantage that turbulence is lower, in principle, but recording steady behaviors is much more difficult.

A. Axial Propulsion

Flow fields measured in the wake of swimming eels (*Anguilla*) show that oscillation of the body produces a wake with laterally oriented momentum jets and negligible downstream flow (momentum added opposite the direction of movement) (Figure 11.6; Müller *et al.*, 2001; Tytell and Lauder, 2004). Wake momentum is thus directed to the side and not in the streamwise direction along the axis of travel; each pulse of lateral momentum probably represents jet flow through the center of unlinked vortex rings. These lateral momentum jets are produced almost entirely by the final third of the body (Figure 11.7) and are not convected along the body via undulatory motion (Tytell and Lauder, 2004). Previously, structures termed "proto-vortices" had been observed along the body (Müller *et al.*, 2001). While these structures do exist (Figure 11.7), they do not contain much vorticity and are therefore not responsible for the vortex wake structure (Tytell and Lauder, 2004). As can be seen in Figure 11.7, there is very little vorticity upstream along the body. Instead, the posterior 15% of the body produces lateral jets of fluid, which results in strong (up to 90 s^{-1}) but unstable shear layers in the wake. The shear layer is visible between two regions of flow traveling in opposite directions, as between the first and second jets in Figure 11.6. The final wake forms through rollup of this shear

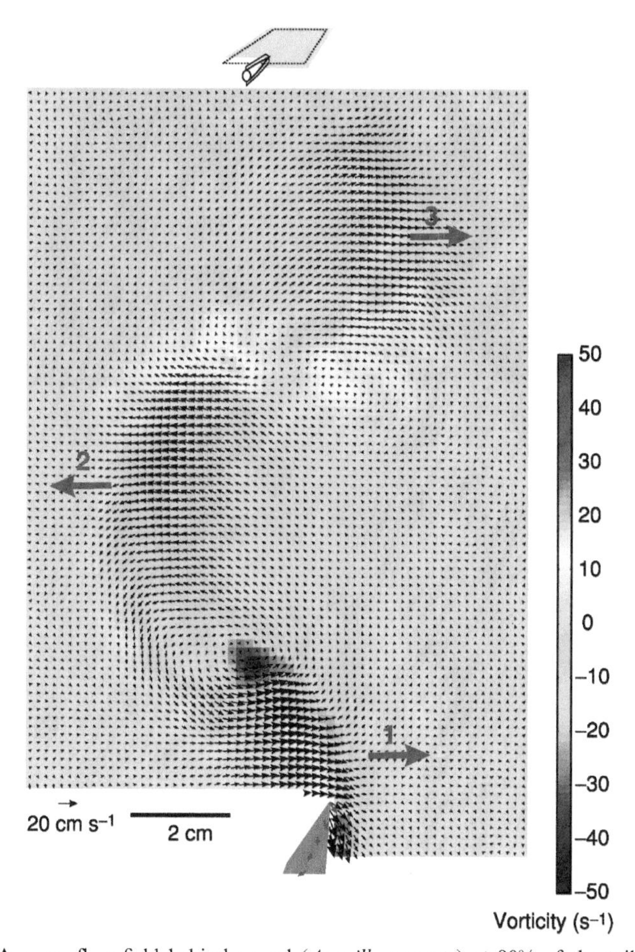

Fig. 11.6. Average flow field behind an eel (*Anguilla rostrata*) at 90% of the tail-beat cycle, recorded in the horizontal plane as shown schematically above and in Figure 11.5C. The vector field shown is a phase average of 14 tail beats. Vorticity is shown in color in the background, and velocity vectors, calculated from high-speed video images of the wake, are shown in black. The eel's tail is in blue at the bottom, with its motion indicated by small red arrows, scaled in the same way as the flow vectors. Freestream flow is from bottom to top and has been subtracted to show the vortex structure in the wake. Vector heads are retained on vectors shorter than 2.5 cm s^{-1} to shown the direction of the flow; otherwise vector head size is scaled with velocity magnitude. Numbered large red arrows show the three regions of lateral momentum in the wake shed by the tail. (Modified from Tytell and Lauder, 2004.)

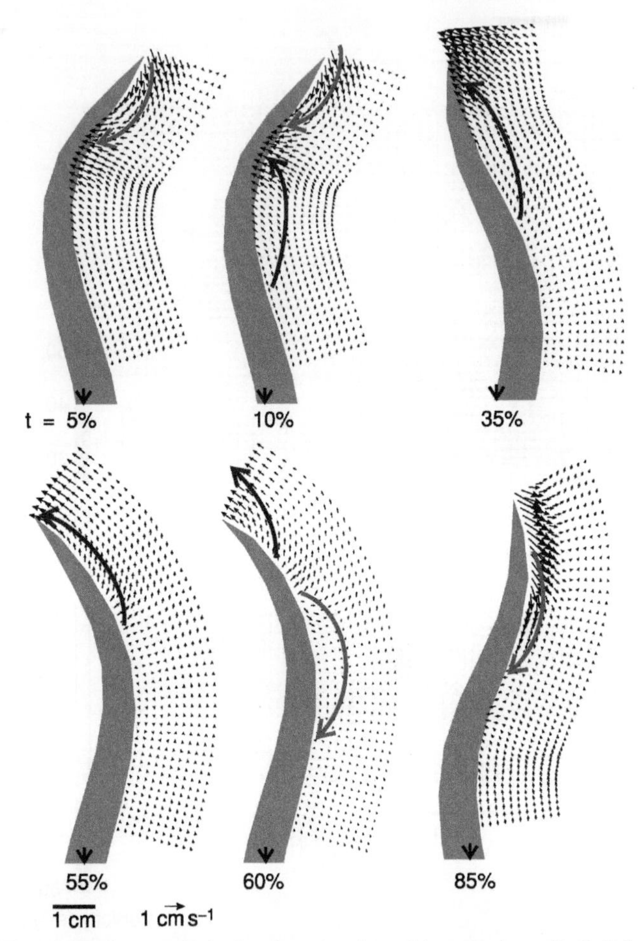

Fig. 11.7. Flow fields close to the body of a swimming eel (gray body outline). Velocity vectors are shown in black for six times during the tail-beat cycle. The lateral position of the eel's snout (off the view below) is shown as a black arrow. Velocities are phase averaged across 14 tail beats by interpolating the normal gridded coordinate system on to a system defined by the distance from the eel's body and the distance along the body from the head. Red and blue arrows indicate the major clockwise and counterclockwise flow directions, respectively, near the body. (Modified from Tytell and Lauder, 2004.)

layer (Figure 11.8), which separates into two or more vortices that are fully developed approximately one tail-beat cycle later. This vortex shedding pattern matches well the pattern of body kinematics (Figure 11.1) with relatively little oscillation of the front half of the body.

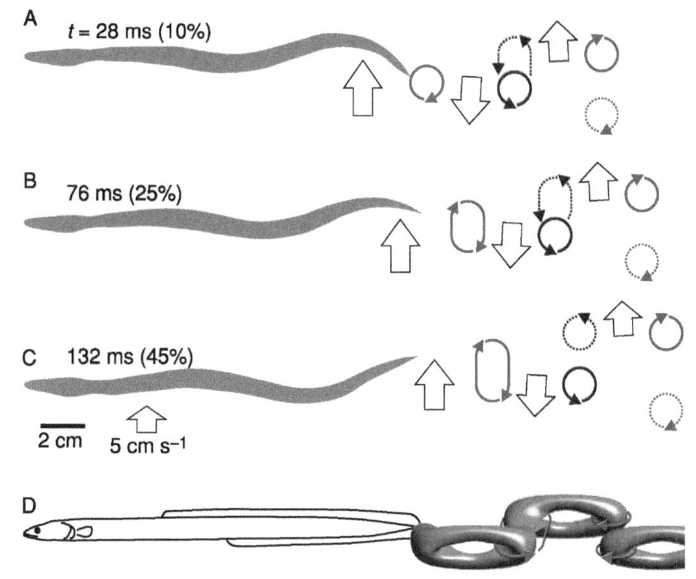

Fig. 11.8. (A–C) Schematic diagram of wake formation behind a steadily swimming eel, *Anguilla*. Vortices shed by the tail are shown at three different times in the locomotor cycle as red and blue arrows; primary vortices are solid, while secondary vortices formed by roll-up of the shear layer shed by the tail are shown as dotted lines. Lateral momentum jets to each side are indicated by block arrows. (Modified from Tytell and Lauder, 2004.) (D) Hypothesized three-dimensional flow, based on (A–C). Lateral vortex rings (shown in blue) are partially linked at the shear layer, but separate completely in the far wake.

As eels increase their swimming speed over a range from 0.5 to 2.0 Ls^{-1}, the fundamental character of the wake does not change (Tytell, 2004a). As long as swimming is steady, there are large lateral momentum jets and little streamwise momentum added to the wake.

In contrast to the wake structure of eels, the wake of fishes such as trout, mackerel, or sunfish shows considerable streamwise momentum (Lauder *et al.*, 2002, 2003; Nauen and Lauder, 2002a,b). Figure 11.9 shows a horizontal slice through the wake of a swimming trout, *Oncorhynchus*. Two distinct centers of vorticity are present in this mid-body wake section, as is a substantial central momentum jet with a strong downstream component. A vertical slice through the wake flow (see Figure 11.5A) of mackerel reveals distinct tip vortices and provides another view of the downstream component of the tail momentum jet (Figure 11.10A). Taken together, the horizontal and vertical wake slices suggest that mackerel, trout, and bluegill produce a linked vortex ring structure in their wakes (Figure 11.10B). Also, note that

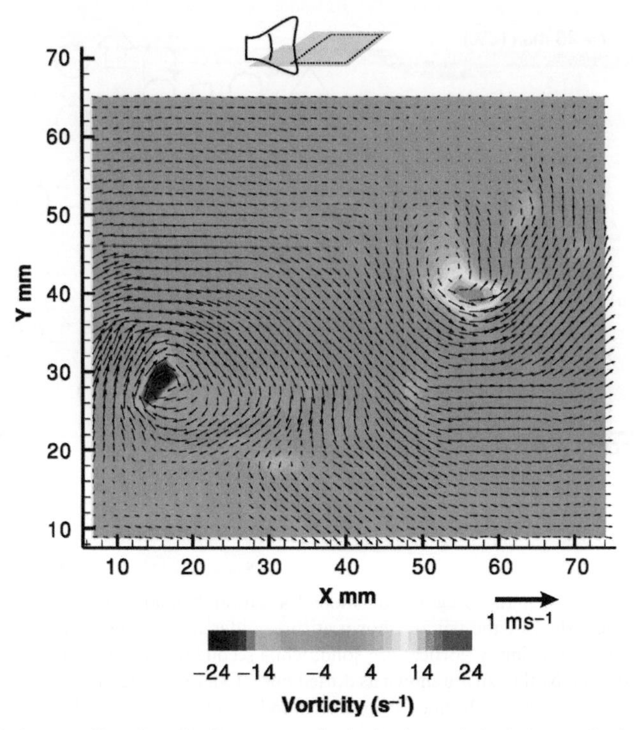

Fig. 11.9. "Subcarangiform" wake flow pattern in the horizontal plane from a freely swimming trout (*Oncorhynchus mykiss*) swimming steadily at 1.0 Ls^{-1}. Light sheet orientation is shown schematically above. Freestream flow is from left to right and has been subtracted to reveal two vortex centers produced by the caudal fin. These two vortex centers represent a planar section through a vortex ring with a high velocity central momentum jet directed downstream and laterally. (Modified from Lauder *et al.*, 2002.)

although downstream momentum is present in fishes as diverse as bluegill sunfish and mackerel, lateral forces are almost always at least equal to streamwise forces and are often as much as two to three times larger (Figure 11.10B) (Lauder and Drucker, 2002; Nauen and Lauder, 2002a).

While this wake structure is different from that of eels, the vorticity pattern in a horizontal section along the body of fishes with distinct tails is similar to that observed in eels (compare Figure 11.11 to Figure 11.7). In both cases, virtually all wake vorticity is developed along the posterior third of the body.

Anderson *et al.* (2000) performed a much more detailed analysis of flow close to the body than is shown in Figures 11.7 and 11.11. By examining the boundary layer, a region close to the body where fluid velocities change from the free stream flow velocity to the body velocity, they estimated skin friction

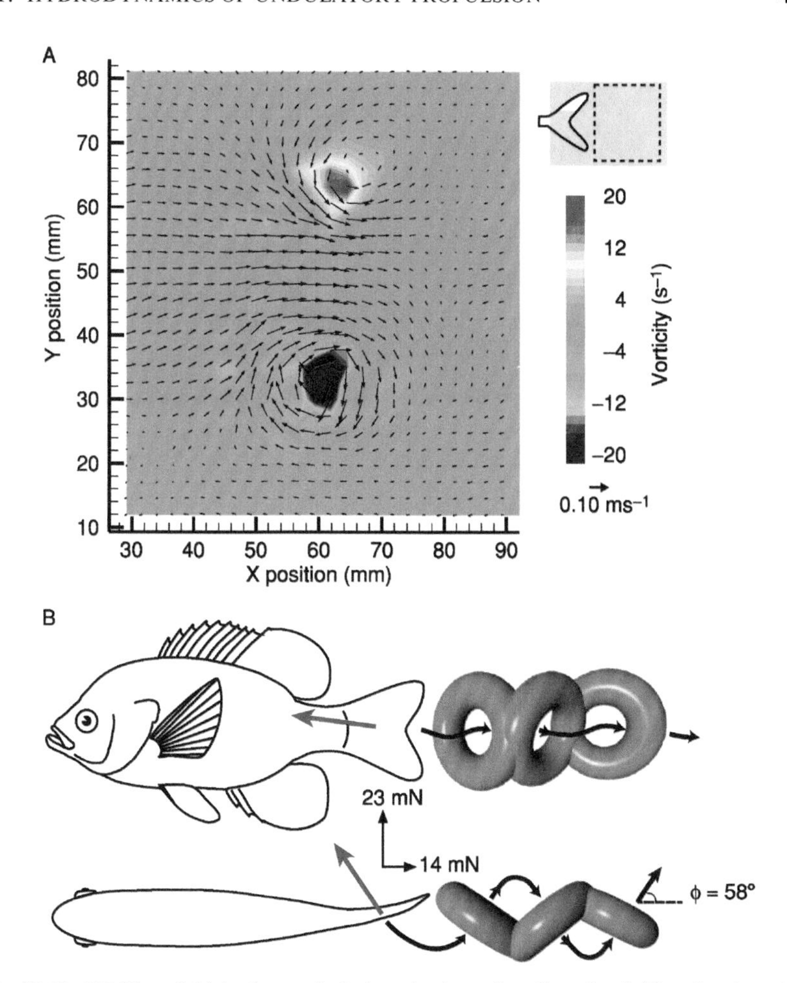

Fig. 11.10. (A) Flow field in the vertical plane in the wake of mackerel (*Scomber japonicus*) swimming steadily at 1.2 Ls^{-1}. Two centers of vorticity are clearly present that were shed by the dorsal and ventral tips of the tail, and the central momentum jet through the tail vortex ring is visible as accelerated flow between the vortex centers. Light sheet orientation is shown schematically above. (Modified from Nauen and Lauder, 2002a.) (B) Schematic model of vortex rings (shown in blue) in the wake of fishes with discrete caudal fins, showing both lateral and dorsal views. During steady locomotion at 1.5 Ls^{-1} the tail sheds linked vortex rings with a central jet of high momentum flow, shown as thick black arrows weaving through the vortex ring centers. For bluegill sunfish (*Lepomis macrochirus*) swimming at this speed, the high-velocity jet flow makes an average angle of 58° with the axis of progression, and the tail generates a mean thrust force of 14 mN, with a mean side force of 23 mN. The reaction force on the tail is shown as a red arrow. (From Lauder and Drucker, 2002.)

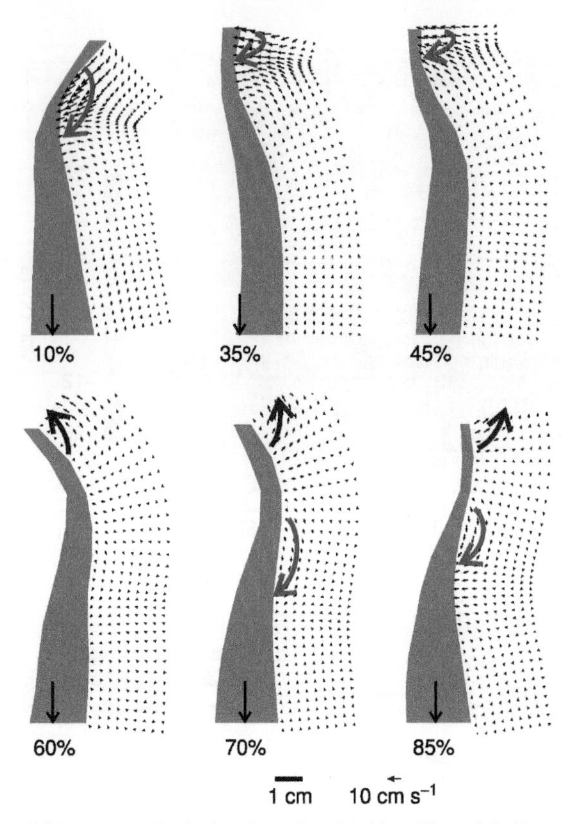

Fig. 11.11. Flow fields close to the body of a swimming bluegill sunfish (*Lepomis macrochirus*, gray body outline) swimming in a "subcarangiform" mode at 1.65 Ls^{-1}. Velocity vectors are shown in black for six times during the tail beat cycle. The lateral position of the sunfish's snout (off the view below) is shown as a large black arrow. Red and blue arrows indicate the major clockwise and counterclockwise flow directions, respectively, near the body.

drag on the bodies of freely swimming scup (*Stenotomus chrysops*) and smooth dogfish (*Mustelus canis*). The magnitude of the drag force on a swimming body has been long debated, with some suggesting that fishes can reduce their drag below that of a stretched straight body (Gray, 1936; Barrett *et al.*, 1999) and others suggesting it should be higher (Lighthill, 1971; Webb, 1975). The Anderson *et al.* (2000) boundary layer measurements clearly indicate that undulatory motion increases drag on both carangiform and anguilliform swimmers.

The differences between the wakes from these two types of swimmers may result less from differences in their midline kinematics (Figure 11.1), and more from differences in the body shapes between eels and other fishes, particularly differences in their tails (see Figure 11.2). Midline kinematics between these fishes are not identical, but the differences are small, whereas their body shapes are very different. The wake differences indicate an important physical difference, as well. When fishes swim steadily, the total drag force must balance thrust over a tail-beat cycle, leading Schultz and Webb (2002) to suggest that there should be no thrust signature in the wake. Indeed, this is what is observed for eel locomotion (Figure 11.6) but not for data from fish such as trout, bluegill, and mackerel that have discrete caudal fin propulsors (Figures 11.9 and 11.10). Tytell and Lauder (2004) and Fish and Lauder (2005) suggested that the tails of such fishes function as a discrete propeller, separate anatomically and functionally from the major areas incurring drag during undulatory locomotion. In a boat with an external motor, the hull of the boat incurs the majority of drag, while the propeller is a separate thrust-generating structure. Sections of a boat wake behind a propeller show a clear thrust signature, just as the wake behind fishes with discrete caudal fins demonstrates thrust. By analogy, the caudal fin of fishes functions as a propeller, with its own intrinsic musculature powering movement. It is not surprising, then, that a clear thrust signature is evident in both experimental wake flow visualizations and in three-dimensional computational models that include a distinct tail propulsor (e.g., Zhu *et al.*, 2002).

B. Function of the Tail

Many fishes have exquisite control over the motion of their tail, allowing them to vary the relative magnitudes of thrust, lateral forces, and lift forces. Even fishes with externally symmetrical (homocercal) tail structures can control the shape and motion of the dorsal and ventral lobes, which usually results in asymmetrical motion and influences the direction of wake flow (Lauder, 1989, 2000; Lauder *et al.*, 2003). Figure 11.12 shows the shape of the caudal fin during one fin beat by a steadily swimming bluegill sunfish: the dorsal lobe of the tail may undergo greater lateral excursion than the ventral lobe, and the tail expands and contracts vertically during the tail beat. Intrinsic tail muscles actively produce these changes in fin conformation (Lauder, 1982, 1989).

Wilga and Lauder's (2002, 2004) studies on sharks indicate some possible hydrodynamic effects from this asymmetrical motion. They performed DPIV in the wake of sharks, which also move their tails asymmetrically. In this case, the inclined trailing edge of the morphologically asymmetrical (heterocercal) caudal fin results in a ring-within-a-ring vortex structure in the wake (compare Figure 11.13 to Figure 11.10B). It is currently unknown

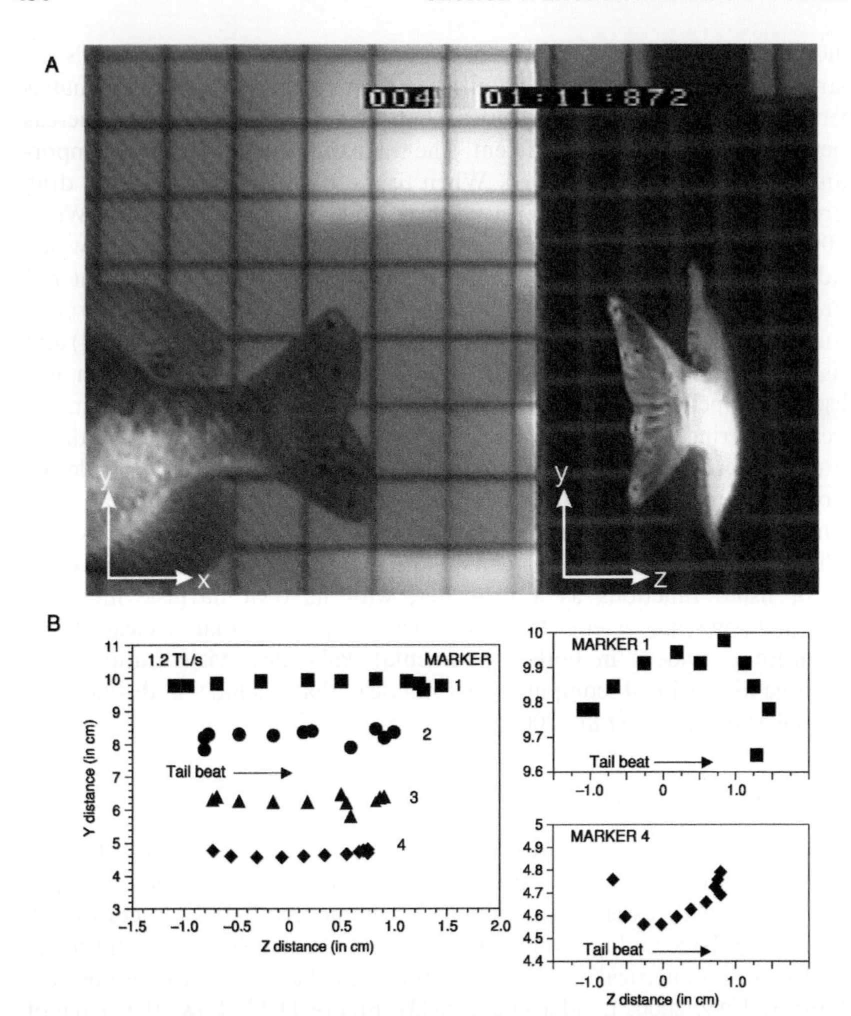

Fig. 11.12. Tail movement in fishes with a homocercal (externally symmetrical) tail need not be symmetrical. (A) Two synchronized video frames showing tail position in bluegill sunfish (*Lepomis macrochirus*) during steady rectilinear locomotion at 1.5 Ls^{-1}. The left panel shows the lateral (xy) view, while the right panel shows a posterior (yz) view through a mirror located in the flow, as illustrated in Figure 11.5. The tail trailing edge has been marked on each side with four small black markers to facilitate digitizing trailing edge angles and excursions. (B) Plot of yz excursions of four tail markers. Tail motion is from left to right. Note that the dorsal tail lobe can undergo considerably greater excursions than the ventral lobe. The two panels to the right show expanded views of dorsal (marker 1) and ventral (marker 4) excursions to demonstrate how the fish expands its tail during the beat. (Modified from Lauder, 2000.)

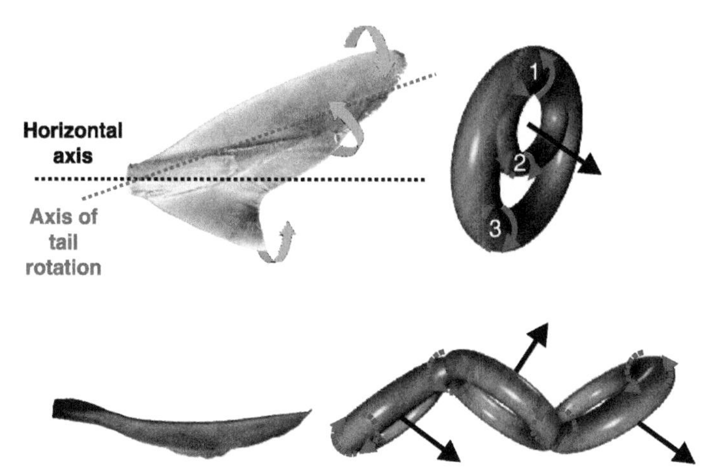

Fig. 11.13. Schematic ring-within-a-ring vortex ring structure in lateral and dorsal view shed by the heterocercal (externally asymmetrical) tail of a swimming spiny dogfish shark (*Squalus acanthias*). Three centers of vorticity are seen in the wake (red curved arrows, numbered 1 to 3 from dorsal to ventral). The central momentum jet through the upper ring is shown as black arrows. Curved green arrows show fluid separating around the tail tips and inclined trailing edge. (Modified from Wilga and Lauder, 2004.)

whether the asymmetric motion described above in fishes such as sunfish that have homocercal tails affects the flow in a similar way; the conformational changes these fishes use may also introduce additional hydrodynamic complexity.

Thus, the tail should not be viewed as a simple extension of the body, a view encouraged by kinematic and hydrodynamic models of fish propulsion based on horizontal mid-body sections (Figures 11.1, 11.2, and 11.4). Rather, the fish tail functions as a independent propulsive surface with distinct three-dimensional anatomy, shedding vortices and used to adjust overall body trim (Liao and Lauder, 2000; Wilga and Lauder, 2002).

Finally, it should be emphasized that flow passing over the tail is affected by the body upstream of it. In mackerel, for example, flow over the body converges toward the tail (Figure 11.14A and B), producing a complex flow around the caudal peduncle. Quantification of particle motion in this region suggests that overall flow results from a combination of two patterns: converging flow along the body, and flow wrapping around the peduncle as it moves from side to side (Figure 11.14C and D) (Nauen and Lauder, 2001). The tail thus does not see freestream flow, but rather encounters flow altered by the body and caudal peduncle. Further alteration of incident tail flow occurs by the dorsal and anal fins, as is discussed in the next section.

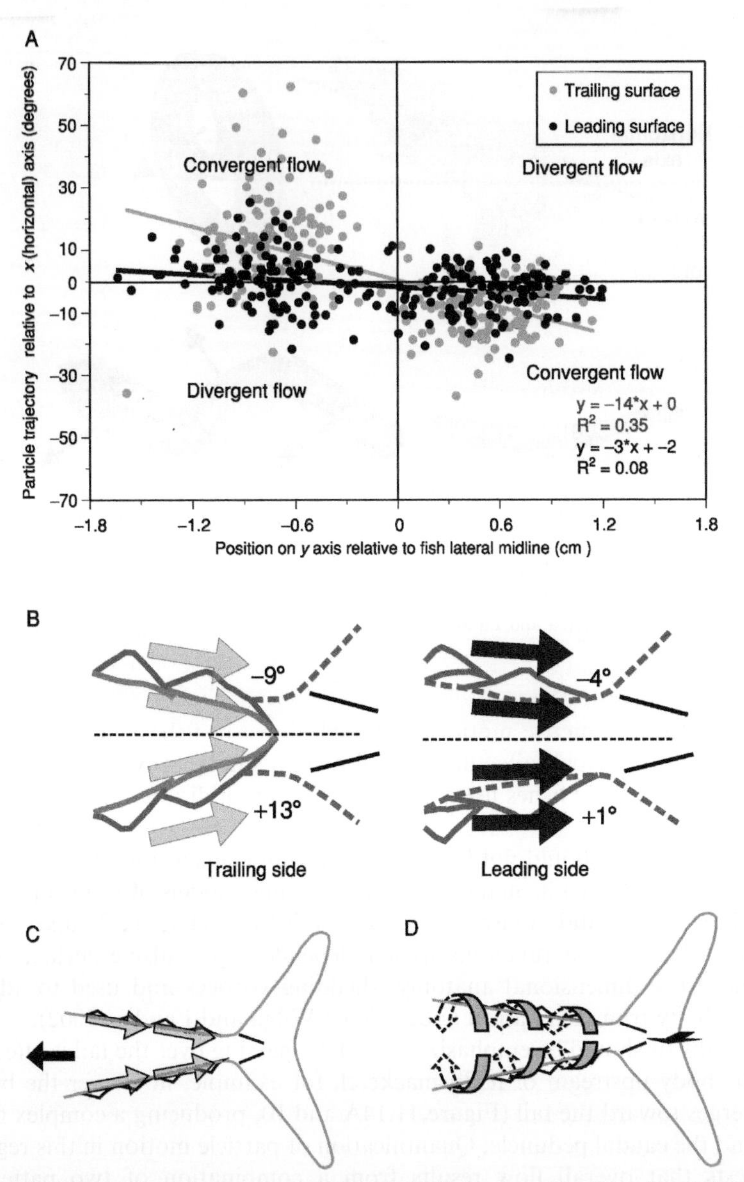

Fig. 11.14. Flow patterns in the region of the caudal peduncle and leading edge of the tail in mackerel (*Scomber japonicus*) swimming steadily at 1.2 Ls^{-1}, measured by manually tracking particle motion. Each symbol represents the position of an individual particle in the flow field. (A) Particle trajectories relative to the horizontal for flow on the leading peduncular surface (black dots) and trailing surface (gray dots). Linear regressions for each flow type show negative slopes, indicating that flow is converging toward the midline of the tail. (B) Schematic summary

C. Dorsal and Anal Fin Function

The dorsal and anal fins of fishes are important fluid control structures with their own intrinsic musculature (Gosline, 1971; Geerlink and Videler, 1974; Winterbottom, 1974; Jayne *et al.*, 1996), ability to exert force on the fluid environment (Arreola and Westneat, 1997; Hove *et al.*, 2001; Standen and Lauder, 2005), and discrete sharp trailing edges at which flow separation may occur (Drucker and Lauder, 2001a; Lauder and Drucker, 2004). However, these fins have not received much attention in the hydrodynamic literature on undulatory propulsion, although some computational models do include passive median fins (Wolfgang *et al.*, 1999; Zhu *et al.*, 2002). Webb (1978) and Webb and Keyes (1981) also discussed vortex sheets that could be shed from median fin trailing edges. In part this lack of attention may be due to the difficulty of studying fin function experimentally and to the primary focus of many studies on undulation of the body midline itself.

Key evidence that dorsal and anal fins play an important role in the overall force balance during locomotion is provided by DPIV analyses of their wakes (Figure 11.15). By recording dorsal fin wake flow patterns in the horizontal plane (Figure 11.16A, planes 1, 2), it is clear that streamwise momentum is added to the wake and that substantial side forces are produced by oscillating dorsal fins (Lauder and Drucker, 2004). In bluegill sunfish, the soft dorsal fin contributes nearly 12% of overall locomotor thrust force at a swimming speed of 1.1 Ls^{-1}, and dorsal fin lateral force is nearly twice thrust force (Drucker and Lauder, 2001a). In trout, lateral forces are even higher relative to thrust force (greater than three times), because the dorsal fin wake oscillates laterally, adding little downstream momentum (Figure 11.15B). Dorsal fins in bluegill and trout thus actively generate considerable side force, which may contribute to body stability during locomotion.

The dorsal and ventral lobes of the caudal fin pass through the wake shed by dorsal and anal fins. The hydrodynamic environment experienced by the tail is thus substantially different than the freestream flow that might be assumed in the absence of other median fins. In bluegill, dorsal fin wake vortices encounter the tail and accelerate flow over the tail surface (note the velocity vectors around the caudal fin in Figure 11.15A) (Drucker and

---

of mean particle tracks in (A) (large grey and black arrows representing trailing and leading flows) giving the mean angle of convergent flow on both sides of the tail. (C and D) Schematic hypothesis of the two peduncular flow patterns that combine to produce the pattern seen in panels (A) and (B): observed flow is the result of convergence along the body surface as it slopes toward the tail midline (C), and divergence results from side-to-side motion in which flow encircles the peduncle (D). (Modified from Nauen and Lauder, 2001.)

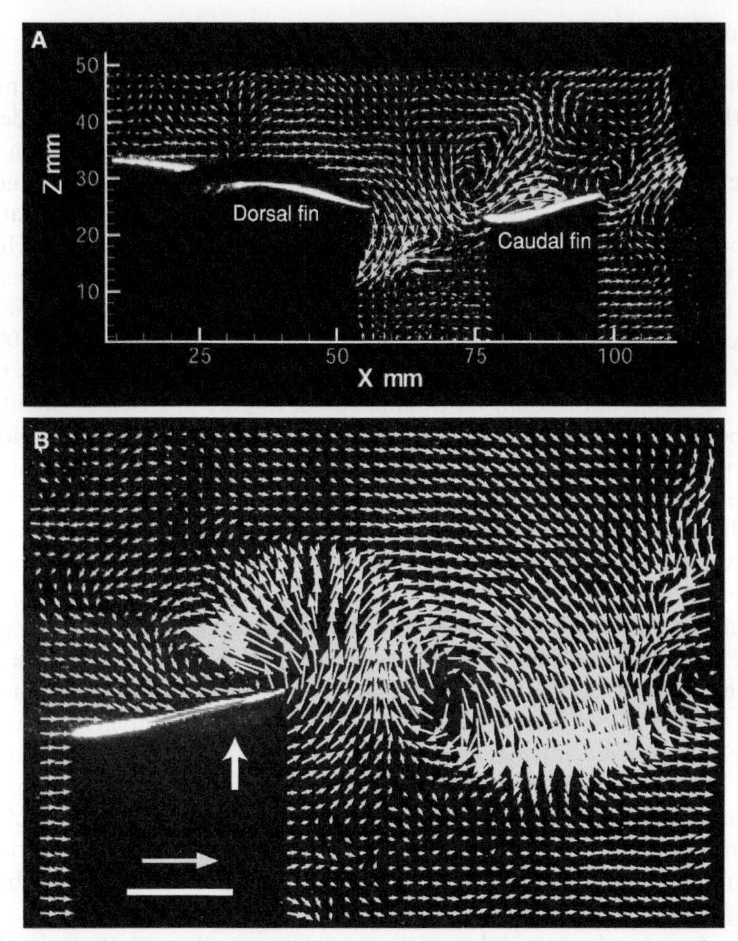

Fig. 11.15. The dorsal fin of bony fishes is an active component of locomotor design and sheds a discrete wake that alters incident flow at the tail. Both panels show flow as viewed from above (*xz* plane). (A) Dorsal fin wake in bluegill sunfish, *Lepomis macrochirus*. The dorsal fin has accelerated flow in the gap between the dorsal fin and tail, and has shed a vortex located just above the tail in this image. (B) The dorsal fin wake in rainbow trout (*Oncorhynchus mykiss*) sheds a linear array of vortex centers with strong lateral jets to each side. The tail passes through these jets and through the centers of the dorsal fin vortices. White arrow pointing up in panel (B) shows the direction of trout dorsal fin motion. Scales: arrow = 10 cm/s; bar = 1 cm. Mean freestream flow has been subtracted from each image. (Modified from Drucker and Lauder, 2001a; Lauder and Drucker, 2004.)

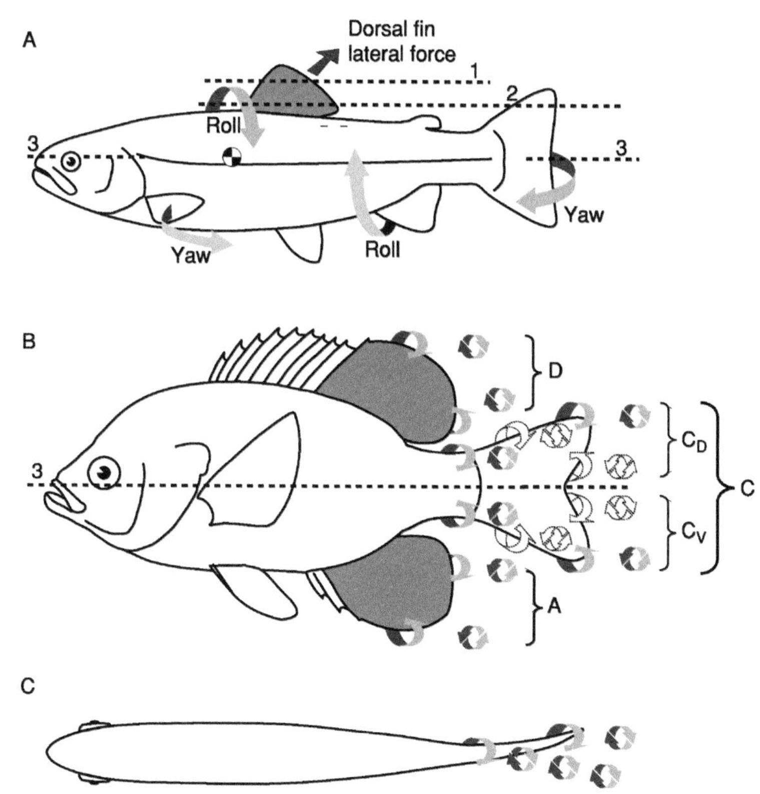

Fig. 11.16. (A) Schematic summary of the overall force balance on freely swimming fishes based on recent particle image velocimetry data. The dorsal fin generates lateral forces (see Figure 11.15) with accompanying yaw and roll torques around the center of mass. These torques must be countered by the action of the caudal, pectoral, and anal fins. The three numbered lines show alternative horizontal slices through the fish at which the wake shed by the body and fins has been analyzed: 1, wake of the dorsal fin alone; 2, wake of the dorsal fin and tail; 3, midline flow of the body and tail (Drucker and Lauder, 2001a; Lauder *et al.*, 2002; Lauder and Drucker, 2004; Tytell, 2004a; Tytell and Lauder, 2004). (B) Hypothesis of streamwise vortical separation from the caudal, dorsal, and anal fins and the caudal peduncle in lateral view. Vortices are shed from the dorsal (D), anal (A), and caudal (C) fins. The dorsal and ventral lobes of the caudal fin may shed discrete vortices also (C_D and C_V). (C) Horizontal section through the middle of the fish at the level of line 3 in (B). Because of undulatory motion, flow rolling up around the caudal peduncle may not be destroyed by the tail, but may instead result in discrete vortices in the wake. Only the upper vortices shed by the caudal peduncle and tail are shown. [Panel (A) modified from Lauder and Drucker (2004).]

Lauder, 2001a). This suggests that dorsal fin vortices might increase tail thrust if caudal fin surface pressures are reduced below what they would be without the dorsal fin. Computational fluid dynamic modeling of dorsal-caudal fin wake interactions corroborates this view, and indicates that vorticity shed from the dorsal fin promotes more rapid growth of the leading edge vortex on the caudal fin, enhancing thrust (Mittal, 2004).

Some fishes such as tuna swim routinely with the dorsal fin retracted (except during maneuvering). In such cases, the tail encounters flow altered only by the body, and roll stability must be achieved by adjusting the position of pectoral fins, which are held in a partially extended position during most routine locomotion (Magnuson, 1978).

D. Overall Force Balance and Three-Dimensional Flow

The large lateral momentum produced by the dorsal fin in ray-finned fishes has important consequences for the overall force balance during locomotion (Figure 11.16A). This lateral force is generated above and posterior to the center of mass and must thus generate both roll and yaw torques. In order for fishes to swim steadily without roll and yaw, other fins must compensate. To counter roll torques induced by the dorsal fin, the anal fin could generate an opposing torque by generating lateral momentum to the same side as the dorsal fin (Figure 11.16A). In addition, the tail and pectoral fins may act in concert to correct any yaw torques.

Overall, the picture that emerges from experimental studies of median fin function is that a complex force balance exists among all fins and the body and tail, even during steady rectilinear locomotion. Both median and paired fins act in concert with the tail to stabilize the body. In light of these experimental data, the classical categorization of locomotor modes as BCF (body-caudal fin) as distinct from MPF (median-paired fin) seems extremely simplistic, if not directly misleading, as median fins clearly are hydrodynamically active during body and caudal fin locomotion, and may in fact play a crucial role in body stability.

Understanding the overall force balance on swimming fishes requires data on buoyancy, body shape, body position, and the direction of forces produced by the fins and body. Despite these complications, progress has been made by quantifying body angles and flow over the body, and by experimental investigation of individual fin forces. Many fishes swim with a positive angle of attack to the body (He and Wardle, 1986; Webb, 1993a; Wilga and Lauder, 1999; Liao and Lauder, 2000; Wilga and Lauder, 2000; Nauen and Lauder, 2002a; Svendsen *et al.*, 2005). This appears to be critical to the overall force balance because body angle changes as swimming speed increases, altering the lift force due to flow over the body.

An additional complication for understanding the overall force balance on swimming fishes is the three-dimensional nature of flow around the body and tail. Figure 11.16B illustrates one possible pattern of flow separation from the trailing edge of the dorsal, anal, and caudal fins, as well as the caudal peduncle in ray-finned fishes. The homocercal tail itself could generate additional small vortices due to rollup at the inclined edges near the midline, by analogy with flow around the heterocercal shark tail (Figure 11.13) (Wilga and Lauder, 2004), although these have yet to be seen experimentally. As seen in a posterior (*yz*) view, then, the vortex wake could consist of as many as 10 vortex filaments: two pairs shed by the tail, one pair by the caudal peduncle due to flow separation at the upper and lower edges, and one pair each by the dorsal and anal fins. How these link up to form discrete vortex rings (Figure 11.10B) is not currently known. Even if the tail moves so that only a single large vortex ring is shed by the caudal fin, additional vortex rings from the dorsal and anal fin may still exist in the wake, producing complex interactions among them. These speculations remain to be examined experimentally, because the structure of streamwise vorticity near the tail and caudal peduncle has not yet been observed. But the importance of three-dimensional effects for understanding fish undulatory hydrodynamics is clear.

E. Undulatory Locomotion in Turbulence

In nature, fishes rarely swim in controlled flows like those produced in laboratory flow tanks. Indeed, one might argue that the vast majority of the locomotor time budget for fishes is spent in turbulent flows. From rapidly flowing streams to strong ocean currents passing over an uneven benthic habitat, most fish habitat is turbulent. While it is certainly understandable that the vast majority of research has used controlled microturbulent flows in laboratory flow tanks to study locomotion, fishes exhibit a wide array of locomotor behaviors such as drafting or entraining on obstacles in turbulent flows (Sutterlin and Waddy, 1975; Gerstner and Webb, 1998; McLaughlin and Noakes, 1998; Webb, 1998). In addition, the use of much larger "natural" flow environments to study fish locomotion (Castro-Santos, 2004, 2005; Haro *et al.*, 2004) reveals new strategies and levels of performance that are not accessible in laboratory settings (Chapter 12).

One example of an unusual locomotor behavior in larger-scale turbulent flows has been termed the "Kármán gait" by Liao *et al.* (2003a,b). By using a D-section cylinder to generate a regular array of alternating vortices in a Kármán vortex street (Figure 11.17), Liao *et al.* observed that a variety of fish species radically alter their locomotor kinematics when swimming in vortical flows. During locomotion in freestream microturbulent flows, fishes

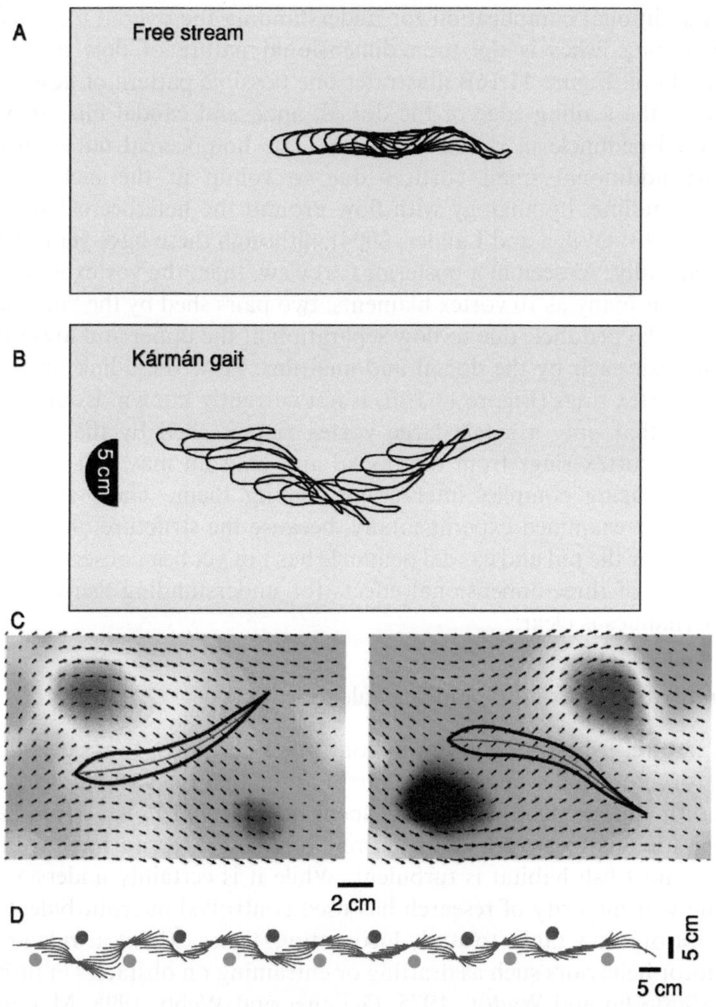

Fig. 11.17. The Kármán gait. (A and B) Trout (*Oncorhynchus mykiss*) swimming in standard freestream microturbulent flow (A) exhibit small lateral oscillations of their center of mass compared to locomotion in a Kármán vortex street behind a D-section cylinder (B). The presence of Kármán vortices radically alters locomotor kinematics in fishes. Fishes in the Kármán street swam three to four body lengths downstream from the D-section cylinder (5 cm wide), well downstream of the suction region behind the cylinder. (C) Quantification of trout body position relative to vortex centers. The trout body outline is shown in black, the midline as a red line, and background flow vorticity in color with superimposed black velocity vectors. (D) During locomotion in a vortex street, trout weave in between vortex centers: trout midlines are shown as red lines during locomotion from right to left, and centers of positive and negative vorticity are indicated by red and blue dots. Flow is from left to right in all plots. (Modified from Liao *et al.*, 2003a,b.)

exhibited the usual pattern of body undulation with only small oscillations of the center of mass (Figure 11.17A; also compare to Figures 11.1 and 11.2). However, in a Kármán vortex street, fishes substantially altered the pattern of body undulation (Figure 11.17B) and displayed very large center of mass oscillations and large-amplitude variations in lateral excursion along the body. In addition, by using DPIV to record simultaneous flow and kinematic patterns during the Kármán gait, Liao *et al.* (2003a) showed that fishes weave in between oncoming vortices (Figure 11.17C and D). The pattern of body bending appears to be primarily passive; both red and white myotomal musculature are largely inactive during the Kármán gait (Liao, 2004). Fishes may use pectoral fins or anterior red myotomal musculature to adjust the body angle of attack to oncoming vortices, but otherwise de-recruit middle and posterior body musculature.

F. Acceleration and Maneuvering

Very few studies have quantified either linear or angular acceleration in fishes as part of the normal undulatory locomotor repertoire (see Webb, 1991). Of course, high-speed unsteady C-start movements with high accelerations and rapid turns have been studied frequently (see Chapter 9), but we know little about routine (low-speed) linear acceleration and turning by fishes and their hydrodynamic causes. Drucker and Lauder (2001b) analyzed the wake of the dorsal fin during controlled yawing maneuvers, and studies by Wolfgang *et al.* (1999), Anderson (1996), and Anderson and Chhabra (2002), and Müller *et al.* (2002) provided some data on the vortices formed by turning fish. But these data lack accompanying detailed kinematics to provide a link between the pattern of body undulation and wake hydrodynamics.

Tytell (2004b) studied controlled linear accelerations in eels (*Anguilla*) under conditions that allowed quantification of both kinematics and the hydrodynamic wake from the same individual locomotor sequences. Eels accelerated or decelerated at rates ranging from 1.3 to $-1.4\ Ls^{-2}$, starting from steady undulatory locomotion at speeds that varied from 0.6 to 1.9 Ls^{-1}. These experiments showed that eels primarily varied tail tip velocity to power acceleration, and that acceleration was accompanied by the addition of significant streamwise momentum to the wake (Tytell, 2004b). The wake of an accelerating eel thus takes on the appearance of the standard undulatory wake in fishes with discrete caudal fins and a downstream momentum jet (e.g., Figures 11.9 and 11.10). Changes in wake structure during linear acceleration in fishes with distinct median and caudal fins have yet to be studied.

V. INTEGRATING THEORY AND EXPERIMENTAL DATA

Comparison of elongated body theory to experimental data (and validation of the model) has mostly used energetic data on swimming efficiency, as this was the only available body of data that permits comparison to theory (Webb, 1975, 1978). Webb (1978, p. 215) noted that "Ideally, validation would be based on observations of flow about swimming fish . . ." and indicated that such observations were problematic at that time. However, the ability in recent years to directly quantify water flow patterns over the body and fins of fishes using DPIV provides a new opportunity to compare experimental hydrodynamic data on fish locomotion to predictions of theoretical models (such as that of Lighthill, 1960), computational models (Carling *et al.*, 1998; e.g., Fauci and Peskin, 1988; Wolfgang *et al.*, 1999; Liu, 2002; Zhu *et al.*, 2002; Cortez *et al.*, 2004; Mittal, 2004), and the results of experiments by fluid engineers on heaving and pitching foils (Triantafyllou *et al.*, 1993; Gopalkrishnan *et al.*, 1994; Anderson *et al.*, 1998; Hover *et al.*, 1998, 2004; Read *et al.*, 2003; Triantafyllou *et al.*, 2004). Both avenues promise to provide considerable insight into the mechanistic bases of force generation during undulatory locomotion, but are still relatively unexplored.

Tytell and Lauder (2004) and Tytell (2004a) compared several parameters measured from DPIV analyses of the wake of swimming eels to predictions of wake impulse derived from Lighthill's elongated body theory (EBT). They found good agreement in estimates of power, but poor agreement in force and impulse estimates. Mean lateral wake power calculated from DPIV was not statistically different from the value calculated from EBT. The overall shape of the wake power profile as it varies over time is also similar for both DPIV and EBT, although DPIV methods generate higher peak power estimates than EBT. However, EBT generally underestimates the wake forces as calculated from DPIV, by as much as 50%.

It is somewhat surprising that agreement was observed in power but not in force. One might expect that either both or neither would agree. However, fluid dynamic effects not included in EBT may affect the wake power without changing the force output substantially (Lighthill, 1971; Webb, 1975), and vice versa. Additionally, discrepancies between DPIV and EBT may come from the assumptions required in DPIV force and power estimates about the three-dimensional structure of the wake. DPIV measures only a single slice of the wake and thus requires assumptions about the flow outside of this slice. Errors in three-dimensional reconstruction may hence under- or over-estimate total wake vorticity. Finally, measuring wake power accurately with DPIV through an entire control volume is difficult.

VI. PROSPECTUS

Over the past 10 years, the development of new video and laser imaging technologies has permitted advances in the study of undulatory hydrodynamics in fishes that were not anticipated at the time the previous major reviews of fish propulsion were written (Lindsey, 1978; Magnuson, 1978; Webb, 1978; Blake, 1983; Webb and Weihs, 1983; Videler, 1993). Detailed analyses of fin and body wake flow patterns are now possible, and recent developments allow measurements of all three velocity components (Nauen and Lauder, 2002b), quantification of undulatory profiles under controlled swimming conditions (Dewar and Graham, 1994; Jayne and Lauder, 1995; Gillis, 1998; Donley and Dickson, 2000; Donley and Shadwick, 2003), and correlations between kinematic and hydrodynamic patterns (Tytell, 2004b). These advances have suggested the need for reevaluating previous classification schemes for fish swimming, and have provided a wealth of new information on how the body and fins interact with the fluid environment.

However, there is still much to learn. We see the following five areas as key directions for future progress. First, to date there has been only limited interaction between researchers developing computational fluid dynamic (CFD) models of undulatory propulsion (e.g., Fauci and Peskin, 1988; Williams *et al.*, 1995; Liu *et al.*, 1996; Carling *et al.*, 1998; Liu, 2002; Cortez *et al.*, 2004; Mittal, 2004) and workers conducting the experimental hydrodynamic studies described here. Additional boundary layer measurements, like those of Anderson *et al.* (2000), are especially key, as they allow direct comparisons of theoretical and empirical estimates of drag forces, which are not measurable through wake flow visualization (Tytell and Lauder, 2004). Future experimental work could focus on testing specific predictions of computational models (as was done for eels; Tytell and Lauder, 2004), on validating those models, and on collaborative studies to investigate new phenomena such as dorsal-caudal fin wake interactions.

Second, we have very little experimental data on three-dimensional flow patterns over the body and fins in swimming fishes. Such data are critical for understanding the vortical structure of the wake, for understanding how fins interact with each other hydrodynamically, and for testing assumptions of theoretical models of fish propulsion. In addition, quantifying three-dimensional flow patterns is necessary to understand the hydrodynamic significance of differences among fish species in body shape, and, specifically, to determine why eel-shaped fish bodies produce a wake different from fishes with discrete caudal fins.

Third, there are currently few data on maneuvering and locomotion in turbulent flows. Both of these locomotor modes are in need of substantial experimental study if we are to understand more completely the locomotor repertoire of fishes.

Fourth, the function of fins and their hydrodynamic interactions with movement of the body is still a largely unexplored area. A hallmark of fish functional design is the use of multiple control surfaces projecting from the body (Lauder and Drucker, 2004), yet we have only modest data on dorsal fin function, a smattering of data on the anal fin, and effectively no data on the function of pelvic fins.

Finally, studies of external hydrodynamics need to be better connected to internal body mechanics, muscle activity and, ultimately, the neural circuitry that controls locomotion. The flow patterns discussed in this chapter develop due to a complex balance between the forces they apply to a fish's body and the forces a fish's muscles apply back to the fluid. The dynamics of this coupling have been preliminarily approached by Cortez *et al.* (2004), but substantially more research must be done to understand it better. Ultimately, these forces are controlled and modulated by the nervous system (Grillner, 2003), which itself responds to feedback from the same hydrodynamic forces. Integrating experimental and computational studies of hydrodynamic flow patterns with research on the mechanics of the body and the neural control of locomotion (Williams *et al.*, 1995; Ekeberg and Grillner, 1999) to understand the process of locomotion from nervous system to fluid, remains a key challenge for the future.

ACKNOWLEDGMENTS

We thank all members of the Lauder Lab for many discussions regarding fish locomotion, especially Paul Webb and Bill Schultz for stimulating debate on the balance of thrust and drag. Kathy Dickson kindly supplied data on scombrid locomotion for Figure 11.1, and Jacquan Horton, Adam Summers, and Eliot Drucker provided data for Figure 11.3B. This research was supported by the National Science Foundation under grant IBN0316675 to G.V.L.

REFERENCES

Aleev, Y. G. (1977). "Nekton." Junk Publishers, The Hague.

Anderson, E. J., McGillis, W. R., and Grosenbaugh, M. A. (2000). The boundary layer of swimming fish. *J. Exp. Biol.* **204,** 81–102.

Anderson, J. (1996). Vorticity control for efficient propulsion. Ph.D. thesis, MIT/WHOI 96–02.

Anderson, J. M., and Chhabra, N. (2002). Maneuvering and stability performance of a robotic tuna. *Integr. Comp. Biol.* **42,** 118–126.
Anderson, J. M., Streitlein, K., Barrett, D., and Triantafyllou, G. S. (1998). Oscillating foils of high propulsive efficiency. *J. Fluid Mech.* **360,** 41–72.
Arnold, G. P., Webb, P. W., and Holford, B. H. (1991). The role of the pectoral fins in station-holding of Atlantic salmon parr (*Salmo salar L.*). *J. Exp. Biol.* **156,** 625–629.
Arreola, V., and Westneat, M. W. (1997). Mechanics of propulsion by multiple fins: Kinematics of aquatic locomotion in the burrfish (*Chilomycterus schoepfi*). *Phil. Trans. Roy. Soc. Lond. B* **263,** 1689–1696.
Barrett, D., Triantafyllou, M. S., Yue, D. K. P., Grosenbaugh, M. A., and Wolfgang, M. J. (1999). Drag reduction in fish-like locomotion. *J. Fluid Mech.* **392,** 183–212.
Batchelor, G. K. (1973). "An Introduction to Fluid Mechanics." Cambridge Univ. Press, Cambridge, UK.
Blake, R. W. (1983). "Fish Locomotion." Cambridge Univ. Press, Cambridge, UK.
Bone, Q. (1978). Locomotor muscle. *In* "Fish Physiology. Vol. VII. Locomotion" (Hoar, W. S., and Randall, D. J., Eds.), pp. 361–424. Academic Press, New York.
Breder, C. M. (1926). The locomotion of fishes. *Zoologica N. Y.* **4,** 159–256.
Carling, J. C., Williams, T. L., and Bowtell, G. (1998). Self-propelled anguilliform swimming: Simultaneous solution of the two-dimensional Navier-Stokes equations and Newton's laws of motion. *J. Exp. Biol.* **201,** 3143–3166.
Castro-Santos, T. (2004). Quantifying the combined effects of attempt rate and swimming capacity on passage through velocity barriers. *Can. J. Fish. Aq. Sci.* **61,** 1602–1615.
Castro-Santos, T. (2005). Optimal swim speeds for traversing velocity barriers: An analysis of volitional high-speed swimming behavior of migratory fishes. *J. Exp. Biol.* **208,** 421–432.
Childress, S. (1981). "Mechanics of Flying and Swimming." Cambridge Univ. Press, Cambridge, UK.
Cortez, R., Fauci, L., Cowen, N., and Dillon, R. (2004). Simulation of swimming organisms: Coupling internal mechanics with external fluid dynamics. *Comp. Sci. Eng.* **6,** 38–45.
Daniel, T. L. (1984). Unsteady aspects of aquatic locomotion. *Amer. Zool.* **24,** 121–134.
Daniel, T. L., and Webb, P. W. (1987). Physics, design and locomotor performance. *In* "Comparative Physiology: Life in Water and on Land" (Dejours, P., Bolis, L., Taylor, C. R., and Weibel, E. R., Eds.), pp. 343–369. Liviana Press, Springer Verlag, New York.
Denny, M. W. (1993). "Air and Water. The Biology and Physics of Life's Media." Princeton Univ. Press, Princeton, NJ.
Dewar, H., and Graham, J. B. (1994). Studies of tropical tuna swimming performance in a large water tunnel. III. Kinematics. *J. Exp. Biol.* **192,** 45–59.
Donley, J., and Dickson, K. A. (2000). Swimming kinematics of juvenile Kawakawa tuna (*Euthynnus affinis*) and chub mackerel (*Scomber japonicus*). *J. Exp. Biol.* **203,** 3103–3116.
Donley, J., and Shadwick, R. (2003). Steady swimming muscle dynamics in the leopard shark *Triakis semifasciata*. *J. Exp. Biol.* **206,** 1117–1126.
Drucker, E. G., and Lauder, G. V. (2000). A hydrodynamic analysis of fish swimming speed: Wake structure and locomotor force in slow and fast labriform swimmers. *J. Exp. Biol.* **203,** 2379–2393.
Drucker, E. G., and Lauder, G. V. (2001a). Locomotor function of the dorsal fin in teleost fishes: Experimental analysis of wake forces in sunfish. *J. Exp. Biol.* **204,** 2943–2958.
Drucker, E. G., and Lauder, G. V. (2001b). Wake dynamics and fluid forces of turning maneuvers in sunfish. *J. Exp. Biol.* **204,** 431–442.
Ekeberg, Ö. (1993). A combined neuronal and mechanical model of fish swimming. *Biol. Cybernet.* **69,** 363–374.

Ekeberg, Ö., and Grillner, S. (1999). Simulations of neuromuscular control in lamprey swimming. *Phil. Trans. Roy. Soc. Lond. B* **354,** 895–902.
Faber, T. E. (1995). "Fluid Dynamics for Physicists." Cambridge Univ. Press, Cambridge, UK.
Fauci, L. J., and Peskin, C. S. (1988). A computational model of aquatic animal locomotion. *J. Comp. Phys.* **77,** 85–108.
Ferry, L. A., and Lauder, G. V. (1996). Heterocercal tail function in leopard sharks: A three-dimensional kinematic analysis of two models. *J. Exp. Biol.* **199,** 2253–2268.
Fish, F., and Lauder, G. V. (2005). Passive and active flow control by swimming fishes and mammals. *Ann. Rev. Fluid Mech.* **38,** 193–224.
Geerlink, P. J., and Videler, J. (1974). Joints and muscles of the dorsal fin of *Tilapia nilotica* L. (Fam. Cichlidae). *Neth. J. Zool.* **24,** 279–290.
Gerstner, C. L., and Webb, P. W. (1998). The station-holding performance of the plaice *Pleuronectes platessa* on artificial substratum ripples. *Can. J. Zool.* **76,** 260–268.
Gillis, G. B. (1996). Undulatory locomotion in elongate aquatic vertebrates: Anguilliform swimming since Sir. James Gray. *Amer. Zool.* **36,** 656–665.
Gillis, G. B. (1998). Environmental effects on undulatory locomotion in the American eel *Anguilla rostrata*: Kinematics in water and on land. *J. Exp. Biol.* **201,** 949–961.
Gopalkrishnan, R., Triantafyllou, M. S., Triantafyllou, G. S., and Barrett, D. (1994). Active vorticity control in a shear flow using a flapping foil. *J. Fluid Mech.* **274,** 1–21.
Gosline, W. A. (1971). "Functional Morphology and Classification of Teleostean Fishes." Univ. of Hawaii Press, Honolulu, HI.
Graham, J. B., and Dickson, K. A. (2004). Tuna comparative physiology. *J. Exp. Biol.* **207,** 4015–4024.
Gray, J. (1933). Studies in animal locomotion. I. The movement of fish with special reference to the eel. *J. Exp. Biol.* **10,** 88–104.
Gray, J. (1936). Studies in animal locomotion. VI. The propulsive powers of the dolphin. *J. Exp. Biol.* **13,** 170–199.
Grillner, S. (2003). The motor infrastructure: From ion channels to neuronal networks. *Nat. Rev. Neurosci.* **4,** 573–586.
Haro, A., Castro-Santos, T., Noreika, J., and Odeh, M. (2004). Swimming performance of upstream migrant fishes in open-channel flow: A new approach to predicting passage through velocity barriers. *Can. J. Fish. Aq. Sci.* **61,** 1590–1601.
He, P., and Wardle, C. S. (1986). Tilting behavior of the Atlantic mackerel, *Scomber scombrus*, at low swimming speeds. *J. Fish. Biol.* **29,** 223–232.
Hertel, H. (1966). "Structure, Form and Movement." Reinhold, New York, NY.
Hess, F., and Videler, J. J. (1984). Fast continuous swimming of saithe (*Pollachius virens*): A dynamic analysis of bending moments and muscle power. *J. Exp. Biol.* **109,** 229–251.
Horton, J. M., Drucker, E., and Summers, A. (2003). Swiftly swimming fish show evidence of stiff spines. *Integ. Comp. Biol.* **43,** 905.
Hove, J. R., O'Bryan, L. M., Gordon, M. S., Webb, P. W., and Weihs, D. (2001). Boxfishes (Teleostei: Ostraciidae) as a model system for fishes swimming with many fins: Kinematics. *J. Exp. Biol.* **204,** 1459–1471.
Hover, F. S., Haugsdal, O., and Triantafyllou, M. S. (2004). Effect of angle of attack profiles in flapping foil propulsion. *J. Fluids Struct.* **19,** 37–47.
Hover, F. S., Techet, A. H., and Triantafyllou, M. S. (1998). Forces on oscillating uniform and tapered cylinders in crossflow. *J. Fluid Mech.* **363,** 97–114.
Jayne, B. C., and Lauder, G. V. (1995). Speed effects on midline kinematics during steady undulatory swimming of largemouth bass, *Micropterus salmoides*. *J. Exp. Biol.* **198,** 585–602.
Jayne, B. C., Lozada, A., and Lauder, G. V. (1996). Function of the dorsal fin in bluegill sunfish: Motor patterns during four locomotor behaviors. *J. Morphol.* **228,** 307–326.

Lauder, G. V. (1982). Structure and function of the caudal skeleton in the pumpkinseed sunfish, *Lepomis gibbosus*. *J. Zool. (Lond.)* **197,** 483–495.
Lauder, G. V. (1989). Caudal fin locomotion in ray-finned fishes: Historical and functional analyses. *Amer. Zool.* **29,** 85–102.
Lauder, G. V. (2000). Function of the caudal fin during locomotion in fishes: Kinematics, flow visualization, and evolutionary patterns. *Amer. Zool.* **40,** 101–122.
Lauder, G. V. (2005). Locomotion. *In* "The Physiology of Fishes" (Evans, D. H., and Claiborne, J. B., Eds.). pp. 3–46. 3rd edn, CRC Press, Boca Raton, FL.
Lauder, G. V., and Drucker, E. (2002). Forces, fishes, and fluids: Hydrodynamic mechanisms of aquatic locomotion. *News Physiolog. Sci.* **17,** 235–240.
Lauder, G. V., and Drucker, E. G. (2004). Morphology and experimental hydrodynamics of fish fin control surfaces. *IEEE J. Oceanic Eng.* **29,** 556–571.
Lauder, G. V., Drucker, E. G., Nauen, J., and Wilga, C. D. (2003). Experimental hydrodynamics and evolution: Caudal fin locomotion in fishes. *In* "Vertebrate Biomechanics and Evolution" (Bels, V., Gasc, J.-P., and Casinos, A., Eds.), pp. 117–135. Bios Scientific Publishers, Oxford.
Lauder, G. V., Nauen, J., and Drucker, E. G. (2002). Experimental hydrodynamics and evolution: Function of median fins in ray-finned fishes. *Integr. Comp. Biol.* **42,** 1009–1017.
Lauder, G. V., and Tytell, E. D. (2004). Three Gray classics on the biomechanics of animal movement. *J. Exp. Biol.* **207,** 1597–1599.
Liao, J. (2004). Neuromuscular control of trout swimming in a vortex street: Implications for energy economy during the Karman gait. *J. Exp. Biol.* **207,** 3495–3506.
Liao, J., and Lauder, G. V. (2000). Function of the heterocercal tail in white sturgeon: Flow visualization during steady swimming and vertical maneuvering. *J. Exp. Biol.* **203,** 3585–3594.
Liao, J., Beal, D. N., Lauder, G. V., and Triantafyllou, M. S. (2003a). Fish exploiting vortices decrease muscle activity. *Science* **302,** 1566–1569.
Liao, J., Beal, D. N., Lauder, G. V., and Triantafyllou, M. S. (2003b). The Kármán gait: Novel body kinematics of rainbow trout swimming in a vortex street. *J. Exp. Biol.* **206,** 1059–1073.
Lighthill, J. (1960). Note on the swimming of slender fish. *J. Fluid Mech.* **9,** 305–317.
Lighthill, J. (1969). Hydromechanics of aquatic animal propulsion: A survey. *Ann. Rev. Fluid Mech.* **1,** 413–446.
Lighthill, J. (1970). Aquatic animal propulsion of high hydromechanical efficiency. *J. Fluid Mech.* **44,** 265–301.
Lighthill, J. (1971). Large-amplitude elongated body theory of fish locomotion. *Proc. Roy. Soc. Lond. B* **179,** 125–138.
Lindsey, C. C. (1978). Form, function, and locomotory habits in fish. *In* "Fish Physiology. Vol. VII. Locomotion" (Hoar, W. S., and Randall, D. J., Eds.), pp. 1–100. Academic Press, New York.
Liu, H. (2002). Computational biological fluid dynamics: Digitizing and visualizing animal swimming and flying. *Integr. Comp. Biol.* **42,** 1050–1059.
Liu, H., Wassersug, R. J., and Kawachi, K. (1996). A computational fluid dynamics study of tadpole swimming. *J. Exp. Biol.* **199,** 1245–1260.
Magnuson, J. J. (1978). Locomotion by scombrid fishes: Hydromechanics, morphology, and behavior. *In* "Fish Physiology. Vol. VII. Locomotion" (Hoar, W. S., and Randall, D. J., Eds.), pp. 239–313. Academic Press, New York.
McCutchen, C. W. (1977). Froude propulsive efficiency of a small fish, measured by wake visualization. *In* "Scale Effects in Animal Locomotion" (Pedley, T. J., Ed.), pp. 339–363. Academic Press, London.

McLaughlin, R. L., and Noakes, D. L. G. (1998). Going against the flow: An examination of the propulsive movements made by young brook trout in streams. *Can. J. Fish. Aq. Sci.* **V55,** 853–860.

Mittal, R. (2004). Computational modeling in biohydrodynamics: Trends, challenges, and recent advances. *IEEE J. Oceanic Eng.* **29,** 595–604.

Müller, U. K., Smit, J., Stamhuis, E. J., and Videler, J. J. (2001). How the body contributes to the wake in undulatory fish swimming: Flow fields of a swimming eel (*Anguilla anguilla*). *J. Exp. Biol.* **204,** 2751–2762.

Müller, U. K., Stamhuis, E., and Videler, J. (2002). Riding the waves: The role of the body wave in undulatory fish swimming. *Integr. Comp. Biol.* **42,** 981–987.

Müller, U. K., Van den Heuvel, B., Stamhuis, E. J., and Videler, J. J. (1997). Fish foot prints: Morphology and energetics of the wake behind a continuously swimming mullet (*Chelon labrosus* Risso). *J. Exp. Biol.* **200,** 2893–2906.

Murray, M., and Howle, L. E. (2003). Spring stiffness influence of an oscillating propulsor. *J. Fluids Struct.* **17,** 915–926.

Nauen, J. C., and Lauder, G. V. (2001). Locomotion in scombrid fishes: Visualization of flow around the caudal peduncle and finlets of the Chub mackerel *Scomber japonicus*. *J. Exp. Biol.* **204,** 2251–2263.

Nauen, J. C., and Lauder, G. V. (2002a). Hydrodynamics of caudal fin locomotion by chub mackerel, *Scomber japonicus* (Scombridae). *J. Exp. Biol.* **205,** 1709–1724.

Nauen, J. C., and Lauder, G. V. (2002b). Quantification of the wake of rainbow trout (*Oncorhynchus mykiss*) using three-dimensional stereoscopic digital particle image velocimetry. *J. Exp. Biol.* **205,** 3271–3279.

Noca, F., Shiels, D., and Jeon, D. (1999). A comparison of methods for evaluating time-dependent fluid dynamic forces on bodies, using only velocity fields and their derivatives. *J. Fluids Struct.* **13,** 551–578.

Pedley, T. J. (1977). "Scale Effects in Animal Locomotion." Academic Press, London.

Raffel, M., Willert, C., and Kompenhans, J. (1998). "Particle Image Velocimetry: A Practical Guide." Springer-Verlag, Heidelberg.

Read, D. A., Hover, F. S., and Triantafyllou, M. S. (2003). Forces on oscillating foils for propulsion and maneuvering. *J. Fluids Struct.* **17,** 163–183.

Rohr, J., and Fish, F. (2004). Strouhal numbers and optimization of swimming by odontocete cetaceans. *J. Exp. Biol.* **207,** 1633–1642.

Rosen, M., Spedding, G. R., and Hedenstrom, A. (2004). The relationship between wingbeat kinematics and vortex wake of a thrush nightingale. *J. Exp. Biol.* **207,** 4255–4268.

Rosen, M. W. (1959). Water flow about a swimming fish. *Naval Ordnance Test Station.* Technical Paper. **2298,** 1–96.

Sakakibara, J., Nakagawa, M., and Yoshida, M. (2004). Stereo-PIV study of flow around a maneuvering fish. *Exp. Fluid* **36,** 282–293.

Schultz, W. W., and Webb, P. W. (2002). Power requirements for swimming: Do new methods resolve old questions? *Integr. Comp. Biol.* **42,** 1018–1025.

Standen, E. M., and Lauder, G. V. (2005). Dorsal and anal fin function in bluegill sunfish (*Lepomis macrochirus*): Three-dimensional kinematics during propulsion and maneuvering. *J. Exp. Biol.* **205,** 2753–2763.

Sutterlin, A. M., and Waddy, S. (1975). Possible role of the posterior lateral line in obstacle entrainment by Brook trout (*Salvelinus fontinalis*). *J. Fish. Res. Bd. Can.* **32,** 2441–2446.

Svendsen, J. C., Koed, A., and Lucas, M. C. (2005). The angle of attack of the body of common bream while swimming at different speeds in a flume tank. *J. Fish. Biol.* **66,** 572–577.

Taylor, G. I. (1952). Analysis of the swimming of long and narrow animals. *Proc. Roy. Soc. Lond. A* **214,** 158–183.

Taylor, G. K., Nudds, R. L., and Thomas, A. (2003). Flying and swimming animals cruise at a Strouhal number tuned for high power efficiency. *Nature* **425,** 707–711.

Triantafyllou, M. S., and Triantafyllou, G. S. (1995). An efficient swimming machine. *Sci. Am.* **272,** 64–70.

Triantafyllou, G. S., Triantafyllou, M. S., and Grosenbaugh, M. A. (1993). Optimal thrust development in oscillating foils with application to fish propulsion. *J. Fluids Struct.* **7,** 205–224.

Triantafyllou, M. S., Techet, A. H., and Hover, F. S. (2004). Review of experimental work in biomimetic foils. *IEEE J. Oceanic Eng.* **29,** 585–594.

Tytell, E. D. (2004a). The hydrodynamics of eel swimming II. Effect of swimming speed. *J. Exp. Biol.* **207,** 3265–3279.

Tytell, E. D. (2004b). Kinematics and hydrodynamics of linear acceleration in eels, *Anguilla rostrata*. *Proc. Roy. Soc. Lond. B* **271,** 2535–2540.

Tytell, E. D., and Lauder, G. V. (2004). The hydrodynamics of eel swimming. I. Wake structure. *J. Exp. Biol.* **207,** 1825–1841.

Veerman, H. P., and Den Boer, R. (2000). PIV measurements in presence of a large out of plane component. *In* "Particle Image Velocimetry. Progress toward Industrial Application" (Stanislas, M., Kompenhans, J., and Westerweel, J., Eds.), pp. 217–225. Kluwer Academic, Dordrecht.

Videler, J. J. (1993). "Fish Swimming." Chapman and Hall, New York.

Videler, J. J., and Hess, F. (1984). Fast continuous swimming of two pelagic predators, saithe (*Pollachius virens*) and mackerel (*Scomber scombrus*): A kinematic analysis. *J. Exp. Biol.* **109,** 209–228.

Webb, P. W. (1975). Hydrodynamics and energetics of fish propulsion. *Bull. Fish. Res. Bd. Can.* **190,** 1–159.

Webb, P. W. (1978). Hydrodynamics: Nonscombroid fish. *In* "Fish Physiology. Vol. VII. Locomotion" (Hoar, W. S., and Randall, D. J., Eds.), pp. 189–237. Academic Press, New York.

Webb, P. W. (1991). Composition and mechanics of routine swimming of rainbow trout, *Oncorhynchus mykiss*. *Can. J. Fish. Aquat. Sci.* **48,** 583–590.

Webb, P. W. (1993a). Is tilting behavior at low speed swimming unique to negatively buoyant fish? Observations on steelhead trout, *Oncorhynchus mykiss*, and bluegill, *Lepomis macrochirus*. *J. Fish. Biol.* **43,** 687–694.

Webb, P. W. (1993b). Swimming. *In* "The Physiology of Fishes" (Evans, D. H., Ed.), pp. 47–73. CRC Press, Boca Raton, FL.

Webb, P. W. (1998). Entrainment by river chub *Nocomis micropogon* and smallmouth bass *Micropterus dolomieu* on cylinders. *J. Exp. Biol.* **201,** 2403–2412.

Webb, P. W., and Blake, R. W. (1985). Swimming. *In* "Functional Vertebrate Morphology" (Hildebrand, M., Bramble, D. M., Liem, K. F., and Wake, D. B., Eds.), pp. 110–128. Harvard Univ. Press, Cambridge, MA.

Webb, P. W., and Keyes, R. S. (1981). Division of labor between median fins in swimming dolphin (Pisces: Coryphaenidae). *Copeia* **1981,** 901–904.

Webb, P. W., and Weihs, D. (1983). "Fish Biomechanics." Praeger Publishers, New York.

Weihs, D. (1972). A hydrodynamic analysis of fish turning manoeuvres. *Proc. Roy. Soc. Lond. B* **182B,** 59–72.

Westerweel, J., and Oord, J.v. (2000). Stereoscopic PIV measurements in a turbulent boundary layer. *In* "Particle Image Velocimetry. Progress toward Industrial Application" (Stanislas, M., Kompenhans, J., and Westerweel, J., Eds.), pp. 459–478. Kluwer Academic, Dordrecht.

Wilga, C. D., and Lauder, G. V. (1999). Locomotion in sturgeon: Function of the pectoral fins. *J. Exp. Biol.* **202,** 2413–2432.

Wilga, C. D., and Lauder, G. V. (2000). Three-dimensional kinematics and wake structure of the pectoral fins during locomotion in leopard sharks *Triakis semifasciata*. *J. Exp. Biol.* **203,** 2261–2278.

Wilga, C. D., and Lauder, G. V. (2002). Function of the heterocercal tail in sharks: Quantitative wake dynamics during steady horizontal swimming and vertical maneuvering. *J. Exp. Biol.* **205,** 2365–2374.

Wilga, C. D., and Lauder, G. V. (2004). Hydrodynamic function of the shark's tail. *Nature* **430,** 850.

Willert, C. (1997). Stereoscopic digital particle image velocimetry for application in wind tunnel flows. *Meas. Sci. Technol.* **8,** 1465–1479.

Willert, C. E., and Gharib, M. (1991). Digital particle image velocimetry. *Exp. Fluid* **10,** 181–193.

Williams, T. L., Bowtell, G., Carling, J. C., Sigvardt, K. A., and Curtin, N. A. (1995). Interactions between muscle activation, body curvature and the water in the swimming lamprey. *In* "Biological Fluid Dynamics" (Ellington, C. P., and Pedley, T. J., Eds.), pp. 49–59. Company of Biologists, Cambridge.

Winterbottom, R. (1974). A descriptive synonymy of the striated muscles of the Teleostei. *Proc. Acad. Nat. Sci. Phila.* **125,** 225–317.

Wolfgang, M. J., Anderson, J. M., Grosenbaugh, M., Yue, D., and Triantafyllou, M. (1999). Near-body flow dynamics in swimming fish. *J. Exp. Biol.* **202,** 2303–2327.

Wu, Y. T. (1971). Hydromechanics of swimming propulsion. Part I. Swimming of a two-dimensional flexible plate at variable forward speeds in an inviscid fluid. *J. Fluid Mech.* **46,** 337–355.

Zhu, Q., Wolfgang, M. J., Yue, D. K. P., and Triantafyllou, G. S. (2002). Three-dimensional flow structures and vorticity control in fish-like swimming. *J. Fluid Mech.* **468,** 1–28.

12

BIOMECHANICS AND FISHERIES CONSERVATION

THEODORE CASTRO-SANTOS
ALEX HARO

I. INTRODUCTION

The field of biomechanics tends by its very nature to focus on suborganismal processes. Muscle kinetics, body and fin morphologies, sensory systems, etc., characterize the mechanical interactions between fishes and their habitat. These in turn define the fishes' ability to exploit their environment, influencing not only home range, but also reproductive success, energetics, trophic interactions, and the scope of available habitats.

The field of conservation, by contrast, is intrinsically integrative. Fisheries managers must take into account more than ecology and life history theory when developing policy: physiology, energetics, behavior, and stochastic habitat and ecosystem-level processes must all be considered. The objective of this chapter is to discuss the current state of the field of fish biomechanics in a management context, identifying advances and gaps in knowledge that may both aid the development of sound management policies and provide helpful directions for future research.

When developing conservation policy, managers often focus on critical life history stages: those parts of the life cycle that determine population-level processes. This is not necessarily when individual animals are most

Fish Biomechanics: Volume 23
FISH PHYSIOLOGY

DOI: 10.1016/S1546-5098(05)23012-1

vulnerable—depending on species and population, stock recruitment may be most affected by larval survival, juvenile growth, adult survival, or spawning success. Many factors influence whether and in what condition fishes progress through these life stages. In some cases, biomechanics can help us understand how such factors act, and identify both developing problems and likely solutions.

For example, we know that salmon construct redds of gravel to ensure adequate flushing and ventilation for egg and larval development (Devries, 1997; Guerrin and Dumas, 2001). This has clear implications for topsoil erosion control and watershed-level management issues (Soulsby *et al.*, 2001; Moir *et al.*, 2002). Similarly, selectivity of fishing gear can be explained largely from biomechanical principles (Chopin and Arimoto, 1995; Huse *et al.*, 2000), and both fishing and management practices can have broad-scale influences on habitat and species characteristics, often with biomechanical implications (Rijnsdorp, 1993; Jennings *et al.*, 1998; Grift *et al.*, 2003).

Biomechanics are particularly important to the management of migratory species. Where opportunities to take advantage of available but remote habitat exist, migration can expand a lifetime track to include such habitat, with consequent benefits to fitness and recruitment (Dingle, 1996). But migrations are often costly, both energetically and physiologically. They can be accompanied by dramatic morphological changes (Tesch, 1977; Klemetsen *et al.*, 2003; Martinez and Bolker, 2003), and are sustainable only when sufficient benefit is gained (Alexander, 1998). Migrations are also associated with the critical life history stages mentioned previously. For this reason, this behavior can put species at risk to both predation and commercial exploitation; and obstacles to migration can cause dramatic changes to populations, even extinction (Holcik, 2001).

The complex interactions between biomechanics and other biological issues make it difficult, if not impossible, to summarize their management implications in isolation. Instead, we have organized this chapter along management-based themes representing important issues confronted by fisheries managers today. Each theme has relevance to multiple life history stages, and the biomechanical issues are varied. We hope that this structure provides a useful context for understanding the intersection of these two disciplines.

We begin with a description of effects of obstacles to riverine migrations, both up- and downstream, and methods employed to mitigate for those obstacles. This leads to a discussion of commercial fishing trawls and the role of biomechanics in both capture efficiency and the problem of bycatch. Next we discuss issues of intraspecific diversity: the importance of local adaptation, and how management practices can interact with biomechanical characteristics of target species. This is followed by a discussion of bioenergetic modeling and the importance of both appropriate energetics equations and accurate behavioral data to this rapidly developing field. Finally, we present a summary of gaps in current knowledge, with suggestions as to how future biomechanics research might help to improve practices in both fisheries management and conservation.

II. RIVERINE MIGRATIONS

A. Upstream Migrations and Passage

Concern over vulnerability of commercially and ecologically important fish stocks of during migratory phases prompted much early interest in fish biomechanics. This focused on locomotor ability, and its obvious importance to the design of both fishing gear and fish passage structures (Beamish, 1966a,b, 1978; Blaxter and Parrish, 1966; Wardle, 1975, 1980). At the same time that Breder (1926) was formalizing the correlations between body forms, swimming modes, and life history traits, others were working to mitigate for massive dam-building projects that had created new barriers to the migrations of numerous species (Denil, 1909, 1937; Stringham, 1924). An ancient concept, dams were one of the primary power sources of the Industrial Revolution. Dam development accelerated rapidly following the introduction of the electric hydro-turbine. By the mid 20th century, they were being built in the United States at a rate of thousands per year; currently there are over 77,000 known high-head dams within the continental United States alone, and when low-head dams are included, this number reaches into the millions (Smith, 1971; U.S. Army Corps of Engineers, 2001).

Construction of these dams has long been known to affect populations of migratory fishes, and by the late 19th century, anadromous salmon stocks had begun to plummet (Orsborn, 1987). This caught people's attention: throughout their range, salmon provided a major source of human dietary protein during that period. Clearly, there was a need to provide passage to this culturally, economically, and ecologically important resource. In order to pass fishes over dams, hydraulic bypass structures (fishways) had to be engineered that could sufficiently dissipate the energy (head) stored by the dam such that flow velocities were low enough to permit passage of upstream migrants. Although the earliest known engineered fishway dates to the 17th century (Clay, 1995), little was known about the ability of fishes to pass such structures. The need to quantify this ability precipitated a flurry of research into both the hydraulic properties of fish passage structures (McLeod and Nemenyi, 1940; Nemenyi, 1941; British Institution of

Civil Engineers, 1942), and the ability of fishes to pass them (Denil, 1937; Bainbridge, 1960; Brett, 1962).

Much of the early work on swimming capacity focused on maximum abilities, specifically, maximum attainable speeds and leaping abilities (Stringham, 1924; Gray, 1936; Stuart, 1962). Bainbridge (1960), however, deserves credit for formalizing the relationship between stamina and fishway design. Contemporaneously with Brett (1964), he characterized the relationship between swimming speed and fatigue time, and identified appropriate methods for quantifying this relationship. Brett (1964) identified the relationship between swim speed and fatigue time as log-linear, i.e.,

$$\ln T = a + bU_s. \tag{1}$$

Fish swimming at speed U_s fatigue at time T, qualified by coefficients a and b (Figure 12.1). Bainbridge (1960) further pointed out that, by taking the fatigue time of a given through-water speed and subtracting the speed of flow, one could predict maximum traversable distance through a velocity barrier (Figure 12.2; see also McCleave, 1980).

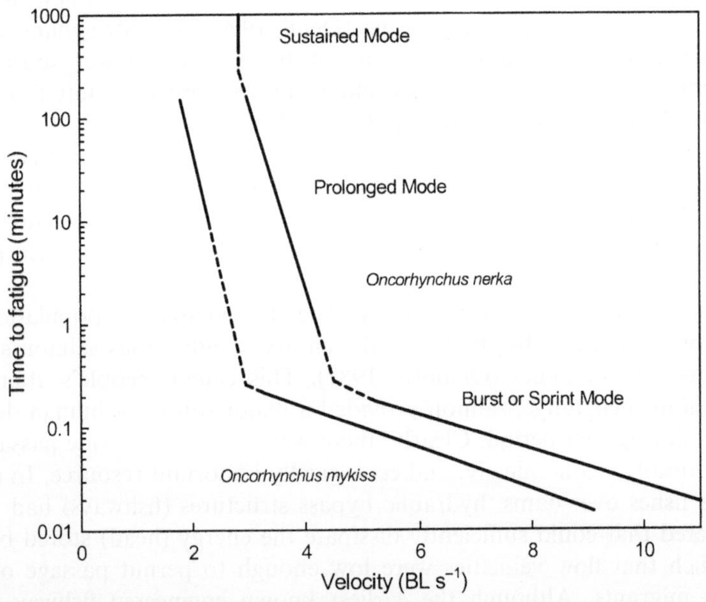

Fig. 12.1. Relationship between swim speed (in body lengths, BL) and fatigue time, redrawn from Brett (1964) and Webb (1975). Note that swim speed is a linear function of the log of fatigue time at speeds greater than the maximum sustained speed. Note also the presence of two swimming modes (prolonged and burst) at these unsustainable speeds.

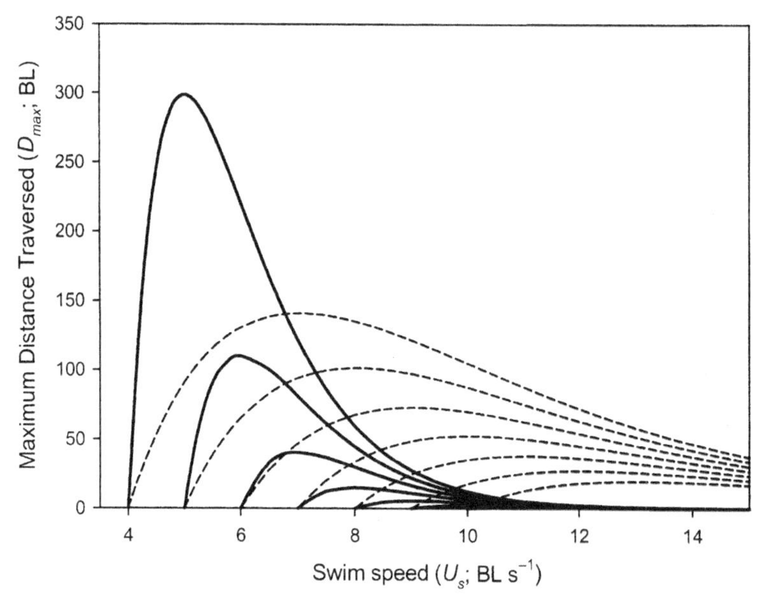

Fig. 12.2. Effect of flow velocity and swim speed on maximum distance traversed (D_{max}, in body lengths, BL), given swim speed-fatigue time relationships for prolonged (solid lines) and burst (sprint, dashed lines) modes. Contours represent D_{max} values for flow velocities of 4–10 BL s^{-1}; note that where fishes swim at the speed of flow, $D_{\text{max}} = 0$. Data are for American shad. (Redrawn from Castro-Santos, 2005).

Oddly, this observation has been largely ignored until recently. Instead, most fish passage manuals, when addressing swimming capacity, provide estimates from Beach (1984). This report incorporates Zhou's (1982) approach of relating performance to a limited energy store, and Brett's (1965, 1972) estimates of maximum aerobic capacity to estimate total available aerobic power as:

$$P_r = 48.11L^{2.964} \tag{2}$$

where P_r = available aerobic power (W), and L = body lengths; the energy available to power sprints (E) is estimated from mean glycogen stores and expected muscle mass as:

$$E = 19396L^{2.964}; \tag{3}$$

and the chemical power (P_c, W) available to the fish is taken from Zhou (1982) for a specific temperature and length as

$$P_c = 0.9751 \times e^{-0.00522M_T} \times U_s^{2.8} \times L^{-1.15}, \tag{4}$$

where M_T = muscle temperature (°C).

Maximum swimming speed is estimated from Wardle (1975, 1980) as

$$U_s = 0.7L/2t, \tag{5}$$

where t is the minimum muscle contraction time, calculated as

$$t = 0.1700L^{0.4288} + 0.0028\ \ln M_T - 0.0425L^{0.4288} \times \ln M_T - 0.0077. \tag{6}$$

Substituting U_s into Eq. (4) yields an estimate of endurance time (T) at maximum swim speed as

$$T = E/[P_c - P_r]. \tag{7}$$

Thus, the time to fatigue is estimated based on both aerobic and anaerobic energy sources, assuming maximum power delivery from both.

Although the effort to predict passage success from purely biomechanical and biochemical principles is commendable, the required assumptions hardly seem reasonable. Glycogen stores are influenced by both condition and previous experience of the fish (Connor *et al.*, 1964; Jonsson *et al.*, 1997); different species have differing muscle composition, with varying mixtures of red and white muscle (Alexander, 1969), presumably indicating differing contributions of aerobic metabolism to high-speed swimming among taxa; and broad inter- and intraspecific variation in both maximum tail-beat frequency and stride length are known to exist (Bainbridge, 1958; Videler, 1993; Coughlin, 2002). Nonetheless, the simplicity of the equations and their proclaimed universality have made them very popular with fishway engineers: Beach's (1984) figures have been reproduced in every major fishway design manual written in the past 20 years (Powers *et al.*,1985; Bell, 1991; Clay, 1995; Vigneux and Larinier, 2002).

Concurrent with the work of Bainbridge and Brett, a more direct approach relating swimming performance to passage success was being developed at a unique facility on the Columbia River (Figure 12.3A). This facility, designed to test prototype fishway models with live, actively migrating fishes, lent itself readily to the study of swimming performance. In a pair of papers, Weaver (1963, 1965) proposed a new method of quantifying swimming performance whereby capacity was measured as distance traversed up a large, open-channel flume against a velocity challenge, i.e., the empirical output of Bainbridge's model. The work of Weaver and his colleagues has largely been forgotten, however: a literature search of peer-reviewed publications from 1978 to 2003 (ISI Web of Science, 2003) indicated that, whereas Bainbridge (1960) was cited 103 times, and Brett (1964) 676 times, Weaver (1963) was cited only 8 times, and Weaver (1965) only twice.

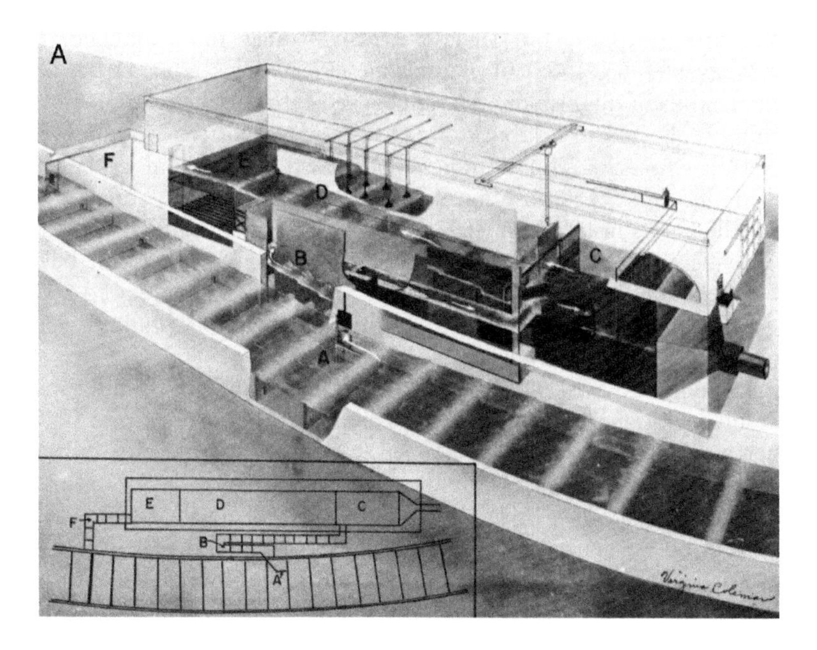

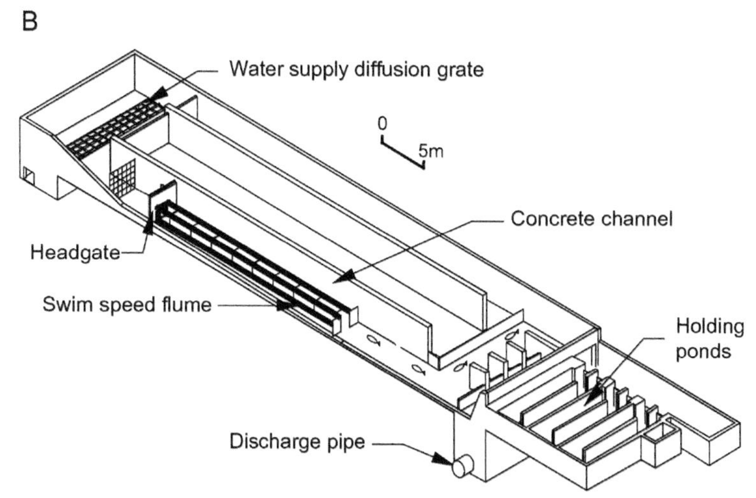

Fig. 12.3. Flume facilities of the Bonneville Fisheries Research Laboratory (A) and the S.O. Conte Anadromous Fish Research Center (B). These facilities were designed to conduct fish passage studies requiring large scales and high flows, and use the head pond stored by adjacent dams for a water supply.

The paper by Brett (1964) has clearly had much more influence than either Weaver (1963, 1965) or Bainbridge (1960). This is in part because Brett's methods of quantifying the swim speed-fatigue time relationship can be readily replicated in a laboratory environment (but see Hammer, 1995 and Nelson *et al.*, 2002 for critiques of the method). Several studies have extended this work, establishing estimates of stamina for a broad range of taxa (references in Beamish, 1978; Videler, 1993). Typically, these studies focus on identifying maximum sustained speeds (speeds that can be sustained indefinitely, or at least for 200 min) and on the swim speed-fatigue time relationship for prolonged swimming (continuous swimming speeds that result in fatigue in greater than 20 s).

Some studies do exist on sprinting as well (continuous swimming speeds that result in fatigue in less than 20 s); however, it is sometimes unclear whether authors really mean prolonged swimming when using this term (e.g., Peake *et al.*, 1997a; McDonald *et al.*, 1998). One of the limitations of the methods described by Brett (1964) and others (Beamish, 1978) is that it can be difficult to generate reliable estimates of endurance at such high speeds, particularly sprinting. Indeed, relatively few authors have attempted to quantify the swim speed-fatigue time relationship for sprinting. Videler (1993, Table 10.1) presents sprinting data from only six studies, of which only two (Bainbridge, 1960; McCleave, 1980) quantified the relationship between sprint speed and fatigue time. Beamish (1978) cited 21 studies of sprint swimming, but of these, many were observations of single individuals in uncontrolled environments; only three papers reported data with sample sizes of 20 or more.

Despite the limitations of these models, they continue to receive attention as methods for characterizing and identifying viable fishway designs for a range of species. Colavecchia *et al.* (1998), Peake *et al.* (1995, 1997a,b, 2000), and Bunt *et al.* (1999, 2000, 2001) have used fixed velocity tests to quantify the swim speed-fatigue time relationship of Atlantic salmon (*Salmo salar*), brown trout (*S. trutta*), white sucker (*Catostomus commersoni*), walleye (*Sander vitreus,* formerly *Stizostedion vitreum*), smallmouth bass (*Micropterus dolomieu*), and lake sturgeon (*Acipenser fulvescens*). The intent of these authors was to define engineering criteria for effective fish passage structures for these species.

Sometimes, conservation interests are best served by excluding, rather than passing, alien or undesirable species from critical habitat (Holloway, 1991; Baxter *et al.*, 2003). In an effort to control populations of introduced sea lamprey (*Petromyzon marinus*) in the Great Lakes, management agencies are exploring the feasibility of using velocity barriers to impede their spawning migrations (Hunn and Youngs, 1980; McCauley, 1996). Estimates of the swim speed-fatigue time relationship are used to predict maximum distance of ascent attainable by this species. Ideally, this value will be less than that of other migratory species that might also need to use these same migratory corridors as part of their life history.

The conflicts arising from such competing requirements of passage and exclusion are characteristic of the types of decisions managers must often make. They also help illustrate the limitations of using the swim speed-fatigue time relationship in developing engineering criteria. For example, while the swim speed-fatigue time relationship may well define a theoretical maximum distance of ascent, the actual distance ascended will depend heavily on the behaviors or strategies fishes employ when attempting to traverse a velocity barrier. Bainbridge's (1960) Figure 15 shows a clear optimum swim speed that maximizes distance against a given velocity of flow (Figure 12.2), but if a fish were to deviate substantially from this optimum, then its maximum distance of ascent may be only a small fraction of its potential (Castro-Santos, 2002, 2005).

This effect is compounded by the fact that swim speeds above maximum sustained speed are not described by the single, log-linear relationship described in Eq. (1). Instead, as mentioned previously, there exist at least two modes of steady state, unsustainable swimming speeds: prolonged and sprint modes. These modes are characterized by distinct parameters of slope and intercept, with the consistent pattern that the slope of the sprint mode relationship is less steep than that for the prolonged mode (Figure 12.1).

The presence of these two modes of unsustainable steady swimming means that competing distance maximizing optima exist. Castro-Santos (2002, 2005) has shown that the distance maximizing swim speed within a given mode is equal to $U_f - 1/b$, where U_f is the flow velocity, and b is the slope coefficient in Eq. (1). This means that the distance maximizing swim speed is a constant groundspeed, regardless of the speed of flow. Castro-Santos (2005) further demonstrated that at a certain critical speed of flow (U_{fcrit}), fishes should switch modes, if possible, from the optimal ground-speed for prolonged mode to that for sprint mode. Failure to make this switch prevents fishes from realizing their maximum potential traversed distance (D_{max}).

The ability to select the best mode in which to swim may be determined by underlying causal mechanisms. Although these mechanisms have yet to be determined, two hypotheses seem reasonable. First, the modes might be a result of a discrete shift from unsustainable use of aerobic (red) musculature to primary recruitment of anaerobic (white) musculature (Brett, 1964). Jayne and Lauder (1994), however, found that, for bluegill sunfish (*Lepomis macrochirus*) at least, the pattern of white muscle recruitment is not discrete, but instead is incremental with speed.

A second hypothesis for the observed pattern is the presence of a kinematic gait shift at the transition from prolonged to sprint speeds. The shift from median and paired fin swimming to body/caudal fin swimming has been shown to exert a similar influence on swimming energetics (Drucker and Jensen, 1996; Webb and Gerstner, 2000; Korsmeyer *et al.*, 2002) and facultative gait selection is widely described for other taxa (Alexander, 1989; Weinstein and Full, 2000). Recent evidence supports this hypothesis, suggesting that the transition from sustained to prolonged mode reflects the switch to a burst-coast gait (Peake and Farrell, 2004). It seems likely that the onset of sprint mode reflects a return to steady-state swimming, powered exclusively and continuously by anaerobic processes (Castro-Santos, 2005; see also Brett, 1964, Figure 20). Under either hypothesis, fishes should always be able to switch from prolonged to sprint mode when traversing velocity barriers. Whether they are able to choose the slower mode depends on flow velocity, but also on the extent to which swimming mode is under volitional control.

The scope of variables detailed previously means that probability of passage or maximum distance swum cannot be reliably predicted from swimming capacity alone. Traversed distance ultimately results from both physiological capacity and behavioral strategies employed by the animals in question. These strategies may themselves reflect life history or biomechanical constraints. Castro-Santos (2005) found that, while anadromous clupeids pursued a distance maximizing strategy, striped bass, walleye, and white sucker did not. Thus, these non-anadromous, but nevertheless migratory species failed to realize their potential D_{max}. Calculating from the data reported in Castro-Santos (2005, Table 2) and Haro *et al.* (2004, Figure 5), it appears that the median distance of ascent was less than the maximum predicted value by about half.

The error in the estimate of realized D_{max} is a result not only of the failure to adopt an optimal swim speed, but also of the error in the model itself. Neither of these two sources of error is considered in published guidelines (Beach, 1984; Powers *et al.*, 1985; Bell, 1991; Clay, 1995) or in more recent recommendations (Peake *et al.*, 1997a,b, 2000; Peake, 2004).

In an effort to bypass the uncertainties inherent in the methods described previously, we sought to quantify D_{max} empirically, not only measuring the average distance traversed, but also describing the distribution for the entire population. At a facility inspired by the one used by Weaver and his colleagues (Figure 12.3B), we adopted and improved upon their methods. A suite of six species was allowed to volitionally ascend a large-scale (24 m long $\times$ 1 m^2), open-channel flume, against regulated flow velocities of 1.5–4.5 m s^{-1}. Progress of fishes up the flume was monitored with uniquely coded passive integrated transponder (PIT) tags (Figure 12.4; Castro-Santos

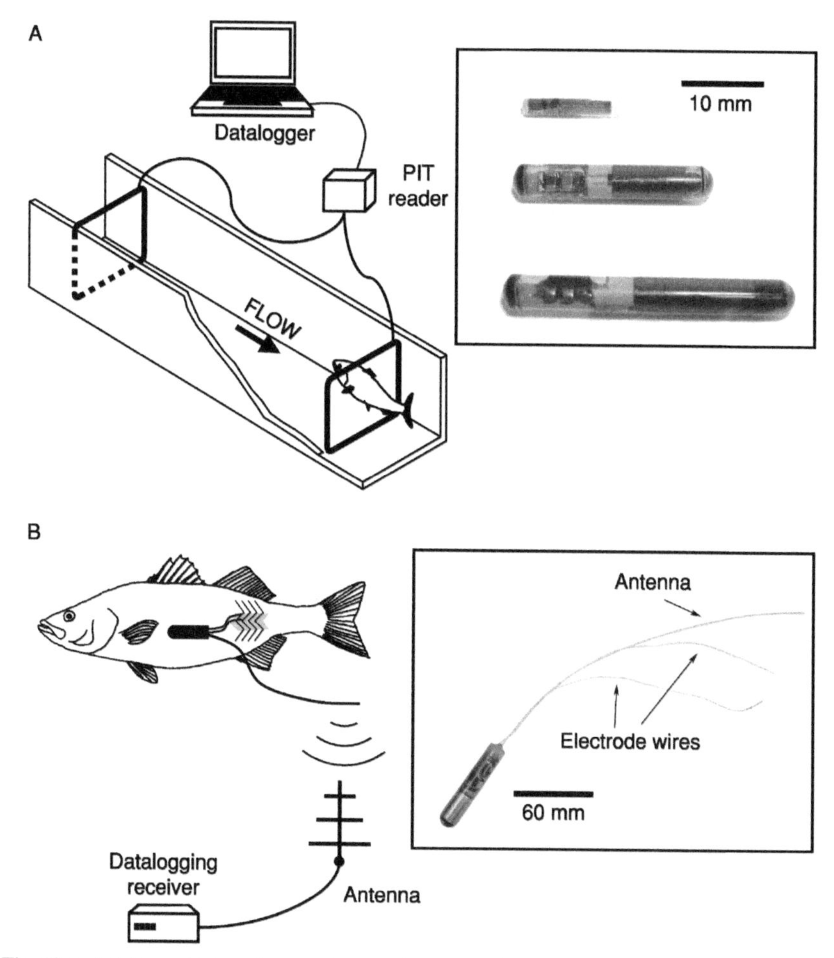

Fig. 12.4. (A) Typical laboratory application for passive integrated transponder (PIT) tags used in fish swimming studies. Tags are detected only when fishes swim through loop coil antennas and fish position (specific antenna) and time are logged. Data are later processed to determine over-ground and through-water swimming speeds, as well as distance of ascent. Inset: commercial PIT tags of various sizes; detection range is typically dependent on tag size. (B) Radio electromyography (EMG) tag application in studies of free-swimming fishes. An EMG tag is typically implanted into peritoneal cavity of fish; electrodes are inserted into fish musculature. EMG data are processed by the tag and transmitted as a pulse rate to receiver/datalogger. Inset: commercial implantable EMG tag showing electrode wires and dipole antenna.

et al., 1996), and regression methods adapted from the survival analysis literature were used to generate estimates of distance of ascent as a function of flow velocity and other biological and environmental variables (Figure 12.5; Castro-Santos, 2002; Castro-Santos and Haro, 2003).

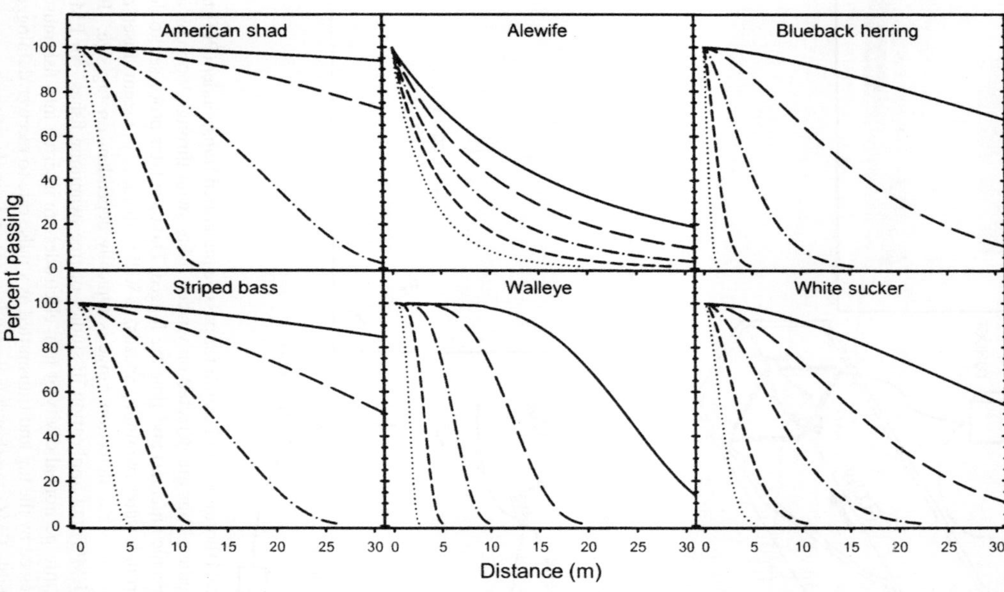

Fig. 12.5. Model curves predicting maximum distance of ascent (D_{max}) of American shad, alewife, blueback herring, striped bass, walleye, and white sucker swimming against velocities of 1–5 m s^{-1} in a large, smooth-surfaced, open-channel flume. Contours indicate flow speeds of 1 (solid), 2 (long dash), 3 (dash-dot), 4 (short dash), and 5 (dotted) m s^{-1} (Haro *et al.*, 2004).

Although failure to adopt an optimal swim speed reduces the maximum traversable distance, fishes can dramatically improve their chances of successfully negotiating a velocity barrier by staging multiple attempts. This is because, having once tried and failed to pass a barrier, those fishes that remain are still free to stage additional attempts. Provided that passage models like those presented in Haro *et al.* (2004) adequately account for the variability of the population in question, those fishes that fail to pass should retain the same probability of passage as those that succeed (i.e., the variance of D_{max} is random). This means that the same equations that were applied to the population on the initial attempt are still relevant, and of those fishes that fail on the first attempt, a similar proportion may be expected to succeed on the second or later attempts (perhaps as a result of improved approximation of U_{opt}). In this way, the likelihood of successfully traversing a barrier increases exponentially with each attempt (Figure 12.6A).

This suggests that the frequency of attempts, the amount of time required for recovery from each attempt, and the consistency of distance traversed among attempts may all be nearly as important as endurance in determining passage rates through velocity barriers (Castro-Santos 2002, 2004; see also Weinstein and Full, 2000, for an interesting discussion of the advantages of intermittent locomotion). Conversely, many fishways are constructed of

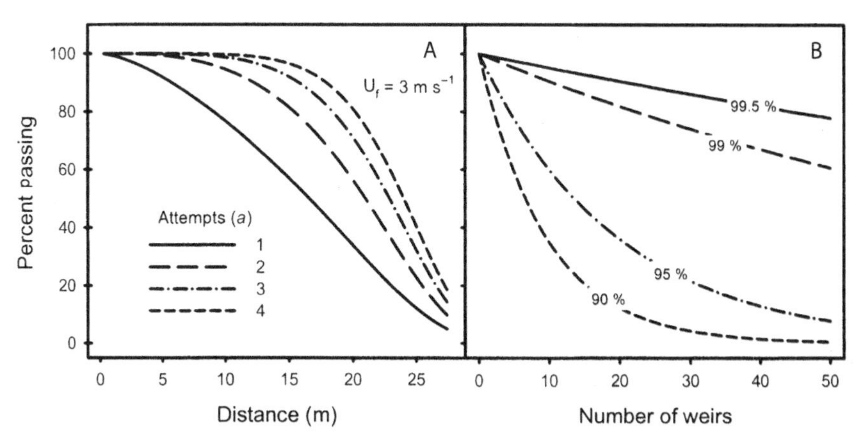

Fig. 12.6. Effect of multiple attempts on expected passage success. (A) When traversing a single velocity barrier, the proportion of population passing increases exponentially with attempt number as $1\text{-}(F(D)^a)$. Estimates are for the predicted distance of ascent by American shad against 3 m s^{-1} flow (Castro-Santos, 2002, 2004). (B) When fishes must traverse multiple weirs to ascend a fishway, likelihood of successful passage declines as P^w, where P is the probability of successfully passing a single weir and w is the number of weirs. Shown are hypothetical isopleths of $P = 0.90$ to 0.995 over 50 weirs (a moderate-sized fishway).

Fig. 12.7. Hydroelectric facility with multiple-pool fishway of the Ice Harbor type (on left). (Photograph by A. Haro.)

multiple pools (Figure 12.7), with sequential velocity challenges numbering in the tens to hundreds. These arrangements *require* fishes to stage multiple attempts in order to pass. Even low rates of failure to pass the individual weirs, chutes, or slots that separate the pools can have dramatic effects on overall passage success (Figure 12.6B). Design criteria call for inclusion of larger resting pools within these fishways, spaced so as to prevent fishes from fatiguing (Connor *et al.*, 1964). To date, very few studies that evaluate the effectiveness of these designs have been published, but there is evidence to suggest that the designs may be inadequate, especially for nonsalmonid species (Rajaratnam *et al.*, 1997; Bunt *et al.*, 1999, 2000; Peake *et al.*, 2000; Sullivan, 2004).

It is a particularly striking feature of fishway design criteria that they are almost exclusively based on the swimming performance and behaviors of anadromous salmonids. Increasing awareness of the prevalence and importance of movements of riverine species has led to exploration of alternative fishway designs (Lucas and Batley, 1996; Lucas and Frear, 1997; Lucas *et al.*, 1999; Gehrke *et al.*, 2002; Oldani and Baigun, 2002; Ovidio and

Philippart, 2002; Stuart and Berghuis, 2002). These may include submerged routes of passage for benthic-oriented species that avoid the surface (Tsunesumi, 2002), fish lifts (elevators) for weakly motivated species and for high dams, or novel structures for passing climbing species. These climbers are generally small (<10 cm in total length) and may employ oral or pelvic suckers (gobies: Schoenfuss and Blob, 2003) or simply utilize water surface tension acting along a long, thin body to hold position on steeply inclined surfaces (galaxiids: Allibone and Townsend, 1997; juvenile anguillid eels: Legault, 1988). Specialized passage structures resembling pegboards have been developed to enhance passage of some these species (Porcher, 2002). Apparently, these researchers were unaware of available biomechanical literature when developing these fishways. Instead, a trial-and-error approach was used, with no reference to previous work on the subject (e.g., Gray, 1946, 1957; Rosen, 1959; Gillis, 1998).

Although it may seem axiomatic that engineers should take advantage of innate behaviors to optimize passage success, this is not always the case. For example, hydraulic jumps (large, horizontal axis eddies, usually forming standing waves or "rollers") commonly occur below waterfalls and are known to trigger leaping behavior by salmon, assisting their ascent. Stuart (1962, 1964), who described this behavior, encouraged engineers to exploit it when designing fishways. In practice, however, fishes leaping in fishways risk injury from striking concrete baffles or from leaping out of the fishway altogether. Because this risk usually outweighs the benefits of the innate behavior, fishways are now typically designed and operated to encourage all species to swim through without leaping (Jonas *et al.*, 2004).

In other situations leaping is clearly a benefit, allowing species access to habitat that has become unavailable to non-leapers. An interesting case in point is poorly designed culverts, such as those often found at road crossings. These frequently scour out streambeds on their downstream end, resulting in a culvert "perched," sometimes at considerable height, above the stream (Roni *et al.*, 2002; Gibson *et al.*, 2005). The only species that are able to gain access to these elevated culverts are leapers (e.g., salmonids). Of course, even then there are other problems, such as insufficient depth of flow and impassibly high velocities, and there is a growing concern that the ubiquitous habitat fragmentation caused by road crossings poses a major threat to freshwater ecosystems (Warren and Pardew, 1998; Roni *et al.*, 2002). As with fishway design, the biological basis for culvert design is weak at best, and there is a pressing need for more data on attraction and entry rates, as well as the effects of such factors as depth of flow and turbulence on the ability of a broad range of species to pass these structures.

Recently, much attention has focused on the fish passage potential of nature-like fishways (e.g., several references in Jungwirth *et al.*, 1998), which typically consist of artificial channels that use boulders in place of weirs to dissipate head. It is often assumed that by attempting to mimic the local environment, these artificial streams will inevitably pass any species that requires passage. In some cases this assumption appears reasonable. Eberstaller *et al.* (1998) found distributions of native species within their fishways to be similar to those in the surrounding habitat, with free movement through the artificial reach of both resident and migratory species.

The success of nature-like fishways ultimately rests on the ability of managers to accurately identify those features that are most conducive to passage. In some cases the required conditions may be inconsistent with the nature-like approach. Fishes that naturally avoid shallow or turbulent channels (e.g., those that are found in large, low-gradient rivers) may not find these structures attractive, and if attracted, may lack the capacities or behaviors required to pass them. The development of nature-like fishways would benefit from research that describes the optimal conditions for both attraction and passage of a range of taxa—including invertebrates (Holmquist *et al.*, 1998; Fievet, 1999).

As the need for fish passage has extended beyond Europe and North America, the salmonid fishway designs have become even less appropriate. A recent study documented 361 species and subspecies of fishes in the Yangtze River alone, of which 177 are endemic and 25 are endangered. Existing and planned hydrologic alterations, including construction of Three Gorges Dam—the largest dam in the world—are expected to further degrade biodiversity within this system (Fu *et al.*, 2003). In Japan and much of south Asia, dozens of species of fishes have taken advantage of traditional wet-field rice culture. These fishes use rice fields as spawning and nursery habitat as well as adult growth during the immersed phase, migrating to and from nearby rivers with the going and coming of the wet season. In Japan, current agricultural practices require that fish passes be provided for these species to access their habitat (Hata, 2002a,b). The presence of fishes in rice paddies provides not only an important ecological role, but also an important protein source for human consumption, and can help reduce dependence on fertilizers and pesticides (Halwart *et al.*, 1996).

Swimming abilities of southern hemisphere and tropical ichthyofauna are even less well-documented than those of their north temperate counterparts. Morphologies differ, and catadromy is much more common than in the north (McDowall, 1997, 1998, 1999; Gross *et al.*, 1998). This means that, instead of prespawning adults, juveniles migrate upriver to grow and feed. These juveniles may measure as little as 10 cm total length (Mallen-Cooper, 1999), and because swimming performance scales with size (Brett, 1965; Brett and Glass, 1973; Goolish, 1991) fishways designed to pass these species must meet criteria very different from existing models (Mallen-Cooper, 1992,

1994). Such passage will become increasingly vital as ever more impoundments are created for both agriculture and hydroelectric power (World Commision on Dams, 2000; Bartle, 2002; Fu *et al.*, 2003).

There is a pressing need for more information on capabilities, behaviors (including the scope of migratory movements), and life histories of migratory fishes worldwide. While some inference across populations and species will inevitably be necessary, the reliability of these inferences remains poorly studied, and much more work needs to be done before such models can in any way be considered risk averse (Caddy, 1999). In any case, the focus on anadromous salmonid passage is no longer a viable paradigm, and the risks of continuing current practices are considerable.

B. Downstream Migrations, Passage, and Entrainment

The importance of swimming mechanics and performance to upstream migration is obvious, but they can also play an important role during downstream movements. Fishes may encounter man-made structures such as dams or water diversion intakes that can pose a serious risk to their survival, either from physical damage (i.e., turbines, screens, or spill), entrapment (impoundments or agricultural diversions), or thermal stress (cooling water structures). In the case of fishes that migrate downstream as part of their life history, passage requirements are similar to those of upstream passage: minimize delay, mortality, injury, and stress while passing the maximum numbers of fishes possible. Passage for diadromous downstream migrants is critical in that the entire upstream population must negotiate the barrier or else become landlocked. The range of size and life stages of fishes that are subject to conditions of downstream passage entrainment is extensive; individuals at risk can be very small (millimeters, in the case of larvae) to large (1–2 m, in the case of post-spawning adult salmonids, catadromous eels, sturgeons, etc.). Protection of downstream migrants is commonly implemented by diverting them away from water intakes and toward "safe" routes, usually either bypass channels or spillways. This approach requires fish to (a) be physically diverted to a safe route by a solid or impermeable structure, such as a wall or fine mesh screen, (b) be passively entrained by a flow field to a safe route, or (c) actively orient and swim away from intakes and toward safe routes.

Physical exclusion by a barrier or screen usually requires that fishes be able to swim as fast or faster than the velocity of water approaching the structure (approach velocity), usually for an extended period of time, while they search for an alternate route. If approach velocities exceed sustained swimming speeds, fishes may become impinged, resulting in injury or death when intake screen or trashrack structures are smaller than the body size of the fish (Figure 12.8). Larger screen or rack structures allow fishes to pass

Fig. 12.8. Downstream migrant adult American shad (post-spawning) swimming above exclusion racks at a hydro-electric facility. Fishes swim actively, holding station above the racks, for extended periods. Eventually, they fatigue and are impinged upon the racks. (Photograph by Boyd Kynard.)

through into the intake (i.e., become entrained) and so incur risk of turbine mortality. Screen or rack structures can have increased diversion performance if they are angled relative to the direction of approaching flow (e.g., Ducharme, 1972; Nettles and Gloss, 1987). Some fine-mesh screen designs (Eicher or wedgewire screen; Amaral *et al.*, 1999) are oriented at a steep angle relative to approaching fishes and permit the fishes to physically come in contact with the screen itself without injury. Because these screens are smooth and employ a high water velocity along rather than through the screen itself (sweeping velocity), fishes are physically swept along the face of the screen to a bypass entrance.

The ability of physical barriers to effectively exclude fishes from intakes is generally related to fish size and swimming performance. As with upstream passage, most of the data used to establish current screening criteria come from Brett-type respirometer studies using juvenile salmonids, with the presumption that a fish will not be entrained if intake velocity does not exceed its sustained swimming ability (Swanson *et al.*, 2004). Pearce and Lee

(1991) documented criteria for maximum approach velocities of <150 mm s^{-1} for salmonid fry (<60 mm in length) and <300 mm s^{-1} for fingerlings (>60 mm in length). Other studies have defined more specific screening criteria (maximum 100 mm s^{-1} approach velocity, with sweeping velocity >100 mm s^{-1} for juvenile Pacific Northwest salmonids; National Marine Fisheries Service, 1997; Bates, 1998). Swimming performance criteria for screen designs have been determined for only a few smaller life stages or weaker-swimming non-salmonid species. These can therefore be at higher risk of entrainment and impingement at screens (Swanson *et al.*, 1998, 2004).

Delta smelt (*Hypomesus transpacificus*) tested in a Brett-type respirometer displayed several velocity-dependent swimming gaits within the range of test velocities, including a "stroke-and-glide" swimming behavior at swimming velocities below 10 cm s^{-1}, continuous swimming above 15 cm s^{-1} up to U_{crit}, and discontinuous "burst-and-glide" swimming at velocities above U_{crit}. Delta smelt also displayed high variability in endurance and significant swimming failure at the transition point between stroke-and-glide and continuous swimming (Swanson *et al.*, 1998). Hayes *et al.* (2000) and Swanson *et al.* (2004) used an innovative circular "treadmill" device that created both approach and sweeping velocities that was used to determine swimming velocity criteria for smaller species including delta smelt, splittail (*Pogonichthys macrolepidotus*), juvenile American shad, and Chinook salmon (*Oncorhynchus tshawytscha*). Their findings indicated that impingement was dependent on both types of water velocity vectors, rheotaxis, illumination, and fatigue rates of fish. Although lower approach velocities did not result in direct impingement, contact with the screen was frequent at lower velocities and was not related exclusively to swimming capability. These results can be used in concert to estimate screen design parameters such as maximum screen length or bypass entrance spacing. Divergent results of data from Brett respirometer and treadmill devices coupled with the tendency for investigators to omit data from individual fishes that fail to swim at test velocities suggest that existing screening criteria may be inadequate, especially for smaller nonsalmonids.

An often-overlooked aspect of physical barriers is that fishes must be able to appropriately orient to the barrier, an ability that may be dependent on sensory mechanisms. For example, fishes that use vision to avoid contacting a barrier during the day may not be able to do so at night. Conversely, if fishes are able to detect a barrier using other means, those cues can be used to direct fish away from harmful routes. Potential stimuli such as turbulence, flow acceleration, pressure changes, and sound may play important roles in avoidance behavior (Coutant and Whitney, 2000; Coutant, 2001; Electric Power Research Institute, 2002). Detection of these cues is mediated largely by the lateral line and inner ear organs (Mogdans and

Bleckmann, 2001), the sensitivities of which are influenced by the habitat an animal occupies: fishes that typically occupy turbulent stream-type habitat will tend to have different stimulus thresholds and different responses to turbulent noise than those occupying lentic habitats (Lannoo and Lannoo, 1996; Coombs *et al.*, 2002; Engelmann *et al.*, 2003; Mogdans *et al.*, 2003; Montgomery *et al.*, 2003). How and whether the structure and function of this system change in association with migratory movements, particularly those where animals move between lotic and lentic habitats, are unknown, however. More work is needed on the ontogeny of the lateral line system to determine the potential effectiveness of these cues.

While behavioral barriers may be effective at some sites, they are prone to reduced effectiveness when environmental conditions change (i.e., increased approach velocities, turbidity, etc.). They might also be effective on only a narrow range of species that respond to specific stimuli and can elicit conflicting responses among multiple species. For example, downstream bypass entrances illuminated at night may be attractive to some species (e.g., juvenile salmonids: Gosset and Travade, 1999) but repellent to others (e.g., eels: Hadderingh *et al.*,1999).

Effective bypass operation requires that fishes either voluntarily swim downstream through the bypass, or are unable to swim back upstream through high-velocity bypass flows. This latter condition usually mandates high water velocities in bypass transit channels or pipes (>3 m s^{-1} for juvenile salmonids). Once diverted to a bypass entrance, fishes may be reluctant to enter a zone of increasing water velocity. Downstream migrant juvenile salmonids have been shown to pass entrances characterized by uniform and gradual water acceleration more readily than those with abrupt accelerations (Brett and Alderdice, 1958; Haro *et al.*, 1998).

Both behavioral and physical barriers often incur some delay as fishes locate and pass an appropriate passage route (Castro-Santos and Haro, 2003). This delay is usually associated with active swimming in an upstream-oriented direction. Downstream migrant fishes often hold station just upstream of intakes or bypasses, often swimming vigorously for long periods (Kynard and O'Leary, 1993). As fishes ultimately pass downstream over a dam crest, they typically continue high-speed swimming in an upstream direction, often in sprint mode, even though they are unable to make progress against the flow (Haro *et al.*, 1998). Energy expended on swimming for protracted periods while fishes search for or are diverted to an exit can be significant, especially if approach velocities (and hence swimming speeds) approach or exceed maximum sustained swim speed (U_{ms}). Such energy expenditures can have important implications for survival, particularly when several obstacles must be passed. Further, having passed a site, fatigued fishes may be more vulnerable to predation downstream.

A new approach to fish protection at hydroelectric projects considers minimizing harm to fishes that pass through turbine units, rather than diverting fishes from them. The premise of this concept is that turbine structures can be designed to pass fishes directly with little damage, stress, delay, or mortality, thus eliminating the need for elaborate and expensive diversion structures (Çada *et al.*, 1997). The hydraulic environment within operating turbine structures can be extreme (Figure 12.9), subjecting fishes to high water velocities (>10 m s^{-1}), high (>3 bar) and low (<0.5 bar) pressures (often with very rapid transitions between the two as fish pass through the turbine runner), high shear stresses (several thousand Nm^{-2}), cavitation, and turbulence (Turnpenny, 1998). The potential for injury or mortality (e.g., the probability of strike by a rotating turbine blade, or hazardous levels of pressure or strain rate) can be estimated using physical and numerical hydraulic modeling (Turnpenny, 1998). Conditions representative of turbine environments are difficult to replicate in the laboratory, however, and we know of only two published attempts to study these effects directly (Odeh *et al.*, 2002; Neitzel *et al.*, 2004). Relationships of these physical forces with attributes of fish biomechanics (i.e., body size, flexibility, skin toughness and resistance to puncture, bruising, or abrasion) need to be determined and require additional research under controlled conditions.

Because through-turbine water velocities exceed maximum swimming capabilities, fishes passing through a turbine were once thought to be passively transported. However, there is some evidence that fishes swim against the flow as they are drawn into the turbine structure (Coutant and Whitney, 2000), although they may be oriented randomly by the time they encounter turbine blades (Montén, 1985). The orientation of the fish may affect the risk of turbine blade strike. Other variables include fish size and point of entry into the turbine structure, the thickness and shape of the leading edge of the turbine blade, blade spacing and speed, number of blades, and water velocity (Montén, 1985; Turnpenny, 1998). Newer turbine designs attempt to reduce passage-induced mortalities by controlling these factors (Çada, 2001). It may be some time before the benefits of these new designs is widespread, however: the life expectancy of a turbine runner is on the order of 50–100 years, and existing powerhouse designs are often not compatible with the new technologies.

Large pumps, such as those used to supply cooling water at power plants, or other types used to deliberately transport fishes from collection facilities, often create hydraulic effects similar to those found in turbines. Sources of injury and mortality within pumping devices are unknown, but are probably similar to those found in turbines. These hazards may be even greater in pumps, the impellers and blades of which are usually smaller and operate at a higher speed than large turbine runners. Tests of Hidrostal pumps and

A

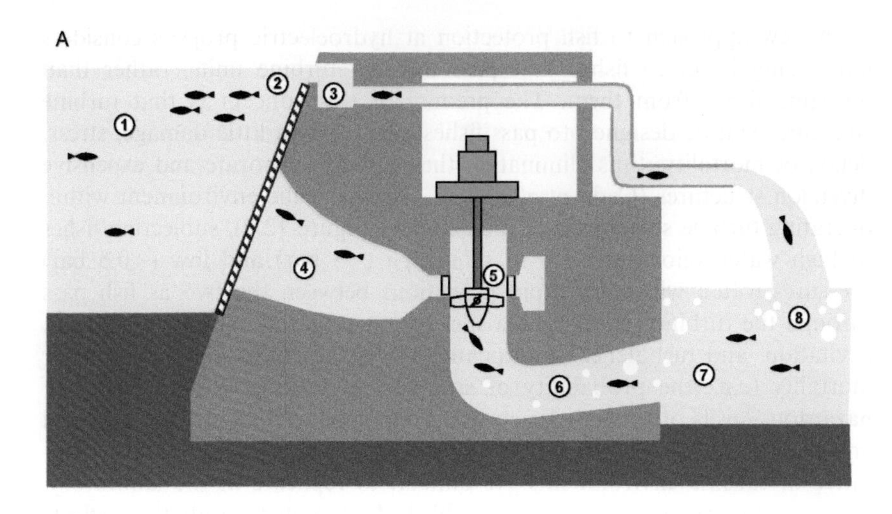

B

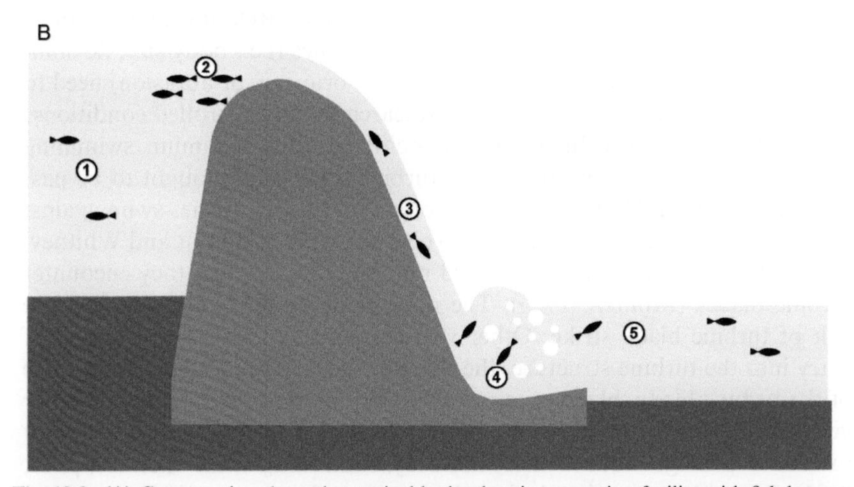

Fig. 12.9. (A) Cross-section through a typical hydroelectric generation facility with fish bypass showing routes of passage and zones of varying hydraulic conditions (circled numbers). 1: *Forebay*; low velocity (1 m s^{-1} or less) and mildly turbulent, directional flow. 2: *Bypass entrance*; accelerating flow, narrowing depth or width of passage channel. 3: *Bypass channel*; moderate to high velocity (2 to 10^{+} m s^{-1}), moderate turbulence. 4: *Penstock*; accelerating flow, moderate turbulence, increasing pressure. 5: *Scroll case, wicket gates, turbine runner*; high velocity (3 to 15^{+} m s^{-1}), potential for strike on gates or turbine blades, sudden drop in pressure, high shear stress near static surfaces. 6: *Draft tube*: increasing turbulence, gas formation or supersaturation. 7: *Tailrace*: reduced velocity (<2 m s^{-1}) but high turbulence, air entrainment. 8: *Bypass outfall*: high velocity (5 to 15 m s^{-1}), high shear stress at outfall jet/tailrace interface, high turbulence and air entrainment. Zones 3–6 are often dark and afford few visual cues for orientation. (B) Cross-section through a typical dam crest, spill bay, or spill gate. 1: *Forebay*; low velocity (1 m s^{-1}or less) and mildly turbulent, directional flow. 2: *Dam crest or gate opening*; accelerating

Archimedes lifts, however, have measured passage survival of >95% for juvenile Chinook salmon and other species up to 500 mm in length (Rodgers and Patrick, 1985; McNabb *et al.*, 2003).

Downstream passage via spill (i.e., over dam crests, spill gates, and bypasses) has historically been considered to pose little risk. Recent work, however, has shown that spillway mortality can be as high as turbine mortality, whether from fishes striking sills, aprons, rocks, or other obstructions on the downstream face of the dam, from abrasion on dam faces or sluiceway surfaces, or from high levels of supersaturated gases in plunge pools below the spillway (Heisey *et al.*, 1996; Figure 12.9). Entry of fishes into near-static water via high velocity outfall jets also has the potential to induce injury or mortality via turbulence or high shear stresses at the jet-tailrace water interface. This risk may not always be high, however: injuries and mortalities have been shown to be minimal for juvenile salmonids at jet velocities below 9 m s^{-1} in simulated low-discharge outfalls (<2.8 m^3 s^{-1}; Neitzel *et al.*, 2004) and below 15 m s^{-1} in high-discharge outfalls (>28 m^3 s^{-1}; Johnson *et al.*, 2003). Injuries to fishes are thought to be less frequent in high-discharge outfalls because the boundary layer of the jet where shear stresses are highest occupies a smaller proportion of the jet's total volume. Because available studies have used a qualitative approach to characterizing injury and mortality, they do little to enhance understanding of the biomechanical basis of these injuries. Downstream passage technologies would benefit from more detailed analyses on the effects of size, body form, abrasion resistance, etc., on survival and injury rates.

Turbulence in areas downstream of dams and hydroelectric projects (i.e., draft tubes, spillways, and bypass outfalls) can be extreme and often is much greater than in most regions of a natural river (Coutant and Whitney, 2000). In addition to high turbulence, these regions may also contain areas of high shear stress, air entrainment, and supersaturated gases, which can be debilitating or lethal to fishes. Mortality incurred from high levels of turbulence has been measured in egg, larval, and small juvenile stages of a few species (Killgore *et al.*, 1987; McEwen and Scobie, 1992) but has not been extensively evaluated for larger stages. Some evidence exists that exposure to turbulent environments can debilitate fishes by eliciting intensive swimming. This leads to exhaustion and decreased sensitivity to external stimuli, including increased susceptibility to predation (Odeh *et al.*, 2002).

---

flow, narrowing depth or width of passage channel. 3: *Dam face*: high velocity (5–10 m s^{-1}), high shear stress near static face of dam. 4: *Plunge pool*: lower velocity, high shear stress, turbulence, and air entrainment/gas supersaturation. 5: *Tailrace*: reduced velocity (<2 m s^{-1}) but high turbulence, air entrainment. Zones 2–4 in both figures represent areas where fishes may spend considerable energy swimming at high speeds for significant durations.

III. TOWED FISHING GEAR

Mechanisms of fish capture by towed fishing gear in lentic environments can be quite similar to the processes influencing entrainment at fixed structures in flowing water. Even though the barrier in the case of towed gear is a net moving through the water, both capture and entrainment "efficiencies" are functions of biomechanics, energetics, and behavior. For example, although mesh geometry of these gears is designed to retain target species and size classes, avoidance and net entry are governed largely by visual cues. Many species avoid contact with nets, even though they may be able to physically pass through meshes (Glass *et al.*, 1993). In an interesting fusion of behavior and kinematics, Ozbilgin and Glass (2004) found that net penetration was influenced by social learning: haddock (*Melanogrammus aegelfinus*) that at first avoided the net were more likely to attempt to pass through the mesh after having observed successful penetration by their conspecifics.

Wardle (1993) reviewed the development of various types of seine and trawl gear and systematically characterized a sequence of typical responses of fishes to towed gears with respect to visual and other cues (Figure 12.10). Fishes generally swim at the minimum speeds necessary to maintain a safe distance between themselves and towed objects (i.e., otter boards, sweeps, and trawl net), and vision is the primary sense driving reactions at close range to the net. Once herded to the mouth of the net, larger individuals often maintain position just ahead of the trawl. This is a behavior one would expect of fishes swimming in unsustainable modes and trying to maximize time to fatigue (Castro-Santos, 2005)—an appropriate response, perhaps, for attempting to outdistance a predator, but not always effective for escaping trawls. Fishes are entrained into the net after they become exhausted or else volitionally turn toward the cod end of the net, seeking an alternate escape route (Walsh and Hickey, 1993).

New technological developments of towed fishing gear attempt to take advantage of fish behavior and morphology to increase gear efficiency and selectivity (Herrmann, 2005a,b). Herding and capture efficiency of trawls are largely a function of fish size, swimming endurance, temperature, and mesh selectivity (He, 1993; Winger *et al.*, 1999, 2000). Although probability of capture can be estimated from these variables, behavioral reactions to towed gear (e.g., optomotor and erratic responses) may be as important to gear efficiency as performance and physiological constraints (Kim and Wardle, 2003). Ideally, net designs capitalize on kinematic and behavioral differences among species to improve efficiency, segregate catch, and reduce bycatch. Although there has been some success on this front (e.g., Broadhurst, 2000),

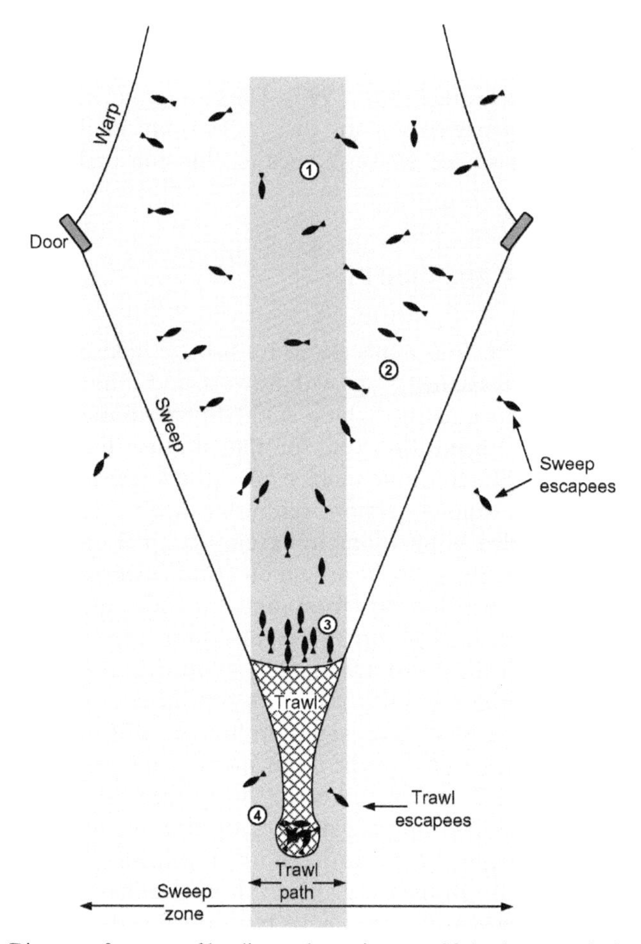

Fig. 12.10. Diagram of process of herding and entrainment of fishes into a typical otter trawl. 1: Fish unaware of approaching trawl exhibit normal schooling and low speed swimming behavior. 2: Approach of doors and sweeps induces avoidance reaction and herding of fish into trawl path. 3: Herded fish congregate near mouth of net and swim at relatively high speeds, often for extended periods of time. 4: Fish turn into net cod end, either volitionally or after exhaustion, and are either captured or escape through net mesh. (Adapted from Winger *et al.*, 1999; see also Wordle, 1993; Godø, 1994.)

there remains much room for improvement, and bycatch continues to be a major source of waste and inefficiency among commercial fisheries. Furthermore, even the better net designs do not always eliminate fatigue among the excluded nontarget species. Having escaped the trawl, these species may still suffer from post-exercise mortality or increased predation risk

(Ryer *et al.*, 2004; Ryer, 2004). Post-escapement survival can vary widely among species, and smaller escaped individuals sometimes sustain higher injury than larger ones (Soldal *et al.*, 1993). The biomechanical basis of these escapement and injury patterns is not always clear, but may be determined by factors such as resistance of fish bodies to abrasion or bending (Revill *et al.*, 2005).

IV. INTRASPECIFIC DIVERSITY

In addition to the obvious implications for fisheries conservation of such anthropogenic factors as dams, water withdrawals, and commercial fisheries, biomechanical concerns underlie habitat suitability at all life stages, regardless of the migratory disposition (or, for that matter, the taxon) of the resident population. Salmon and sticklebacks (*Gasterosteus* spp.) provide instructive examples. Among salmon, renowned fidelity to natal streams (philopatry) has enabled populations to develop adaptations that are very locally specific. For example, sockeye salmon (*Oncorhynchus nerka*) spawn in a range of habitat types that vary from sand to cobble, in riffles, pools, or lakes, and are characterized by a broad range of discharge and temperature. Redds are constructed to ensure adequate ventilation and flushing, appropriate to the habitat, and providing necessary conditions for egg and larval development. Timing of spawning, egg morphology, and growth rates are synchronized such that fry emerge the following spring when food is readily available (Quinn *et al.*, 1995; Hilborn *et al.*, 2003). Depending on spawning location, fry might need to migrate up- or downstream to arrive at their lacustrine nursery habitat, a behavior that is under genetic control (Brannon, 1972; *sensu* Quinn and Dittman, 1990). Analogous behaviors, along with functionally significant adaptations, have been documented for other salmonid species as well (Northcote *et al.*, 1970; Tsuyuki and Williscroft, 1977; Kelso *et al.*, 1981; Northcote and Kelso, 1981; Taylor and Larkin, 1986; Nemeth *et al.*, 2003).

After a variable period of freshwater growth, young sea-run salmon metamorphose into smolts and migrate to the marine habitat where they will experience much greater growth rates. The transition from fresh- to saltwater is preceded by morphological, behavioral, and physiological changes. Smolts become longer and leaner (Hoar, 1939; Winans and Nishioka, 1987) and their swim bladders increase in volume (Saunders, 1965). These changes prepare the young salmon for their downstream migration. The increased buoyancy accompanies a switch from solitary, benthic-oriented territorial behavior to active, surface-oriented schooling behavior (Hoar, 1976). The functional importance of the length change is somewhat confusing. Because length-specific swimming ability decreases among smolts (Coughlin, 2002), many authors have interpreted this as reducing their ability to stem currents, thereby facilitating passive downstream transport (various references in McCormick and Saunders, 1987; Peake and McKinley, 1998). In fact, however, absolute swimming ability increases with the increased length, and it may be that ability to capture prey and escape predators is actually enhanced by these changes (Peake and McKinley, 1998). Downstream migration, therefore, does not result from reduced swimming ability, but rather appears to be an endocrine-mediated volitional response (Specker *et al.*, 2000).

Concurrent with these morphological changes, the structure of the osmoregulatory apparatus is reorganized. For a brief period (perhaps as little as 2 weeks), salmon are able to maintain osmotic homeostasis in either fresh- or saltwater. Failure to arrive at the marine environment within the so-called "smolt window" can have catastrophic consequences for survival, owing both to reduced osmoregulatory ability (McCormick *et al.*, 1998), and, perhaps consequentially, to reduced ability to detect and avoid predators (Handeland *et al.*, 1996). Similar limitations appear to be present for other anadromous species (Zydlewski and McCormick, 1997a,b).

The timing of these migrations is influenced by both environmental and genetic factors (Whalen *et al.*, 1999; Nielsen *et al.*, 2001). Presumably, this timing reflects long-term average optima, allowing the fishes to arrive at the marine environment both when they are best able to withstand the physiological stress of the transition to seawater and when the marine environment is most likely to be conducive to rapid growth (Hvidsten *et al.*, 1998; Friedland *et al.*, 2000, 2003). Rapid growth upon entry to seawater is considered vital for survival: larger size reduces risk of predation, increases the range of prey items available, and appears to be a prerequisite for some populations that undergo lengthy marine migrations (Schaffer and Elson, 1975; Roff, 1988, 1991; Alexander, 1998). In summary, downstream migrations involve a complex interplay of behavior, physiology, and biomechanics, the intricacies of which are still not well understood but which do show evidence of an important role for local adaptation.

Natal habitat also exerts influence over the shape of the returning adults. For example, male salmon with deeper bodies may have greater access to females (Quinn and Foote, 1994), but deeper-bodied individuals of both sexes are more vulnerable to predation by bears and can be less able to traverse or exploit shallow spawning habitats (Quinn and Kinnison, 1999; Quinn *et al.*, 2001a,b; Figure 12.11). In some cases, fishes with longer freshwater migrations have greater length and less depth than those with shorter migrations. This is true of some salmon populations and also of threespine sticklebacks (*Gasterosteus aculeatus*) (Schaffer and Elson, 1975;

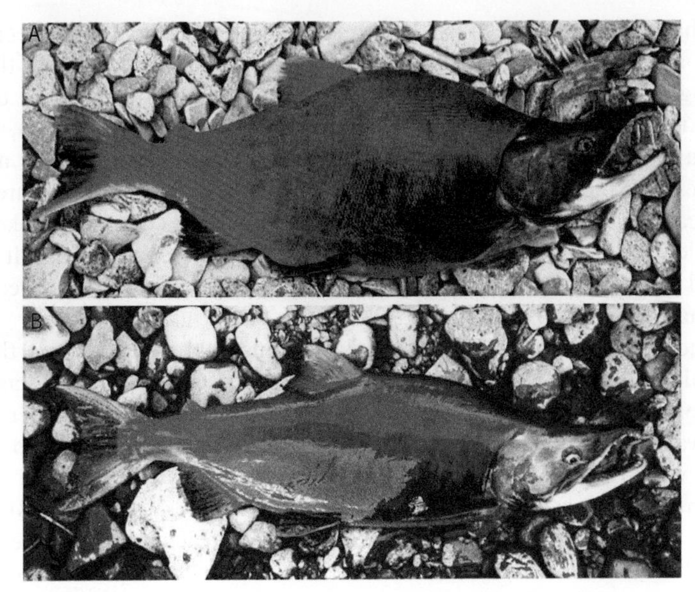

Fig. 12.11. Sockeye salmon vary widely in body morphology. Shown are photos of typical males spawning in beach (A) and river habitats (B). (Photographs by Andrew Hendry.)

Taylor and Mcphail, 1985a; Labeelund, 1991; Schaarschmidt and Jurss, 2003), although it may not be true of Pacific salmon (Fleming and Gross, 1989; Roni and Quinn, 1995). Where this effect occurs, it appears that the more fusiform body shapes are adaptive, providing improved sustained swimming stamina and energetic efficiency over the nonmigratory populations (Taylor and Mcphail, 1985b). The nonmigrants, in turn, have deeper bodies and superior fast-start performance. This performance measure has obvious implications, both for ability to capture prey and to avoid predators, and Taylor and McPhail (1985b, 1986) interpreted the deeper-bodied form as being adaptive for fast starts. Hydrodynamic theory, however, suggests otherwise. Fast-start specialists typically have a fusiform shape coupled with posterior positioning of the median fins (Webb, 1988, 1998), and Walker (1997) argued from habitat, morphometric, hydromechanical, and behavioral data that the deeper body morphologies are more likely to be primarily adaptive for maneuverability. Under this interpretation, the observed differences may have more to do with habitats occupied and the need to exploit highly structured environments.

These two interpretations are not mutually exclusive. Maneuverability may still provide good predator protection, and the improved fast-start performance may be a result of the morphological changes, even if they are not optimized for this behavior (Walker, 1997). Moreover, deeper body forms are often more difficult for predators to process, and may function as a gape-limiting predator defense mechanism (bear predation on sockeye being a notable exception). In any case, predator–prey relations can certainly be a key feature influencing life history and morphology of local populations (Magnhagen and Heibo, 2001, 2004; Doucette *et al.*, 2004; Langerhans *et al.*, 2004).

Wherever morphological diversity reflects locally important adaptations, there are significant implications for fisheries management. Meristic and morphological differences have long been used in stock identification (Hill, 1959; Ihssen *et al.*, 1981a,b; Winans, 1984), and local adaptation may be the rule, rather than the exception. Proper understanding of the importance of these adaptations requires integrated studies, however, synthesizing morphological, physiological, ecological, and behavioral data and identifying key performance measures that define niche characteristics of populations and the extent to which they are locally specific. A convincing case for the importance of biomechanics in these characters requires a functionally rigorous analysis of the relationship between observed behaviors and both body and fin morphologies.

Even where such work is achieved, its relevance may still be questioned, as the complexities of various ecological interactions might preclude ready extrapolation to generalities (Kerr, 1976). Nevertheless, it is the outcome of these interactions that are of interest to managers, and some trends in available data suggest that these issues cannot be ignored. For example, while philopatric adaptations presumably benefit local populations, they can also put them at risk. The extent of specialization is often so great that individuals that stray to nearby spawning sites experience substantial reductions in fitness (Tallman and Healey, 1994). Similarly, and perhaps because they often originate from remote locations, fishes of hatchery origin have reduced fitness in their introduced range (Verspoor, 1997; McLean *et al.*, 2003).

Growing awareness of the importance of local adaptations has caused researchers and managers to reevaluate stocking practices. Whereas supplementation of native stocks through hatchery rearing was once viewed as entirely beneficial, the practice is now known to hold considerable risk. Despite the reduced fitness mentioned previously, substantial introgression of nonnative genes has occurred as a direct result of introductions from foreign stocks. In some cases, these may have contributed to the very declines they were intended to prevent (Anonymous, 1991; Ryman *et al.*, 1995). This situation could be substantially aggravated by the recent growth of the aquaculture industry, particularly of Atlantic salmon (Hindar *et al.*, 1991; Nash and Waknitz, 2003; Waknitz *et al.*, 2003).

In New England, efforts to restore the Atlantic salmon have intentionally randomized matings of returning adults in order to avoid inbreeding depression (Moring *et al.*, 1995; Connecticut River Atlantic Salmon Commission, 1998). With improved genetic techniques, however, the ability to assess the risk of inbreeding has improved (King *et al.*, 2000; Spidle *et al.*, 2003), and broodstock are now segregated by river in Maine (Moring *et al.*, 1995). Future techniques may allow for even finer-scale management of broodstock, and there is cause for optimism that management practices may be tailored to match the phenotypic requirements of populations on a subdrainage scale (Letcher and King, 2001; Youngson *et al.*, 2003).

Because philopatry is so well documented among freshwater species, they provide excellent examples of the importance of intraspecific phenotypic diversity to management and conservation strategies. This diversity has long been attributed to biogeography and the intrinsic fragmentation of freshwater environments (Darwin, 1859). Recent evidence, however, suggests that it is not exclusively the province of freshwater and anadromous fishes. Ruzzante *et al.* (1996a,b) provided evidence that North Atlantic cod stocks are much more heterogeneous, both genetically and morphologically, than previously thought. Adaptations may exist on an even finer scale: Pepin and Carr (1993) described morphological differences associated with both depth and latitudinal gradients, and Ruzzante *et al.* (2000) found differences in size among bays within the near-shore stock. The persistence and geographical distribution of these genotypic and phenotypic differences suggest that recovery of cod stocks may depend on recruitment from a smaller spatial scale than provided for in current management divisions (Ruzzante *et al.*, 1996a, 1997, 2000, 2001).

These findings appear to be part of a pattern, in which species believed to be panmictic across their range actually exhibit metapopulation structure (Ryman *et al.*, 1995; Thorrold *et al.*, 2001; Wirth and Bernatchez, 2001). Although the adaptive value of philopatry and population substructure is often unknown and difficult to demonstrate, adaptation remains a likely explanation for this emergent pattern. Management strategies should strive to maintain diversity where it exists; conservation measures that overlook important local adaptations may actually work to depress those stocks they are designed to maintain (Booke, 1999; Conover and Munch, 2002; Hilborn *et al.*, 2003).

V. BIOENERGETICS MODELING

Much of the preceding discussion has focused on swimming performance and how it affects the ability of fishes to traverse barriers, escape entrapment, or adapt to local environments. Integral to swimming capacity are the mechanics of thrust generation and the associated energetic costs. The accuracy of swimming cost estimates is of considerable interest to fishery managers. Bioenergetics models (Winberg, 1956; Brett and Groves, 1979; Kitchell, 1983; Adams and Breck, 1990) provide a theoretical framework for relating rates of feeding, activity, and growth, and can provide insights into causal mechanisms among these variables. They have been used to estimate effects of predators on prey base (Rand *et al.*, 1993, 1995; Petersen and Ward, 1999), of prey on predator growth and recruitment (Yako *et al.*, 2000); and of thermal regimes on life history strategies (Glebe and Leggett, 1981a,b; McCann and Shuter, 1997), including the consequences of climate change (Linton *et al.*, 1997; Welch *et al.*, 1998; Petersen and Kitchell, 2001). These models have also been used to characterize habitat quality (Letcher *et al.*, 1996; Nislow *et al.*, 2000) and are a core component of ecosystem-based management (Latour *et al.*, 2003).

Interestingly, models describing cost of activity (COA) measured using respirometry (oxygen consumption: mg O_2 kg^{-1} h^{-1}) often do not agree mathematically with those based on hydrodynamic principles. Brett (1964) noted that hydrodynamic theory predicts that COA should be a power function of U_s:

$$COA = a + bU_s^c, \tag{8}$$

where c is expected to take on a value of about 3 (Webb, 1975, 1993; Videler, 1993). Brett (1964), however, determined that for his dataset, an exponential relationship better described the effect of swim speed on oxygen consumption:

$$\log(COA) = a + bU_s. \tag{9}$$

Oxygen consumption is a meaningful measure of activity costs, but not necessarily of mechanical power produced (Brett, 1962). This is because it reflects the actual energy requirements of locomotion: variance in the efficiencies of muscle contraction, in the contributions of aerobic and anaerobic metabolism and in the endpoints of metabolic byproducts, means that the cost of locomotion will always be greater than the work performed, and the association may not scale according to hydrodynamic principles alone (Driedzic and Hochachka, 1978; Jones and Randall, 1978; Moyes and West, 1995). Webb (1993) and Korsmeyer *et al.* (2002) maintained that the power function [Eq. (8)] was still preferable. Webb further argued that residual analysis did not indicate sufficient improvement in model fit to justify the exponential model over the power function model. Acknowledging the complexity of the underlying biological processes, however, he concluded that, although the power function model maximizes hydrodynamic verisimility, sufficient uncertainty exists surrounding these other components to justify using whichever model produces the best fit.

The choice of which energetic model to use may be largely academic over the sustainable range of swimming speeds. Important inaccuracies arise primarily when extrapolating beyond the observed data, i.e., to prolonged and burst speeds. Figure 12.12 is derived from Webb (1971), in which COA (oxygen consumption) of rainbow trout (*Oncorhynchus mykiss*) was modeled using Eq. (9), and a curve fit to the data using Eq. (8). The two models agree almost exactly ($r^2 > 0.999$) over the modeled range of swim speeds ($10.1 \text{ cm s}^{-1} < U_s < U_{\text{crit}} = 58.1 \text{ cm s}^{-1} = 2$ lengths (L) per second; Webb, 1971). The models diverge, however, at higher speeds. At twice the observed U_{crit} ($4\ L\ \text{s}^{-1}$), the exponential model exceeds the power model by a factor of 2.1 (note that this is still within the range of expected critical swim speeds for this species [Beamish, 1978]). At burst speeds of $10\ L\ \text{s}^{-1}$, this error has increased to a factor of 158.

Respirometry data can also be used to estimate standard (resting) metabolic rate (SMR) by extrapolating to zero swim speed. In the example

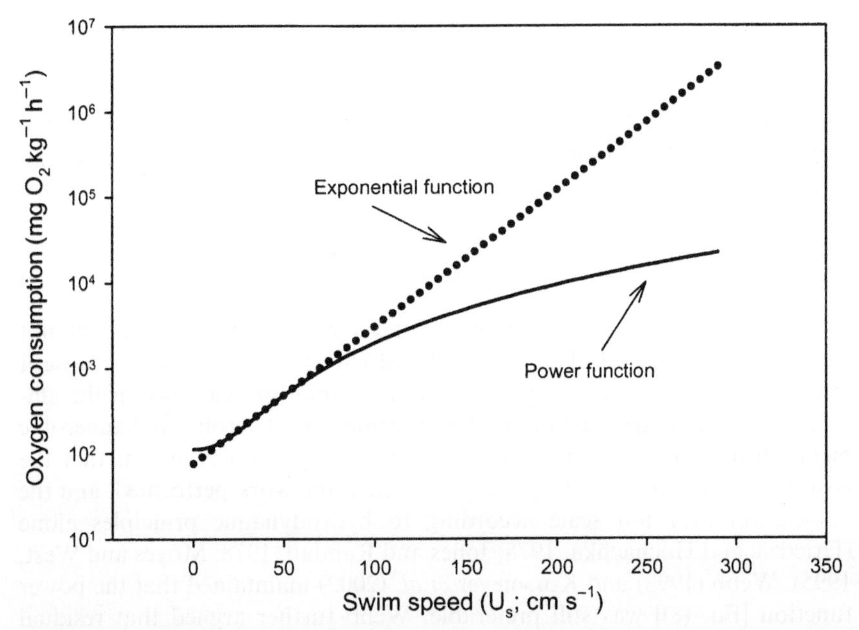

Fig. 12.12. Comparison of extrapolated values of activity costs (COA) from power function model (solid curve) and exponential function model (dotted line). Data are from Webb (1971), where a power function was fit to the exponential function provided over the observed range of swim speeds (10.2–58.1 cm s^{-1}). Note that curves agree well within the observed range, but diverge widely for extrapolated values.

presented here, the power function model exceeds the exponential model by 48%. An alternative method for estimating SMR is to measure oxygen consumption of fishes that have been immobilized with a neuroblocker. Leonard *et al.* (1999) used this approach for American shad, and found that the results agreed well (within 13%) with those generated by extrapolating to zero swim speed using the exponential model [Eq. (9)]. An equivalent power function would have overestimated the directly measured SMR by 93%. Thus, although there appears to be little cost in choosing the wrong model over the range of swim speeds typically observed in laboratories, extrapolations outside the observed range should be viewed with caution. Substantial errors can result if inaccurate models are applied to a field setting, where swim speeds may often fall outside the modeled range; more work is needed to verify the accuracy of these projections.

Accurate and realistic bioenergetics models can provide important insights into the costs of migration, as well as the effects of barriers to migration. Bernatchez and Dodson (1987) used them to test for optimizing behavior among migrant fishes, and found that only those with the most arduous migrations swam at speeds that minimized cost of transport, suggesting that energetic concerns are not necessarily limiting for the species tested. Among those populations with difficult migrations, however, the energetic consequences of challenging hydraulic and other environmental conditions can be dramatic (Leggett and Whitney, 1972; Leggett and Carscadden, 1978; Jonsson *et al.*, 1997; Leonard and McCormick, 1999), and in some years may even lead to stock failure (Rand and Hinch, 1998).

An important restriction on the reliability of bioenergetics models is that they involve so many estimated parameters that broad assumptions are required about the behaviors of individual fish. The form of the generalized model is:

$$C = M + G, \tag{10}$$

where C is the assimilated portion of consumed food, G is growth, either somatic or allotted to reproduction, and M is the total energy budget of the fish, including standard metabolic rate, costs of food processing (digestion, etc.), and costs of activity (Figure 12.13). Of these, G is readily measured in the field, and components of C and M can be estimated in the laboratory. Activity costs can be measured in the laboratory as a function of swim speed, but the speed at which wild fishes actually swim is largely unknown. For this reason, bioenergetic models typically require the assumption that activity costs take on some fixed value, usually a multiple of the standard metabolic rate (e.g., Winberg, 1956; Kitchell *et al.*, 1977). Commonly, this is set to a value believed to optimize some energetic feature, such as minimizing swimming costs (Weihs, 1973) or maximizing growth rates (Ware, 1975, 1978).

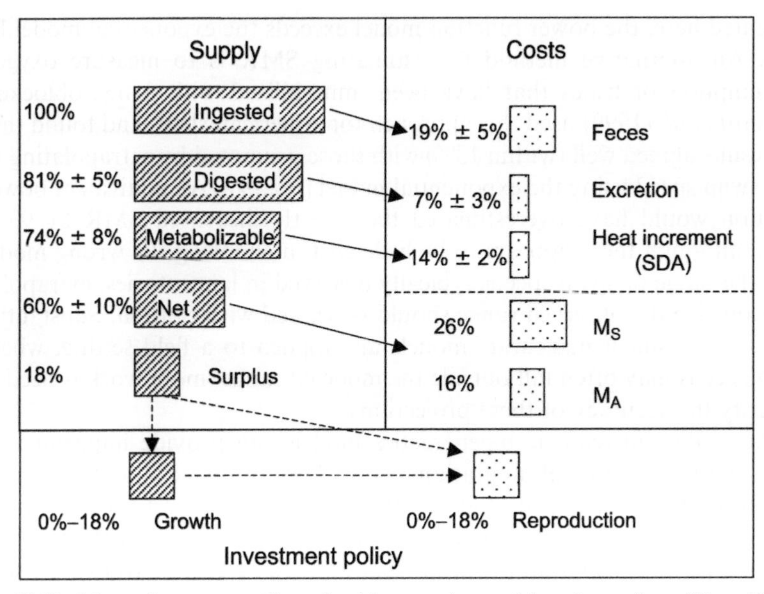

Fig. 12.13. Schematic representation of a bioenergetics model, redrawn from Ware (1982). Components of Eq. (10) are broken down to supply, costs, and investment policy. Assimilated food consumed [C in Eq. (10)] is the net supply; standard (M_s, i.e., SMR) and active (M_a) metabolism comprise the total energy budget [M in Eq. (10)]; and investment policy [G in Eq. (10)] is allotted to growth and reproduction. Numeric values are percentages, thought to represent a "typical carnivorous fish."

This practice, though convenient, can introduce substantial bias into bioenergetics models, even to the point of compromising their legitimacy. Boisclair and Leggett (1989) found that, contrary to common practices and assumptions, activity composed a major component of the overall energy budget (as much as 40% in their study) and was highly variable among populations. This contrasts starkly with the expectations based on optimal swim speeds (Ware, 1980; Figure 12.13), and reflects variability, not only in mean activity level, but also in its variance.

The importance of variability in activity to overall energy budgets bears on the preceding discussion of appropriate energetics models. Even relatively rare bouts of burst swimming can have a substantial effect on daily energy budgets (e.g., Tang *et al.*, 1994; Reidy *et al.*, 1995; Enders and Herrmann, 2003; Figures 12.12 and 12.13), and understanding the magnitude and frequency of this effect should be an objective for those engaged in bioenergetics modeling (Hansen *et al.*, 1993; Comeau and Boisclair, 1998; Aubin-Horth *et al.*, 1999; Boisclair, 2001).

A primary reason why simplifying assumptions about fish activity patterns are often made when modeling energetics is the difficulty inherent in acquiring real-time data from free-ranging animals. Recent developments in telemetry technology promise to remove this barrier, however. Early use of custom-engineered accelerometer and heart rate transmitters (Johnstone *et al.*, 1992; Lucas *et al.*, 1993) has given way to commercially available electromyogram (EMG)-transmitting radio tags (McKinley and Power, 1992; Hinch and Rand, 1998; Hinch *et al.*, 2002; Standen *et al.*, 2002; Cooke *et al.*, 2004). These tags are equipped with electrodes that are placed about 10 mm apart, usually within the red (aerobic) axial musculature of the fish (Figure 12.4B). The tags transmit a filtered signal, representing the rectified EMG integrated over several tail beats. With appropriate calibration, these transmitters can provide detailed information on activity patterns of free-swimming animals in near-real time. Because of the attenuating properties of salt water, radio applications are restricted to freshwater environments. Alternative methods described by Dewar *et al.* (1999) and Lowe *et al.* (1998), however, promise to expand this capability to the marine environment.

An important limitation to the commercially available units used by McKinley and Power (1992) and Hinch and Rand (1998) is that they need to be individually calibrated (preferably in the field; Geist *et al.*, 2002). Placement of electrodes, both with respect to each other and longitudinally, can affect data output (Beddow and McKinley, 1999), as can changes in muscle composition, suggesting that even individually calibrated applications can change over time (Beddow and McKinley, 1998).

A further limitation of these tags is that they may not provide accurate data at high swim speeds. Typically, the EMG signals are calibrated at sustained swim speeds, and at these speeds the data they provide can be quite accurate (Hinch *et al.*, 1996). Reliability of the relationship between the EMG signal and tail-beat frequency, however, will be compromised at speeds well outside of the calibrated range. One way to circumvent this problem is to transmit the raw EMG (Dewar *et al.*, 1999), or, perhaps more intuitively, to transmit the actual motion of the tail, which can be accomplished using some form of accelerometer (Lowe *et al.*, 1998). These tags may allow more precise monitoring of rapid, powerful movements, the frequency and magnitude of which could have substantial energetic consequences. In any case, data of unprecedented accuracy and detail are now available to describe activity patterns of individual free-ranging fishes, thus eliminating one of the major weaknesses of bioenergetic models. Coupled with emerging technologies and data describing large-scale movements of free-ranging fishes (Block *et al.*, 1998, 2002; Welch and Eveson, 1999; Welch *et al.*, 2002), bioenergetic modeling, particularly when individual based, is poised to become a very valuable management tool (Rand *et al.*, 1997; Rand and Hinch, 1998; Latour *et al.*, 2003).

VI. CONCLUSIONS AND RECOMMENDATIONS

From the preceding sections, one might get the impression that fisheries management is a deterministic discipline, and that biomechanics and other organismic disciplines alone can provide the basis for sound management. This is clearly not the case. Population dynamics are notoriously stochastic processes, and it is often impossible to draw generalities from a few specifics. Such complex systems are said to have "emergent properties," characteristics that could not be predicted from the sum of their parts; reductionist methods of analysis are therefore unlikely to yield reliable predictions unless preceded by evaluation of the system as a whole (Kerr, 1982). Thus, unanticipated changes to the environment, whether from climate change, introduction or removal of a predator, or the invasion of exotic species, may produce effects not predictable from autecological arguments alone.

This does not mean that management solutions or predictions from biomechanical principles are futile. On the contrary, as we have shown, biomechanics plays an integral role in resolving engineering problems posed by conservation objectives. The value of biomechanics in addressing broader management goals may be less intuitive but can still be important. Retrospective analyses of existing datasets can inform predictions of future patterns, and biomechanics can comprise an important part of these predictions; invasive zebra mussels, with their profligate pumping, filtering, and reproductive potential provide instructive examples (Ricciardi and Rasmussen, 1998; Madenjian *et al.*, 2002; Lammens *et al.*, 2004).

The integrative nature of fisheries management and conservation means that any direction of biological endeavor holds the potential to reveal relevant information. This chapter has focused largely on issues and species that are of major ecological and economic interest, especially to north temperate habitats. In doing so, we have largely ignored other taxa and locations that may be of equal or greater interest from an evolutionary, a biodiversity, or an indigenous perspective (e.g., Arthington, 1991; Ogutu-Ohwayo and Hecky, 1991; Concepcion and Nelson, 1999; Foster *et al.*, 2003).

We hope, however, that readers will recognize in our summaries a greater relevance to conservation issues generally. Dams pose barriers to upstream migration because the hydraulic and physical conditions they create exceed the capabilities (both mechanical and behavioral) of fishes to pass them. Mitigation for these projects will be best served by a holistic understanding of the factors that limit these capacities. The search for limits need not itself be limited by the search for mechanical or physiological maxima or minima. Rather, these ultimate boundaries need to take individual, behavioral, and environmental diversity into account. Predictions of performance based only on data collected under controlled conditions may be very inaccurate, and can under- as well as overestimate performance (Hinch and Bratty, 2000; Hinch and Rand, 2000; Castro-Santos, 2002, 2005).

These inaccuracies could be overcome, provided a better understanding of the complexities underlying swimming performance is gained. This objective will be best served by greater integration between laboratory and field observations. Descriptions of actual performance (especially when that performance does not match predicted maxima or minima) are very helpful to engineers, whether they are designing fish passage structures or fishing gear. Explanations of how and under what conditions fishes exceed expected limits are of substantial interest to biologists and engineers alike. In any case, an improved understanding of the effects of behavioral variability on swimming performance could greatly improve the accuracy of predictions of upstream passage success, as well as the effectiveness of diversion structures and bycatch reduction devices.

One area in which very little basic research has been performed is the biomechanics and energetics of swimming in turbulent flows. Laboratory observations of swimming capacity typically attempt to minimize turbulence, but turbulence often dominates the hydraulic environment in which fishes spend their lives, particularly those that live in lotic environments (Pavlov *et al.*, 2000). It is well known that the ability to detect and respond to turbulence varies by native habitat (Mogdans and Bleckmann, 2001; Coombs *et al.*, 2002; Engelmann *et al.*, 2003; Montgomery *et al.*, 2003). We know of no study, however, that describes how the structure of the lateral line might change for fishes that switch between lotic and lentic environments as part of their life history. Might it be that the relatively poor performance of some species in ascending fishways could be a result of stress induced by the turbulent environments? What changes, if any, must the lateral line of diadromous fishes undergo when they transition to and from the marine environment? If changes occur, are they synchronous with the restructuring of the osmoregulatory apparatus? What are the potential implications for restocking and restoration efforts?

Similarly, can the ability of fishes to tune to turbulence be used to engineer improved fishway designs? Pavlov *et al.* (2000) found that swimming performance decreased with increasing turbulence, particularly at sustained swim speeds, and Enders *et al.* (2003) showed that COA increased with increasing turbulence. These results were for highly chaotic forms of turbulence, however; where turbulence is more structured, the energetic effects may be quite different. Recent work by Liao *et al.* (2003a,b) suggested that fishes are able to achieve dramatic reductions in energy use by tuning the bending of their bodies to vortices shed by objects in flow. Although much more work needs to be done before realizing any practical application of this information, the potential benefits to fishway design (especially of nature-like fishways) are at once obvious and intriguing.

Responses to hydraulic environments are relevant to downstream passage as well, and better information is needed to understand the relative risks of turbine, bypass, and spill passage. Similarly, the effects of the severe turbulence and air entrainment that fishes encounter just after passage remain poorly understood. Development of improved (i.e., "fish-friendly") hydro-turbine design will depend on better understanding of these factors and how they can be expected to affect actively migrating fish. Development of improved guidance systems might be aided by an improved understanding of which hydraulic conditions are most attractive (or repulsive) to the target species (Pavlov *et al.*, 2000), and why.

Also, there are exciting developments in the area of bioenergetics modeling. New technologies for monitoring activities of free-ranging fish are changing the way modelers think about the importance of the activity component of the daily energy budget. Here again, behavioral diversity appears to be important. Activity patterns vary by life history, but also among individuals with similar life histories and in similar environments. This variation can have a dramatic effect on our understanding of trophic interactions, growth, and recruitment. As more information becomes available, we look forward to the prospect of improved reliability of these models as both a research and management tool.

Whether the issue is energetics, migratory timing, survival, growth, or reproduction, increasing evidence shows that local adaptations can be very important to fish species occupying a broad range of habitats. Although managers are well aware of this in general, it can be difficult to act on the apparent principles, especially when these principles are not themselves well understood. More work needs to be done to document and understand the functional importance of intraspecific diversity and to do so in a way that is rigorous with respect to both biomechanics and behavior. The data produced by such efforts will be a significant contribution, not only to management, but also to ecological and evolutionary theory; and improved integration of these disciplines in support of conservation efforts is a worthy goal indeed.

ACKNOWLEDGMENTS

Several people provided useful feedback and suggestions during the preparation of this manuscript. Paul Webb was particularly helpful, offering a detailed and careful review that resulted in substantial improvements to the final product. Helpful reviews and comments were also provided by Bob Shadwick, George Lauder, Kitty Griswold, Les Kauffman, Tom Quinn, Andrew Hendry, and several members of the staff of the S.O. Conte Anadromous Fish Research Center.

REFERENCES

Anonymous (1991). International Symposium on "The Ecological and Genetic Implications of Fish Introductions (FIN)" *Can. J. Fish. Aquat. Sci.* **48**(Suppl. 1), 1–181.

Adams, S. M., and Breck, J. E. (1990). Bioenergetics. *In* "Methods for Fish Biology" (Schreck, C. B., and Moyle, P. B., Eds.), pp. 389–415. American Fisheries Society, Bethesda, Maryland, USA.

Alexander, R. M. (1969). The orientation of muscle fibres in the myotomeres of fishes. *J. Mar. Biol. Assoc. UK* **49,** 263–290.

Alexander, R. M. (1989). Optimization and gaits in the locomotion of vertebrates. *Physiological Reviews* **69,** 1199–1227.

Alexander, R. M. (1998). When is migration worthwhile for animals that walk, swim or fly? *J. of Avian Biol.* **29,** 387–394.

Allibone, R. M., and Townsend, C. R. (1997). Distribution of four recently discovered Galaxiid species in the Taieri River, New Zealand: The role of macrohabitat. *J. Fish Biol.* **51,** 1235–1246.

Amaral, S., Taft, N., Winchell, F. C., Plizga, A., Paolini, E., and Sullivan, C. W. (1999). Fish diversion effectiveness of a modular inclined screen system. *In* "Innovations in Fish Passage Technology" (Odeh, M., Ed.), pp. 61–78. American Fisheries Society, Bethesda, Maryland.

Arthington, A. H. (1991). Ecological and genetic impacts of introduced and translocated freshwater fishes in Australia. *Can. J. Fish. Aquat. Sci.* **48,** 33–43.

Aubin-Horth, N., Gingras, J., and Boisclair, D. (1999). Comparison of activity rates of 1+ yellow perch (*Perca flavescens*) from populations of contrasting growth rates using underwater video observations. *Can. J. Fish. Aquat. Sci.* **56,** 1122–1132.

Bainbridge, R. (1958). The speed of swimming of fish as related to size and to the frequency and amplitude of the tail beat. *J. Exp. Biol.* **35,** 109–133.

Bainbridge, R. (1960). Speed and stamina in three fish. *J. Exp. Biol.* **37,** 129–153.

Bartle, A. (2002). World atlas and industry guide. *The International Journal on Hydropower and Dams* **2**(4), 3–6.

Bates, K. (1998). "Screen Criteria for Juvenile Salmon." Washington State Department of Fisheries, Seattle, Washington.

Baxter, J. S., Birch, G. J., and Olmsted, W. R. (2003). Assessment of a constructed fish migration barrier using radio telemetry and floy tagging. *N. Am. J. Fish. Mgt.* **23,** 1030–1035.

Beach, M. H. (1984). "Fish pass design–criteria for the design and approval of fish passes and other structures to facilitate the passage of migratory fish in rivers." Ministry of Agriculture, Fisheries, and Food; Directorate of Fisheries Research; Fisheries Research Technical Report, Lowestoft.

Beamish, F. W. H. (1966a). Muscular fatigue and mortality in haddock *Melanogrammus aeglevinus*, caught by otter trawl. *J. Fish. Res. Bd. Canada* **23,** 1507–1519.

Beamish, F. W. H. (1966b). Swimming endurance of some Northwest Atlantic fishes. *J. Fish. Res. Bd. Canada* **23,** 341–347.

Beamish, F. W. H. (1978). Swimming capacity. *In* "Fish Physiology, Vol. VII, Locomotion" (Hoar, W. S., and Randall, D. J., Eds.), pp. 101–187. Academic Press, London.

Beddow, T. A., and McKinley, R. S. (1998). Effects of thermal environment on electromyographical signals obtained from Atlantic salmon (*Salmo salar* L.) during forced swimming. *Hydrobiologia* **372,** 225–232.

Beddow, T. A., and McKinley, R. S. (1999). Importance of electrode positioning in biotelemetry studies estimating muscle activity in fish. *J. Fish Biol.* **54,** 819–831.

Bell, M. C. (1991). "Fisheries Handbook of Engineering Requirements and Biological Criteria." U.S. Army Corps of Engineers, Portland, OR.

Bernatchez, L., and Dodson, J. J. (1987). Relationship between bioenergetics and behavior in anadromous fish migrations. *Can. J. Fish. Aquat. Sci.* **44,** 399–407.

Blaxter, J. H. S., and Parrish, B. B. (1966). The reactions of marine fish to moving netting and other devices in tanks. *Marine Research* **1,** 1–5.

Block, B. A., Costa, D. P., Boehlert, G. W., and Kochevar, R. E. (2002). Revealing pelagic habitat use: The tagging of Pacific pelagics program. *Oceanologica Acta* **25,** 255–266.

Block, B. A., Dewar, H., Farwell, C., and Prince, E. D. (1998). A new satellite technology for tracking the movements of Atlantic bluefin tuna. *Proc. Natl. Acad. Sci.* **95,** 9384–9389.

Boisclair, D. (2001). Fish habitat modeling: From conceptual framework to functional tools. *Can. J. Fish. Aquat. Sci.* **58,** 1–9.

Boisclair, D., and Leggett, W. C. (1989). The importance of activity in bioenergetics models applied to actively foraging fishes. *Can. J. Fish. Aquat. Sci.* **46,** 1859–1867.

Booke, H. E. (1999). The stock concept revisited: Perspectives on its history in fisheries. *Fish. Res.* **43,** 9–11.

Brannon, E. L. (1972). Mechanisms controlling migration of sockeye salmon fry. *International Pacific Salmon Fisheries Commission Bulletin* **21,** 1–86.

Breder, C. M. (1926). The locomotion of fishes. *Zoologica* **4,** 159–256.

Brett, J. R. (1962). Some considerations in the study of respiratory metabolism in fish, particularly salmon. *J. Fish. Res. Bd. Canada* **19,** 1025–1038.

Brett, J. R. (1964). The respiratory metabolism and swimming performance of young sockeye salmon. *J. Fish. Res. Bd. Canada* **21,** 1183–1226.

Brett, J. R. (1965). The relations of size to the rate of oxygen consumption and sustained swimming speeds of sockeye salmon (*Oncorhynchus nerka*). *J. Fish. Res. Bd. Canada* **22,** 1491–1501.

Brett, J. R. (1972). Metabolic demand for oxygen in fish, particularly salmonids, and a comparison with other vertebrates. *Respiration Physiology* **14,** 151–170.

Brett, J. R., and Alderdice, D. F. (1958). Research on guiding young salmon at two British Columbia field stations. *Fish Res. Bd. Can.* **117,** 1–75.

Brett, J. R., and Glass, N. R. (1973). Metabolic rates and critical swimming speeds of sockeye salmon (*Oncorhynchus nerka*) in relation to size and temperature. *J. Fish. Res. Bd. Canada* **30,** 379–387.

Brett, J. R., and Groves, T. D. D. (1979). Physiological energetics. *In* "Fish Physiology Volume 8: Bioenergetics and Growth" (Hoar, W. S., Randall, D. J., and Brett, J. R., Eds.), pp. 279–352. Academic Press, New York.

British Institution of Civil Engineers (1942). "Report of the committee on fish-passes." William Clowes and Sons, London (UK)(Reprinted 1948).

Broadhurst, M. K. (2000). Modifications to reduce bycatch in prawn trawls: A review and framework for development. *Reviews in Fish Biology and Fisheries* **10,** 27–60.

Bunt, C. M., Cooke, S. J., and McKinley, R. S. (2000). Assessment of the Dunnville fishway for passage of walleyes from Lake Erie to the Grand River, Ontario. *Journal of Great Lakes Research* **26,** 482–488.

Bunt, C. M., Katopodis, C., and McKinley, R. S. (1999). Attraction and passage efficiency of white suckers and smallmouth bass by two Denil fishways. *N. Am. J. Fish. Mgt.* **19,** 793–803.

Bunt, C. M., van Poorten, B. T., and Wong, L. (2001). Denil fishway utilization patterns and passage of several warmwater species relative to seasonal, thermal and hydraulic dynamics. *Ecol. Freshw. Fish* **10,** 212–219.

Çada, G. (2001). The development of advanced hydroelectric turbines to improve fish passage survival. *Fisheries* **26,** 14–23.

Çada, G., Coutant, C. C., and Whitney, R. R. (1997). "Development of biological criteria for the design of advanced hydropower turbines." U.S. Department of Energy, Report , Idaho Operations Office, Idaho Falls.

Caddy, J. F. (1999). Fisheries management in the twenty-first century: Will new paradigms apply? *Reviews in Fish Biology & Fisheries* **9,** 1–43.

Castro-Santos, T. (2002). "Swimming performance of upstream migrant fishes: New methods, new perspectives." Ph.D. thesis, University of Massachusetts, Amherst.

Castro-Santos, T. (2004). Quantifying the combined effects of attempt rate and swimming performance on passage through velocity barriers. *Can. J. Fish. Aquat. Sci.* **61,** 1602–1615.

Castro-Santos, T. (2005). Optimal swim speeds for traversing velocity barriers: An anlaysis of volitional high-speed swimming behavior of migratory fishes. *J. Exp. Biol.* **208,** 421–432.

Castro-Santos, T., and Haro, A. (2003). Quantifying migratory delay: A new application of survival analysis methods. *Can. J. Fish. Aquat. Sci.* **60,** 986–996.

Castro-Santos, T., Haro, A., and Walk, S. (1996). A passive integrated transponder (PIT) tagging system for monitoring fishways. *Fish. Res.* **28,** 253–261.

Chopin, F. S., and Arimoto, T. (1995). The condition of fish escaping from fishing gears - a review. *Fish. Res.* **21,** 315–327.

Clay, C. H. (1995). "Design of Fishways and Other Fish Facilities." Second ed., Lewis Publishers, Boca Raton.

Colavecchia, M., Katopodis, C., Goosney, R., Scruton, D. A., and McKinley, R. S. (1998). Measurement of burst swimming performance in wild Atlantic salmon (*Salmo salar* L.) using digital telemetry. *Regul. Rivers: Res. Mgmt.* **14,** 41–51.

Comeau, S., and Boisclair, D. (1998). Day-to-day variation in fish horizontal migration and its potential consequence on estimates of trophic interactions in lakes. *Fish. Res.* **35,** 75–81.

Concepcion, G. B., and Nelson, S. G. (1999). Effects of a dam and reservoir on the distributions and densities of macrofauna in tropical streams of Guam (Mariana islands). *Journal of Freshwater Ecology* **14,** 447–454.

Connecticut River Atlantic Salmon Commission (1998). "Strategic plan for the restoration of Atlantic salmon to the Connecticut River." US Fish and Wildlife Service, Sunderland, MA (USA).

Connor, A. R., Elling, C. H., Black, E. C., Collins, G. B., Gauley, J. R., and Trevor-Smith, E. (1964). Changes in glycogen and lactate levels in migrating salmonid fishes ascending experimental "endless" fishways. *J. Fish. Res. Bd. Canada* **21,** 255–290.

Conover, D. O., and Munch, S. B. (2002). Sustaining fisheries yields over evolutionary time scales. *Science* **297,** 94–96.

Cooke, S. J., Thorstad, E. B., and Hinch, S. G. (2004). Activity and energetics of free-swimming fish: Insights from electromyogram telemetry. *Fish and Fisheries* **5,** 1–32.

Coombs, S., New, J. G., and Nelson, M. (2002). Information-processing demands in electrosensory and mechanosensory lateral line systems. *Journal of Physiology* **96,** 341–354.

Coughlin, D. J. (2002). A molecular mechanism for variations in muscle function in rainbow trout. *Integrative & Comparative Biology* **42,** 190–198.

Coutant, C. C. (2001). Integrated, multi-sensory, behavioral guidance systems for fish diversions. *Am. Fish. Soc. Symp.* **26,** 105–113.

Coutant, C. C., and Whitney, R. R. (2000). Fish behavior in relation to passage through hydropower turbines: A review. *Trans. Am. Fish. Soc.* **129,** 351–380.

Darwin, C. (1859). "On the origin of species." John Murray, London.

Denil, G. (1909). Les eschelles a poissons et leru application aux barrages des Meuse et d'Ourthe. *Ann. Trav. Publ. Belg.* **2,** 1–152.

Denil, G. (1937). La mecanique du poisson de riviere: les capacites mecaniques de la truite et du saumon. *Ann. Trav. Publ. Belg.* **38,** 412–423.

Devries, P. (1997). Riverine salmonid egg burial depths - review of published data and implications for scour studies. *Can. J. Fish. Aquat. Sci.* **54,** 1685–1698.

Dewar, H., Deffenbaugh, M., Thurmond, G., Lashkari, K., and Block, B. A. (1999). Development of an acoustic telemetry tag for monitoring electromyograms in free-swimming fish. *J. Exp. Biol.* **202,** 2693–2699.

Dingle, H. (1996). "Migration: The Biology of Life on the Move." Oxford University Press, New York.

Doucette, L. I., Skulason, S., and Snorrason, S. S. (2004). Risk of predation as a promoting factor of species divergence in threespine sticklebacks (*Gasterosteus aculeatus* L.). *Biological Journal of the Linnean Society* **82,** 189–203.

Driedzic, W. R., and Hochachka, P. W. (1978). Metabolism in fish during exercise. *In* "Fish Physiology, Vol. VII: Locomotion" (Hoar, W. S., and Randall, D. J., Eds.), pp. 425–501. Academic Press, New York.

Drucker, E. G., and Jensen, J. S. (1996). Pectoral fin locomotion in the striped surfperch. II. Scaling swimming kinematics and performance at a gait transition. *J. Exp. Biol.* **199,** 2243–2252.

Ducharme, L. J. A. (1972). An application of louver deflectors for guiding Atlantic salmon (Salmo salar) smolts from power turbines. *J. Fish. Res. Bd. Can.* **29,** 1397–1404.

Eberstaller, J., Hinterhofer, M., and Parasiewicz, P. (1998). The effectiveness of two nature-like bypass channels in an upland Austrian river. *In* "Fish migration and fish bypasses" (Jungwirth, M., Schmutz, S., and Weiss, S. E., Eds.), pp. 363–383. Fishing News Books, Oxford.

Electric Power Research Institute (2002). "Upstream and downstream fish passage and protection technologies for hydroelectric application: A fish passage and protection manual." Palo Alto, California.

Enders, E. C., Boisclair, D., and Roy, A. G. (2003). The effect of turbulence on the cost of swimming for juvenile Atlantic salmon (*Salmo salar*). *Can. J. Fish. Aquat. Sci.* **60,** 1149–1160.

Enders, E. C., and Herrmann, J. P. (2003). Energy costs of spontaneous activity in horse mackerel quantified by a computerised imaging analysis. *Archive of Fishery & Marine Research* **50,** 205–219.

Engelmann, J., Krother, S., Bleckmann, H., and Mogdans, J. (2003). Effects of running water on lateral line responses to moving objects. *Brain, Behavior & Evolution* **61,** 195–212.

Fievet, E. (1999). An experimental survey of freshwater shrimp upstream migration in an impounded stream of Guadeloupe Island, Lesser Antilles. *Archiv fur Hydrobiologie* **144,** 339–355.

Fleming, I. A., and Gross, M. R. (1989). Evolution of adult female life history and morphology in a Pacific salmon (Coho: *Oncorhynchus kisutch*). *Evolution* **43,** 141–157.

Foster, S. A., Baker, J. A., and Bell, M. A. (2003). The case for conserving threespine stickleback populations: Protecting an adaptive radiation. *Fisheries* **28,** 10–18.

Friedland, K. D., Hansen, L. P., Dunkley, D. A., and Maclean, J. C. (2000). Linkage between ocean climate, post-smolt growth, and survival of Atlantic salmon (*Salmo salar* L.) in the North Sea area. *ICES Journal of Marine Science* **57,** 419–429.

Friedland, K. D., Reddin, D. G., McMenemy, J. R., and Drinkwater, K. F. (2003). Multidecadal trends in North American Atlantic salmon (*Salmo salar*) stocks and climate trends relevant to juvenile survival. *Can. J. Fish. Aquat. Sci.* **60,** 563–583.

Fu, C. Z., Wu, J. H., Chen, J. K., Qu, Q. H., and Lei, G. C. (2003). Freshwater fish biodiversity in the Yangtze River basin of China: Patterns, threats and conservation. *Biodiversity & Conservation* **12,** 1649–1685.

Gehrke, P. C., Gilligan, D. M., and Barwick, M. (2002). Changes in fish communities of the Shoalhaven River 20 years after construction of Tallowa Dam, Australia. *River Research & Applications* **18,** 265–286.

Geist, D. R., Brown, R. S., Lepla, K., and Chandler, J. (2002). Practical application of electromyogram radiotelemetry: The suitability of applying laboratory-acquired calibration data to field data. *N. Am. J. Fish. Mgt.* **22,** 474–479.

Gibson, R. J., Haedrich, R. L., and Wernerheim, C. M. (2005). Loss of fish habitat as a consequence of inappropriately constructed stream crossings. *Fisheries* **30,** 10–17.

Gillis, G. B. (1998). Environmental effects on undulatory locomotion in the American eel Anguilla rostrata - kinematics in water and on land. *J. Exp. Biol.* **201,** 949–961.

Glass, C. W., Wardle, C. S., and Gosden, S. J. (1993). Behavioural studies of the principles underlying mesh penetration by fish. *ICES Marine Science Symposium* **196,** 92–97.

Glebe, B. D., and Leggett, W. C. (1981a). Latitudinal differences in energy allocation and use during the freshwater migrations of American shad (*Alosa sapidissima*) and their life history consequences. *Can. J. Fish. Aquat. Sci.* **38,** 806–820.

Glebe, B. D., and Leggett, W. C. (1981b). Temporal, intra-population differences in energy allocation and use by American shad (*Alosa sapidissima*) during the spawning migration. *Can. J. Fish. Aquat. Sci.* **38,** 795–805.

Godø, O. R. (1994). Factors affecting the reliability of groundfish abundance estimates from bottom trawl surveys. *In* "Marine Fish Behavior in Capture and Abundance Estimation" (Fernö, A., and Olsen, S., Eds.), pp. 169–199. Fishing News Books, Oxford.

Goolish, E. M. (1991). Aerobic and anaerobic scaling in fish. *Biol. Rev.* **66,** 33–56.

Gosset, C., and Travade, F. (1999). Devices to aid downstream salmonid migration: Behavioral barriers [French]. *Cybium* **23,** 45–66.

Gray, J. (1936). Studies in animal locomotion. *J. Exp. Biol.* **13,** 192–199.

Gray, J. (1946). The mechanism of locomotion in snakes. *J. Exp. Biol.* **23,** 101–123.

Gray, J. (1957). How Fishes Swim. *Scientific American* **197,** 48.

Grift, R. E., Rijnsdorp, A. D., Barot, S., Heino, M., and Dieckmann, U. (2003). Fisheries-induced trends in reaction norms for maturation in North Sea plaice. *Marine Ecology-Progress Series* **257,** 247–257.

Gross, M. R., Coleman, R. M., and McDowall, R. M. (1998). Aquatic productivity and the evolution of diadromy. *Science* **239,** 1291–1293.

Guerrin, F., and Dumas, J. (2001). Knowledge representation and qualitative simulation of salmon redd functioning. Part I: Qualitative modeling and simulation. *Biosystems* **59,** 75–84.

Hadderingh, R. H., Van Aerssen, G. H. F. M., De Beijer, R. F. L. J., and Van d, V. (1999). Reaction of silver eels to artificial light sources and water currents: An experimental deflection study. *Regul. Rivers: Res. Mgmt.* **15,** 365–371.

Halwart, M., Borlinghaus, M., and Kaule, G. (1996). Activity pattern of fish in rice fields. *Aquaculture* **145,** 159–170.

Hammer, C. (1995). Fatigue and exercise tests with fish. *Comparative Biochemistry & Physiology* **112A,** 1–20.

Handeland, S. O., Jarvi, T., Ferno, A., and Stefansson, S. O. (1996). Osmotic stress, antipredator behaviour, and mortality of Atlantic salmon (*Salmo salar*) smolts. *Can. J. Fish. Aquat. Sci.* **53,** 2673–2680.

Hansen, M. J., Boisclair, D., Brandt, S. B., Hewett, S. W., Kitchell, J. F., Lucas, M. C., and Ney, J. J. (1993). Applications of bioenergetics models to fish Ecology and management - where do we go from here. *Trans. Am. Fish. Soc.* **122,** 1019–1030.

Haro, A., Castro-Santos, T., Noreika, J., and Odeh, M. (2004). Swimming performance of upstream migrant fishes in open-channel flow: A new approach to predicting passage through velocity barriers. *Can. J. Fish. Aquat. Sci.* **61,** 1590–1601.

Haro, A., Odeh, M., Noreika, J., and Castro-Santos, T. (1998). Effect of water acceleration on downstream migratory behavior and passage of Atlantic salmon smolts and juvenile American shad at surface bypasses. *Trans. Am. Fish. Soc.* **127,** 118–127.

Hata, K. (2002a). Field experiment on the migration of fishes to a paddy field with a small fishway. *Japan Agricultural Research Quarterly* **36,** 219–225.

Hata, K. (2002b). Perspectives for fish protection in Japanese paddy field irrigation systems. *Japan Agricultural Research Quarterly* **36,** 211–218.

Hayes, D. E., Mayr, S. D., Kawas, M. L., Chen, Z. Q., Velagic, E., Karakas, A., Bandeh, H., Dogrul, E. C., Cech, J. J., Jr., Swanson, C., and Young, P. S. (2000). Fish screen velocity criteria development using a screened, circular swimming channel. *In* "Advances in fish passage technology: Engineering design and biological evaluation." (Odeh, M., Ed.), pp. 137–147. American Fisheries Society, Bethesda, Maryland.

He, P. (1993). Swimming speeds of marine fish in relation to fishing gears. *ICES Mar. Sci. Symp.* **196,** 183–189.

Heisey, P. G., Mathur, D., and Euston, E. T. (1996). Passing fish safely: A closer look at turbine vs. spillway survival. *Hydro Review* **15,** 42–50.

Herrmann, B. (2005a). Effect of catch size and shape on the selectivity of diamond mesh cod-ends I. Model development. *Fish. Res.* **71,** 1–13.

Herrmann, B. (2005b). Effect of catch size and shape on the selectivity of diamond mesh cod-ends II. Theoretical study of haddock selection. *Fish. Res.* **71,** 15–26.

Hilborn, R., Quinn, T. P., Schindler, D. E., and Rogers, D. E. (2003). Biocomplexity and fisheries sustainability. *Proc. Natl. Acad. Sci.* **100,** 6564–6568.

Hill, D. R. (1959). Some uses of statistical analysis in classifying races of American shad (*Alosa sapidissima*). *Fish. Bull.* **59,** 269–286.

Hinch, S. G., and Bratty, J. (2000). Effects of swim speed and activity pattern on success of adult sockeye salmon migration through an area of difficult passage. *Trans. Am. Fish. Soc.* **129,** 598–606.

Hinch, S. G., Diewert, R. E., Lissimore, T. J., Prince, A. M. J., Healey, M. C., and Henderson, M. A. (1996). Use of electromyogram telemetry to assess difficult passage areas for river-migrating adult sockeye salmon. *Trans. Am. Fish. Soc.* **125,** 253–260.

Hinch, S. G., and Rand, P. S. (1998). Swim speeds and energy use of up-river migrating sockeye salmon (*Oncorhynchus nerka*): Role of local environment and fish characteristics. *Can. J. Fish. Aquat. Sci.* **55,** 1821–1831.

Hinch, S. G., and Rand, P. S. (2000). Optimal swimming speeds and forward-assisted propulsion: Energy-conserving behaviours of upriver-migrating adult salmon. *Can. J. Fish. Aquat. Sci.* **57,** 2470–2478.

Hinch, S. G., Standen, E. M., Healey, M. C., and Farrell, A. P. (2002). Swimming patterns and behaviour of upriver-migrating adult pink (*Oncorhynchus gorbuscha*) and sockeye

(*O. nerka*) salmon as assessed by EMG telemetry in the Fraser River, British Columbia, Canada. *Hydrobiologia* **483,** 147–160.
Hindar, K., Ryman, N., and Utter, F. (1991). Genetic effects of cultured fish on natural fish populations. *Can. J. Fish. Aquat. Sci.* **48,** 945–957.
Hoar, W. S. (1939). The length-weight relationship of the Atlantic salmon. *J. Fish. Res. Bd. Canada* **4,** 441–460.
Hoar, W. S. (1976). Smolt transformation - evolution, behavior, and physiology. *J. Fish. Res. Bd. Canada* **33,** 1233–1252.
Holcik, J. (2001). The impact of stream regulations upon the fish fauna and measures to prevent it. *Ekologia*-Bratislava **20,** 250–262.
Holloway, G. A. (1991). The Brule River sea lamprey barrier and fish ladder, Wisconsin. *Am. Fish. Soc. Symp.* **10,** 264–267.
Holmquist, J. G., Schmidtgengenbach, J. M., and Yoshioka, B. B. (1998). High dams and marine-freshwater linkages - effects on native and introduced fauna in the Caribbean. *Conservation Biology* **12,** 621–630.
Hunn, J. B., and Youngs, W. D. (1980). Role of physical barriers in the control of sea lamprey (*Petromyzon marinus*). *Can. J. Fish. Aquat. Sci.* **37,** 2118–2122.
Huse, I., Lokkeborg, S., and Soldal, A. V. (2000). Relative selectivity in trawl, longline and gillnet fisheries for cod and haddock. *ICES Journal of Marine Science* **57,** 1271–1282.
Hvidsten, N. A., Heggberget, T. G., and Jensen, A. J. (1998). Sea water temperatures at Atlantic salmon smolt entrance. *Nordic Journal of Freshwater Research* **74,** 79–86.
Ihssen, P. E., Booke, H. E., Casselman, J. M., Mcglade, J. M., Payne, N. R., and Utter, F. M. (1981a). Stock identification—materials and methods. *Can. J. Fish. Aquat. Sci.* **38,** 1838–1855.
Ihssen, P. E., Evans, D. O., Christie, W. J., Reckahn, J. A., and Desjardine, R. L. (1981b). Life-history, morphology, and electrophoretic characteristics of 5 allopatric stocks of lake whitefish (*Coregonus clupeaformis*) in the Great Lakes Region. *Can. J. Fish. Aquat. Sci.* **38,** 1790–1807.
ISI Web of Science (2003). Science Citation Index. Thompson Scientific, Philadelphia.
Jayne, B. C., and Lauder, G. V. (1994). How swimming fish use slow and fast muscle fibers: Implications for models of vertebrate muscle recruitment. *J. Comp. Physiol. A* **175,** 123–131.
Jennings, S., Reynolds, J. D., and Mills, S. C. (1998). Life history correlates of responses to fisheries exploitation. *Proceedings of the Royal Society of London - Series B: Biological Sciences* **265,** 333–339.
Johnson, G. D., Ebberts, B. D., Dauble, D. D., Giorgi, A. E., Heisey, P. G., Mueller, R. P., and Neitzel, D. A. (2003). Effects of jet entry at high-flow outfalls on juvenile Pacific salmon. *N. Am. J. Fish. Mgt.* **23,** 441–449.
Johnstone, A. D. F., Lucas, M. C., Boylan, P., and Carter, T. J. (1992). Telemetry of tail-beat frequency of Atlantic salmon (*Salmo salar* L.) during spawning. *In* "Wildlife Telemetry: Remote Monitoring and Tracking of Animals" (Priede, I. G., and Swift, S. M., Eds.), pp. 456–465. Horwood, New York.
Jonas, M. R., Dalen, J. T., Jones, S. T., and Madson, P. L. (2004). "Evaluation of the John Day Dam south fish ladder modification 2003." U.S. Army Corps of Engineers CENWP-OP-SRF, Cascade Locks, Oregon.
Jones, D. R., and Randall, D. J. (1978). The respiratory and circulatory systems during exercise. *In* "Fish Physiology, Vol. VII: Locomotion" (Hoar, W. S., and Randall, D. J., Eds.), pp. 425–501. Academic Press, New York.
Jonsson, N., Jonsson, B., and Hansen, L. P. (1997). Changes in proximate composition and estimates of energetic costs during upstream migration and spawning in Atlantic salmon *Salmo salar*. *Journal of Animal Ecology* **66,** 425–436.

Jungwirth, M., Schmutz, S., and Weiss, S. (1998). "Fish migration and fish bypasses." Fishing News Books, Cambridge.

Kelso, B. W., Northcote, T. G., and Wehrhahn, C. F. (1981). Genetic and environmental aspects of the response to water current by rainbow-trout (*Salmo gairdneri*) originating from inlet and outlet streams of 2 lakes. *Canadian Journal of Zoology-Revue Canadienne de Zoologie* **59,** 2177–2185.

Kerr, S. R. (1976). Ecological analysis and Fry Paradigm. *J. Fish. Res. Bd. Canada* **33,** 329–335.

Kerr, S. R. (1982). The role of external analysis in fisheries science. *Trans. Am. Fish. Soc.* **111,** 165–170.

Killgore, K. J., Miller, A. C., and Conley, K. C. (1987). Effects of turbulence on yolk-sac larvae of paddlefish. *Trans. Am. Fish. Soc.* **116,** 670–673.

Kim, Y.-H., and Wardle, C. S. (2003). Optomotor response and erratic response: Quantitative analysis of fish reaction to towed fishing gears. *Fish. Res.* **60,** 455–470.

King, T. L., Spidle, A. P., Eackles, M. S., Lubinski, B. A., and Schill, W. B. (2000). Mitochondrial DNA diversity in North American and European Atlantic salmon with emphasis on the Downeast rivers of Maine. *J. Fish Biol.* **57,** 614–630.

Kitchell, J. F. (1983). Energetics. *In* "Fish Biomechanics" (Webb, P. W., and Weihs, D., Eds.), pp. 312–338. Praeger Scientific, New York.

Kitchell, J. F., Stewart, D. J., and Weininger, D. (1977). Application of a bioenergetics model to yellow perch (*Perca flavescens*), and walleye (*Stizostedion vitreum vitreum*). *J. Fish. Res. Bd. Canada* **34,** 1922–1935.

Klemetsen, A., Amundsen, P. A., Dempson, J. B., Jonsson, B., Jonsson, N., O'Connell, M. F., and Mortensen, E. (2003). Atlantic salmon *Salmo salar* L., brown trout *Salmo trutta* L. and Arctic charr *Salvelinus alpinus* L.: A review of aspects of their life histories. *Ecol. Freshw. Fish* **12,** 1–59.

Korsmeyer, K. E., Steffensen, J. F., and Herskin, J. (2002). Energetics of median and paired fin swimming, body and caudal fin swimming, and gait transition in parrotfish (*Scarus schlegeli*) and triggerfish (*Rhinecanthus aculeatus*). *J. Exp. Biol.* **205,** 1253–1263.

Kynard, B., and O'Leary, J. (1993). Evaluation of a bypass system for spent American shad at Holyoke Dam, Massachusetts. *North Am. J. Fish. Manage.* **13,** 782–789.

Labeelund, J. H. (1991). Variation within and between rivers in adult size and sea age at maturity of anadromous brown trout, salmo-trutta. *Can. J. Fish. Aquat. Sci.* **48,** 1015–1021.

Lammens, E. H. R. R., van Nes, E. H., Meijer, M. L., and van den Berg, M. S. (2004). Effects of commercial fishery on the bream population and the expansion of *Chara aspera* in Lake Veluwe. *Ecological Modelling* **177,** 233–244.

Langerhans, R. B., Layman, C. A., Shokrollahi, A. M., and Dewitt, T. J. (2004). Predator-driven phenotypic diversification in *Gambusia affinis*. *Evolution* **58,** 2305–2318.

Lannoo, M. J., and Lannoo, S. J. (1996). Development of the electrosensory lateral line lobe in the channel catfish, *Ictalurus punctatus*, with reference to the onset of swimming and feeding behaviors. *Marine Behavior & Physiology* **28,** 45–53.

Latour, R. J., Brush, M. J., and Bonzek, C. F. (2003). Toward ecosystem-based fisheries management: Strategies for multispecies modeling and associated data requirements. *Fisheries* **28,** 10–22.

Legault, A. (1988). The dam clearing of eel by climbing study in Sevre Niortaise. *Bulletin Francais de la Peche et de la Pisciculture* 1–10.

Leggett, W. C., and Carscadden, J. E. (1978). Latitudinal variation in reproductive characteristics of American shad (*Alosa sapidissima*): Evidence for population specific life history strategies in fish. *Can. J. Fish. Aquat. Sci.* **35,** 1469–1478.

Leggett, W. C., and Whitney, R. R. (1972). Water temperature and the migrations of American shad. *Fish. Bull.* **70,** 659–670.

Leonard, J. B. K., and McCormick, S. D. (1999). Effects of migration distance on whole-body and tissue-specific energy use in American shad (*Alosa sapidissima*). *Can. J. Fish. Aquat. Sci.* **56,** 1159–1171.

Leonard, J. B. K., Norieka, J. F., Kynard, B., and McCormick, S. D. (1999). Metabolic rates in an anadromous clupeid, the American shad (*Alosa sapidissima*). *J. Comp. Physiol. B* **169,** 287–295.

Letcher, B. H., and King, T. L. (2001). Parentage and grandparentage assignment with known and unknown matings: Application to Connecticut River Atlantic salmon restoration. *Can. J. Fish. Aquat. Sci.* **58,** 1812–1821.

Letcher, B. H., Rice, J. A., Crowder, L. B., and Rose, K. A. (1996). Variability in survival of larval fish: Disentangling components with a generalized linear individual-based model. *Can. J. Fish. Aquat. Sci.* **53,** 787–801.

Liao, J. C., Beal, D. N., Lauder, G. V., and Triantafyllou, M. S. (2003a). Fish exploiting vortices decrease muscle activity. *Science* **302,** 1566–1569.

Liao, J. C., Beal, D. N., Lauder, G. V., and Triantafyllou, M. S. (2003b). The Karman gait: Novel body kinematics of rainbow trout swimming in a vortex street. *J. Exp. Biol.* **206,** 1059–1073.

Linton, T. K., Reid, S. D., and Wood, C. M. (1997). The metabolic costs and physiological consequences to juvenile rainbow trout of a simulated summer warming scenario in the presence and absence of sublethal ammonia. *Trans. Am. Fish. Soc.* **126,** 259–272.

Lowe, C. G., Holland, K. N., and Wolcott, T. G. (1998). A new acoustic tailbeat transmitter for fishes. *Fish. Res.* **36,** 275–283.

Lucas, M. C., and Batley, E. (1996). Seasonal movements and behaviour of adult barbel *Barbus barbus*, a riverine cyprinid fish: implications for river management. *Journal of Applied Ecology* **33,** 1345–1358.

Lucas, M. C., and Frear, P. A. (1997). Effects of a flow-gauging weir on the migratory behaviour of adult barbel, a riverine, cyprinid. *J. Fish Biol.* **50,** 382–396.

Lucas, M. C., Johnstone, A. D. F., and Priede, I. G. (1993). Use of physiological telemetry as a method of estimating metabolism of fish in the natural-environment. *Trans. Am. Fish. Soc.* **122,** 822–833.

Lucas, M. C., Mercer, T., Armstrong, J. D., McGinty, S., and Rycroft, P. (1999). Use of a flat-bed passive integrated transponder antenna array to study the migration and behaviour of lowland river fishes at a fish pass. *Fish. Res.* **44,** 183–191.

Madenjian, C. P., Fahnenstiel, G. L., Johengen, T. H., Nalepa, T. F., Vanderploeg, H. A., Fleischer, G. W., Schneeberger, P. J., Benjamin, D. M., Smith, E. B., Bence, J. R., Rutherford, E. S., Lavis, D. S., Robertson, D. M., Jude, D. J., and Ebener, M. P. (2002). Dynamics of the Lake Michigan food web, 1970–2000. *Can. J. Fish. Aquat. Sci.* **59,** 736–753.

Magnhagen, C., and Heibo, E. (2001). Gape size allometry in pike reflects variation between lakes in prey availability and relative body depth. *Functional Ecology* **15,** 754–762.

Magnhagen, C., and Heibo, E. (2004). Growth in length and in body depth in young-of-the-year perch with different predation risk. *J. Fish Biol.* **64,** 612–624.

Mallen-Cooper, M. (1992). Swimming ability of juvenile Australian bass, *Macquaria novemaculeata* (Steindachner), and juvenile barramundi, *Lates calcarifer* (Bloch), in an experimental vertical-slot fishway. *Australian Journal of Marine and Freshwater Research* **43,** 823–834.

Mallen-Cooper, M. (1994). Swimming ability of adult golden perch, *Macquaria ambigua* (Percichthyidae), and adult silver perch, *Bidyanus bidyanus* (Teraponidae), in an experimental vertical-slot fishway. *Australian Journal of Marine and Freshwater Research* **45,** 191–198.

Mallen-Cooper, M. (1999). Developing fishways for nonsalmonid fishes: A case study from the Murray River in Australia. *In* "Innovations in Fish Passage Technology" (Odeh, M., Ed.), pp. 173–196. American Fisheries Society, Bethesda, Maryland.

Martinez, G. M., and Bolker, J. A. (2003). Embryonic and larval staging of summer flounder (*Paralichthys dentatus*). *J. Morph.* **255,** 162–176.

McCann, K., and Shuter, B. (1997). Bioenergetics of life history strategies and the comparative allometry of reproduction [review]. *Can. J. Fish. Aquat. Sci.* **54,** 1289–1298.

McCauley, T. C. (1996). "Development of an Instream Velocity Barrier to Stop Sea Lamprey (*Petromyzon marinus*) Migrations in Great Lakes Streams." M.S. thesis, University of Manitoba.

McCleave, J. D. (1980). Swimming performance of European eel (*Anguilla anguilla* (L.)) elvers. *J. Fish Biol.* **16,** 445–452.

McCormick, S. D., Hansen, L. P., Quinn, T. P., and Saunders, R. L. (1998). Movement, migration, and smolting of Atlantic salmon (*Salmo salar*). *Can. J. Fish. Aquat. Sci.* **55,** 77–92.

McCormick, S. D., and Saunders, R. L. (1987). Preparatory physiological adaptations for marine life of salmonids: Osmoregulation, growth, and metabolism. *Am. Fish. Soc. Symp.* **1,** 211–229.

McDonald, D. G., McFarlane, W. J., and Milligan, C. L. (1998). Anaerobic capacity and swim performance of juvenile salmonids. *Can. J. Fish. Aquat. Sci.* **55,** 1198–1207.

McDowall, R. M. (1997). The evolution of diadromy in fishes (revisited) and its place in phylogenetic analysis. *Reviews in Fish Biology & Fisheries* **7,** 443–462.

McDowall, R. M. (1998). Driven by diadromy: its role in the historical and ecological biogeography of the New Zealand freshwater fish fauna. *Italian Journal of Zoology* **65,** 73–85.

McDowall, R. M. (1999). Different kinds of diadromy: Different kinds of conservation problems. *ICES Journal of Marine Science* **56,** 410–413.

McEwen, D., and Scobie, G. (1992). "Estimation of the Hydraulic Conditions Relating to Fish Passage through Turbines." National Engineering Laboratory, East Kilbride, Glasgow.

McKinley, R. S., and Power, G. (1992). Measurement of activity and oxygen consumption for adult lake sturgeon (*Acipenser fulvescens*) in the wild using radio-transmitted EMG signals. *In* "Wildlife Telemetry: Remote Monitoring and Tracking of Animals" (Priede, I. G., and Swift, S. M., Eds.), pp. 307–318. Horwood, New York.

McLean, J. E., Bentzen, P., and Quinn, T. P. (2003). Differential reproductive success of sympatric, naturally spawning hatchery and wild steelhead trout (*Oncorhynchus mykiss*) through adult stage. *Can. J. Fish. Aquat. Sci.* **60,** 433–440.

McLeod, A. M., and Nemenyi, P. (1940). "An Investigation of Fishways." State University of Iowa, Iowa City.

McNabb, C. D., Liston, C. R., and Borthwick, S. M. (2003). Passage of juvenile Chinook salmon and other fish species through Archimedes lifts and a Hidrostal pump at Red Bluff, California. *Trans. Am. Fish. Soc.* **132,** 326–334.

Mogdans, J., and Bleckmann, H. (2001). The mechanosensory lateral line of jawed fishes. *In* "Sensory Biology of Jawed Fishes: New Insights" (Kapoor, B. G., and Hara, T. J., Eds.), pp. 181–213. Science Publishers, Inc., Plymouth.

Mogdans, J., Engelmann, J., Hanke, W., and Krother, S. (2003). The fish lateral line: How to detect hydrodynamic stimuli. *In* "Sensors and Sensing in Biology and Engineering" (Barth, F. G., Friedrich, G., Humphrey, J. A. C., and Secomb, T. W., Eds.), pp. 173–185. Springer, New York.

Moir, H. J., Soulsby, C., and Youngson, A. F. (2002). Hydraulic and sedimentary controls on the availability and use of Atlantic salmon (*Salmo salar*) spawning habitat in the River Dee system, north-east Scotland. *Geomorphology* **45,** 291–308.

Montén, E. (1985). "Fish and Turbines." Vattenfall, Stockholm.

Montgomery, J. C., McDonald, F., Baker, C. F., Carton, A. G., and Ling, N. (2003). Sensory integration in the hydrodynamic world of rainbow trout. *Proceedings of the Royal Society of London - Series B: Biological Sciences* **270,** S195–S197.

Moring, J. R., Marancik, J., and Griffiths, F. (1995). Changes in stocking strategies for Atlantic salmon restoration and rehabilitation in Maine, 1871–1993. *Am. Fish. Soc. Symp.* **15,** 38–46.

Moyes, C. D., and West, T. G. (1995). Exercise metabolism of fish. *In* "Biochemistry and Molecular Biology of Fishes" (Hochachka, P. W., and Mommsen, T. P., Eds.), pp. 367–392. Elsevier Science, Amsterdam.

Nash, C. E., and Waknitz, F. W. (2003). Interactions of Atlantic salmon in the Pacific Northwest I. Salmon enhancement and the net-pen farming industry. *Fish. Res.* **62,** 237–254.

National Marine Fisheries Service (1997). "Fish Screening Criteria for Anadromous Salmonids." Sacramento, California.

Neitzel, D. A., Dauble, D. D., Cada, G. F., Richmond, M. C., Guensch, G. R., Mueller, R. R., Abernethy, C. S., and Amidan, B. (2004). Survival estimates for juvenile fish subjected to a laboratory-generated shear environment. *Trans. Am. Fish. Soc.* **133,** 447–454.

Nelson, J. A., Gotwalt, P. S., Reidy, S. P., and Webber, D. M. (2002). Beyond U-crit: Matching swimming performance tests to the physiological ecology of the animal, including a new fish 'drag strip.' *Comp. Biochem. Physiol. A* **133,** 289–302.

Nemenyi, P. (1941). "An Annotated Bibliography of Fishways." State University of Iowa, Iowa City.

Nemeth, M. J., Krueger, C. C., and Josephson, D. C. (2003). Rheotactic response of two strains of juvenile landlocked Atlantic salmon: implications for population restoration. *Trans. Am. Fish. Soc.* **132,** 904–912.

Nettles, D. C., and Gloss, S. P. (1987). Migration of landlocked Atlantic salmon smolts and effectiveness of a fish bypass structure at a small-scale hydroelectric facility. *N. Am. J. Fish. Mgt.* **7,** 562–568.

Nielsen, C., Holdensaard, G., Petersen, H. C., Bjornsson, B. T., and Madsen, S. S. (2001). Genetic differences in physiology, growth hormone levels and migratory behaviour of Atlantic salmon smolts. *J. Fish Biol.* **59,** 28–44.

Nislow, K. H., Folt, C. L., and Parrish, D. L. (2000). Spatially explicit bioenergetic analysis of habitat quality for age-0 Atlantic salmon. *Trans. Am. Fish. Soc.* **129,** 1067–1081.

Northcote, T. G., and Kelso, B. W. (1981). Differential response to water current by 2 homozygous LDH phenotypes of young rainbow trout (*Salmo gairdneri*). *Can. J. Fish. Aquat. Sci.* **38,** 348–352.

Northcote, T. G., Williscr, S. N., and Tsuyuki, H. (1970). Meristic and lactate dehydrogenase genotype differences in stream populations of rainbow trout below and above a waterfall. *J. Fish. Res. Bd. Canada* **27,** 1987–R.

Odeh, M., Noreika, J., Haro, A., Maynard, A., Castro-Santos, T., and Çada, G. (2002). Evaluation of the Effects of Turbulence on the Behavior of Migratory Fish. Report to the Bonneville Power Administration SOE/BP-00000022-1. US Department of Energy, Portland OR.

Ogutu-Ohwayo, R., and Hecky, R. E. (1991). Fish introductions in Africa and some of their implications. *Can. J. Fish. Aquat. Sci.* **48,** 8–12.

Oldani, N. O., and Baigun, C. R. M. (2002). Performance of a fishway system in a major South American dam on the Parana River (Argentina-Paraguay). *River Research & Applications* **18,** 171–183.

Orsborn, J. F. (1987). Fishways – historical assesment of design practices. *Am. Fish. Soc. Symp.* **1,** 122–130.

Ovidio, M., and Philippart, J. C. (2002). The impact of small physical obstacles on upstream movements of six species of fish – Synthesis of a 5-year telemetry study in the River Meuse basin. *Hydrobiologia* **483,** 55–69.

Ozbilgin, H., and Glass, C. W. (2004). Role of learning in mesh penetration behaviour of haddock (*Melanogrammus aeglefinus*). *ICES Journal of Marine Science* **61,** 1190–1194.

Pavlov, D. S., Lupandin, A. I., and Skorobogatov, M. A. (2000). The effects of flow turbulence on the behavior and distribution of fish. *Journal of Ichthyology* **40**(Suppl. 2), S232–S261.

Peake, S. (2004). An evaluation of the use of critical swimming speed for determination of culvert water velocity criteria for smallmouth bass. *Trans. Am. Fish. Soc.* **133,** 1472–1479.

Peake, S., Beamish, F. W. H., McKinley, R. S., Katopodis, C., and Scruton, D. A. (1995). Swimming performance of lake sturgeon, *Acipenser fulvescens.* **2063,** 1–30.

Peake, S., Beamish, F. W. H., McKinley, R. S., Scruton, D. A., and Katopodis, C. (1997a). Relating swimming performance of lake sturgeon, *Acipenser fulvescens*, to fishway design. *Can. J. Fish. Aquat. Sci.* **54,** 1361–1366.

Peake, S., and McKinley, R. S. (1998). A re-evaluation of swimming performance in juvenile salmonids relative to downstream migration. *Can. J. Fish. Aquat. Sci.* **55,** 682–687.

Peake, S., McKinley, R. S., and Scruton, D. A. (1997b). Swimming performance of various freshwater Newfoundland salmonids relative to habitat selection and fishway design. *J. Fish Biol.* **51,** 710–723.

Peake, S., McKinley, R. S., and Scruton, D. A. (2000). Swimming performance of walleye (*Stizostedion vitreum*). *Can. J. Zool.* **78,** 1686–1690.

Peake, S. J., and Farrell, A. P. (2004). Locomotory behaviour and post-exercise physiology in relation to swimming speed, gait transition and metabolism in free-swimming smallmouth bass (*Micropterus dolomieu*). *J. Exp. Biol.* **207,** 1563–1575.

Pearce, R. O., and Lee, R. T. (1991). Some design considerations for approach velocities at juvenile salmonid screening facilities. *In* "American Fisheries Society Symposium 10" (Colt, J., and White, R. J., Eds.), pp. 237–248. American Fisheries Society, Bethesda, Maryland.

Pepin, P., and Carr, S. M. (1993). Morphological, meristic, and genetic analysis of stock structure in juvenile Atlantic cod (*Gadus morhua*) from the Newfoundland shelf. *Can. J. Fish. Aquat. Sci.* **50,** 1924–1933.

Petersen, J. H., and Kitchell, J. F. (2001). Climate regimes and water temperature changes in the Columbia River: Bioenergetic implications for predators of juvenile salmon. *Can. J. Fish. Aquat. Sci.* **58,** 1831–1841.

Petersen, J. H., and Ward, D. L. (1999). Development and corroboration of a bioenergetics model for northern pikeminnow feeding on juvenile salmonids in the Columbia River. *Trans. Am. Fish. Soc.* **128,** 784–801.

Porcher, J. P. (2002). Fishways for eels. *Bulletin Francais de la Peche et de la Pisciculture* 147–155.

Powers, P. D., Orsborn, J. F., Bumstead, T. W., Klinger-Kingsley, S., and Mih, W. C. (1985). "Fishways–An assessment of their Development and Design." Bonneville Power Administration, US Department of Energy.

Quinn, T. P., and Dittman, A. H. (1990). Pacific salmon migrations and homing: Mechanisms and adaptive significance. *Trends in Ecology & Evolution* **5,** 174–177.

Quinn, T. P., and Foote, C. J. (1994). The effects of body size and sexual dimorphism on the reproductive behaviour of sockeye salmon, *Oncorhynchus nerka*. *Animal Behaviour* **48,** 751–761.

Quinn, T. P., Hendry, A. P., and Buck, G. B. (2001a). Balancing natural and sexual selection in sockeye salmon: Interactions between body size, reproductive opportunity and vulnerability to predation by bears. *Evolutionary Ecology Research* **3,** 917–937.

Quinn, T. P., Hendry, A. P., and Wetzel, L. A. (1995). The influence of life history trade-offs and the size of incubation gravels on egg size variation in sockeye salmon (*Oncorhynchus nerka*). *Oikos* **74,** 425–438.

Quinn, T. P., and Kinnison, M. T. (1999). Size-selective and sex-selective predation by brown bears on sockeye salmon. *Oecologia* **27,** 3–282.

Quinn, T. P., Wetzel, L., Bishop, S., Overberg, K., and Rogers, D. E. (2001b). Influence of breeding habitat on bear predation and age at maturity and sexual dimorphism of sockeye salmon populations. *Can. J. Zool.* **79,** 1782–1793.

Rajaratnam, N., Katopodis, C., Wu, S., and Sabur, M. A. (1997). Hydraulics of resting pools for Denil fishways. *Journal of Hydraulic Engineering* **123,** 632–638.

Rand, P. S., and Hinch, S. G. (1998). Swim speeds and energy use of upriver-migrating sockeye salmon (*Oncorhynchus nerka*) – simulating metabolic power and assessing risk of energy depletion. *Can. J. Fish. Aquat. Sci.* **55,** 1832–1841.

Rand, P. S., Scandol, J. P., and Walter, E. E. (1997). Nerkasim - a research and educational tool to simulate the marine life history of Pacific salmon in a dynamic environment. *Fisheries* **22,** 6–13.

Rand, P. S., Stewart, D. J., Lantry, B. F., Rudstam, L. G., Johannsson, O. E., Goyke, A. P., Brandt, S. B., O'Gorman, R., and Eck, G. W. (1995). Effect of lake-wide planktivory by the pelagic prey fish community in Lakes Michigan and Ontario. *Can. J. Fish. Aquat. Sci.* **52,** 1546–1563, 1995.

Rand, P. S., Stewart, D. J., Seelbach, P. W., Jones, M. L., and Wedge, L. R. (1993). Modeling steelhead population energetics in Lakes Michigan and Ontario. *Trans. Am. Fish. Soc.* **122,** 977–1001.

Reidy, S. P., Nelson, J. A., Tang, Y., and Kerr, S. R. (1995). Post-exercise metabolic rate in Atlantic cod and its dependence upon the method of exhaustion. *J. Fish Biol.* **47,** 377–386.

Revill, A. S., Dulvy, N. K., and Holst, R. (2005). The survival of discarded lesser-spotted dogfish (*Scyliorhinus canicula*) in the Western English Channel beam trawl fishery. *Fish. Res.* **71,** 121–124.

Ricciardi, A., and Rasmussen, J. B. (1998). Predicting the identity and impact of future biological invaders – a priority for aquatic resource management. *Can. J. Fish. Aquat. Sci.* **55,** 1759–1765.

Rijnsdorp, A. D. (1993). Fisheries as a large-scale experiment on life-history evolution – disentangling phenotypic and genetic effects in changes in maturation and reproduction of North-Sea plaice, *Pleuronectes platessa. Oecologia* **96,** 391–401.

Rodgers, D. W., and Patrick, P. H. (1985). Evaluation of a Hidrostal pump fish return system. *N. Am. J. Fish. Mgt.* **5,** 393–399.

Roff, D. A. (1988). The evolution of migration and some life history parameters in marine fishes. *Environmental Biology of Fishes* **22,** 133–146.

Roff, D. A. (1991). Life history consequences of bioenergetic and biomechanical constraints on migration. *Amer. Zool.* **31,** 205–215.

Roni, P., Beechie, T. J., Bilby, R. E., Leonetti, F. E., Pollock, M. M., and Pess, G. R. (2002). A review of stream restoration techniques and a hierarchical strategy for prioritizing restoration in Pacific northwest watersheds. *N. Am. J. Fish. Mgt.* **22,** 1–20.

Roni, P., and Quinn, T. P. (1995). Geographic variation in size and age of North American chinook salmon. *North Am. J. Fish. Manage.* **15,** 325–345.

Rosen, M. W. (1959). "Water Flow about a Swimming Fish." University of California, Los Angeles.

Ruzzante, D. E., Taggart, C. T., and Cook, D. (1996a). Spatial and temporal variation in the genetic composition of a larval cod (*Gadus morhua*) aggregation: Cohort contribution and genetic stability. *Can. J. Fish. Aquat. Sci.* **53,** 2695–2705.

Ruzzante, D. E., Taggart, C. T., Cook, D., and Goddard, S. (1996b). Genetic differentiation between inshore and offshore Atlantic cod (*Gadus morhua*) off Newfoundland: microsatellite DNA variation and antifreeze level. *Can. J. Fish. Aquat. Sci.* **53,** 634–645.

Ruzzante, D. E., Taggart, C. T., Cook, D., and Goddard, S. V. (1997). Genetic differentiation between inshore and offshore Atlantic cod (*Gadus morhua*) off Newfoundland: A test and evidence of temporal stability. *Can. J. Fish. Aquat. Sci.* **54,** 2700–2708.

Ruzzante, D. E., Taggart, C. T., Doyle, R. W., and Cook, D. (2001). Stability in the historical pattern of genetic structure of Newfoundland cod (Gadus morhua) despite the catastrophic decline in population size from 1964 to 1994. *Conservation Genetics* **2,** 257–269.

Ruzzante, D. E., Wroblewski, J. S., Taggart, C. T., Smedbol, R. K., Cook, D., and Goddard, S. V. (2000). Bay-scale population structure in coastal Atlantic cod in Labrador and Newfoundland, Canada. *J. Fish Biol.* **56,** 431–447.

Ryer, C. H., Ottmar, M. L., and Sturm, E. A. (2004). Behavioral impairment after escape from trawl codends may not be limited to fragile fish species. *Fish. Res.* **66,** 261–269.

Ryer, C. H. (2004). Laboratory evidence for behavioural impairment of fish escaping trawls: A review. *ICES Journal of Marine Science* **61,** 1157–1164.

Ryman, N., Utter, F., and Laikre, L. (1995). Protection of intraspecific biodiversity of exploited fishes. *Reviews in Fish Biology and Fisheries* **5,** 417–446.

Saunders, R. L. (1965). Adjustment of buoyancy in young Atlantic salmon and brook trout by changes in swimbladder volume. *J. Fish. Res. Bd. Canada* **22,** 335–352.

Schaarschmidt, T., and Jurss, K. (2003). Locomotory capacity of Baltic Sea and freshwater populations of the threespine stickleback (*Gasterosteus aculeatus*). *Comparative Biochemistry and Physiology A-Molecular & Integrative Physiology* **135,** 411–424.

Schaffer, W. M., and Elson, P. F. (1975). The adaptive significance of variations in life history among local populations of Atlantic salmon in North America. *Ecology* **56,** 577–590.

Schoenfuss, H. L., and Blob, R. W. (2003). Kinematics of waterfall climbing in Hawaiian freshwater fishes (Goblidae): Vertical propulsion at the aquatic-terrestrial interface. *Journal of Zoology, London* **261,** 191–205.

Smith, N. (1971). "A History of Dams." Peter Davies, London.

Soldal, A. V., Engas, A., and Isaksen, B. (1993). Survival of gadoids that escape from a demersal trawl. *ICES Marine Science Symposium* **196,** 122–127.

Soulsby, C., Youngson, A. F., Moir, H. J., and Malcolm, I. A. (2001). Fine sediment influence on salmonid spawning habitat in a lowland agricultural stream: A preliminary assessment. *Science of the Total Environment* **265,** 295–307.

Specker, J. L., Eales, J. G., Tagawa, M., and Tyler, W. A. (2000). Parr-smolt transformation in Atlantic salmon: Thyroid hormone deiodination in liver and brain and endocrine correlates of change in rheotactic behavior. *Can. J. Zool.* **78,** 696–705.

Spidle, A. P., Kalinowski, S. T., Lubinski, B. A., Perkins, D. L., Beland, K. F., Kocik, J. F., and King, T. L. (2003). Population structure of Atlantic salmon in Maine with reference to populations from Atlantic Canada. *Trans. Am. Fish. Soc.* **132,** 196–209.

Standen, E. M., Hinch, S. G., Healey, M. C., and Farrell, A. P. (2002). Energetic costs of migration through the Fraser River Canyon, British Columbia, in adult pink (*Oncorhynchus gorbuscha*) and sockeye (*Oncorhynchus nerka*) salmon as assessed by EMG telemetry. *Can. J. Fish. Aquat. Sci.* **59,** 1809–1818.

Stringham, E. (1924). The maximum speed of fresh-water fishes. *Am. Nat.* **18,** 156–161.

Stuart, I. G., and Berghuis, A. P. (2002). Upstream passage of fish through a vertical-slot fishway in an Australian subtropical river. *Fish. Mgt. Ecol.* **9,** 111–122.

Stuart, T. A. (1962). "The Leaping Behavior of Salmon and Trout at Falls and Obstructions." Department of Agriculture and Fisheries for Scotland, Freshwater and Salmon Fisheries Research, His Majesty's Stationery Office, Edinburgh.

Stuart, T. A. (1964). Biophysical aspects of leaping behaviour in salmon and trout. *Annals of Applied Biology* **53,** 503–505.

Sullivan, T. J. (2004). "Evaluation of the Turner's Falls Fishway Complex and Potential Improvements for Passing Adult American Shad." M.S. thesis, University of Massachusetts, Amherst.

Swanson, C., Young, P. S., and Cech, J. J. (1998). Swimming performance of delta smelt – maximum performance, and behavioral and kinematic limitations on swimming at submaximal velocities. *J. Exp. Biol.* **201,** 333–345.

Swanson, C., Young, P. S., and Cech, J. J., Jr. (2004). Swimming in two-vector flows: Performance and behavior of juvenile Chinook salmon near a simulated screened water diversion. *Trans. Am. Fish. Soc.* (in press).

Tallman, R. F., and Healey, M. C. (1994). Homing, straying, and gene flow among seasonally separated populations of chum salmon (*Oncorhynchus keta*). *Can. J. Fish. Aquat. Sci.* **51,** 577–588.

Tang, Y., Nelson, J. A., Reidy, S. P., Kerr, S. R., and Boutilier, R. G. (1994). A reappraisal of activity metabolism in Atlantic cod (*Gadus morhua*). *J. Fish Biol.* **44,** 1–10.

Taylor, E. B., and Larkin, P. A. (1986). Current response and agonistic behavior in newly emerged fry of Chinook salmon, *Oncorhynchus tshawytscha*, from ocean-type and stream-type populations. *Can. J. Fish. Aquat. Sci.* **43,** 565–573.

Taylor, E. B., and Mcphail, J. D. (1985a). Variation in body morphology among British Columbia populations of coho salmon, *Oncorhynchus kisutch. Can. J. Fish. Aquat. Sci.* **42,** 2020–2028.

Taylor, E. B., and Mcphail, J. D. (1985b). Variation in burst and prolonged swimming performance among British Columbia populations of coho salmon, *Oncorhynchus kisutch. Can. J. Fish. Aquat. Sci.* **42,** 2029–2033.

Taylor, E. B., and Mcphail, J. D. (1986). Prolonged and burst swimming in anadromous and fresh-water threespine stickleback, *Gasterosteus aculeatus. Canadian Journal of Zoology-Revue Canadienne de Zoologie* **64,** 416–420.

Tesch, F. W. (1977). "The Eel: Biology and Management of Anguillid Eels." Chapman and Hall, London (UK).

Thorrold, S. R., Latkoczy, C., Swart, P. K., and Jones, C. M. (2001). Natal homing in a marine fish metapopulation. *Science* **291,** 297–299.

Tsunesumi, N. (2002). Development of a new fishway for various fish species and cost reduction. *Japan Agricultural Research Quarterly* **36,** 201–209.

Tsuyuki, H., and Williscroft, S. N. (1977). Swimming stamina differences between genotypically distinct forms of rainbow (*Salmo gairdneri*) and steelhead trout. *J. Fish. Res. Bd. Canada* **34,** 996–1003.

Turnpenny, A. W. H. (1998). Mechanisms of fish damage in low-head turbines: an experimental appraisal. *In* "Fish Migration and Fish Bypasses" (Jungwirth, M., Schmutz, S., and Weiss, S., Eds.), pp. 300–314. Fishing News Books, Cambridge.

U.S.Army Corps of Engineers (2001). "U.S. National Inventory of Dams." Alexandria, VA (USA).

Verspoor, E. (1997). Genetic diversity among Atlantic salmon (*Salmo salar* L.) populations. *ICES Journal of Marine Science* **54,** 965–973.

Videler, J. J. (1993). "Fish Swimming." Chapman & Hall, London.

Vigneux, E., and Larinier, M. (2002). Fishways: biological basis, design criteria and monitoring. *Bulletin Francais de la Peche et de la Pisciculture* **364,** (suppl), 1–205.

Waknitz, F. W., Iwamoto, R. N., and Strom, M. S. (2003). Interactions of Atlantic salmon in the Pacific northwest IV. Impacts on the local ecosystems [Review]. *Fish. Res.* **62,** 307–328.

Walker, J. A. (1997). Ecological morphology of lacustrine threespine stickleback *Gasterosteus aculeatus* L. (Gasterosteidae) body shape. *Biological Journal of the Linnean Society* **61,** 3–50.

Walsh, S. J., and Hickey, W. M. (1993). Behavioural reactions of demersal fish to bottom trawls at various light conditions. *ICES Mar. Sci. Symp.* **196,** 68–78.

Wardle, C. S. (1975). Limit of fish swimming speed. *Nature* **255,** 725–727.

Wardle, C. S. (1980). Effects of temperature on the maximum swimming speed of fishes. *In* "Environmental Physiology of Fishes" (Ali, M. A., Ed.), pp. 519–531. Plenum, New York.

Wardle, C. S. (1993). Fish behaviour and fishing gear. *In* "Behaviour of Teleost Fishes" (Pitcher, T. J., Ed.), pp. 609–643. Chapman and Hall, London.

Ware, D. M. (1975). Growth, metabolism, and optimal swimming speed of a pelagic fish. *J. Fish. Res. Bd. Canada* **32,** 33–41.

Ware, D. M. (1978). Bioenergetics of pelagic fish - theoretical change in swimming speed and ration with body size. *Journal of the Fisheries Research Board of Canada* **35,** 220–228.

Ware, D. M. (1980). Bioenergetics of stock and recruitment. *Can. J. Fish. Aquat. Sci.* **37,** 1012–1024.

Ware, D. M. (1982). Power and evolutionary fitness of teleosts. *Can. J. Fish. Aquat. Sci.* **39,** 3–13.

Warren, M. L., and Pardew, M. G. (1998). Road crossings as barriers to small-stream fish movement. *Trans. Am. Fish. Soc.* **127,** 637–644.

Weaver, C. R. (1963). Influence of water velocity upon orientation and performance of adult migrating salmonids. *Fish. Bull.* **63,** 97–121.

Weaver, C. R. (1965). Observations on the swimming ability of adult American shad (*Alosa sapidissima*). *Trans. Am. Fish. Soc.* **94,** 382–385.

Webb, P. W. (1971). The swimming energetics of trout: II. Oxygen consumption and swimming capacity. *J. Exp. Biol.* **55,** 521–540.

Webb, P. W. (1975). Hydrodynamics and energetics of fish propulsion. *Bull. Fish. Res. Bd. Canada* **190,** 1–158.

Webb, P. W. (1988). Simple physical principles and vertebrate aquatic locomotion. *Amer. Zool.* **28,** 709–725.

Webb, P. W. (1993). Swimming. *In* "The Physiology of Fishes" (Evans, D. H., Ed.), pp. 47–73. CRC Press, Boca Raton.

Webb, P. W. (1998). Swimming. *In* "The Physiology of Fishes" (Evans, D. H., Ed.), Second ed., pp. 3–24. CRC Press, Boca Raton.

Webb, P. W., and Gerstner, C. L. (2000). Fish swimming behaviour: Predictions from physical principles. *In* "Biomechanics in Animal Behaviour" (Domenici, P., and Blake, R. W., Eds.), pp. 59–77. BIOS Scientific Publishing, Ltd., Oxford.

Weihs, D. (1973). Optimal fish cruising speed. *Nature* **245,** 48–50.

Weinstein, R. B., and Full, R. J. (2000). Intermittent locomotor behavior alters total work. *In* "Biomechanics in Animal Behaviour" (Domenici, P., and Blake, R. W., Eds.), pp. 33–48. BIOS Scientific Publishers Ltd, Oxford.

Welch, D. W., Boehlert, G. W., and Ward, B. R. (2002). POST – the Pacific Ocean salmon tracking project. *Oceanologica Acta* **25,** 243–253.

Welch, D. W., and Eveson, J. P. (1999). An assessment of light-based geoposition estimates from archival tags. *Can. J. Fish. Aquat. Sci.* **56,** 1317–1327.

Welch, D. W., Ishida, Y., and Nagasawa, K. (1998). Thermal limits and ocean migrations of sockeye salmon (*Oncorhynchus nerka*) – long-term consequences of global warming. *Can. J. Fish. Aquat. Sci.* **55,** 937–948.

Whalen, K. G., Parrish, D. L., and McCormick, S. D. (1999). Migration timing of Atlantic salmon smolts relative to environmental and physiological factors. *Trans. Am. Fish. Soc.* **128,** 289–301.

Winans, G. A. (1984). Multivariate morphometric variability in Pacific salmon – technical demonstration. *Can. J. Fish. Aquat. Sci.* **41,** 1150–1159.

Winans, G. A., and Nishioka, R. S. (1987). A multivariate description of change in body shape of coho salmon (*Oncorhynchus kisutch*) during smoltification. *Aquaculture* **66,** 235–245.

Winberg, G. G. (1956). "Rate of Metabolism and Food Requirements of Fishes." Belarussian University, Minsk, USSR (Transl. from Russian by Fish. Res. Bd. Can. Transl. Ser. No. 194, 1967).

Winger, P. D., He, P., and Walsh, S. J. (1999). Swimming endurance of American plaice (*Hippoglossoides platessoides*) and its role in fish capture. *ICES Journal of Marine Science* **56,** 252–265.

Winger, P. D., He, P. G., and Walsh, S. J. (2000). Factors affecting the swimming endurance and catchability of Atlantic cod (*Gadus morhua*). *Can. J. Fish. Aquat. Sci.* **57,** 1200–1207.

Wirth, T., and Bernatchez, L. (2001). Genetic evidence against panmixia in the European eel. *Nature* **409,** 1037–1040.

World Commision on Dams (2000). "Dams and Development: a New Framework for Decision-Making." Earthscan Publications, LTD, London.

Yako, L. A., Mather, M. E., and Juanes, F. (2000). Assessing the contribution of anadromous herring to largemouth bass growth. *Trans. Am. Fish. Soc.* **129,** 77–88.

Youngson, A. F., Jordan, W. C., Verspoor, E., McGinnity, P., Cross, T., and Ferguson, A. (2003). Management of salmonid fisheries in the British Isles: Towards a practical approach based on population genetics. *Fish. Res.* **62,** 193–209.

Zhou, Y. (1982). The swimming behaviour of fish in towed gears; a reexamination of the principles. *Work. Pap., Dept. Agric. Fish. Scotl.* **4,** 1–55.

Zydlewski, J., and McCormick, S. D. (1997a). The ontogeny of salinity tolerance in the American shad, *Alosa sapidissima*. *Can. J. Fish. Aquat. Sci.* **54,** 182–189.

Zydlewski, J., and McCormick, S. D. (1997b). The loss of hyperosmoregulatory ability in migrating juvenile American shad *Alosa sapidissima*. *Can. J. Fish. Aquat. Sci.* **54,** 2377–2387.

INDEX

C

F

G

J

K

L

N

O

P

Q

R

S

T

Y

Z

OTHER VOLUMES IN THE FISH PHYSIOLOGY SERIES

VOLUME 1 — Excretion, Ionic Regulation, and Metabolism
Edited by W. S. Hoar and D. J. Randall

VOLUME 2 — The Endocrine System
Edited by W. S. Hoar and D. J. Randall

VOLUME 3 — Reproduction and Growth: Bioluminescence, Pigments, and Poisons
Edited by W. S. Hoar and D. J. Randall

VOLUME 4 — The Nervous System, Circulation, and Respiration
Edited by W. S. Hoar and D. J. Randall

VOLUME 5 — Sensory Systems and Electric Organs
Edited by W. S. Hoar and D. J. Randall

VOLUME 6 — Environmental Relations and Behavior
Edited by W. S. Hoar and D. J. Randall

VOLUME 7 — Locomotion
Edited by W. S. Hoar and D. J. Randall

VOLUME 8 — Bioenergetics and Growth
Edited by W. S. Hoar, D. J. Randall, and J. R. Brett

VOLUME 9A — Reproduction: Endocrine Tissues and Hormones
Edited by W. S. Hoar, D. J. Randall, and E. M. Donaldson

VOLUME 9B — Reproduction: Behavior and Fertility Control
Edited by W. S. Hoar, D. J. Randall, and E. M. Donaldson

VOLUME 10A — Gills: Anatomy, Gas Transfer, and Acid-Base Regulation
Edited by W. S. Hoar and D. J. Randall

VOLUME 10B — Gills: Ion and Water Transfer
Edited by W. S. Hoar and D. J. Randall

VOLUME 11A The Physiology of Developing Fish: Eggs and Larvae
Edited by W. S. Hoar and D. J. Randall

VOLUME 11B The Physiology of Developing Fish: Viviparity and Posthatching Juveniles
Edited by W. S. Hoar and D. J. Randall

VOLUME 12A The Cardiovascular System
Edited by W. S. Hoar, D. J. Randall, and A. P. Farrell

VOLUME 12B The Cardiovascular System
Edited by W. S. Hoar, D. J. Randall, and A. P. Farrell

VOLUME 13 Molecular Endocrinology of Fish
Edited by N. M. Sherwood arid C. L. Hew

VOLUME 14 Cellular and Molecular Approaches to Fish Ionic Regulation
Edited by Chris M. Wood and Trevor J. Shuttleworth

VOLUME 15 The Fish Immune System: Organism, Pathogen, and Environment
Edited by George Iwama and Teruyuki Nakanishi

VOLUME 16 Deep Sea Fishes
Edited by D. J. Randall and A. P. Farrell

VOLUME 17 Fish Respiration
Edited by Steve F. Perry and Bruce Tufts

VOLUME 18 Muscle Growth and Development
Edited by Ian A. Johnson

VOLUME 19 Tuna: Physiology, Ecology, and Evolution
Edited by Barbara A. Block and E. Donald Stevens

VOLUME 20 Nitrogen Excretion
Edited by Patricia A. Wright and Paul M. Anderson

VOLUME 21 The Physiology of Tropical Fishes
Edited by Adalberto L. Val, Vera Maria F. De Almeida-Val, and David J. Randall

VOLUME 22 The Physiology of Polar Fishes
Edited by Anthony P. Farrell and John F. Steffensen

VOLUME 23 Fish Biomechanics
Edited by Robert E. Shadwick and George V. Lauder

VOLUME 24 Behaviour and Physiology of Fish
Edited by Katherine A. Sloman, Rod W. Wilson, and Sigal Balshine

Printed and bound by CPI Group (UK) Ltd, Croydon, CR0 4YY

08/05/2025

01864951-0003